Synthesis Green Metrics
Problems, Exercises, and Solutions

Synthesis Green Metrics
Problems, Exercises, and Solutions

By
John Andraos

CareerChem
Toronto, ON
Canada
c1000@careerchem.com

CRC Press is an imprint of the
Taylor & Francis Group, an **informa** business

CRC Press
Taylor & Francis Group
6000 Broken Sound Parkway NW, Suite 300
Boca Raton, FL 33487-2742

© 2019 by Taylor & Francis Group, LLC
CRC Press is an imprint of Taylor & Francis Group, an Informa business

No claim to original U.S. Government works

Printed on acid-free paper

International Standard Book Number-13: 978-0-367-00226-8 (Hardback)
International Standard Book Number-13: 978-0-367-00225-1 (Paperback)

This book contains information obtained from authentic and highly regarded sources. Reasonable efforts have been made to publish reliable data and information, but the author and publisher cannot assume responsibility for the validity of all materials or the consequences of their use. The authors and publishers have attempted to trace the copyright holders of all material reproduced in this publication and apologize to copyright holders if permission to publish in this form has not been obtained. If any copyright material has not been acknowledged, please write and let us know so we may rectify in any future reprint.

Except as permitted under U.S. Copyright Law, no part of this book may be reprinted, reproduced, transmitted, or utilized in any form by any electronic, mechanical, or other means, now known or hereafter invented, including photocopying, microfilming, and recording, or in any information storage or retrieval system, without written permission from the publishers.

For permission to photocopy or use material electronically from this work, please access www.copyright.com (http://www.copyright.com/) or contact the Copyright Clearance Center, Inc. (CCC), 222 Rosewood Drive, Danvers, MA 01923, 978-750-8400. CCC is a not-for-profit organization that provides licenses and registration for a variety of users. For organizations that have been granted a photocopy license by the CCC, a separate system of payment has been arranged.

Trademark Notice: Product or corporate names may be trademarks or registered trademarks, and are used only for identification and explanation without intent to infringe.

Visit the Taylor & Francis Web site at
http://www.taylorandfrancis.com

and the CRC Press Web site at
http://www.crcpress.com

Dedication

Dr. Christopher Randall Schmid

(1959–2007)

Inaugural Editor of Organic Process Research & Development

Contents

Preface xi

Chapter 1 Introduction 1
        1.1   What This Book Is About 1
        1.2   Structure 2
        1.3   How to Use the Book 3

Chapter 2 Synthesis Plan Analysis 5
        2.1   Synthesis Tree Diagrams 5
              2.1.1   Terms, Definitions, and Examples 5
              2.1.2   Problems 13
        2.2   Linear Synthesis Plans 15
              2.2.1   Terms, Definitions, and Examples 15
              2.2.2   Problems 16
        2.3   Convergent Synthesis Plans 19
              2.3.1   Terms, Definitions, and Examples 19
              2.3.2   Problems 34
        2.4   Reaction Networks 36
              2.4.1   Terms, Definitions, and Examples 36
              2.4.2   Problems 41
        2.5   Ranking Synthesis Plans According to Material
            Efficiency 45
              2.5.1   Terms, Definitions, and Examples 45
        2.6   Excel Spreadsheet Tools: Part 1 82
              2.6.1   Synthesis-Plan-Analysis-Basis-Materials-Metrics.xls 82
              2.6.2   Template-SYNTHESIS.xls 82
              2.6.3   Calculator-Cumulative-PMI.xls 83
        2.7   Problems 83
        References 147

Chapter 3 Environmental Impact Metrics 153
        3.1   Life Cycle Assessment 153
              3.1.1   Terms, Definitions, and Examples 155
        3.2   Multi-Compartment Model 166
              3.2.1   Terms, Definitions, and Examples 166
        3.3   Environmental Impact Potentials 169
              3.3.1   Terms, Definitions, and Examples 169
        3.4   Benign Index 191
              3.4.1   Terms, Definitions, and Examples 191

	3.5	Excel Spreadsheet Tools: Part 2 196
		3.5.1 Descriptions of Spreadsheets 196
		3.5.2 Assumptions and Dealing with Missing Data 198
	3.6	Problems 200
	References 208

Chapter 4 Safety-Hazard Impact Metrics 215

    4.1 Parameters 215
        4.1.1 Terms, Definitions, and Examples 215
    4.2 Safety-Hazard Impact Potentials 223
        4.2.1 Terms, Definitions, and Examples 223
    4.3 Safety-Hazard Index 241
        4.3.1 Terms, Definitions, and Examples 241
    4.4 Excel Spreadsheet Tools: Part 3 247
    4.5 Problems 248
    References 251

Chapter 5 Energy Metrics 253

    5.1 Thermodynamic Preliminaries 255
        5.1.1 Terms, Definitions, and Examples 255
        5.1.2 Predicting Thermochemical Properties 274
    5.2 Thermodynamic Calculations 285
        5.2.1 Determination of Phases of Input Materials in a Given Chemical Reaction 285
        5.2.2 Determination of Molar Volume of a Substance at a Given Pressure and Temperature for a Given Cubic Equation of State 285
        5.2.3 Derivation of $\int pdV$ Expressions for Various Cases in P-V Diagrams Related to Determining Work Done by Fluids During Reversible Isothermal Expansion or Contraction 289
        5.2.4 Derivation of Expressions for $\int p(V)dV$ for Various Equations of State 295
            5.2.4.1 van der Waals EOS 295
            5.2.4.2 Redlich–Kwong EOS 296
            5.2.4.3 Peng–Robinson EOS 296
        5.2.5 Determination of Expressions for Enthalpy Change due to Temperature and Pressure Contributions 297
            5.2.5.1 Case I: State 1 (gas) to State 2 (gas) 297
            5.2.5.2 Case II: State 1 (liquid) to State 2 (gas) 301
            5.2.5.3 Case III: State 1 (liquid) to State 2 (liquid) 302
            5.2.5.4 Case IV: State 1 (gas) to State 2 (liquid) 303
        5.2.6 Worked Examples 307

Contents ix

	5.3	Types of Energy Metrics 319
		5.3.1 Terms, Definitions, and Examples 319
	5.4	Excel Spreadsheet Tools: Part 4 335
	5.5	Problems 337
	References 341

Chapter 6 Algorithms 345

    6.1 Environmental Assessment Tool for Organic Synthesis 345
        6.1.1 Description of Program 345
        6.1.2 Instructions for Using Program 348
    6.2 EcoScale 355
        6.2.1 Description of Algorithm 355
        6.2.2 EcoScale-Template.xls 355
    6.3 Edwards–Lawrence Algorithm 357
        6.3.1 Description of Algorithm 357
        6.3.2 Edwards-Lawrence-Template.xls 358
    6.4 GreenStar 361
        6.4.1 Description of Algorithm 361
        6.4.2 GreenStar-Template.xls 362
    6.5 Multivariate Method 362
        6.5.1 Description of Algorithm 362
        6.5.2 Multivariate-Template.xls 368
    6.6 UTGCI Method 368
        6.6.1 Description of Algorithm 368
        6.6.2 UTGCI-Template.xls 370
    6.7 Andraos Algorithm 370
        6.7.1 Description of Algorithm 370
    6.8 Rowan Solvent Greenness Index 377
        6.8.1 Description of Algorithm 377
        6.8.2 Rowan-Solvent-Greenness-Index.xls 380
    6.9 Epilogue 385
    6.10 Problems 387
    References 389

Chapter 7 Examples from the Chemical Industry 391

    7.1 Terms, Definitions, and Examples 391
    7.2 Problems 392
    References 499

Appendix: Other Terminologies 503

Index 509

Preface

> An investment in knowledge pays the best interest.
>
> – Benjamin Franklin

> Education is what you remember after you have forgotten what you learned in school.
>
> – Albert Einstein

> We don't need no education. We don't need no thought control. No dark sarcasm in the classroom. Teachers leave them kids alone.
>
> – Pink Floyd

This book and its prequel volume, *Reaction Green Metrics: Problems, Exercises, and Solutions,* are the culmination of my 15 years of independent study of the now mature field of green chemistry since 2004. As with all research endeavors, the seed from which this effort emerged is teaching, as was the case when Dmitry Mendeleev came up with the idea of periodicity of the elements or when Gilbert Lewis came up with the idea of drawing dot diagrams for chemical structures to keep track of bonding and non-bonding electrons between atoms. My real education in chemistry began the day I first taught it to others. Up to that point nearly everything I learned from my professors was presented as a set of disjointed facts that needed to be committed to memory for the sole purpose of recall for examination purposes and degree accreditation. So, in order to really understand my subject, I had to relearn it from the beginning. In doing so, two essential tools needed to be mastered. The first was the mapping of the scientific genealogy of the people and ideas that shaped the field. This exercise revealed the connections between those people as a roadmap for career development and it also revealed connections between their ideas, which essentially elucidated the trajectory of important research directions. The second was the fortuitous emergence of green chemistry as a unifying force that linked many scientific disciplines (organic and inorganic chemistries, toxicology, environmental studies, mathematics, computer science, and optimization research) together for a common purpose; namely, to invent improved syntheses as an intellectual endeavor that would have great impact both on preserving and utilizing our planet's finite resources and also on the quality of human life. A great advantage in mastering these skills is quantitative reasoning, which was the key missing ingredient when the 12 Principles of Green Principles were announced in 1998 as merely a set of qualitative guiding principles for chemists to think about in the back of their minds. Quantitative reasoning moves that thinking to the front of the mind where decision-making actions are initiated and ultimately carried out.

No accomplishment is made single-handedly. I thank Dr. Floyd H. Dean and his son, Jason, for being intellectual sounding boards on matters of science. I thank Dr. Thuy van Pham for this generous gift of a laptop. I thank Andrei Hent for gathering some of the references from which problems were posed. I thank Dr. Laura B. Hoch and Dr. Melanie L. Mastronardi, inaugural founders of the University of Toronto Green Chemistry Initiative who helped to develop the UTGCI algorithm and who invited me to participate in Green Chemistry Workshops at the University of Toronto for the benefit of graduate students in the Departments of Chemistry and Chemical Engineering. I thank Hilary LaFoe for her patience as Editor at CRC Press/Taylor & Francis Group throughout this campaign over the last three years. Lastly, I thank my family for their emotional support during the course of my lifelong study of chemistry.

John Andraos
Toronto, Canada

1 Introduction

1.1 WHAT THIS BOOK IS ABOUT

This book builds on the topics covered in *Reaction Green Metrics: Problems, Exercises, and Solutions* which covered parameterization of material efficiency of individual reactions. Here, we extend the same analysis to linear and convergent synthesis plans that are composed of sequences of individual reactions. We also expand the range of metrics applied to both individual reactions and synthesis plans to include energy consumption, environmental impact, and safety-hazard impact. The result is a comprehensive suite of tools used to analyze a set of synthesis plans to a given target molecule with a view to rank them in a comparative sense so that relative strengths and weaknesses can be pinpointed and then best candidate plans can be selected for further optimization. In order to successfully navigate the topics presented it is expected that the reader has an adequate background knowledge in the following subjects as necessary prerequisites. In chemistry, the reader should have already taken an introductory organic chemistry course, has familiarity with reaction mechanism, has a basic library of named organic reactions committed to memory, and is familiar with drawing and interpreting catalytic cycles. Most importantly, balancing chemical equations and drawing synthesis plans following the principle of conservation of structural aspect are highly emphasized. These skills will be the most important to master for a student or a professional, but they are a mandatory starting point for solving every example and question posed in this book. No metrics analysis of any kind can begin without first establishing a balanced chemical equation for a given chemical transformation. Since this book combines chemistry with quantitative reasoning there are certain topics in mathematical science that the reader is expected to be fluent in. Specifically, these include basic arithmetic, basic algebra, and elementary calculus for dealing with topics related to thermodynamics. In order for the reader to take full advantage of the accompanying online suite of spreadsheet algorithms (available at https://www.crcpress.com/9780367002251), mastery of Microsoft Excel spreadsheet software and facility with graphing are highly recommended. These spreadsheets and calculators were designed to remove the tedious task of computing metrics by hand, particularly for assessing environmental and safety-hazard impacts, which requires assembling and collating large amounts of data, particularly for synthesis plan analysis. In addition, extensive databases of key physical and toxicological properties of commonly used industrial chemicals are given in one place so the reader does not have to scour the literature to find them, thereby speeding up the process of conducting calculations and interpreting the results of those calculations.

1.2 STRUCTURE

As in the case of the previous book, *Reaction Green Metrics: Problems, Exercises, and Solutions*, the topics in this book are also presented in a ladder approach where each successive topic follows logically from the previous one. Chapter 2 introduces synthesis tree diagrams to streamline the display and analysis of synthesis plans, particularly convergent ones. These useful diagrams greatly facilitate the computation of step and overall material efficiency metrics for synthesis plans. Particular attention is paid to establishing relationships between the overall material efficiency metrics for a given plan (overall atom economy, overall E-factor, and overall PMI (process mass intensity)) with the corresponding step material efficiency metrics pertaining to each reaction step in a given plan. The analysis of reaction networks involving multiple synthetic routes to a given target molecule using elementary graph theoretical concepts is also examined. Chapter 3 introduces life cycle assessment analysis based on the Mackay compartment environmental model and the Guinée environmental impact potentials formalism. The benign index (BI) is discussed as a unifying parameter that describes environmental impact. Assumptions, limitations, and dealing with missing data are also given special attention. Various Microsoft Excel spreadsheets are introduced to automate the analyses. Chapter 4 introduces safety-hazard impacts using the same kind of potential analysis as in Chapter 3. The safety-hazard index (SHI) is introduced as a second unifying parameter. Chapter 5 introduces in-depth analysis of energy metrics based on fundamental thermodynamics principles. The key operative energy metric is energy consumption per unit mass of target product, where energy consumption refers to the amount of input energy that is required (i.e., enthalpy) to go from an initial state of ambient conditions (25°C, 1 atm) to a final state of reaction temperature and pressure conditions. Therefore, a thorough coverage of selected thermodynamic parameters is given that will serve both to refresh intermediate readers on these topics and to clarify much of the complex jargon found in specialized books on chemical thermodynamics for novices. Special attention is paid to equations of state for gases and liquids, and to the determination of enthalpy changes for input materials and heats of reaction. Various methods are also summarized for the estimation of key thermodynamic parameters such as critical constants, boiling and melting points, enthalpies of fusion and vaporization, vapor pressure, and heat capacities at constant pressures when these are not available in property databases. Chapter 6 surveys various algorithms available in the literature including Environmental Assessment Tool for Organic Synthesis (EATOS), EcoScale, Edwards–Lawrence, GreenStar, Multivariate, University of Toronto Green Chemistry Initiative (UTGCI), Andraos, and Rowan Solvent Greenness Index. For clear illustrative purposes the performances of these algorithms are compared head-to-head on the same set of linear and convergent plans. Chapter 7 is the concluding chapter, which is a collection of 40 advanced problems covering all topics introduced in this book and in the prequel *Reaction Green Metrics: Problems, Exercises, and Solutions*. These problems are posed from interesting chemistry found in the organic chemistry, pharmaceutical chemistry, and industrial chemistry literature. Microsoft Excel

Introduction

spreadsheets have been developed throughout to evaluate material efficiency, energy efficiency, environmental impact, and safety-hazard impact of synthesis plans in an automated, reliable, and easy-to-use fashion.

1.3 HOW TO USE THE BOOK

First, it is recommended that this book be used only *after* all of the topics in *Reaction Green Metrics: Problems, Exercises, and Solutions* have been thoroughly covered and mastered. For teaching purposes instructors can use the worked examples in a classroom setting to introduce topics for discussion in lectures and use the problems as homework set exercises for deeper thought. As before, it is also recommended that this book be used in a second one-semester intermediate to advanced level course on green chemistry as a dedicated subject spanning academic and industrial chemistry topics. The style of presentation is brief, where each section of each chapter begins with definitions of terms followed immediately in each case by easy examples that illustrate the implementation of the definitions. Each topic is briefly introduced followed by worked examples to illustrate the ideas for immediate understanding and implementation. Advanced problems are posed at the conclusion of each chapter. Consistent with good pedagogy, all examples and problems posed come directly from literature sources that are cited upfront for both instructors and students to view. Solutions to all problems are available electronically through the publisher's website.

For research purposes, experienced readers may skip topics they are familiar with and focus on those areas they wish to learn more about as needed. The same pedagogical style described above is applicable to experienced chemists for their benefit. Examples and problems of interest to industrial chemists, process chemists, and chemical engineers have been taken from both research journals and patents.

At the end of the journey, the reader should consider themselves to be experienced enough in applying concepts learned to their own research work in chemical science, be confident in practicing green chemistry principles in a rigorous quantitative manner, and to be able to critically evaluate literature procedures when "green" claims are announced.

2 Synthesis Plan Analysis

2.1 SYNTHESIS TREE DIAGRAMS

2.1.1 TERMS, DEFINITIONS, AND EXAMPLES

Degree of Asymmetry of a Synthesis Plan, β^1

A parameter determined from the shape of a synthesis tree diagram for a synthesis plan that describes the degree of skewness of a triangle whose vertices are the target product node, the origin, and the node for the last reagent projected onto the ordinate axis.

The mathematical relationship is given by Equation (2.1).

$$\beta = \frac{2y}{I-1} - 1 \qquad (2.1)$$

where y is the ordinate of the point representing the final target product and I is the number of reactant inputs in the entire synthesis plan. Figure 2.1 shows an example synthesis tree diagram indicating the geometric origin of the degree of asymmetry parameter. A synthesis plan with a symmetric synthesis tree diagram has $\beta=0$ (when $y=0.5*(I-1)$), and one having an asymmetric diagram has $\beta=1$ (when $y=I-1$).

Degree of Convergence, δ^1

A parameter determined from the shape of a synthesis tree diagram for a synthesis plan that describes the ratio of the angle subtended at the actual product node vertex to that at a product node vertex corresponding to the hypothetical case of all reaction substrates in a plan reacting in a single step.

The mathematical relationship is given by Equation (2.2).

$$\delta = \frac{\arctan\left(\frac{I-1-y}{N}\right) + \arctan\left(\frac{y}{N}\right)}{2\arctan\left(\frac{I-1}{2}\right)} \qquad (2.2)$$

where y is the ordinate of the point representing the final target product, I is the number of reactant inputs in the entire synthesis plan, and N is the number of reaction stages. For a linear plan, the number of reaction stages is equal to the number of reaction steps; for a convergent plan, the number of reaction stages is less than the number of reaction steps.

Figure 2.2 shows an example synthesis tree diagram indicating the geometric origin of the degree of convergence parameter. A completely convergent plan has $\delta=1$ where all input materials are used to reach the target product in a single reaction stage such as a one-pot multi-component reaction. A plan with $\delta=1$ corresponds

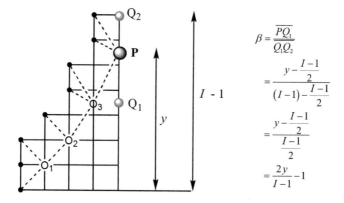

FIGURE 2.1 Synthesis tree diagram showing the geometric origin of Equation (2.1).

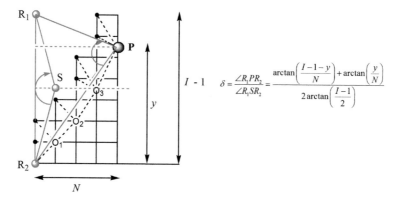

FIGURE 2.2 Synthesis tree diagram showing the geometric origin of Equation (2.2).

to one that is the most linear consisting of sequential transformations over several reaction stages with no input reagents incorporated into the intermediate products. An example of such a synthesis plan is a sequence of rearrangements, electrocyclic, and/or elimination reactions.

Centroid of Product Nodes in Synthesis Tree[1]

The formula for calculating the centroid of input reagents for a reaction step is given by Equation (2.3). From a synthesis tree diagram, essentially it is found by determining the centroid of nodes representing reactants at each reaction stage in a successive manner until the final node representing the final product of the entire synthesis plan is reached.

$$y = \frac{1}{2^{n-1}}\left[\sum_{j=0}^{n-1} a_{j+1} \binom{n-1}{j}\right] = \frac{1}{2^{n-1}}\left[\sum_{j=0}^{n-1} a_{j+1} \left(\frac{(n-1)!}{j!(n-1-j)!}\right)\right] \quad (2.3)$$

Synthesis Plan Analysis

where $n > 2$, n is the number of points corresponding to the number of reactant input structures, a_{j+1} is the ordinate of the $(j+1)$th input, and $0! = 1$ by definition.

Synthesis Tree[1]

A diagram that facilitates the tracking of input reagents, intermediates, and the counting of reaction steps, branches, and points of convergence for a synthesis plan. It is also used to track the mole scales of each reaction from final target product to any intermediate or input material along the various branches. The shape of the diagram also is used to determine the degree of convergence, the degree of asymmetry parameters, and the first moment building up parameter. The diagram is depicted on a grid of squares. Input reagents are depicted as dark dots, intermediates are depicted as open dots, and the final product of the synthesis is depicted as a gray dot. The first input reagent appears as a dark dot situated at the origin. The identities and molecular weights of all input reagents along with their associated stoichiometric coefficients are written out adjacent to the dark dots. Each intermediate dot has a numerical subscript corresponding to its associated reaction step number. For each reaction step dotted lines connect the respective input reagents to the corresponding intermediate product for that step. The ordinate of each intermediate dot is determined using the centroid formula shown in Equation (2.3). Reaction yields for each reaction step are written along the x-axis below the diagram. For convergent plans, convergent branches and their step reaction yields are depicted using a * notation for branch 2, ** notation for branch 3, and so on. Yields pertaining to convergent branches are written below the set of reaction yields for the main branch, which appears immediately below the x-axis.

Example 2.1

Red mercuric oxide is made in a two-step sequence by reacting elemental mercury with chlorine gas and then treating the product with aqueous sodium hydroxide solution. The reaction yields for the two steps are 74% and 48%, respectively.

Part 1

Write out balanced chemical equations for the synthesis plan and draw its corresponding synthesis tree diagram.

SOLUTION

Step 1

$$Hg + Cl_2 \longrightarrow HgCl_2$$
$$200.59 \quad 70.9 \quad\quad\quad 271.49$$

Step 2

$$HgCl_2 + 2\,NaOH \longrightarrow HgO + 2\,NaCl + H_2O$$
$$271.49 \quad 2(40) \quad\quad 216.59 \quad 2\,(58.45) \quad 18$$

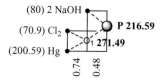

Part 2

From the synthesis tree diagram, determine the ordinate of the final product node, P, the degree of asymmetry and the degree of convergence.

Solution

For the first step, there are 2 input materials ($n = 2$):

$$y_1 = \frac{1}{2^{2-1}} \left[\sum_{j=0}^{2-1} \alpha_{j+1} \frac{(2-1)!}{j!(2-1-j)!} \right]$$

$$= \frac{1}{2} \left[\alpha_1 \frac{1!}{0!(1-0)!} + \alpha_2 \frac{1!}{1!(1-1)!} \right]$$

$$= \frac{1}{2} \left[0(1) + 1(1) \right]$$

$$= 0.5$$

For the second step, there are 2 input materials ($n = 2$):

$$y_2 = \frac{1}{2^{2-1}} \left[\sum_{j=0}^{2-1} \alpha_{j+1} \frac{(2-1)!}{j!(2-1-j)!} \right]$$

$$= \frac{1}{2} \left[\alpha_1 \frac{1!}{0!(2-1-0)!} + \alpha_2 \frac{1!}{1!(2-1-1)!} \right]$$

$$= \frac{1}{2} \left[0.5(1) + 2(1) \right]$$

$$= \frac{1}{2}(2.5)$$

$$= 1.25$$

Degree of asymmetry ($l = 3$ and $y = 1.25$):

$$\beta = \frac{2y}{l-1} - 1 = \frac{2(1.25)}{3-1} - 1 = 0.25$$

Synthesis Plan Analysis

Degree of convergence ($l=3$, $y=1.25$, $N=2$):

$$\delta = \frac{\arctan\left(\frac{l-1-y}{N}\right) + \arctan\left(\frac{y}{N}\right)}{2\arctan\left(\frac{l-1}{2}\right)}$$

$$= \frac{\arctan\left(\frac{3-1-1.25}{2}\right) + \arctan\left(\frac{1.25}{2}\right)}{2\arctan\left(\frac{3-1}{2}\right)}$$

$$= \frac{\arctan(0.375) + \arctan(0.625)}{2\arctan(1)}$$

$$= 0.5840$$

Example 2.2

Part 1

The ordinate values of dots for intermediate I_5 and reagents C, D, E, and F are 4, 8, 9, 10, and 11, respectively, for the sixth reaction in a synthesis tree diagram. Determine the centroid of the product dot pertaining to intermediate I_6 by taking midpoints of successive pairs of dots until one dot remains.

SOLUTION

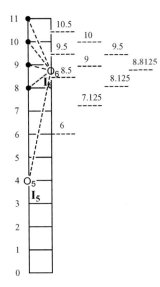

Part 2

Use Equation (2.3) to determine the centroid. Does your answer agree with that found in Part 1?

Solution

There are 5 input materials, hence $n = 5$.

$$y = \frac{1}{2^{5-1}} \left[\sum_{j=0}^{5-1} \alpha_{j+1} \left(\frac{(5-1)!}{j!(5-1-j)!} \right) \right]$$

$$= \frac{1}{16} \left[\sum_{j=0}^{4} \alpha_{j+1} \left(\frac{(4)!}{j!(4-j)!} \right) \right]$$

$$= \frac{3}{2} \left[\alpha_1 \frac{1}{0!(4-0)!} + \alpha_2 \frac{1}{1!(4-1)!} + \alpha_3 \frac{1}{2!(4-2)!} + \alpha_4 \frac{1}{3!(4-3)!} + \alpha_5 \frac{1}{4!(4-4)!} \right]$$

$$= \frac{3}{2} \left[4\left(\frac{1}{24}\right) + 8\left(\frac{1}{6}\right) + 9\left(\frac{1}{4}\right) + 10\left(\frac{1}{6}\right) + 11\left(\frac{1}{24}\right) \right]$$

$$= \frac{3}{2} \left[\left(\frac{1}{24}\right)(15) + \left(\frac{1}{6}\right)(18) + 9\left(\frac{1}{4}\right) \right]$$

$$= \frac{3}{2} \left[\frac{15 + 4(18) + 9(6)}{24} \right]$$

$$= \frac{3}{2} \left[\frac{141}{24} \right]$$

$$= 8.8125$$

Example 2.3

Part 1

For the synthesis tree diagram shown below, write out the synthesis plan as a sequence of reactions in the conventional way.

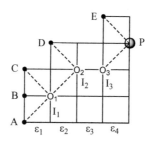

Solution

$$A + B + C \xrightarrow{} I_1 \xrightarrow{D} I_2 \xrightarrow{} I_3 \xrightarrow{E} P$$

Synthesis Plan Analysis

Part 2
How many steps are in the reaction plan? How many intermediates are isolated along the way?

Solution
There are four reaction steps and three intermediates that are isolated.

Part 3
For a linear plan of N steps, how many intermediates are isolated before the final product is reached?

Solution
There are N–1 reaction intermediates.

Part 4
What kind of reaction is involved in the third step if the molecular weight of I_3 is the same as that of I_2?

Solution
A rearrangement reaction.

Part 5
What kind of reaction is involved in the third step if the molecular weight of I_3 is less than that of I_2?

Solution
An elimination (fragmentation) reaction.

Example 2.4[2]

From the synthesis tree diagram shown below, deduce the synthesis plan for L-DOPA by writing out all the balanced chemical equations for each reaction step along with their reaction yields.

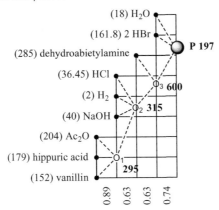

Solution

Note: The reported reaction yield of 63% for the classical resolution step is calculated with respect to the desired enantiomeric stereoisomer starting material. This explains why this number exceeds 50%. The correct yield should be 0.5*63% = 31.5%, which is calculated with respect to the racemic starting material.

Synthesis Plan Analysis

Example 2.5

Rank the shapes of all possible synthesis tree diagrams corresponding to plans involving four input reagents with respect to their degrees of asymmetry and convergence.

SOLUTION

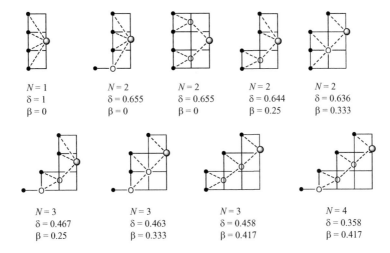

2.1.2 PROBLEMS

PROBLEM 2.1[3-6]

The synthesis of (-)-menthol from myrcene is given below, which involves a stereoselective hydrogenation in the second step using a chiral rhodium catalyst with a loading of 0.1 mol%.

(-)-Menthol synthesis:

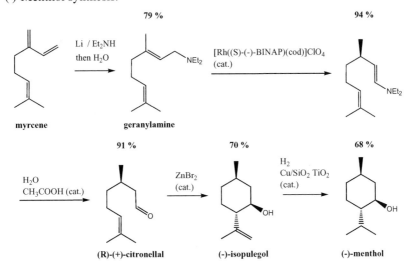

Part 1
Balance all reactions in the synthesis plan.

Part 2
Draw a synthesis tree diagram for the synthesis plan and determine the centroid of the final product node, the degree of convergence, and the degree of asymmetry.

Part 3
If 1 mole of (-)-menthol is synthesized, determine the number of moles of chiral catalyst needed in step 2.

PROBLEM 2.2[4,7,8]
The chiral rhodium catalyst used in the hydrogenation reaction for the menthol synthesis in Problem 2.1 is made according to the following sequence.

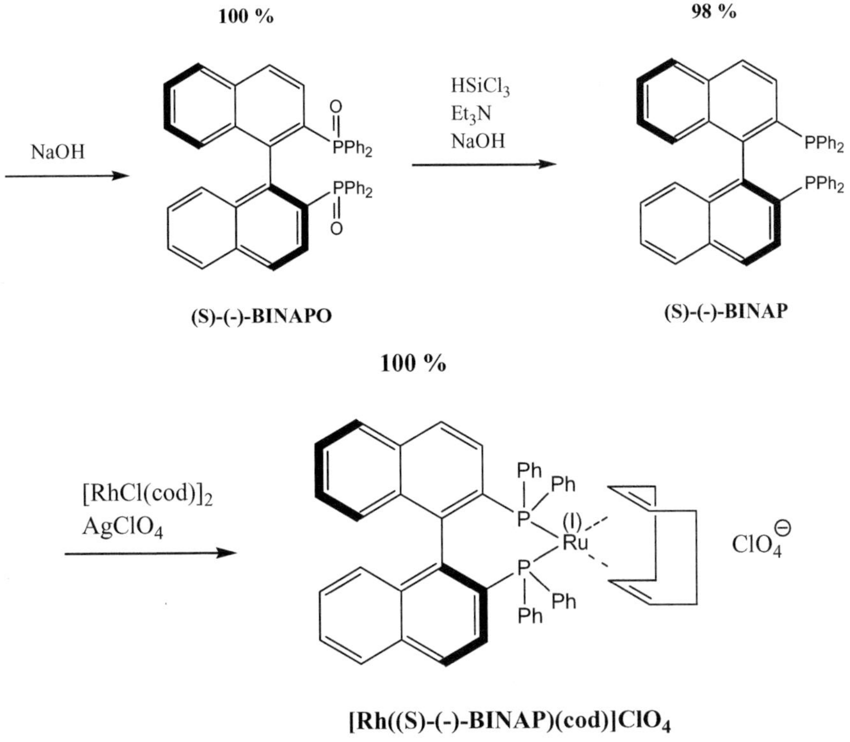

Synthesis Plan Analysis

100 %

(S)-(-)-BINAPO

98 %

(S)-(-)-BINAP

100 %

[Rh((S)-(-)-BINAP)(cod)]ClO$_4$

cod = 1,5-cyclooctadiene
DBT = dibenzoyltartrate

Part 1
Balance all reactions in the synthesis plan.

Part 2
Draw a synthesis tree diagram for the synthesis plan and determine the centroid of the final product node, the degree of convergence, and the degree of asymmetry.

Part 3
Determine the overall kernel process mass intensity (PMI) for the synthesis of 1 mole of (-)-menthol that includes the synthesis of the required amount of rhodium catalyst.

2.2 LINEAR SYNTHESIS PLANS

2.2.1 Terms, Definitions, and Examples

Synthesis: Linear
A synthesis plan that contains one branch made up of a consecutive sequence of reaction steps.

2.2.2 Problems

PROBLEM 2.3[9,10]

Lipases are often used to carry out enzymatic resolutions in selective acylation reactions for racemic alcohols or selective amidations for racemic amines. There are two senses in which they can be employed. In the first case, the unfunctionalized product is the desired product. An example of this is the second step of the synthesis of (R)-fluoxetine where the non-acetylated product is desired. In the second case, the functionalized product is the desired product. An example of this is the fourth step of the synthesis of (S)-dapoxetine where the unwanted amine is not amidated.

(R)-fluoxetine synthesis:

Synthesis Plan Analysis

(S)-dapoxetine synthesis:

Part 1
For each synthesis plan, provide balanced chemical equations for each step.

Part 2

For each synthesis plan, draw its corresponding synthesis tree diagram. Specify whether or not the reactant in the resolution step is included in the diagram.

PROBLEM 2.4[11-15]

The Solvay process to make sodium carbonate was invented in 1864. A series of reactions are shown below.

$$CaCO_3 \rightarrow CaO + CO_2$$

$$NH_3 + H_2O \rightarrow NH_4OH$$

$$NH_4OH + CO_2 \rightarrow NH_4(HCO_3)$$

$$NH_4(HCO_3) + NaCl \rightarrow NaHCO_3 + NH_4Cl$$

$$NaHCO_3 \rightarrow Na_2CO_3 + CO_2$$

Part 1

Balance each chemical equation showing by-products.

Part 2

Determine the overall balanced chemical equation for the process.

Part 3

Represent the entire process in the form of a synthesis tree diagram using general reaction yields.

Part 4

Determine the mass ratio of ammonium chloride to calcium oxide waste by-products. Determine the mass ratio of carbon dioxide input to carbon dioxide output.

Part 5

The following set of recycling reactions is added to convert the waste by-products back into ammonia.

$$CaO + H_2O \rightarrow Ca(OH)_2$$

$$Ca(OH)_2 + 2NH_4Cl \rightarrow 2NH_3 + CaCl_2 + 2H_2O$$

Determine the overall balanced chemical equation for the process including the recycling loop.

Part 6

Draw a synthesis tree diagram using general reaction yields for the recycling loop only.

Part 7
Determine the mass ratio of ammonium chloride to calcium oxide that is needed to match the input mass of ammonia in the main stream to produce sodium carbonate. Compare this value of the ratio with that found in Part 4. What is the condition that makes them equal?

2.3 CONVERGENT SYNTHESIS PLANS

2.3.1 TERMS, DEFINITIONS, AND EXAMPLES

Synthesis: Convergent

A synthesis plan that contains at least two independent branches made up of parallel consecutive sequences of reaction steps leading to intermediates that are used in convergent reaction steps. Such plans are made up of B branches and B−1 points of convergence. The branch with the longest number of reaction steps is the main branch and is the one used to determine the overall yield of the plan.

Synthesis: Divergent

A synthesis plan that contains one branch made up of a consecutive sequence of reaction steps leading to a common hub intermediate, which is then used to make different target products via divergent branches in subsequent steps along parallel but independent paths.

Example 2.6

A convergent synthesis plan according to the following sequence of reactions is shown below.

step 1 $\quad A + B \longrightarrow I_1 + Q_1 \quad$ Yield $= \varepsilon_1$

step 2 $\quad I_1 + C \longrightarrow I_2 + Q_2 \quad$ Yield $= \varepsilon_2$

step 2* $\quad D + E \longrightarrow I_2^* + Q_3 \quad$ Yield $= \varepsilon_2^*$

step 3 $\quad I_2 + I_2^* \longrightarrow I_3 + Q_4 \quad$ Yield $= \varepsilon_3$

step 4 $\quad F + I_3 \longrightarrow P + Q_5 \quad$ Yield $= \varepsilon_4$

This series of reactions can be written as follows to show the convergence of two branches.

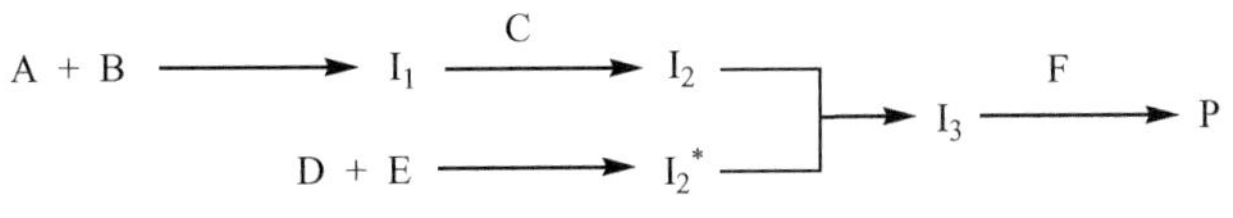

$$A + B \longrightarrow I_1 \xrightarrow{C} I_2 \searrow$$
$$\qquad\qquad\qquad\qquad\qquad I_3 \xrightarrow{F} P$$
$$D + E \longrightarrow I_2^* \nearrow$$

Part 1
Draw the corresponding synthesis tree diagram.

Solution

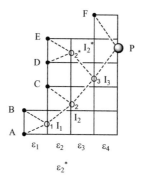

Note that all intermediate open dots lie at the median of their preceding feeder dark dots that represent their precursor reagents. Note also that the second branch depicted by the reaction $D + E \rightarrow I_2^* + Q_3$ occurs in parallel with the second step in the main branch of the synthesis. The asterisk labels refer to this second branch. The convergent step is the third step of the sequence.

This convergent synthesis is composed of two branches and one point of convergence.

Part 2

If a convergent synthesis plan has B branches, how many points of convergence are expected?

Solution

Maximum number of points of convergence $= B - 1$.

The maximum number of points of convergence is equal to the maximum partition number for the integer B. For example, if we have a plan with 3 branches then the maximum partition number is 2. Hence, the number 3 can be partitioned into 1 part, namely 3; and 2 parts, namely, 2 and 1 or 1 and 2 depending on the order. A 1-partition corresponds to 1 point of convergence and a 2-partition corresponds to 2 points of convergence in the plan. We can make the statement that:

Maximum number of points of convergence = maximum partition number for B.

Table 2.1 gives the possible partitions for plans containing up to six branches.

Note that when $B = 1$, we have a linear plan and when $B > 1$ we have convergent plans.

The possible partitions in the table may be given in graphical form as follows. Points of convergence are shown as red dots.

$B = 1$

(1)

$B = 2$

(2)

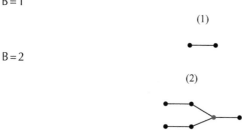

Synthesis Plan Analysis

TABLE 2.1
List of Possible Partitions for Plans Containing up to Six Branches

No. Branches	Maximum No. Partitions	Possible Partitions
1	0	1
2	1	2
3	2	3
		(2,1); (1,2)
4	3	4
		(3,1); (1,3)
		(2.1.1); (1,2,1); (1,1,2)
5	4	5
		(4,1); (1,4)
		(3.2); (2.3)
		(2.2,1); (2,1,2); (1,2,2)
		(3,1,1); (1,3,1); (1,1,3)
		(1,1,1,2); (1,1,2,1); (1,2,1,1); (2,1,1,1)
6	5	6
		(5,1); (1,5)
		(4,2); (2,4)
		(3,3)
		(4,1,1); (1,4,1); (1,1,4)
		(3,2,1); (2,3,1); (2,1,3); (3,1,2); (1,2,3); (1,3,2)
		(2,2,2)
		(1,1,1,3); (1,1,3,1); (1,3,1,1); (3,1,1,1)
		(2,2,1,1); (2,1,2,1); (1,2,2,1); (2,1,1,2); (1,1,2,2); (1,2,1,2)

B = 3

    (3)   (1,2); (2,1)

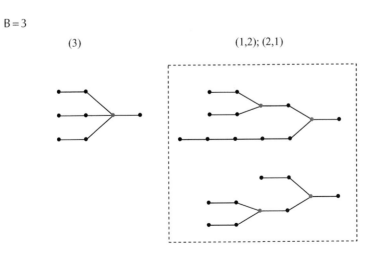

B = 4

(4)　　　　　(3,1); (1,3)　　　　　(2,1,1); (1,2,1); (1,1,2)

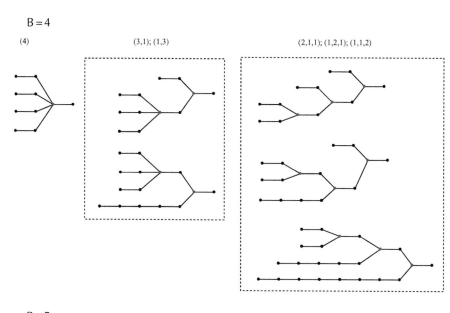

B = 5

(5)　　　　　(4,1); (1,4)　　　　　(3,2); (2,3)

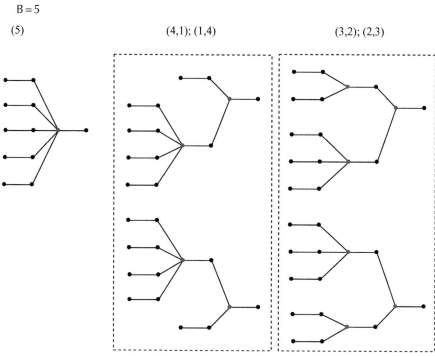

Synthesis Plan Analysis

(2,2,1); (2,1,2); (1,2,2)

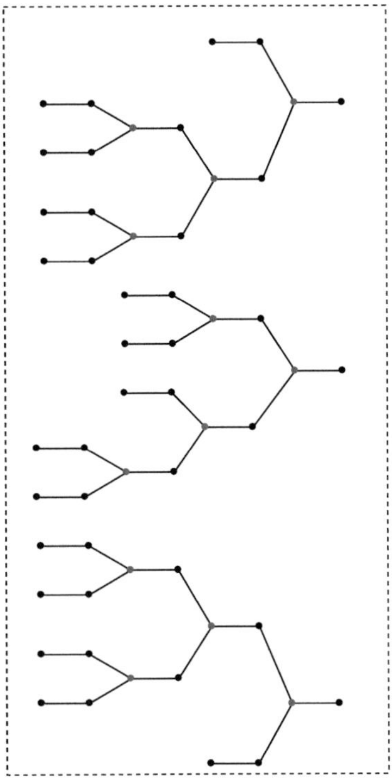

(3,1,1); (1,3,1); (1,1,3)

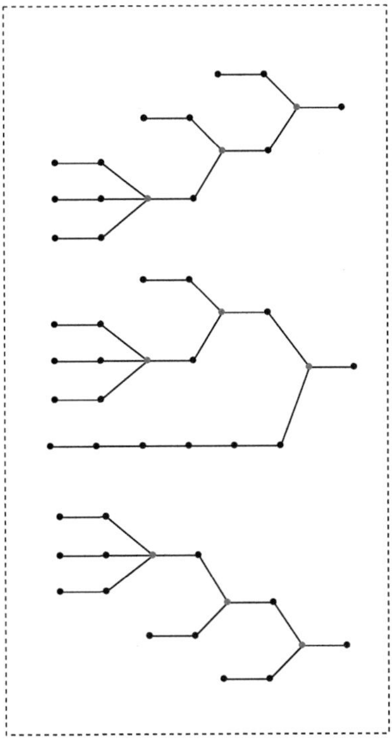

(1,1,1,2); (1,1,2,1); (1,2,1,1); (2,1,1,1)

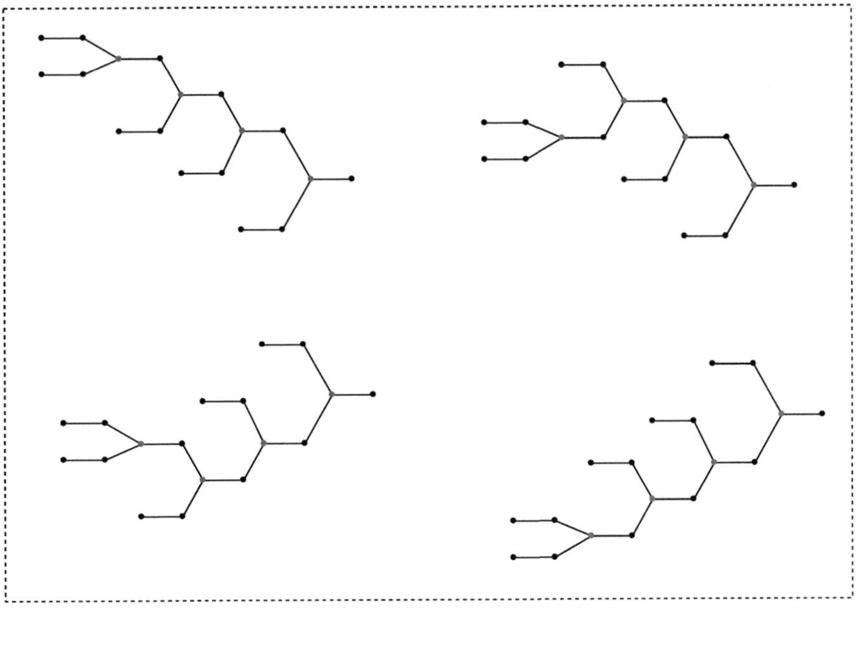

B = 6

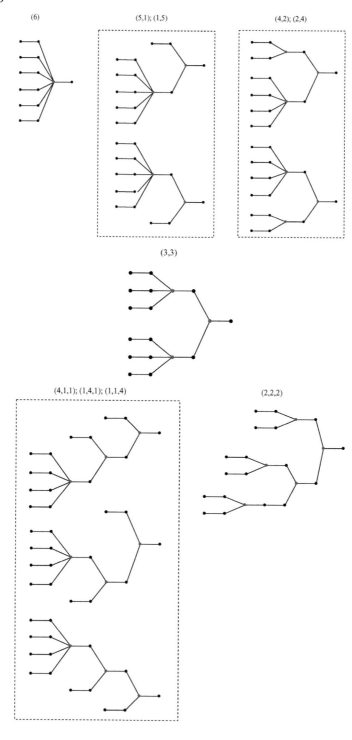

Synthesis Plan Analysis

(3,2,1); (2,3,1); (2,1,3); (3,1,2); (1,2,3); (1,3,2)

(1,1,1,3); (1,1,3,1); (1,3,1,1); (3,1,1,1)

(2,2,1,1); (2,1,2,1); (1,2,2,1); (2,1,1,2); (1,1,2,2); (1,2,1,2)

Example 2.7

A synthesis tree diagram for a multi-convergent plan is shown below.

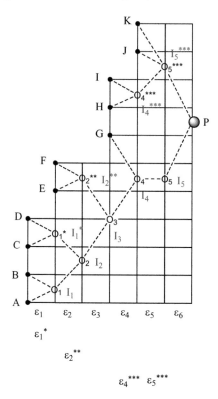

Part 1
How many reaction steps are in the plan?

SOLUTION

Total number of steps = 6 + 1 + 1 + 2 = 10.

Part 2
How many branches are in the plan?

SOLUTION

Number of branches = 4.

Part 3
How many points of convergence are in the plan?

SOLUTION

Number of points of convergence = 3 (at I_2, I_3, and P).

Part 4

If 1 mole of target product P is to be made, what are the number of moles for intermediate pairs I_1 and I_1^*, I_2 and I_2^{**}, and I_5 and I_5^{***}"?"

Solution

Intermediate	Number of Moles
I_1	$1/(\varepsilon_2\, \varepsilon_3\, \varepsilon_4\, \varepsilon_5\, \varepsilon_6)$
I_1^*	$1/(\varepsilon_2\, \varepsilon_3\, \varepsilon_4\, \varepsilon_5\, \varepsilon_6)$
I_2	$1/(\varepsilon_3\, \varepsilon_4\, \varepsilon_5\, \varepsilon_6)$
I_2^*	$1/(\varepsilon_3\, \varepsilon_4\, \varepsilon_5\, \varepsilon_6)$
I_5	$1/(\varepsilon_6)$
I_5^{***}	$1/(\varepsilon_6)$

Part 5

Draw out a conventional synthesis scheme showing clearly the convergent steps.

Solution

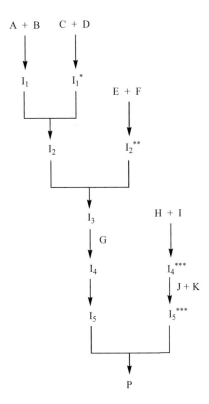

Part 6
What is the overall yield along the branch beginning with reagents E and F?

Solution
Overall yield from reagents E and F = $(\varepsilon_2^{**}) (\varepsilon_3) (\varepsilon_4) (\varepsilon_5) (\varepsilon_6)$.

Example 2.8

A multi-convergent synthesis is composed of five branches. The main (longest) branch has ten reaction steps, the second branch has three steps, the third branch has three steps, the fourth branch has two steps, and the fifth branch has four steps.

Part 1
How many intermediate products are isolated in such a plan before reaching the final product?

Solution

Branch	No. steps	No. intermediates
1	10	9
2	3	3
3	3	3
4	2	2
5	4	4
TOTAL	22	21

Part 2
Derive an expression for the number of intermediates isolated in a general multi-convergent plan that depends on the number of branches and the number of reaction steps in each branch.

Solution
Let the main branch (B1) have N steps. Therefore, the number of intermediates in the main branch is $N-1$.
Let the second branch (B2) have M_2 steps, the third branch have M_3 steps, etc. The total number of intermediates is given by $N-1+M_2+M_3+M_4+\ldots$
We may write the result as

$$N-1+\sum_{j=2}^{B-1} M_j,$$

where N is the number of steps in the main branch and B is the total number of branches in the plan.

Synthesis Plan Analysis

Example 2.9[16]

A generic linear route containing five reaction steps and a generic convergent route also containing five reaction steps are shown below.

Linear route:

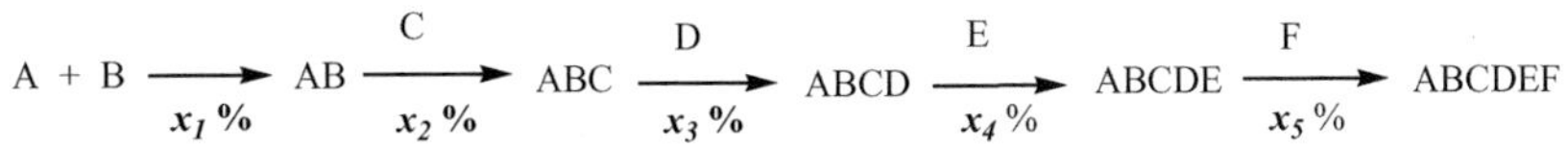

$$A + B \xrightarrow{x_1\%} AB \xrightarrow[x_2\%]{C} ABC \xrightarrow[x_3\%]{D} ABCD \xrightarrow[x_4\%]{E} ABCDE \xrightarrow[x_5\%]{F} ABCDEF$$

Convergent route:

$$\begin{array}{c} A + B \xrightarrow{y_1\%} AB \\ C + D \xrightarrow{y_2\%} CD \end{array} \xrightarrow{y_3\%} ABCD$$

$$E + F \xrightarrow{y_4\%} EF$$

combining to give ABCDEF with yield $y_5\%$.

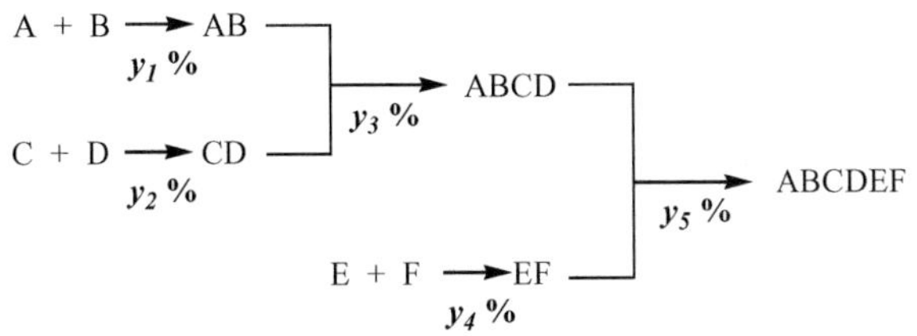

Part 1
For each plan, determine the overall yield.

SOLUTION

For the linear route, the overall yield is $x_1\% * x_2\% * x_3\% * x_4\% * x_5\%$.

For the convergent, the overall yield is defined along the longest linear path. In the scheme shown there are two such paths of equal length: $y_1\% * y_3\% * y_5\%$ and $y_2\% * y_3\% * y_5\%$.

Part 2
Under what conditions will the overall yield of the linear plan exceed that of the convergent plan? What about the other way around?

SOLUTION

Linear plan exceeding the convergent plan:
$x_1\% * x_2\% * x_3\% * x_4\% * x_5\% > y_1\% * y_3\% * y_5\%$
and
$x_1\% * x_2\% * x_3\% * x_4\% * x_5\% > y_2\% * y_3\% * y_5\%$

Convergent route exceeding linear route:
$x_1\% * x_2\% * x_3\% * x_4\% * x_5\% < y_1\% * y_3\% * y_5\%$
and
$x_1\% * x_2\% * x_3\% * x_4\% * x_5\% < y_2\% * y_3\% * y_5\%$

Part 3
Give numerical examples that satisfy the results of Part 2.

Solution

Linear plan exceeding convergent plan:

90%*90%*100%*90%*90% > 50%*60%*70%, or 65.61% > 21.00%

and

90%*90%*100%*90%*90% > 50%*70%*70%, or 65.61% > 24.50%.

Convergent plan exceeding linear plan:

80%*80%*80%*80%*80% < 90%*90%*90%, or 32.77% < 72.90%

and

80%*80%*80%*80%*80% < 80%*90%*70%, or 32.77% < 50.40%.

Part 4
What can be concluded about the relative efficiencies of linear plans versus convergent ones with respect to the overall yield metric?

Solution
No definite statement can be made because the outcome entirely depends on the number of steps and the yield values for each reaction step.

Example 2.10

Two synthesis tree diagrams are shown below. Figure A represents a three-step linear plan and Figure B represents a three-step convergent plan with two parallel steps.

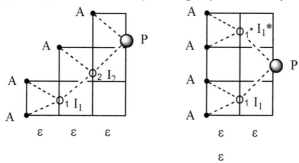

Part 1
From the tree diagrams and the variables shown, determine expressions for the respective kernel reaction mass efficiencies.

Solution

Linear plan: $\text{RME} = \dfrac{p}{\dfrac{2\alpha}{\varepsilon^3} + \dfrac{\alpha}{\varepsilon^2} + \dfrac{\alpha}{\varepsilon}}$

Convergent plan: $\text{RME} = \dfrac{p}{\dfrac{2\alpha}{\varepsilon^2} + \dfrac{2\alpha}{\varepsilon^2}}$

Synthesis Plan Analysis

Part 2

From the two kernel RME expressions in Part 1, plot the respective functions in reaction yield on the same graph. Which function is greater than the other? Plot the difference function as a function of reaction yield. Determine at what reaction yield is the difference between the functions at maximum.

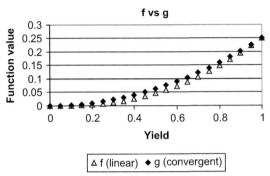

2-step linear function: $f = \dfrac{1}{\dfrac{2}{\varepsilon^3}+\dfrac{1}{\varepsilon^2}+\dfrac{1}{\varepsilon}} = \dfrac{\varepsilon^3}{2+\varepsilon+\varepsilon^2}$

2-step convergent function: $g = \dfrac{1}{\dfrac{2}{\varepsilon^2}+\dfrac{2}{\varepsilon^2}} = \dfrac{1}{\dfrac{4}{\varepsilon^2}} = \dfrac{\varepsilon^2}{4}$

Difference function:

$$g - f = h = \dfrac{\varepsilon^2}{4} - \dfrac{\varepsilon^3}{\varepsilon^2+\varepsilon+2} = \dfrac{\varepsilon^4+\varepsilon^3+2\varepsilon^2-4\varepsilon^3}{4(\varepsilon^2+\varepsilon+2)} = \dfrac{\varepsilon^4-3\varepsilon^3+2\varepsilon^2}{4(\varepsilon^2+\varepsilon+2)}$$

Extrema exist when $\dfrac{dh}{d\varepsilon} = 0$.

$$\dfrac{dh}{d\varepsilon} = \dfrac{(4\varepsilon^3-9\varepsilon^2+4\varepsilon)(4)(\varepsilon^2+\varepsilon+2)-(\varepsilon^4-3\varepsilon^3+2\varepsilon^2)(4)(2\varepsilon+1)}{16(\varepsilon^2+\varepsilon+2)^2}$$

$$= \dfrac{\varepsilon(4\varepsilon^2-9\varepsilon+4)(\varepsilon^2+\varepsilon+2)-\varepsilon^2(\varepsilon^2-3\varepsilon+2)(2\varepsilon+1)}{4(\varepsilon^2+\varepsilon+2)^2}$$

Therefore,

$$\varepsilon(4\varepsilon^2 - 9\varepsilon + 4)(\varepsilon^2 + \varepsilon + 2) = \varepsilon^2(\varepsilon^2 - 3\varepsilon + 2)(2\varepsilon + 1)$$

$$(4\varepsilon^2 - 9\varepsilon + 4)(\varepsilon^2 + \varepsilon + 2) = \varepsilon(\varepsilon^2 - 3\varepsilon + 2)(2\varepsilon + 1)$$

$$4\varepsilon^4 + 4\varepsilon^3 + 8\varepsilon^2 - 9\varepsilon^3 - 9\varepsilon^2 - 18\varepsilon + 4\varepsilon^2 + 4\varepsilon + 8 = \varepsilon(2\varepsilon^3 + \varepsilon^2 - 6\varepsilon^2 - 3\varepsilon + 4\varepsilon + 2)$$

$$4\varepsilon^4 - 5\varepsilon^3 + 3\varepsilon^2 - 14\varepsilon + 8 = \varepsilon(2\varepsilon^3 - 5\varepsilon^2 + \varepsilon + 2)$$

$$4\varepsilon^4 - 5\varepsilon^3 + 3\varepsilon^2 - 14\varepsilon + 8 = 2\varepsilon^4 - 5\varepsilon^3 + \varepsilon^2 + 2\varepsilon$$

$$2\varepsilon^4 + 2\varepsilon^2 - 16\varepsilon + 8 = 0$$

$$\varepsilon^4 + \varepsilon^2 - 8\varepsilon + 4 = 0$$

The roots of the quartic are given by
$\varepsilon = 0.54903903241396$ and $\varepsilon = 1.57181887893436$

Since $\varepsilon < 1$, we take $\varepsilon = 0.54903903241396$ as the root. Hence, the difference between the convergent and linear functions is a maximum at a reaction yield of about 55%.

Example 2.11[17–19]

Creatine is a muscle enhancing supplement used by athletes to bulk up their muscle mass. It is synthesized by the following convergent plan.

Part 1
Rewrite the scheme showing balanced chemical equations.

Solution

Synthesis Plan Analysis

Part 2

Represent this as a synthesis tree diagram and determine the kernel mass efficiency for the plan.

SOLUTION

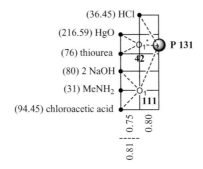

$$RME_{kernel} = \frac{131}{\dfrac{216.59+76}{0.81*0.8} + \dfrac{94.45+31+80}{0.75*0.8} + \dfrac{36.45}{0.8}}$$

$$= \frac{131}{451.528 + 342.417 + 45.563}$$

$$= \frac{131}{839.507}$$

$$= 0.156$$

or 15.6%.

Part 3

Two side products are produced in the convergent step: creatinine and dicyandiamide. Suggest how these could be accounted for using reaction mechanisms.

creatinine **dicyandiamide**

SOLUTION

dicyandiamide

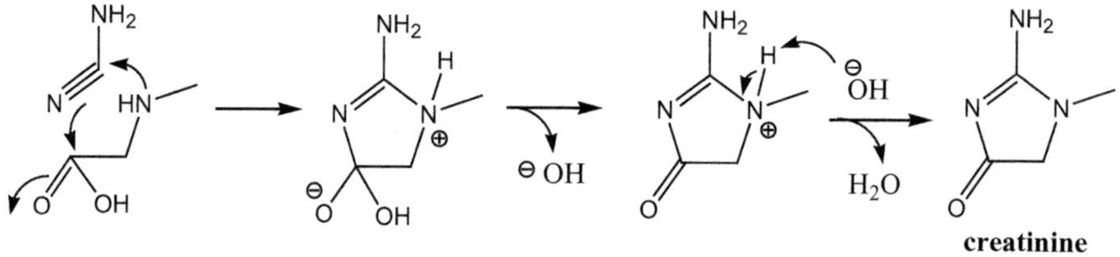

2.3.2 PROBLEMS

PROBLEM 2.5[20–22]

The dyestuff aniline yellow is made according to the synthesis plan shown below.

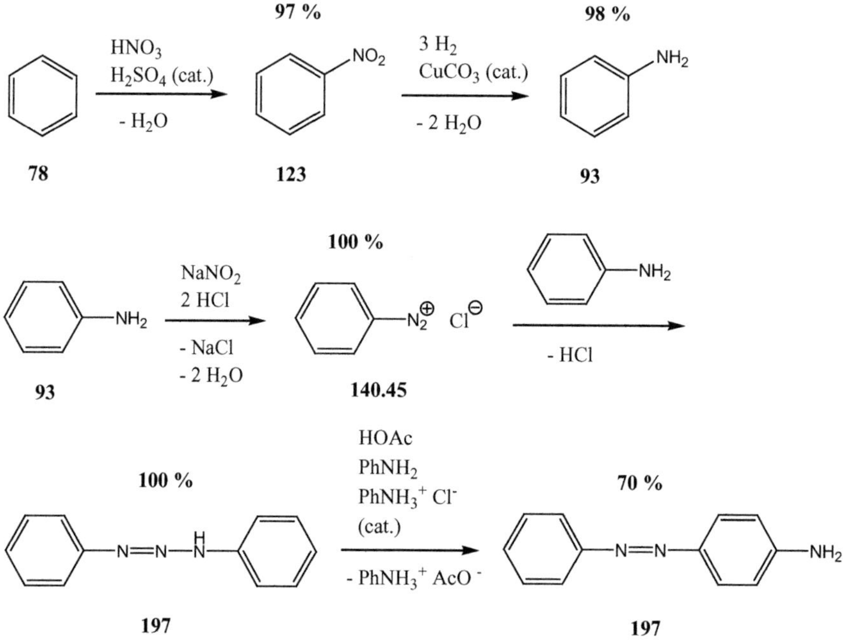

Part 1

Draw a synthesis tree diagram for the plan.

Part 2

Determine the mass of aniline required in each of the places it occurs as a reagent.

Part 3

Determine the mass split ratio pertaining to Part 2.

Synthesis Plan Analysis

Part 4

If the reaction yield for nitration of benzene was reduced by 10%, how would this impact the mass of waste generated from the production of nitrobenzene at each stage it appears in the plan, and on the overall kernel E-factor for the plan?

Part 5

Repeat Parts 2, 3, and 4 if the diazotization and subsequent coupling reactions have yields that are 90% and 80%, respectively.

PROBLEM 2.6

Sodium thiosulfate is produced industrially according to the following process:

$$FeS + 2HCl \rightarrow FeCl_2 + H_2S$$

$$4FeS + 7O_2 \rightarrow 2Fe_2O_3 + 4SO_2$$

$$SO_2 + 2NaOH \rightarrow Na_2SO_3 + H_2O$$

$$SO_2 + 2H_2S \rightarrow 3S + 2H_2O$$

$$Na_2SO_3 + S \rightarrow Na_2S_2O_3$$

Part 1

Draw a synthesis tree diagram for this process using general reaction yields.

Part 2

Determine the ratio of mass of FeS fed to produce SO_2 versus that fed to produce H_2S.

Part 3

Determine the ratio of mass of SO_2 fed to produce Na_2SO_3 versus that fed to produce S.

Part 4

The unit costs for the input materials are shown below.

NaOH	$212.20 per 12 kg
O_2	$171.60 per 56 L at STP (25 C, 1 atm)
FeS	$28.50 per 250 g
HCl	$125.60 per 10 L of 37 wt% solution ($d = 1.18$ g/mL)

Determine the minimum cost to produce 1 kg of sodium thiosulfate by this process.

Determine the general input material cost function for a mass of X grams of sodium thiosulfate using the general yields used in Part 1.

2.4 REACTION NETWORKS

2.4.1 Terms, Definitions, and Examples

Example 2.12

An industrial network for the production of aniline from benzene starting material is shown below.

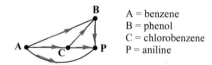

Part 1

Determine the number of routes in the network.

Solution

Reduced digraph representing reaction network

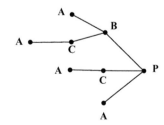

A = benzene
B = phenol
C = chlorobenzene
P = aniline

Path diagram

There are four routes in the network.

Part 2

Enumerate the routes.

Solution

Route #1: A → B → P
Route #2: A → C → B → P
Route #3: A → C → P
Route #4: A → P

Part 3

For each route, write out balanced chemical equations, draw a synthesis tree diagram, and determine the respective atom economies.

Solution

Route #1

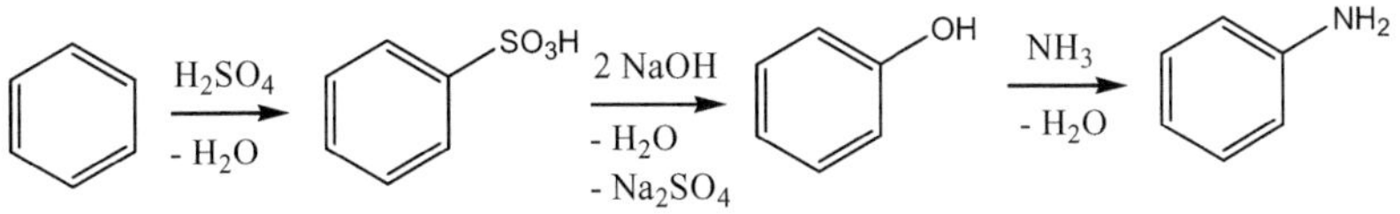
SO3H
H2SO4
- H2O
2 NaOH
- H2O
- Na2SO4
OH
NH3
- H2O
NH2

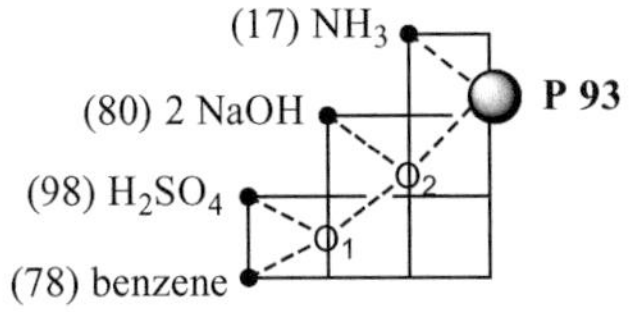
(17) NH3
(80) 2 NaOH
(98) H2SO4
(78) benzene
P 93
1
2

AE = 93/(78 + 98 + 80 + 17) = 0.341.

Route #2

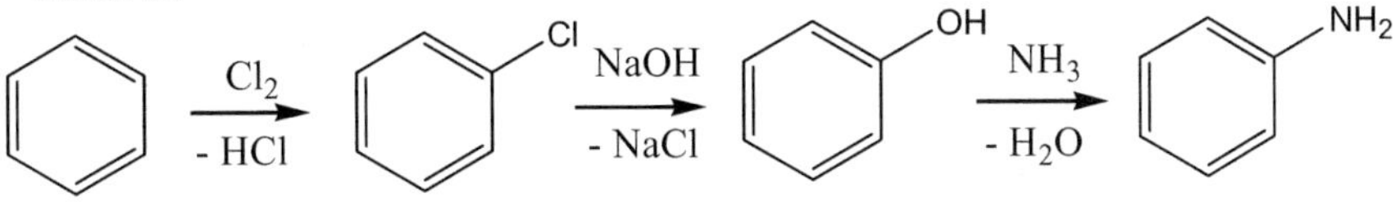
Cl2
- HCl
Cl
NaOH
- NaCl
OH
NH3
- H2O
NH2

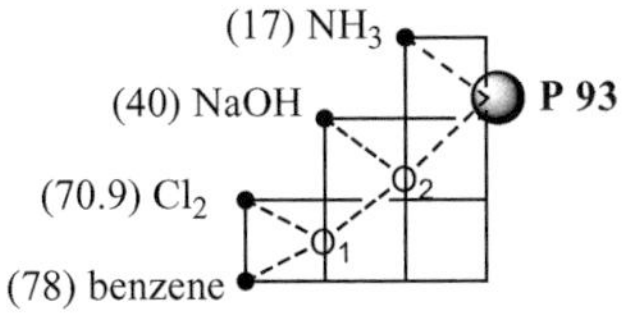
(17) NH3
(40) NaOH
(70.9) Cl2
(78) benzene
P 93
1
2

AE = 93/(78 + 70.9 + 40 + 17) = 0.452.

Route #3

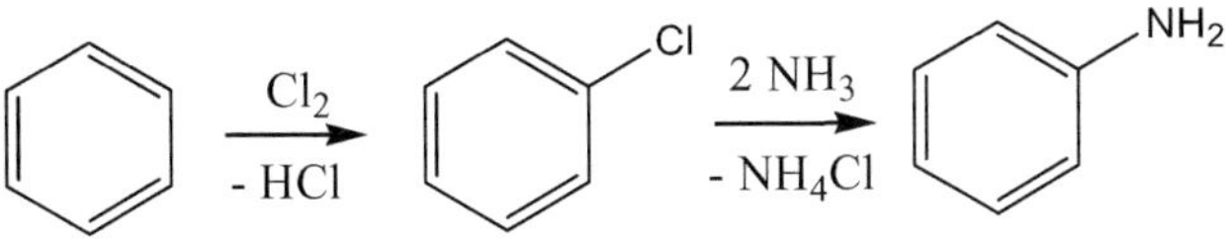
Cl2
- HCl
Cl
2 NH3
- NH4Cl
NH2

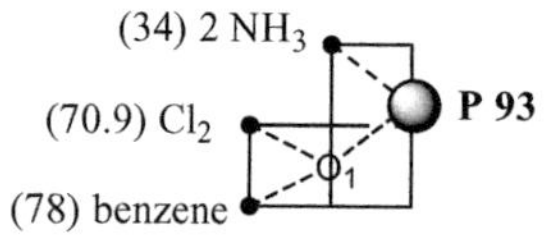

$AE = 93/(78 + 70.9 + 34) = 0.508$.

Route #4

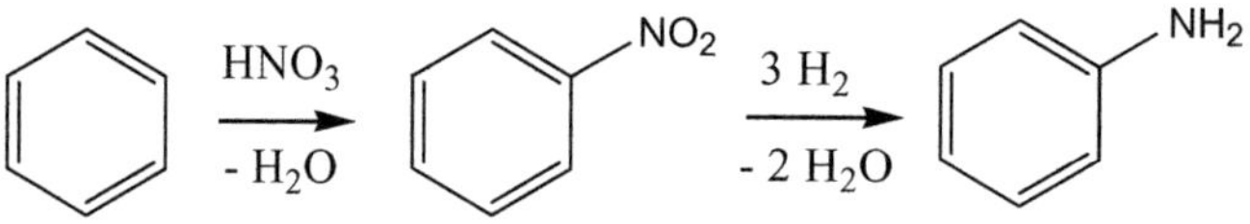

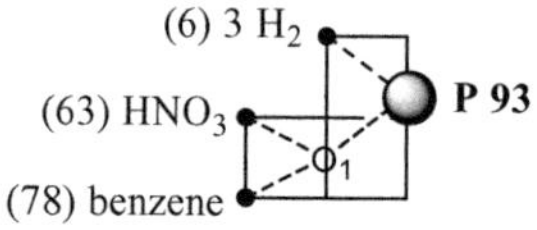

$AE = 93/(78 + 63 + 6) = 0.633$.

Part 4

Which route is the most atom economical?

SOLUTION

Route #4 has the highest atom economy.

Example 2.13

A divergent industrial network for ethylene feedstock is shown below.

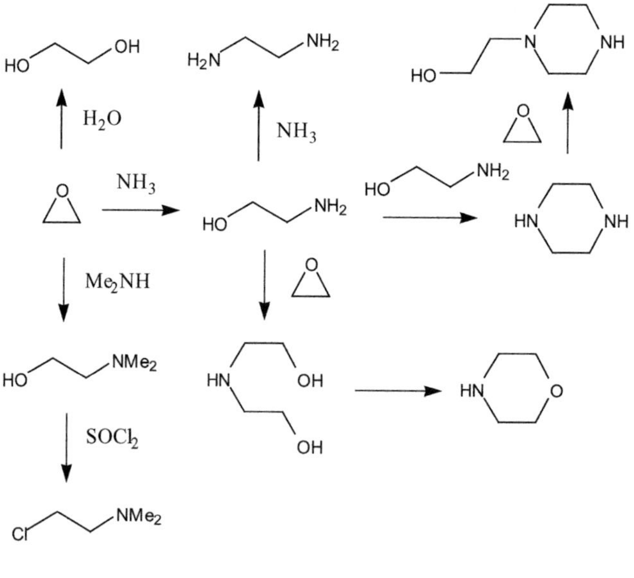

Synthesis Plan Analysis

Part 1
Determine the number of routes in the network.

Solution
Reduced digraph representing reaction network

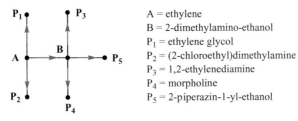

A = ethylene
B = 2-dimethylamino-ethanol
P_1 = ethylene glycol
P_2 = (2-chloroethyl)dimethylamine
P_3 = 1,2-ethylenediamine
P_4 = morpholine
P_5 = 2-piperazin-1-yl-ethanol

There are five routes in the network.

Part 2
Enumerate the routes to each product and write out balanced chemical equations for each route.

Solution
Route #1: $A \to P_1$
Route #2: $A \to P_2$
Route #3: $A \to B \to P_3$
Route #4: $A \to B \to P_4$
Route #5: $A \to B \to P_5$

Route #1

[ethylene oxide] + H$_2$O ⟶ HO–CH$_2$CH$_2$–OH

Route #2

[ethylene oxide] + Me$_2$NH ⟶ HO–CH$_2$CH$_2$–NMe$_2$ $\xrightarrow[\text{– HCl}]{\text{SOCl}_2,\ \text{– SO}_2}$ Cl–CH$_2$CH$_2$–NMe$_2$

Route #3

[ethylene oxide] + NH$_3$ ⟶ HO–CH$_2$CH$_2$–NH$_2$ $\xrightarrow[\text{– H}_2\text{O}]{\text{NH}_3}$ H$_2$N–CH$_2$CH$_2$–NH$_2$

Route #4

[ethylene oxide] + NH$_3$ ⟶ HO–CH$_2$CH$_2$–NH$_2$ $\xrightarrow{\text{[ethylene oxide]}}$ HO–CH$_2$CH$_2$–NH–CH$_2$CH$_2$–OH $\xrightarrow{\text{– H}_2\text{O}}$ morpholine

Route #5

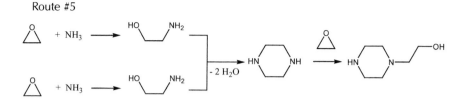

Part 3

If 1 tonne of each product is desired determine the starting mass of ethylene that is needed to synthesize all products in the network. Assume that each reaction yield in the network is 90%.

SOLUTION

Synthesis tree diagrams for products

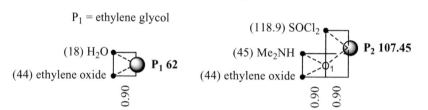

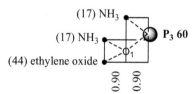

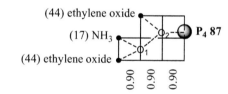

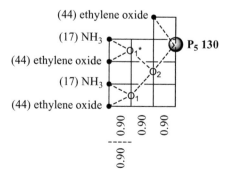

Synthesis Plan Analysis

From the synthesis trees we have the following results.

Product	MW Product (g/mol)	Mass of Product (kg)	Moles of Product	Moles of Ethylene Oxide Required	Mass of Ethylene Oxide Required (kg)
P_1	62	1000	16129.03	16129.03/0.9 = 17921.14	17.92114*44 = 788.53
P_2	107.45	1000	9306.65	9306.65/(0.9)^2 = 11489.69	11.48969*44 = 505.55
P_3	60	1000	16666.67	16666.67/(0.9)^2 = 20576.14	20.57614*44 = 905.35
P_4	87	1000	11494.25	11494.25/(0.9)^3 = 15767.15 11494.25/(0.9)^2 = 14190.43	15.76715*44 = 693.75 14.19043*44 = 624.38
P_5	130	1000	7692.31	7692.31/0.9 = 8547.01 7692.31/(0.9)^3 = 10551.87 7692.31/(0.9)^3 = 10551.87	8.54701*44 = 376.07 10.55187*44 = 464.28 10.55187*44 = 464.28

Total mass of ethylene oxide required for all products
 = 788.53 + 505.55 + 905.35 + 693.75 + 624.38 + 376.07 + 464.28 + 464.28
 = 4822.19 kg
 = 4.8 tonnes

2.4.2 PROBLEMS

PROBLEM 2.7

Part 1

Complete the retrosynthetic plan for the "green" acrylamide substitute compound, vinylformamide, by filling in appropriate reagents as necessary in the forward direction.

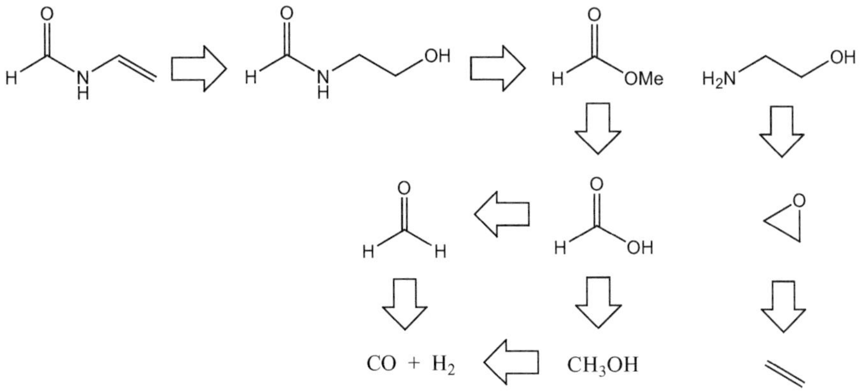

Part 2

Incorporate the above plan with the more complex one shown below. Draw an overall reaction network showing all possible routes. Using synthesis tree diagrams enumerate the routes and determine which one is the most atom economical.

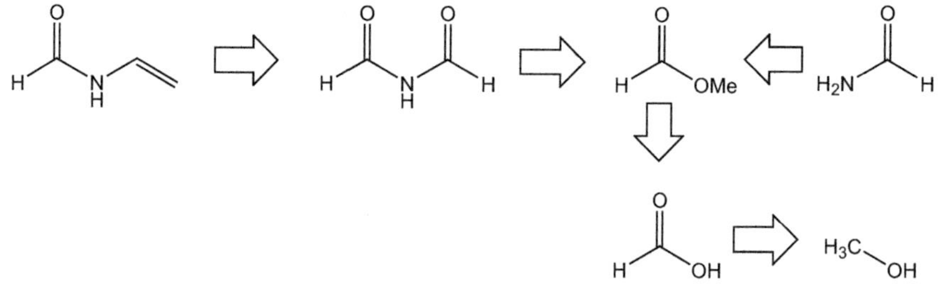

PROBLEM 2.8

An industrial network for the production of vanillin from eugenol and guaiacol starting materials is shown below.

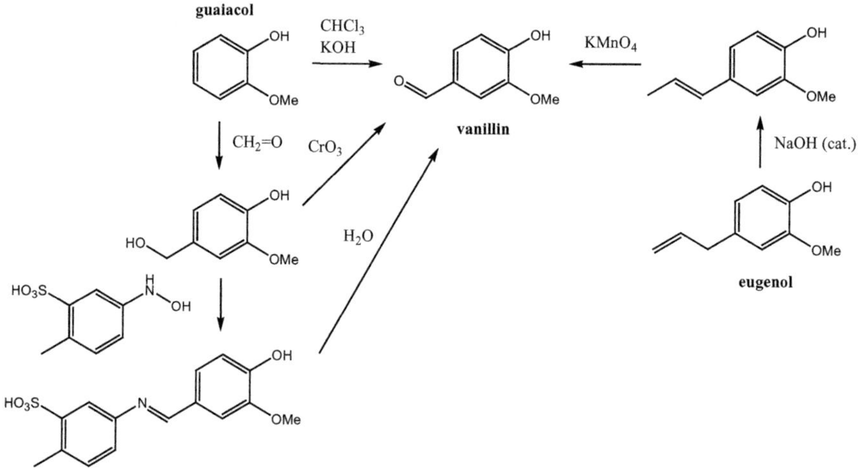

Part 1
Determine the number of routes in the network.

Part 2
Enumerate the routes.

Part 3
For each route, write out balanced chemical equations, draw a synthesis tree diagram, and determine the respective atom economies.

Synthesis Plan Analysis

Part 4
Which route is the most atom economical?

Problem 2.9
An industrial network for the production of *p*-toluidine from benzene starting material is shown below.

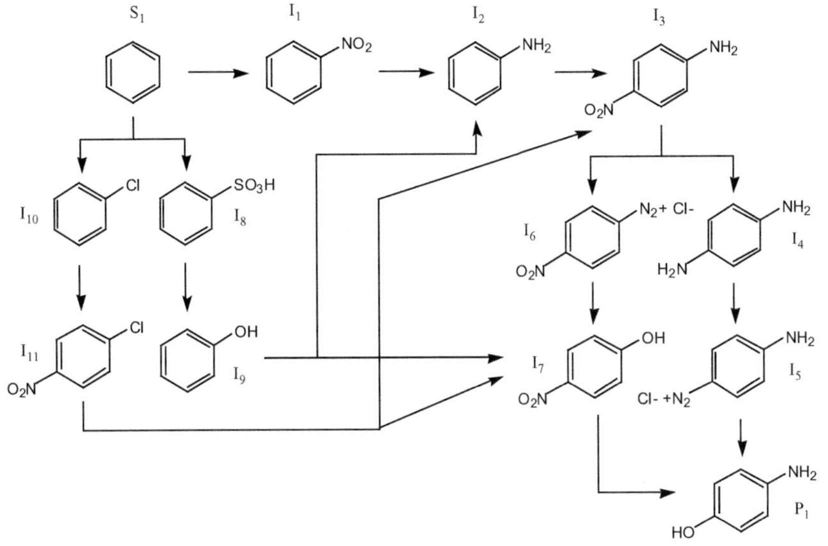

Part 1
Determine the number of routes in the network.

Part 2
Enumerate the routes.

Part 3
For each route, write out balanced chemical equations, draw a synthesis tree diagram, and determine the respective atom economies.

Part 4
Which route is the most atom economical?

PROBLEM 2.10
If 1 kg each of N,N-dimethyl-*p*-toluidine, *p*-toluidine hydrochloride salt, N-acetyltoluidine, and tolylmethylamine are to be prepared from toluene determine the mass of toluene required given the scheme shown below in terms of the reaction yield parameters shown.

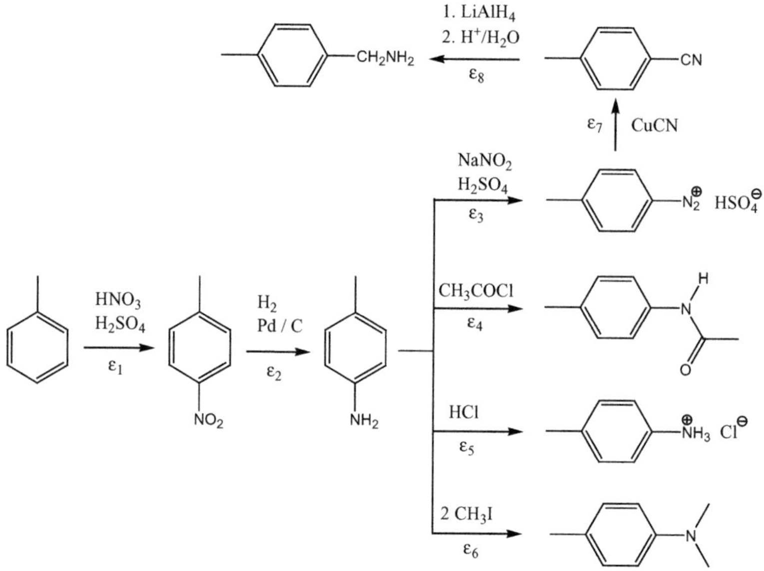

PROBLEM 2.11
An industrial network for the production of glycerol from propylene is shown below.

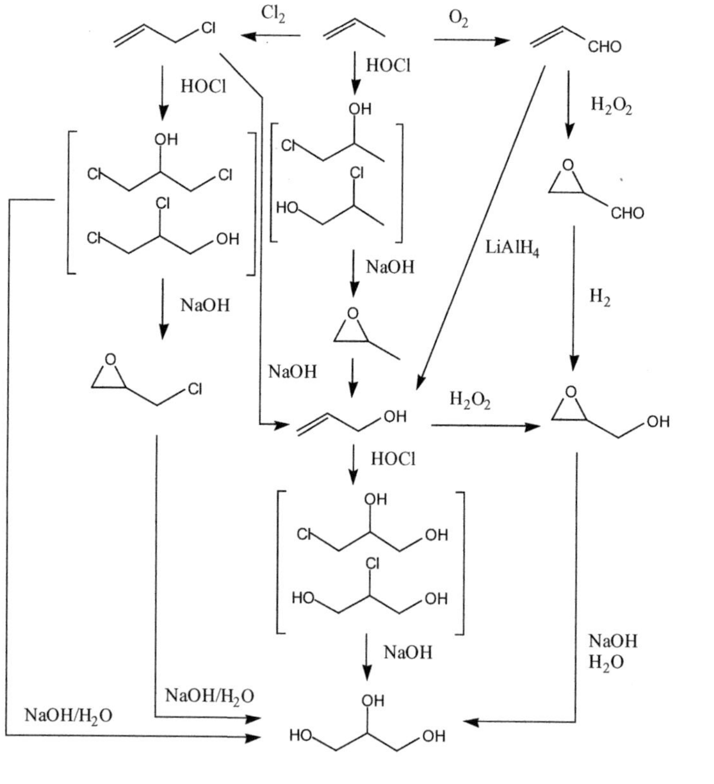

Synthesis Plan Analysis

Part 1

Determine the number of routes in the network.

Part 2

Enumerate the routes.

Part 3

For each route, write out balanced chemical equations, draw a synthesis tree diagram, and determine the respective atom economies.

Part 4

Which route is the most atom economical?

2.5 RANKING SYNTHESIS PLANS ACCORDING TO MATERIAL EFFICIENCY

2.5.1 Terms, Definitions, and Examples

Concession Step[23,24]

A reaction step in a synthesis plan that does not result in the formation of a target bond that appears in the final target structure of the plan. It is typically a redox correction step, a functional group protection step, a functional group deprotection step, or a substitution correction step. Well-strategized synthesis plans have few concession steps.

See *sacrificial reaction or step*.

Concise Synthesis

A phrase often used in the organic synthesis literature to designate a synthesis plan to a known target molecule that has as a key feature a minimum number of reaction steps achieved compared with previously reported syntheses of that compound. It may also be associated with a convergent strategy compared with a linear one with fewer reaction steps. A concise synthesis does *not* automatically imply an atom economical synthesis or a green synthesis.

Construction Step[23,24]

A reaction step in a synthesis plan that results in the formation of a target bond that appears in the final target structure of the plan. Well-strategized synthesis plans have a high proportion of construction steps, often containing construction steps that produce more than one target bond in the same reaction.

Cumulative Material Efficiency Metrics

For a linear N-step plan, the relationship between the overall value of a metric and the values of the individual steps is given by Equations (2.4) to (2.9) for yield, atom economy, E-factor, PMI, gRME, and Lavoisier number, respectively.

$$\varepsilon_T = \prod_{j=1}^{N} \varepsilon_j \qquad (2.4)$$

$$(AE)_T = \cfrac{1}{\cfrac{1}{(MW)_P}\left(\sum_{j=1}^{N-1}(MW)_{Y_j}\left(\cfrac{1}{(AE)_j}-1\right)\right)+\cfrac{1}{(AE)_N}} \qquad (2.5)$$

$$E_T = E_N + \frac{1}{m_P}\sum_{j=1}^{N-1} m_{Y_j}(E_j) \qquad (2.6)$$

$$\mathrm{PMI}_T = E_T + 1 = \mathrm{PMI}_N + \frac{1}{m_P}\sum_{j=1}^{N-1} m_{Y_j}(\mathrm{PMI}_j - 1) \qquad (2.7)$$

$$\mathrm{gRME} = \frac{1}{(\mathrm{PMI})_T} = \cfrac{1}{\cfrac{1}{(\mathrm{RME})_N}+\cfrac{1}{m_P}\sum_{j=1}^{N-1} m_{Y_j}\left(\cfrac{1}{\mathrm{RME}_j}-1\right)} \qquad (2.8)$$

$$(LN)_T = (LN)_N + \frac{1}{(MW)_P}\sum_{j=1}^{N-1}(MW)_{Y_j}\left((LN)_j - 1\right) \qquad (2.9)$$

where the subscripts T and N refer to the total or overall value of the metric and the value of the metric for the Nth step in the linear synthesis plan, respectively; ε is reaction yield; AE is atom economy; MW is molecular weight; E is E-factor; PMI is process mass intensity; gRME is global reaction mass efficiency; LN is Lavoisier number; $m_{Y_j} = \dfrac{(MW)_{Y_j}}{\varepsilon_{j+1}\ldots\varepsilon_N}$, m_{Y_j} and $(MW)_{Y_j}$ represent the mass and molecular weight of intermediate product Y_j along the linear chain, respectively, m_P is the mass of the target product, and $(MW)_P$ is the molecular weight of the target product.

We may also write the following recursive formulas for the same cumulative metrics (Equations (2.10) to (2.15)), which are amenable to calculation by computer programming methods in an iterative sense.

$$\varepsilon_{1\to j+1} = \varepsilon_{1\to j}\varepsilon_{j+1} \qquad (2.10)$$

$$(AE)_{1\to j+1} = \frac{(MW)_{Y_j}}{(MW)_{Y_{j+1}}}\left[(AE)_{1\to j}-1\right]+(AE)_{j+1} \qquad (2.11)$$

$$(E)_{1\to j+1} = \frac{m_{Y_j}}{m_{Y_{j+1}}}(E)_{1\to j}+(E)_{j+1} \qquad (2.12)$$

$$(\text{PMI})_{1\to j+1} = \frac{m_{Y_j}}{m_{Y_{j+1}}}\left[(\text{PMI})_{1\to j} - 1\right] + (\text{PMI})_{j+1} \quad (2.13)$$

$$(\text{RME})_{1\to j+1} = \frac{1}{\dfrac{m_{Y_j}}{m_{Y_{j+1}}}\left[\dfrac{1}{(\text{RME})_{1\to j}} - 1\right] + \dfrac{1}{(\text{RME})_{j+1}}} \quad (2.14)$$

$$(\text{LN})_{1\to j+1} = \frac{1}{\dfrac{(MW)_{Y_j}}{(MW)_{Y_{j+1}}}\left[\dfrac{1}{(\text{LN})_{1\to j}} - 1\right] + \dfrac{1}{(\text{LN})_{j+1}}} \quad (2.15)$$

Example 2.14

Derive Equation (2.7) for the overall PMI for a linear synthesis plan containing N steps in terms of the step PMIs.

Solution

For a linear sequence of N steps involving the production of intermediate products $P_1, P_2, \ldots,$ and P_N in steps 1, 2, …, N where all the preceding intermediates in any given step is committed as a reagent in the next step, we can write the step PMI for step j as shown in Equation (2.16).

$$(\text{PMI})_j = \frac{\phi_j}{m_{P_j}} \quad (2.16)$$

where ϕ_j is the mass of input materials used in step j and m_{P_j} is the mass of intermediate product collected in step j. For the entire synthesis plan covering all N steps, the overall PMI is given by Equation (2.17) where we have substituted the step PMI expression given in Equation (2.16).

$$\begin{aligned}(\text{PMI})_{1\to N} &= \frac{\phi_1 + \phi_2 + \ldots + \phi_N - (m_{P_1} + m_{P_2} + \ldots + m_{P_{N-1}})}{m_{P_N}} \\ &= \frac{1}{m_{P_N}}\left[\sum_{j=1}^{N}\phi_j - \sum_{j=1}^{N-1} m_{P_j}\right] \\ &= \frac{1}{m_{P_N}}\left[\sum_{j=1}^{N} m_{P_j}(\text{PMI})_j - \sum_{j=1}^{N-1} m_{P_j}\right] \\ &= \frac{1}{m_{P_N}}\left[m_{P_N}(\text{PMI})_N + \sum_{j=1}^{N-1} m_{P_j}\left((\text{PMI})_j - 1\right)\right]\end{aligned} \quad (2.17)$$

Example 2.15[25]

From the results found above derive an analogous expression for the overall E-factor for a linear synthesis plan in terms of the step E-factors. Compare your result with the following statement given in the paper: "For a three-step process, the overall E-factor is E(total) = $E_1 + E_2 + E_3$; whereas the overall PMI is PMI(total) = $PMI_1 + E_2 + E_3$." Does your result agree with this statement? Comment on the result.

Solution

Transforming the expression given in Equation (2.17) using the substitution $PMI = E + 1$ yields Equation (2.18).

$$E_{1 \to N} + 1 = \frac{1}{m_{P_N}} \left[m_{P_N}(E_N + 1) + \sum_{j=1}^{N-1} m_{P_j}\left((E_j + 1) - 1\right) \right]$$

$$= \frac{1}{m_{P_N}} \left[m_{P_N}(E_N + 1) + \sum_{j=1}^{N-1} m_{P_j} E_j \right] \quad (2.18)$$

$$E_{1 \to N} = E_N + \frac{1}{m_{P_N}} \sum_{j=1}^{N-1} m_{P_j} E_j$$

The above results show that the overall E-factor for a linear synthesis plan is not the sum of the step E-factors. Therefore, the Sheldon statement is formally incorrect. Based on the formula given in Equation (2.18), $E_{1 \to N} = \sum_{j=1}^{N} E_j$ under the very special condition when the masses of the intermediate products are each identical to the final product of the linear synthesis, that is, $m_{P_N} = m_{P_j}$ for $j = 1, 2, 3, \ldots, N$. Such a scenario implies a linear sequence comprising consecutive rearrangement reactions where the molecular weight of each intermediate product remains the same and each reaction yield is 100% with no loss of mass arising from by-products or unreacted starting materials along the way.

Example 2.16

For a linear synthesis plan, prove the recursive relation shown for the cumulative PMI.

$$(cPMI)_{1 \to i} = \frac{m_{P_{i-1}}}{m_{P_i}} \left[(cPMI)_{1 \to i-1} - 1 \right] + (PMI)_i$$

This expression is identical to Equation (2.13).

Solution

For a three-step plan, the step PMIs for steps 1, 2, and 3 are given by Equations (2.19) to (2.21).

Synthesis Plan Analysis

$$(PMI)_1 = \frac{\phi_1}{m_{P_1}}$$

$$(PMI)_2 = \frac{\phi_2}{m_{P_2}} \quad (2.19\text{–}2.21)$$

$$(PMI)_3 = \frac{\phi_3}{m_{P_3}}$$

The cumulative PMI for steps 1 and 2 is given by Equation (2.22).

$$\begin{aligned}
(cPMI)_{1\to 2} &= \frac{\phi_1 + \phi_2 - m_{P_1}}{m_{P_2}} = \frac{\phi_1}{m_{P_2}} + \frac{\phi_2}{m_{P_2}} - \frac{m_{P_1}}{m_{P_2}} \\
&= \frac{\phi_1}{m_{P_1}}\frac{m_{P_1}}{m_{P_2}} + \frac{\phi_2}{m_{P_2}} - \frac{m_{P_1}}{m_{P_2}} = \frac{m_{P_1}}{m_{P_2}}\left[\frac{\phi_1}{m_{P_1}} - 1\right] + \frac{\phi_2}{m_{P_2}} \\
&= \frac{m_{P_1}}{m_{P_2}}\left[(PMI)_1 - 1\right] + (PMI)_2 \\
&= \frac{m_{P_1}}{m_{P_2}}\left[(cPMI)_{1\to 1} - 1\right] + (PMI)_2
\end{aligned} \quad (2.22)$$

where it is understood that the step PMI for step 1 is identical to the cumulative PMI for step 1. Similarly, the cumulative PMI for steps 1, 2, and 3 is given by Equation (2.23).

$$\begin{aligned}
(cPMI)_{1\to 3} &= \frac{\phi_1 + \phi_2 + \phi_3 - m_{P_1} - m_{P_2}}{m_{P_3}} = \frac{m_{P_2}(cPMI)_{1\to 2} + \phi_3 - m_{P_2}}{m_{P_3}} \\
&= \frac{m_{P_2}}{m_{P_3}}\left[(cPMI)_{1\to 2} - 1\right] + (PMI)_3
\end{aligned} \quad (2.23)$$

It is readily apparent from the emerging pattern of Equations (2.22) and (2.23) that the stated recursive relation is verified for general step i.

Example 2.17

What is the analogous recursive expression for cumulative E-factor?

Solution

Substituting PMI = E + 1 for the step PMIs and cPMI = cE + 1 for the cumulative PMIs yields Equation (2.24).

$$\begin{aligned}
(cE)_{1\to i} + 1 &= \frac{m_{P_{i-1}}}{m_{P_i}}\left[(cE)_{1\to i-1} + 1 - 1\right] + (E)_i + 1 \\
(cE)_{1\to i} &= \frac{m_{P_{i-1}}}{m_{P_i}}(cE)_{1\to i-1} + (E)_i
\end{aligned} \quad (2.24)$$

This expression is identical to Equation (2.12).

Example 2.18

For the synthesis tree diagram shown below, determine recursive expressions for the cumulative PMI in terms of the step PMIs.

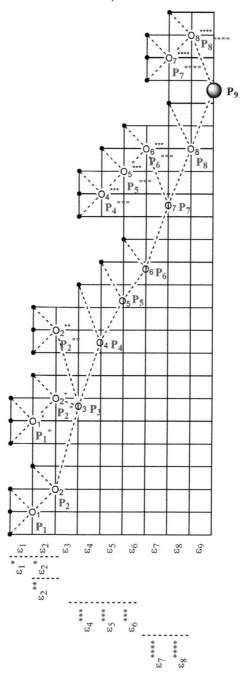

Synthesis Plan Analysis

SOLUTION

Recursive cumulative PMI expressions (S1 to S17):

1. Main Branch

   a. (S1) $(cPMI)_{1\to1} = (PMI)_1$

   b. (S2) $(cPMI)_{1\to2} = \dfrac{m_{P_1}}{m_{P_2}}\left[(cPMI)_{1\to1} - 1\right] + (PMI)_2$

   c. (S3) $(cPMI)_{1\to3} = \dfrac{m_{P_2}}{m_{P_3}}\left[(cPMI)_{1\to2} - 1\right] + (PMI)_3$
   $$+ \dfrac{m_{P_{2^*}}}{m_{P_3}}\left[(cPMI)_{1^*\to2^*} - 1\right] + \dfrac{m_{P_{2^{**}}}}{m_{P_3}}\left[(cPMI)_{2^{**}\to2^{**}} - 1\right]$$

   d. (S4) $(cPMI)_{1\to4} = \dfrac{m_{P_3}}{m_{P_4}}\left[(cPMI)_{1\to3} - 1\right] + (PMI)_4$

   e. (S5) $(cPMI)_{1\to5} = \dfrac{m_{P_4}}{m_{P_5}}\left[(cPMI)_{1\to4} - 1\right] + (PMI)_5$

   f. (S6) $(cPMI)_{1\to6} = \dfrac{m_{P_5}}{m_{P_6}}\left[(cPMI)_{1\to5} - 1\right] + (PMI)_6$

   g. (S7) $(cPMI)_{1\to7} = \dfrac{m_{P_6}}{m_{P_7}}\left[(cPMI)_{1\to6} - 1\right] + (PMI)_7 + \dfrac{m_{P_{6^{***}}}}{m_{P_7}}\left[(cPMI)_{4^{***}\to6^{***}} - 1\right]$

   h. (S8) $(cPMI)_{1\to8} = \dfrac{m_{P_7}}{m_{P_8}}\left[(cPMI)_{1\to7} - 1\right] + (PMI)_8$

   i. (S9) $(cPMI)_{1\to9} = \dfrac{m_{P_8}}{m_{P_9}}\left[(cPMI)_{1\to8} - 1\right] + (PMI)_9$
   $$+ \dfrac{m_{P_{8^{****}}}}{m_{P_9}}\left[(cPMI)_{7^{****}\to8^{****}} - 1\right] = (PMI)_{total}$$

2. Branch 2

   a. (S10) $(cPMI)_{1^*\to1^*} = (PMI)_{1^*}$

   b. (S11) $(cPMI)_{1^*\to2^*} = \dfrac{m_{P_{1^*}}}{m_{P_{2^*}}}\left[(cPMI)_{1^*\to1^*} - 1\right] + (PMI)_{2^*}$

3. Branch 3
   a. (S12) $(cPMI)_{2^{**} \to 2^{**}} = (PMI)_{2^{**}}$

4. Branch 4
   a. (S13) $(cPMI)_{4^{***} \to 4^{***}} = (PMI)_{4^{***}}$

   b. (S14) $(cPMI)_{4^{***} \to 5^{***}} = \dfrac{m_{P_{4^{***}}}}{m_{P_{5^{***}}}}\left[(cPMI)_{4^{***} \to 4^{***}} - 1\right] + (PMI)_{5^{***}}$

   c. (S15) $(cPMI)_{4^{***} \to 6^{***}} = \dfrac{m_{P_{5^{***}}}}{m_{P_{6^{***}}}}\left[(cPMI)_{4^{***} \to 5^{***}} - 1\right] + (PMI)_{6^{***}}$

5. Branch 5
   a. (S16) $(cPMI)_{7^{****} \to 7^{****}} = (PMI)_{7^{****}}$

   b. (S17) $(cPMI)_{7^{****} \to 8^{****}} = \dfrac{m_{P_{7^{****}}}}{m_{P_{8^{****}}}}\left[(cPMI)_{7^{****} \to 7^{****}} - 1\right] + (PMI)_{8^{****}}$

Example 2.19[26]

UM171 was recently discovered as an agonist of hematopoietic stem cell self-renewal, which makes it a potential therapeutic to treat various blood diseases. Use the *calculator-cumulative-PMI.xls* spreadsheet (see Section 2.6) to determine the cumulative PMIs for the synthesis plan depicted in the synthesis tree shown below.

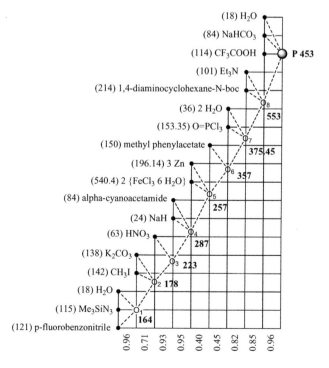

Synthesis Plan Analysis

The step PMIs are as follows:

Step j	Step PMI
1	118.70
2	85.3
3	191.12
4	48.13
5	1210.09
6	57.49
7	1337.22
8	300.26
9	519.33

SOLUTION

Step j	Step PMI	MW Y(j)	Product of step j+1 MW Y(j+1)	Yield of step j+1 ε(j+1)	[MW Y(j)/MW Y(j+1)]/ε(j+1)	Cumulative PMI
1	118.70	164	178	0.71	1.298	118.70
2	85.30	178	223	0.93	0.858	238.04
3	191.12	223	287	0.95	0.818	394.56
4	48.13	287	257	0.4	2.792	370.03
5	1210.09	257	357	0.45	1.600	2240.35
6	57.49	357	375.45	0.82	1.160	3639.89
7	1337.22	375.45	553	0.85	0.799	5556.82
8	300.26	553	453	0.96	1.272	4737.94
9	519.33	453				6542.90

Example 2.20[27]

The anti-coagulant pharmaceutical apixaban developed by Bristol-Myers Squibb has been synthesized using the following convergent strategy.

Synthesis Plan Analysis

The reaction yields for the synthesis are as follows:

Step 1: 76%; Step 2: 92%; Step 3: 89%; Step 4: 85%; Step 5: 75%; Step 6: 95%; and Step 2b: 78%.

The step PMIs are as follows:

Step 1: 45.92; Step 2: 19.19; Step 3: 13.92; Step 4: 25.61; Step 5: 36.03; Step 6: 39.85; and Step 2b: 24.28.

The product of the convergent branch is used in 6.6% excess in the convergent step.
Use the *calculator-cumulative-PMI.xls* spreadsheet (see Section 2.6) to determine the overall PMI for the synthesis plan.

SOLUTION

We first determine the overall balanced chemical equations for each step.

Next, we draw out the synthesis tree diagram for the plan.

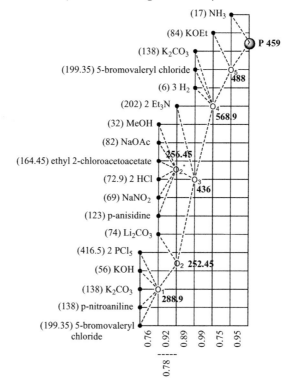

Synthesis Plan Analysis

The results of the cumulative PMI calculator are as follows:

SECTION A

Number of steps along main branch	6
Number of branches	2

SECTION B
Identification of terminal steps of convergent branches

Branch no.	Terminal step no.	SF of terminal product used in convergent step
2	2*	1.066

SECTION C

Convergent branch		Multiply entry in column F (Section C) of terminal step of convergent branch by SF of terminal product (Section B)					
Step j	Step PMI	Product of step j MW Y(j)	Product of step j+1 MW Y(j+1)	Yield of step j+1 ε(j+1)	[MW Y(j)/ MW Y(j+1)]/ε(j+1)	Cumulative PMI	
2*	24.28	256.45 436	436	0.89	**0.706** #DIV/0!	24.28 **16.44**	cPMI(2*=>2*) increment for convergent step 3

SECTION D

Main branch		Add last entry in column G (Section C) to cumulative PMI entry for convergent step (Section D)					
Step j	Step PMI	Product of step j MW Y(j)	Product of step j+1 MW Y(j+1)	Yield of step j+1 ε(j+1)	[MW Y(j)/ MW Y(j+1)]/ ε (j+1)	Cumulative PMI	
1	45.92	288.9	252.45	0.92	1.249	45.92	cPMI(1=>1)
2	19.19	252.45	436	0.89	0.652	75.31	cPMI(1=>2)
3	13.92	436	568.9	0.85	0.898	**78.81**	cPMI(1=>3)
4	25.61	568.9	488	0.75	1.561	95.52	cPMI(1=>4)
5	36.03	488	459	0.95	1.124	183.54	cPMI(1=>5)
6	39.85	459				245.01	cPMI(1=>6)

The convergent step is step 3.
The overall PMI for the plan is 245.01.

Cumulative Yield

The cumulative yield is the multiplicative product of reaction step yields along a linear synthesis sequence from step 1 to any step $j > 1$ downstream along that pathway. If the multiplication is taken along an entire N-step linear sequence, then the cumulative yield corresponds to the overall yield of the linear synthesis. For convergent syntheses, the cumulative yield can be determined along any point on a given linear branch from the beginning of that branch up to the end product of that branch, which corresponds to an intermediate used as input material in a convergent step. Hence, for a convergent synthesis plan containing M branches, there are in principle M cumulative yields that can be determined, one for each branch. One of these M branches corresponds to the main branch having the longest linear sequence from starting materials to final desired product at the terminus of the convergent synthesis.

First Moment Building Up Parameter[1]

This metric belongs to the set of essential synthesis strategy parameters that describes the net building up of a structure from the set of initial input and intermediate structures toward the final target product. The molecular weight first moment per reaction stage about the target product molecular weight in units of grams per mole per reaction stage is given by Equation (2.25).

$$\mu = \frac{\omega_1 - \omega_2}{N+1} \qquad (2.25)$$

Synthesis Plan Analysis

where:

$\omega_1 = \Sigma\,(MW_{\text{intermediates}} + MW_{\text{starting material at beginning of each branch}})$

$\omega_2 = $ (total number of intermediate nodes + starting material nodes at beginning of each branch)$(MW)_{\text{product}}$

$N = $ is the number of reaction stages in a synthesis plan.

The starting materials correspond to those inputs at the beginning stage of each branch that actually get incorporated in the subsequent intermediate product. The zeroth stage representing the starting substrates for the longest branch or root of the synthesis tree is accounted for by the extra stage in the denominator. If a reaction stage has parallel reactions and therefore consists of more than one intermediate product being formed in that stage, then each of their respective molecular weights are included in the first summed term, ω_1. The second term in ω_1 accounts for molecular weights of input starting materials at the beginning of each branch provided they contribute to the structure of the immediately resulting product. A positive value for μ indicates an overall net loss in MW per reaction stage (net degradation) and a negative value indicates an overall net gain in MW per reaction stage (net building up). The larger the magnitude of the first moment the greater is the effect of degradation or building up. Good synthesis plans are characterized by fewer reaction stages, the frequent occurrence of convergent reaction stages (i.e., parallel reactions), and large negative molecular weight first moments per reaction stage. The building up toward the final product structure may be visualized by plotting the differences in molecular weight between all intermediates along the main branch and the final product of the synthesis as a function of the reaction stage in the form of a histogram. The histogram also includes the difference in molecular weight between the starting materials at the zeroth stage that end up in the first intermediate structure and the final product structure.

Example 2.21

Determine the first moment building up parameter for the synthesis of L-DOPA given in Example 2.4. Draw a histogram of $MW(P_j) - MW(P_{\text{final}})$ versus reaction stage.

SOLUTION

MW(L-DOPA) = 197 g/mol
The synthesis tree consists of one branch.
Number of stages = 4
Number of intermediate nodes = 3
Number of starting material nodes at beginning of branch = 2
$\omega_1 = 295 + 315 + 600 + 152 + 179 = 1541$
$\omega_2 = (3+2)*197 = 985$
$\mu = (\omega_1 - \omega_2)/(N+1) = (1541-985)/(4+1) = 111.2$ g/mol/stage

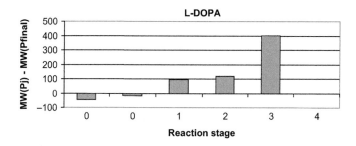

We note that the large positive overshoot in stage 3 is due to the high molecular weight of the diastereomeric salt produced during the classical resolution step. This causes the first moment building up parameter to have a net positive value for the linear synthesis.

Example 2.22

Determine the first moment building up parameter for the synthesis of apixaban given in Example 2.20. Draw a histogram of $MW(P_j) - MW(P_{final})$ versus reaction stage.

SOLUTION

MW(apixaban) = 459 g/mol
The synthesis tree consists of two branches.
Number of stages = 6
Number of intermediate nodes in main branch = 5
Number of intermediate nodes in convergent branch = 1
Number of starting material nodes at beginning of main branch = 3
Number of starting material nodes at beginning of convergent branch = 3

For main branch:

$$\omega_1(\text{main}) = 288.9 + 252.45 + 436 + 568.9 + 488 + 199.35 + 138 + 416.5$$

$$= 2788.1$$

For convergent branch:

$\omega_1(\text{convergent}) = 256.45 + 123 + 69 + 164.45 = 612.9$
$\omega_2 = (5 + 1 + 3 + 3)*459 = 5508$
$\mu = (\omega_1 - \omega_2)/(N+1) = (2788.1 + 612.9 - 5508)/(6+1) = -301$ g/mol/stage

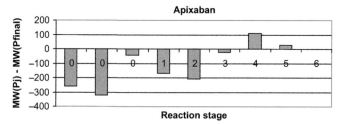

Synthesis Plan Analysis

We note that in this convergent synthesis there is a small overshoot in molecular weight in stage 4, however, there is an overall significant net building up toward the product structure.

Gantt Diagram[28-31]

A Gantt diagram is a chart invented by Henry Gantt in 1903 to streamline the scheduling of work and hence this resulted in increasing the efficiency of production management. The concept of Gantt diagrams has been applied to the chemical industry.

Example 2.23[32]

A synthesis tree diagram may be used to determine the scheduling of reactions in a synthesis plan in the same way as a Gantt diagram. Instead of labeling the bottom of the grid with reaction yields, the labels pertain to the length of time in hours needed to carry out all operations for each reaction step (reaction time, isolation time, and purification time). This kind of diagram is particularly advantageous for working out the scheduling of reactions in a convergent plan.

Part 1

For the synthesis diagram shown below, when should intermediates at stage 4* and stage 5**, be available relative to the final product if there is to be no downtime for executing the entire plan?

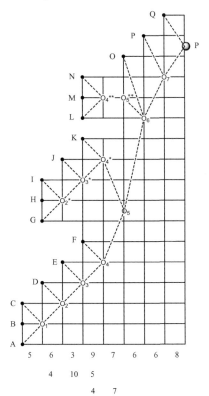

Solution

Drawing a scaled time diagram gives a clear visualization of the problem.

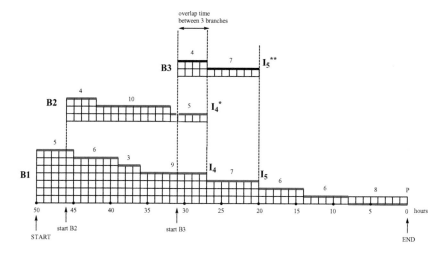

Intermediate I4* should be ready 27 hours before attaining the final product. Its synthesis should begin 4 hours after the start of the main branch.

Intermediate I5** should be ready 20 hours before attaining the final product. Its synthesis should begin 19 hours after the start of the main branch.

Part 2

When should the syntheses of the two convergent branches begin relative to the final product?

Solution

The synthesis of branch 2 should begin 46 hours before attaining the final product. The synthesis of branch 3 should begin 31 hours before attaining the final product. Alternatively, we can state the start times of each convergent branch in the forward sense relative to the beginning of the main branch (B1) as follows: branch 2 to start 4 hours after beginning B1; branch 3 to start 19 hours after beginning B1.

Part 3

What is the overlap time when reactions in all three branches in the synthesis are running simultaneously? When will this period occur relative to the beginning of branch 1? Which reactions and what time periods will this overlap time cover?

Solution

The overlap time when all 3 branches would be running simultaneously is 4 hours.
The overlap period will begin 19 hours after the beginning of branch 1 coinciding with when branch 3 begins.

Synthesis Plan Analysis

The overlap time begins 5 hours after step 4 in the main branch starts; 1 hour after step 3 in the second branch starts; and exactly when the first step of branch 3 begins.

Part 4

Suppose a different procedure was selected for the first step in branch 3, which required a reaction time of 6 hours instead of 4. How would the answers in Part 3 change?

Solution

The following diagram depicts the effect of an additional two-hours for the first step of branch 3.

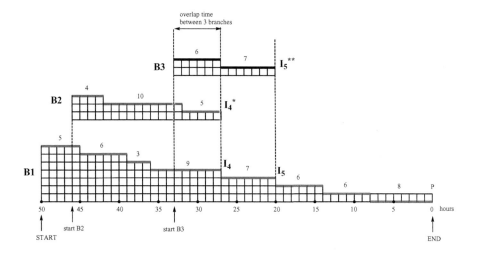

The overlap time when all three branches would be running simultaneously is 6 hours.

The overlap period will begin 17 hours after the beginning of branch 1 coinciding with when branch 3 begins.

The overlap time begins 3 hours after step 4 in the main branch starts; 1 hour before step 3 in the second branch starts (or 9 hours after step 2 in the second branch starts); and exactly when the first step of branch 3 begins.

Part 5

Suppose an unforeseen delay of 2 hours occurred while carrying out the first step in branch 3. How would this impact the schedule of the synthesis?

Solution

The following diagram depicts the effect of a 2-hour delay for the first step of branch 3.

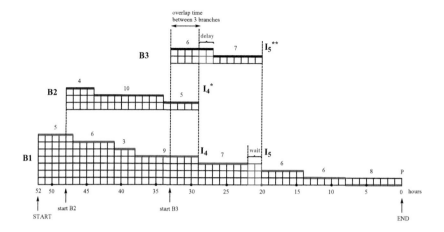

The overlap time when all three branches would be running simultaneously is 4 hours.

The overlap period will begin 19 hours after the beginning of branch 1 coinciding with when branch 3 begins.

The overlap time begins 5 hours after step 4 in the main branch starts; 1 hour after step 3 in the second branch starts; and exactly when the first step of branch 3 begins.

The 2-hour delay in step 1 of branch 3 means that intermediate I5** will be available 2 hours after I5 is ready, that is, 32 hours after the main branch begins instead of 30 hours.

The entire plan will require 52 hours.

Global Atom Economy

This metric pertains to the overall atom economy for a synthesis plan containing at least two reaction steps. The synthesis tree diagram facilitates the computation of the global atom economy for linear and convergent synthesis plans because all necessary input materials are distinguishable from intermediates generated as products from step 1 to step $N-1$ in an N-step plan. The mathematical expression for global atom economy is given by Equation (2.26).

$$\mathrm{GAE} = AE_{\mathrm{overall}} = \frac{MW_{\mathrm{product}}}{\sum_j \nu_j (MW)_{\mathrm{input},j}} \quad (2.26)$$

Global E-Factor

This metric synonymous to overall E-factor is applicable to single balanced chemical reactions and to synthesis plans (linear or convergent) composed of sets of appropriately scaled balanced chemical equations. The mathematical representation of this metric is given by Equation (2.27).

$$gE = E_{\mathrm{overall}} = \frac{\sum_j m_{\mathrm{waste},j}}{m_{\mathrm{product}}} = \frac{\sum_j m_{\mathrm{input},j} - m_{\mathrm{product}}}{m_{\mathrm{product}}} = \mathrm{gPMI} - 1 \quad (2.27)$$

Synthesis Plan Analysis

where the m terms refer to the masses of final target product and all input materials including reagents, catalysts, reaction solvents, workup materials, and purification materials. Masses of intermediates formed as products in step 1 to step $N-1$ are not included since they are made and consumed in consecutive steps in an appropriately scaled synthesis plan.

Global Mass Efficiency
This term is identical in meaning to global process mass intensity.

Global Material Economy
This term is identical in meaning to global process mass intensity.

Global Process Mass Intensity
This metric is synonymous to overall process mass intensity and is applicable to single balanced chemical reactions and to synthesis plans (linear or convergent) composed of sets of appropriately scaled balanced chemical equations. The mathematical representation of this metric is given by Equation (2.28).

$$\text{gPMI} = \frac{\sum_j m_{\text{input},j}}{m_{\text{product}}} \qquad (2.28)$$

where the m terms refer to the masses of final target product and all input materials including reagents, catalysts, reaction solvents, workup materials, and purification materials. Masses of intermediates formed as products in step 1 to step $N-1$ are not included since they are made and consumed in consecutive steps in an appropriately scaled synthesis plan.

Global Reaction Mass Efficiency[33]
This metric is synonymous to overall reaction mass efficiency and is applicable to single balanced chemical reactions and to synthesis plans (linear or convergent) composed of sets of appropriately scaled balanced chemical equations. The mathematical representation of this metric is given by Equation (2.29).

$$\text{gRME} = \frac{m_{\text{product}}}{\sum_j m_{\text{input},j}} = \frac{1}{\text{gPMI}} \qquad (2.29)$$

where the m terms refer to the masses of final target product and all input materials including reagents, catalysts, reaction solvents, workup materials, and purification materials. Masses of intermediates formed as products in step 1 to step $N-1$ are not included since they are made and consumed in consecutive steps in an appropriately scaled synthesis plan.

Hub Intermediate
An intermediate that appears as a multivalent node in a reaction network or that appears more than once in a synthesis tree diagram signifying that such an

intermediate is common to several reaction pathways and therefore has a prominent status. A material efficient synthesis of a hub intermediate is central to achieve material efficient syntheses of other products that originate from it. The chemical structure of a hub intermediate is usually a common substructure of other more advanced products appearing in a reaction network.

Example 2.24[34]

Carpanone is a lignan obtained from the bark of the carpano tree. A synthesis of this compound utilized 6-propenyl-benzo[1,3]dioxol-5-ol as a key hub intermediate. Hub intermediates are used more than once in a synthesis owing to their repeated structural motif in the final target product. It is imperative that the synthesis sequence that makes such a hub intermediate be optimized as far as possible since that sequence is utilized more than once in the overall plan. If done successfully, this technique results in high overall material efficiency performances for the syntheses of complex structures.

carpanone

Part 1
Find where 6-propenyl-benzo[1,3]dioxol-5-ol maps onto the structure of carpanone.

SOLUTION

Part 2
Suggest how 6-propenyl-benzo[1,3]dioxol-5-ol could be made from benzo[1,3]dioxol-5-ol. Show balanced chemical equations for each step.

Synthesis Plan Analysis

SOLUTION
Using data found in references [35, 36] we can suggest the following.

[Reaction scheme: sesamol + allyl chloride with KOH, −H₂O, −KCl → allyl sesamol ether; then heated at 210–220°C → ortho-allyl sesamol; then KO^tBu → propenyl product]

Ideality[23,24,37]
Applied to a synthesis plan, ideality is the fraction of reaction steps in a synthesis that are not concession or sacrificial steps. In other words, it is the fraction of reactions in a synthesis plan that pertains to target bond-forming or construction steps.
  See *concession step; sacrificial step.*

Inventory Analysis
For the analysis of the material efficiency, a synthesis plan inventory analysis involves the determination of the scaled masses of all input materials required in the synthesis including all reagents and all associated auxiliary materials used in each reaction. The scaled masses of all input materials are worked out using appropriate scaling factors based on a given mole scale of target product produced at the end of the synthesis plan.

Maximum Molecular Weight Fraction of Sacrificial Reagents, f(sac)$_{max}$
The maximum value of the molecular weight fraction of sacrificial reagents. This metric is determined from a balanced chemical equation written using Markush structures for reagents and products highlighting structural features that are variable and those that are not. The maximum value of the fraction is determined when the variable groups are set to be as large as possible.

Molecular Weight Fraction of Sacrificial Reagents, f(sac)[38]
This metric belongs to the set of synthesis strategy parameters that measures the MW fraction of reagents that absolutely do not end up in the final target molecule structure; that is, no atom in any of those reagents ends up in the final target molecule. The metric can be applied to a single reaction or a synthesis plan. The mathematical expression is given by Equation (2.30).

$$f(\text{sac}) = \frac{\sum_j (MW)_{\text{reagent not ending up in target product},j}}{\sum_j (MW)_{\text{reagent},j}}. \qquad (2.30)$$

Synthesis Green Metrics

This parameter is more probing than atom economy in that it is intimately connected to the dissection of a target structure that traces the origin of each atom back to its corresponding starting materials.

Example 2.25[39]

(-)-Aromoadendranediol is a widely distributed sesquiterpene that has been isolated from the marine coral *Sinularia mayi* as well as from the leaves of the Amazonian tree *Xylopia brasiliensis*. Extracts containing this substance are used in Chinese and Brazilian folk medicine as analgesics, sedatives, and treatments for lung inflammation. The first step of its total synthesis utilized a one-pot triple cascade reaction that involved a tandem Grubbs cross metathesis, Mukaiyama–Michael addition, and intramolecular aldol reaction. The scheme below shows the synthesis plan.

Part 1

Write out the scheme showing fully balanced equations in each step.

Solution

Synthesis Plan Analysis

Part 2

Based on the synthesis plan draw the target bond map for the product structure and describe the ring construction strategy.

Solution

Ring target map: $[4+1]_A + [4+2+1]_B + [1+1+1]_C$

Ring construction strategy:

$$\xrightarrow{[4+1]} A_1 \xrightarrow{[7+0]} B_7 \xrightarrow{[2+1]} C_8$$

Part 3

Determine the MW fraction of sacrificial reagents used in the plan.

Solution

Reagent	MW (g/mol)
Trimethyl-(5-methyl-furan-2-yloxy)-silane	170
But-2-enal	70
Hex-5-en-2-one	98
H_2O	18
H_2O	18
H_2O	18
$LiAlH_4$	38
$CHBr_3$	252.7
SUM	682.7

Sum of MW of all reagents used = 3968.85
Sum of MW of reagents ending up in product = 682.7
MW fraction of reagents ending up in product = 682.7/3968.85 = 0.172, or 17.2%
MW fraction of sacrificial reagents = 1 − 0.172 = 0.828, or 82.8%

Number of Bonds Made Per Step

The count of the number of target bonds made in a given reaction that appear in the final target product structure of the plan.

Number of Branches

The count of branches in a synthesis plan. A linear plan has one branch. A convergent plan has at least two branches.

Number of Input Materials

The count of input reactants used in a synthesis plan that appears along the ordinate of a synthesis tree diagram.

Number of Nonbonding Steps Per Reaction Stage

The ratio of the number of reaction steps in a synthesis plan that do not produce target bonds in the final product structure of the plan and the number of reaction stages in the plan. In a linear plan, the number of reaction stages is identical to the number of reaction steps. In a convergent plan, the number of reaction stages is less than the number of reaction steps.

Number of Reaction Steps

The count of reaction steps in a synthesis plan where a step constitutes isolation of the intermediate product along the way.

Overall Atom Economy

See *global atom economy*.

Overall E-Factor

See *global E-factor*.

Overall Process Mass Intensity

See *global process mass intensity*.

Overall Reaction Mass Efficiency

See *global reaction mass efficiency*.

Overall Yield

For a linear synthesis plan, the overall yield corresponds to the multiplicative product of the individual step reaction yields. For a convergent plan, the overall yield corresponds to the multiplicative product of the individual step reaction yields along the longest branch of the synthesis plan; that is, the branch having the greatest number of reaction steps.

Reaction Stage

In a linear synthesis plan, the number of reaction stages is equal to the number of reaction steps. In a convergent plan, the number of reaction stages is less than the number of reaction steps. In a convergent plan, a reaction stage can contain at least two reaction steps run in parallel as determined by its synthesis tree diagram.

Reaction Step

A reaction step is defined as a chemical transformation that begins with an isolated set of starting materials and ends up with an isolated reaction product.

Sacrificial Reagent[1]

A reagent used in a sacrificial reaction that is classified as a non-productive reaction in a synthesis plan since it does not form target bonds appearing in the target product structure. None of the atoms in a sacrificial reagent end up in the final product structure. Sacrificial reactions are also known as concession reactions.

Sacrificial Reaction or Step[1]

A reaction step in a synthesis plan that does not result in the formation of a target bond that appears in the final target structure of the plan. It is typically a redox correction step, a function group protection step, a functional group deprotection step, or a substitution correction step. Well-strategized synthesis plans have few sacrificial steps.

See *concession reaction*.

Scaling Factor

A factor used to adjust the mole scale of reagents in a given chemical reaction so it can be linked with other sequential reactions in a synthesis plan to a given target product with a pre-defined mole scale called the basis scale for the entire synthesis. Each reaction step in a synthesis plan will have its own associated scaling factor. The scaling factor for a given reaction step is also applied to all auxiliary materials used in that step. The scaling factors are determined in a backward fashion from the target product of the synthesis plan step-by-step until the starting material for a given branch in the synthesis is reached. If a convergence step is reached in the backward tracing, then a scaling factor is applied to each of the intermediates that constitute the input reactants in that convergence step. These factors are then used to determine the scaling factors for prior steps in each branch separately.

Step Atom Economy

The atom economy pertaining to a particular reaction step in a synthesis plan.

Step Economy[40–43]

A term coined by Paul Wender at Stanford University that states that an ideal synthesis is characterized by the fewest number of reaction steps.

Synthesis Plan Analysis

Step E-Factor
The E-factor pertaining to an individual reaction step in a synthesis plan.

Step Process Mass Intensity
The process mass intensity pertaining to an individual reaction step in a synthesis plan.

Step Reaction Mass Efficiency
The reaction mass efficiency pertaining to an individual reaction step in a synthesis plan.

Synthesis Efficiency
A general term used by classically trained synthetic organic chemists to describe the material efficiency of a synthesis plan usually according to the number of reaction steps and the overall yield only. In reality, the parameterization of synthesis efficiency involves the following suite of metrics applied to each step and to the whole plan: atom economy, yield, excess reagent consumption, E-factor, reaction mass efficiency, and process mass intensity.

Synthesis Strategy
A general term used by classically trained synthetic organic chemists to describe the sequence of strategic steps employed in the total synthesis of a given target molecule. This entails (a) the type of reactions employed that can be in any one of the following broad categories: additions, substitutions, eliminations, rearrangements, redox reactions, or multi-component; (b) the number of those reactions that are target bond forming (construction steps) and those that are sacrificial steps (concession steps); and (c) whether the synthesis plan follows a linear or convergent trajectory. Tools often used to strategize how a given target molecule can be assembled from smaller molecules are retrosynthetic analysis and the large database of known named organic reactions.

Example 2.26[44]

Suggest a linear and a convergent plan for the target molecule shown and compare their atom economy performances. Show by-products in each reaction step.

Solution

Convergent plan

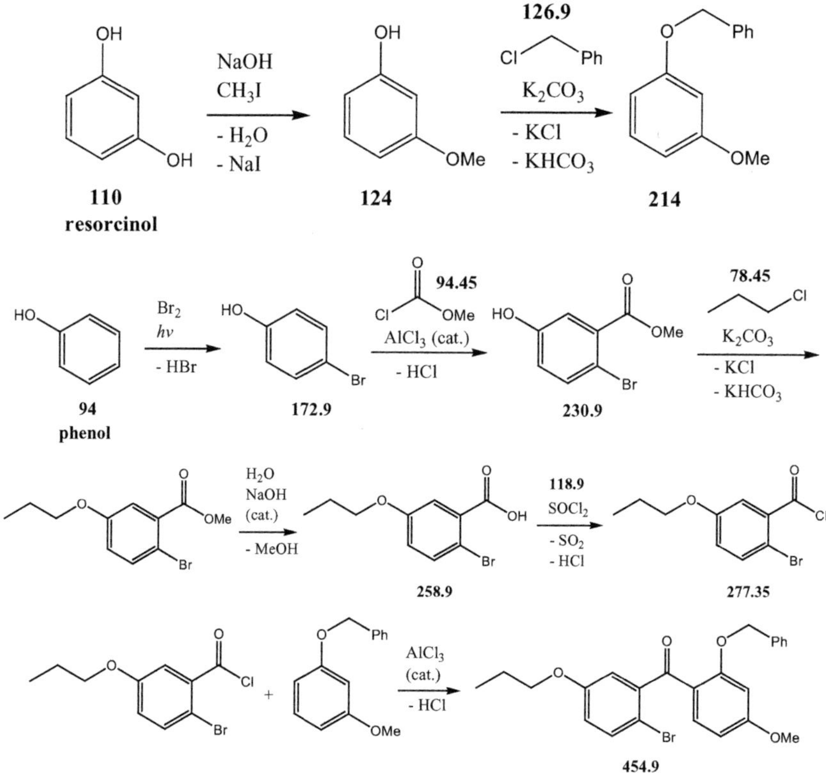

[Scheme: resorcinol (110) → (124) via NaOH/CH₃I, −H₂O, −NaI; then → (214) via ClCH₂Ph (126.9), K₂CO₃, −KCl, −KHCO₃]

[Scheme: phenol (94) → (172.9) via Br₂, hv, −HBr; → (230.9) via ClC(O)OMe (94.45), AlCl₃ (cat.), −HCl; → next via propyl chloride (78.45), K₂CO₃, −KCl, −KHCO₃]

[Scheme: → (258.9) via H₂O, NaOH (cat.), −MeOH; → (277.35) via SOCl₂ (118.9), −SO₂, −HCl]

[Scheme: acid chloride + benzyl ether (214) → (454.9) via AlCl₃ (cat.), −HCl]

Overall AE = 454.9/(110 + 40 + 142 + 126.9 + 138 + 159.8 + 94 + 94.45 + 78.45 + 138 + 18 + 118.9) = 0.361, or 36.1%

Linear plan

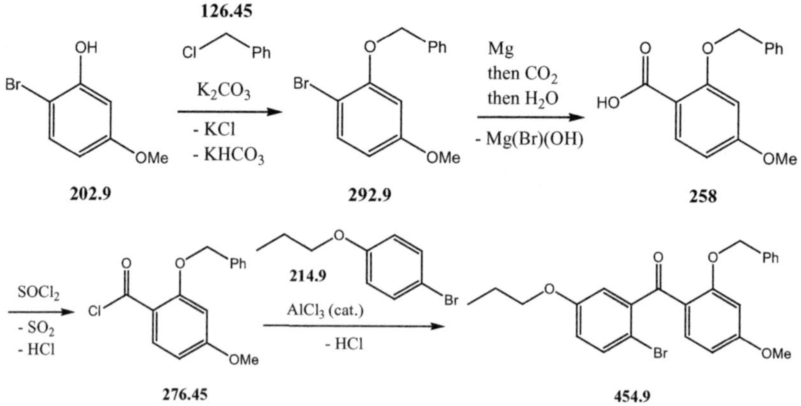

[Scheme: (202.9) → (292.9) via ClCH₂Ph (126.45), K₂CO₃, −KCl, −KHCO₃; → (258) via Mg then CO₂ then H₂O, −Mg(Br)(OH)]

[Scheme: → (276.45) via SOCl₂, −SO₂, −HCl; + (214.9) → (454.9) via AlCl₃ (cat.), −HCl]

Overall AE = 454.9/(202.9 + 126.45 + 138 + 24.3 + 44 + 18 + 118.9 + 214.9) = 0.513, or 51.3%

Synthesis Plan Analysis

Target Bond Forming Reaction or Step[1]

A target bond forming reaction is a reaction step in a synthesis plan that involves forming a target bond that is found in the final product structure.

See *construction step*.

Target Bond Profile

A bar graph or histogram that plots the number of target bonds made in each reaction step. This diagram is useful in categorizing which steps are target bond forming (construction) steps and which steps are sacrificial (concession) steps in a synthesis plan.

Example 2.27[45,46]

The alkaloid (+)-plicamine isolated from the Turkish Amayllidaceae *Glanthus plicatus* subspecies *byzantinus* has a unique tetracyclic ring system containing a 6,6-spirocyclic core defining three of the four stereogenic centers. Ley and co-workers have used the technique of solid-supported reagents and catalysts to carry out total syntheses of complex natural products. This technique has a number of advantages including cleaner work-up of reaction intermediates, facile removal of excess reactant or by-product from the product, easy retrieval and recyclability of reagents and catalysts, elimination of auxiliary material demand normally used in work-up and chromatographic procedures, and immobilization facilitates use of toxic, obnoxious, or volatile compounds.

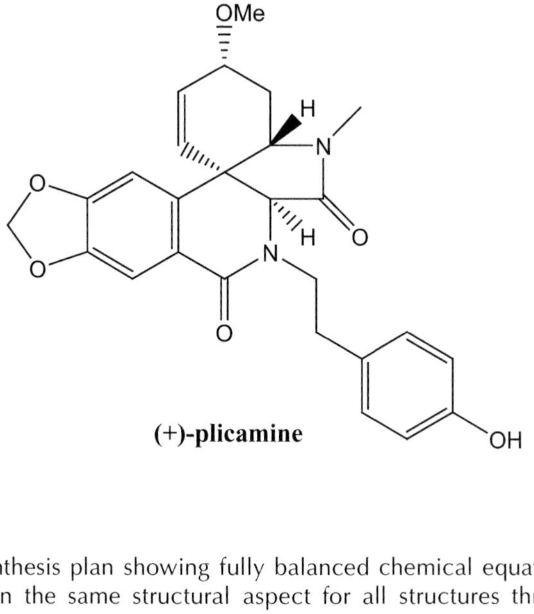

(+)-plicamine

Part 1

Write the synthesis plan showing fully balanced chemical equations in each step. Maintain the same structural aspect for all structures throughout the scheme.

Solution

Synthesis Plan Analysis

410 → (99%) [⬤—SO₃H, Me₃SiCHN₂, MeOH; −N₂, −Me₃SiOMe] → **424** (95%)

→ [⬤—⁺NMe₃ ⁻OH; −⬤—⁺NMe₃ CF₃CO₂⁻] → **328** (96%)

→ [**200.9** Br-CH₂CH₂-C₆H₄-OH, ⬤—⁺NEt₃ NaCO₃⁻; −⬤—⁺NEt₃ Br⁻, −NaHCO₃] → **448** (90%)

→ {4/3 CrO₃, 96 (3,5-dimethyl-1H-pyrazole); 4 ⬤—SO₃H; −3 H₂O, −4/3 [⬤—SO₃⁻]₃ Cr⁺³, −4/3 3,5-Dimethyl-1H-pyrazole} → **462** (70%)

Part 2

Draw the product structure highlighting the target bonds made. Determine the ring construction strategy and the ring mapping. Determine the MW fraction of sacrificial reagents used. Draw a target bond-step profile and determine the fraction of the reaction steps that do not form target bonds.

Solution

Ring construction strategy:

$$\xrightarrow{(3+3)} A_4 \xrightarrow{(5+0)} B_5$$

Ring mapping:

$[3+3]_A + [5+0]_B$

Reagent ending up in product	MW (g/mol)	
Benzo[1,3]dioxole-5-carbaldehyde	150	
2-Amino-2-(4-hydroxy-phenyl)-N-methyl-acetamide	180	
Me$_3$SiCHN$_2$	114	
CrO$_3$	99.996	
MeOH	32	
4-(2-Bromo-ethyl)-phenol	200.9	
R-SO$_3$H	96	R = Me
[R-NMe$_3$] [BH$_4$]	89	R = Me
Total	961.896	

Synthesis Plan Analysis

Assume minimum size of R group in solid-supported resins is methyl.
Sum of MW of all reagents used in plan = 2699.228
MW fraction of reagents ending up in product = 961.896/2699.228 = 0.356, or 35.6%
MW fraction of sacrificial reagents = 1 − 0.356 = 0.644, or 64.4%
Since a minimum sized R group is considered, these fractions are maximum estimates.

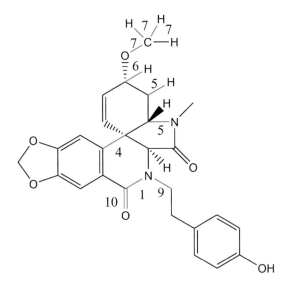

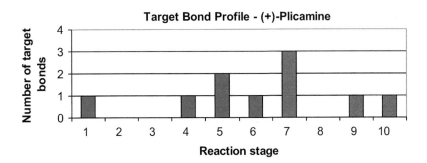

The number of steps in the plan is 10.
The number of target bonds made is 10.
The number of steps producing at least one target bond is 7.
The number of steps producing no target bonds is 3.
Three out of ten or 30% of the synthesis steps produce no target bonds.

Part 3

Draw a synthesis tree diagram for the plan and determine whether it is linear or convergent and the kernel material efficiency metrics.

Solution

The plan is linear.

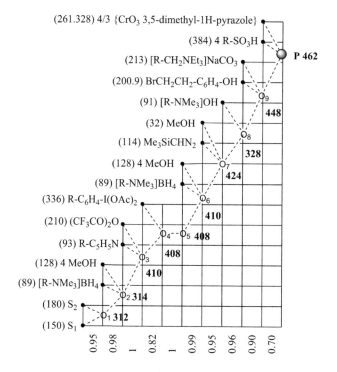

S1 = Benzo[1,3]dioxole-5-carbaldehyde
S2 = 2-Amino-2-(4-hydroxy-phenyl)-N-methyl-acetamide

Assume R=Me. This represents the minimum size of the solid-supported reagent substituent.

Overall yield = 43.4%
Overall AE = 17.1% (maximum value)
Overall kernel RME = 9.3% (maximum value)
Overall kernel PMI = 10.76 (minimum value)

Part 4

For each step in the plan indicate where a solid-supported reagent or catalyst can be recycled or retrieved.

Solution

In step 1, no solid-supported reagents are used.
In steps 2 and 6, the by-product may be converted back to the reagent.

$$\text{⊙–NMe}_3^{\oplus}\ \text{MeO}^{\ominus} \xrightarrow[-\text{NaOMe}]{\text{NaBH}_4} \text{⊙–NMe}_3^{\oplus}\ \text{BH}_4^{\ominus}$$

Synthesis Plan Analysis

In step 3, the by-product may be neutralized back to the reagent.

In step 4, the by-product may be re-oxidized back to the reagent.

In steps 5 and 7, the solid-supported reagents act as catalysts and can be used again.

In steps 8 and 9, the by-products can be converted back to the solid-supported reagents via the following sequences.

In step 10, the chromium bound by-product may be converted back to the solid-supported reagent and the freed metal can be re-oxidized.

[O] $CrCl_3 + 3 H_2O \longrightarrow CrO_3 + 3 Cl^- + 6 H^+ + 3 e$

[R] $KMnO_4 + 4 H^+ + 3 e \longrightarrow K^+ + MnO_2 + 2 H_2O$

Net reaction:

$$CrCl_3 + KMnO_4 + H_2O \rightarrow CrO_3 + KCl + MnO_2 + 2HCl$$

2.6 EXCEL SPREADSHEET TOOLS: PART 1

In this section, we describe three spreadsheet tools that facilitate computation of material efficiency metrics for synthesis plans in an automated way. These spreadsheets complement the ones described in Chapter 8, Section 8.1 for individual reactions in the prequel book *Reaction Green Metrics: Problems, Exercises, and Solutions*.

2.6.1 Synthesis-Plan-Analysis-Basis-Materials-Metrics.xls

This spreadsheet computes the overall AE, overall kernel RME, and overall kernel PMI for a linear synthesis plan. Once the balanced chemical equations for each reaction step are established, the reaction step numbers are first entered in column A beginning with the last step and working backward. Next, the molecular weights of final product and all intermediate products are entered in column C. Then the molecular weights of all input reagents multiplied by their respective stoichiometric coefficients are entered in column B followed by the molecular weights of all reaction by-products also adjusted with their own stoichiometric coefficients. Columns E, F, and G check that the entered molecular weights for each reaction step lead to properly balanced chemical equations. Specifically, if all equations are balanced correctly then all cell entries appearing in column G should be zero. Columns H and I determine the step E-factor and step atom economies based on molecular weight. The step reaction yields are entered in column J, beginning with the last reaction step and working backward toward the first step. Column K calculates the cumulative reaction yields along the chain corresponding to the main branch in a synthesis tree diagram. The last entry at the bottom of column K corresponds to the overall yield for the main branch of a linear synthesis plan. For convergent synthesis plans, the entire block of cells appearing in the spreadsheet is copied and pasted below the first block. The second block of cells corresponds to the convergent branch. The same procedure is followed in filling out the entry data in the second block as is done in the first block working backward from the terminal product of the convergent branch to its first reaction step. The two blocks are numerically linked by multiplying the reaction yield entry in column J for the last step in the convergent branch by the cumulative yield shown in column K corresponding to the convergent reaction step where the terminal product of the convergent branch reacts with the appropriate intermediate in the main branch of the synthesis plan.

2.6.2 Template-SYNTHESIS.xls

This spreadsheet computes the kernel and global quantities for the following metrics for any kind of synthesis plan: AE, RME, and PMI. The computation of all metrics is done using a 1 kg basis mass of final target product; however, this may be adjusted to any value if desired. In addition, partitioning of the global E-factor into its kernel, excess reagent, reaction solvent, catalyst, workup material, and purification material consumptions are also calculated. These are displayed visually as a pie chart. A radial pentagon is also displayed pertaining to the overall quantities of AE, reaction yield, inverse stoichiometric factor, materials recovery parameter, and reaction mass efficiency in the same way as for individual reaction steps determined in the *template-REACTION.xls* spreadsheet. The corresponding synthesis tree diagram is entered to facilitate the tracking of entry data for reaction steps in columns C to H, and column L. The data for these columns are obtained from the results obtained from the *template-REACTION.xls* spreadsheets pertaining to each reaction step in the synthesis plan. The following four histograms display the overall performance of a synthesis plan: (a) auxiliary material consumption as a function of reaction stage with catalyst, reaction solvent, work-up, and purification contributions for each step displayed; (b) AE and reaction yield profile as a function of reaction step; (c) stoichiometric and excess reagent consumption as a function of reaction stage; and (d) PMI reaction profile as a function of reaction step. The spreadsheet is defaulted to plans containing three reaction steps, however, syntheses exceeding three steps may be easily evaluated by simply copying additional reaction step blocks as needed. When extra reaction blocks are entered, the formula cells corresponding to the sums of adjusted molecular weights of reagents (column F), total mass of input reagents (column I), mass proportion (column J), total input kernel mass (column M), total input excess mass (column P), total input unreacted intermediates mass (column S), and total input auxiliary mass (column V) need appropriate adjustment to cover the correct range of values in those columns. Also, similar adjustments are needed for the total masses of catalysts, reaction solvents, workup materials, and purification materials appearing in column AH.

2.6.3 Calculator-Cumulative-PMI.xls

This spreadsheet is composed of two sheets: (a) linear plans, and (b) convergent plans. For both sheets, the synthesis tree diagrams are entered to facilitate tracking of reaction steps in the synthesis plan being evaluated. For linear plans, the total number of steps is entered as well as the step PMIs (determined from the *template-REACTION.xls* spreadsheets), molecular weights of intermediate products, and step reaction yields. The cumulative PMI values appear in column G. The cumulative PMI values are determined using Equation (2.13). The spreadsheet is defaulted to a ten-step linear plan, however, additional steps may be easily incorporated by copying and pasting additional cells in the spreadsheet with embedded formulas. For convergent plans, in Section A the total number of branches is entered in addition to the total number of steps in the main branch. In Section B, the terminal step number, the branch number of the convergent branch, and the stoichimetric factor of the terminal product used in the convergent step are entered. The latter quantity is found from the *template-REACTION.xls* spreadsheet pertaining to the convergent step. Section C refers to the convergent branch and Section D to the main branch. Data entries are made in these sections in the same way as for linear plans. The two sections C and D are linked by following the instruction statements embedded in the spreadsheet as shown by worked Example 2.20 for the convergent synthesis of apixaban.

2.7 PROBLEMS

PROBLEM 2.12
Perusal of the database of named organic reactions shows that there are four named reactions that are used to prepare biphenyl.

Suzuki coupling

⟨⟩—B(OH)₂ + Br—⟨⟩ →[Pd(OAc)₂ (cat.) NaOH] ⟨⟩—⟨⟩

Gomberg–Bachmann reaction

⟨⟩—N₂⁺ Cl⁻ + ⟨⟩ →[NaOH] ⟨⟩—⟨⟩

Ullmann coupling

⟨⟩—Br + Br—⟨⟩ →[Cu, heat] ⟨⟩—⟨⟩

Wurtz–Fittig reaction

⟨⟩—Br + Br—⟨⟩ →[Na] ⟨⟩—⟨⟩

Part 1

Balance each reaction showing reaction by-products and determine the corresponding atom economies. Rank the reactions according to this metric.

Part 2

Each of the starting materials can be traced back to benzene as a common starting material for all routes. Show the precursor reactions and reanalyze the overall atom economies of all routes to biphenyl. Rank the routes according to AE. How does the ranking change compared with Part 1?

PROBLEM 2.13

Given the following schedule for a synthesis plan.

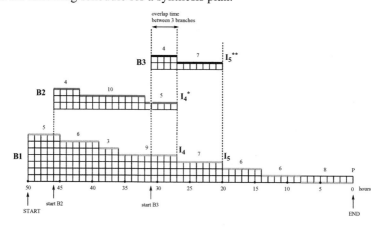

Synthesis Plan Analysis

Part 1
Suppose there is change of protocol in step 4 of B1 that reduces the workup time by 4 hours. How does this impact the schedule of the final product P?

Part 2
Suppose there is an unforeseen delay of 3 hours in the procedure of step 4 in B1 that occurs 1 hour before the synthesis of B3 is scheduled to begin. How does this impact the schedule of the final product P?

Part 3
Suppose there is an unforeseen delay of 3 hours in the procedure of step 4 in B1 that occurs 1 hour after the synthesis of B3 begins. How does this impact the schedule of the final product P?

PROBLEM 2.14[47–49]
Alkane fuels can be synthesized from biofeedstocks as shown by the example below where nonane is made from D-fructose via 5-(hydroxymethyl)furfural (5-HMF).

Experimental Procedure:
Step 1
A 500-mL single-necked, round-bottomed flask equipped with a 3.5-cm ellipsoidal magnetic stir bar is charged with tetraethylammonium bromide (45 g, 214 mmol) distilled water (5 mL) and fructose (**1**) (10 g, 0.056 mol). The flask

is placed in an oil bath at 100°C. The reaction temperature is monitored by a thermometer placed inside the reaction mixture. When the internal temperature reaches 90°C, Amberlyst® 15 (1 g, 10%w/w) is added. The stirring is continued for 15 min, during which time the internal temperature reaches 96–98°C and the color of the reaction mixture turns brown. After this period, the flask is removed from the heat source and the solution separated from the Amberlyst® 15 by decantation of the hot solution into a 2-L single-necked round-bottomed flask. Absolute ethanol (20–30 mL) is used to wash the catalyst after the transfer, and the ethanol is added to the aqueous solution. The reaction mixture is placed on a rotary evaporator for 1 h (15 mmHg, 50°C). A 3.5-cm ellipsoidal magnetic stir bar is added, and the residue is dissolved in hot absolute anhydrous ethanol (25 mL), which is followed by immediate addition of ethyl acetate (1.5 L) under vigorous stirring. Immediate precipitation of the tetraethylammonium bromide is observed and the mixture is cooled down to room temperature. The precipitate is vacuum-filtered through a fritted filter and the cake rinsed with ethyl acetate (100 mL). The collected filtrate is vacuum-filtered through a pad of silica gel (30 g) using a fritted filter and the silica gel is rinsed with ethyl acetate (100 mL). The filtrate is concentrated by rotary evaporation (32°C, 15 mmHg). The oily residue is transferred into a 50 mL single-necked round-bottomed flask and the residual organic solvent is removed *in vacuo* (0.5 mmHg); 5-HMF is obtained as a deeply orange liquid (5.8–6.1 g, 83%–87%) and 90%–92% purity. Flash column chromatography was applied to give the title compound as an orange liquid (5.5 g, 79%) with 97% purity.

Step 2
The reaction was performed in water without adjustment of pH. Piperidine was used in 5 mol%; 5-HMF (0.631 g, 5 mol) was charged in a small round bottom flask with a stirring bar, water (4 mL) was added. Acetone (0.290 g, 5 mmol, 1 equiv.) was added. The mixture was then stirred while piperidine was added at room temperature. The reaction mixture was kept closed with a plastic cap and stirred for 20 hrs. The stirring bar was removed and silica gel added. The mixture was dried with rotary evaporation. The residue was loaded on silica gel and eluted with 50% ethyl acetate in hexanes to provide mono-HMF adduct **A** (0.191 g, 23%) as a light yellow solid.

Step 3
A (830 mg, 5.00 mmol) was dissolved in a 50% vol/vol solution of acetic acid in H_2O (15 mL) and Pd/C added (83 mg of 10 wt% Pd/C, 0.08 mmol Pd, 1.6 mol% Pd relative to substrate). The solution was added to a round bottom flask and put under an atmosphere of H_2 using a balloon and heated at 65°C for 1 h to yield a pale yellow solution. The flask was then equipped with a condenser, opened to the air, and heated at 100°C for 3 h to yield a colorless solution. On cooling, the solution was filtered to remove the Pd/C, the aqueous layer was neutralized with $NaHCO_3$ and extracted with dichloromethane (3 × 5 mL), dried over $MgSO_4$, filtered and solvent removed in vacuo to yield **B** (817 mg, 96% yield).

Step 4

To a 25.00 mL volumetric flask, 851 mg (5 mmol) of **B** was dissolved in glacial acetic acid together with dimethyl sulfone (DMS) (235.3 mg, 2.5 mmol) as an internal standard for GC-MS to yield an orange solution. The solution was sonicated for 5 min and then loaded in the autoclave reactor with Pd/C (170 mg of 5 wt% Pd/C, 0.08 mmol Pd; that is, 1.6 mol% Pd relative to substrate) and La(OTf)$_3$ (426 mg, 0.727 mmol) and sealed. The reactor was pressurized with H$_2$ to 3.45 MPa and vented three times to remove any residual oxygen atmosphere. The reactor was pressurized to 3.45 MPa one final time and heated to the desired reaction temperature (200°C). After 16 h, the reactor was cooled to room temperature and the H$_2$ pressure released. The sole reaction product was identified as *n*-nonane (90% yield).

Part 1
Write out balanced chemical equations for each step and determine the respective atom economies.

Part 2
Draw out a synthesis tree diagram for the plan.

Part 3
Using the *template-REACTION.xls* and *template-SYNTHESIS.xls* spreadsheets, and the experimental details given, determine the material efficiency metrics for each reaction and for the overall synthesis plan. State any assumptions made in the calculations.

PROBLEM 2.15[50,51]
A convergent plan to make 2,3,7,8,12,13,17,18-octaethylporphyrin from ethyl glycinate, 1-nitropropane, and propionaldehyde is shown below.

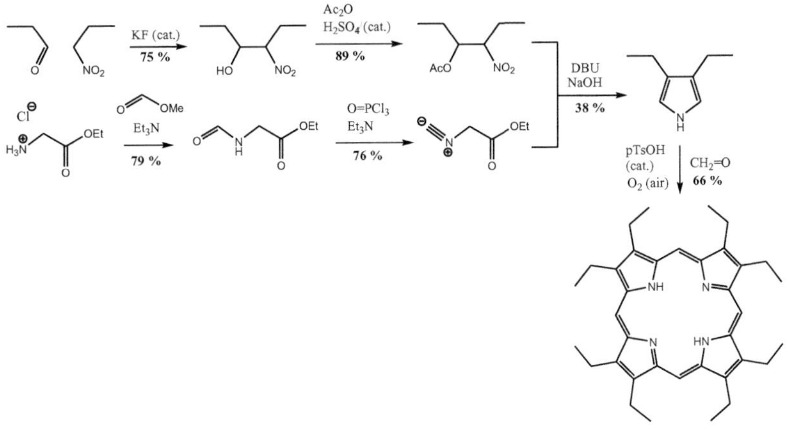

Part 1
Rewrite the scheme showing balanced chemical equations for all steps.

Part 2
Look up the experimental procedures for each step and use the *template-REACTION.xls* and *template-SYNTHESIS.xls* spreadsheets to determine the step and overall values for the following metrics: AE, RME, PMI, and E-factor. State any assumptions made in the calculations.

Part 3
Verify the following connecting relationships between the step and overall metrics.

$$(AE)_T = \frac{p}{\left[\sum_{j=1}^{N-1} I_j \left(\frac{1}{(AE)_j} - 1\right)\right] + \frac{p}{(AE)_N}}$$

$$(RME)_T = \frac{m_p}{\frac{m_p}{(RME)_N} + \sum_{j=1}^{N-1} m_{I_j}\left[\frac{1}{(RME)_j} - 1\right]}$$

$$(PMI)_T = \frac{m_p(PMI)_N + \sum_{j=1}^{N-1} m_{I_j}\left[(PMI)_j - 1\right]}{m_p}$$

$$E_T = \frac{\sum_{j=1}^{N-1} m_{I_j} E_j + m_p E_N}{m_p}$$

Compare the sum of the step E-factors with the total E-factor.
Compare the sum of the step PMIs with the total PMI.

Part 4
Determine the step E-factor contributions to the total E-factor.
  Determine the step PMI contributions to the total PMI.
  What is the relationship between the sum of the step metric contributions and the total metric value in each case?

PROBLEM 2.16
A reaction network showing various routes for the preparation of (R)-N-(1-phenylethyl)-benzamide from acetophenone is shown below. In this network,

Synthesis Plan Analysis

(R)-1-phenylethylamine is a hub intermediate since all routes converge to this product before it is transformed to the final amide product.

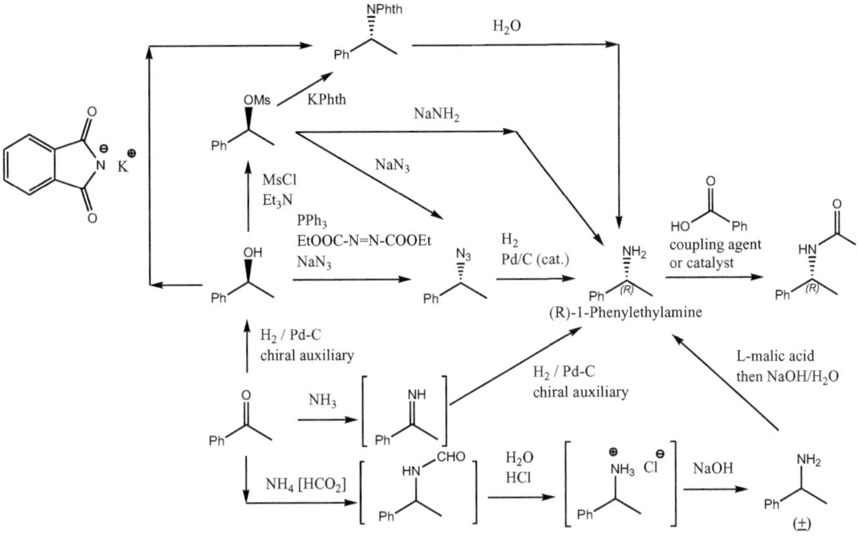

Possible coupling agents for consideration:

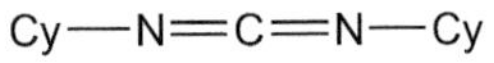

Cy—N=C=N—Cy (DCC)

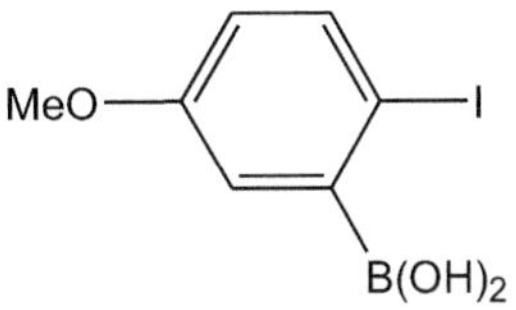

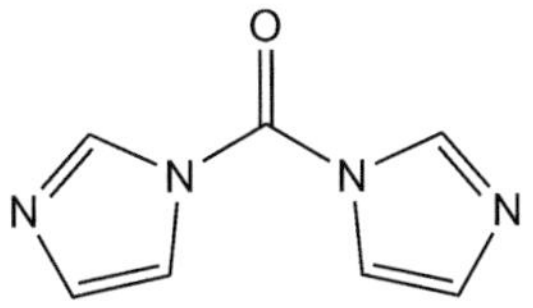

(CDI)

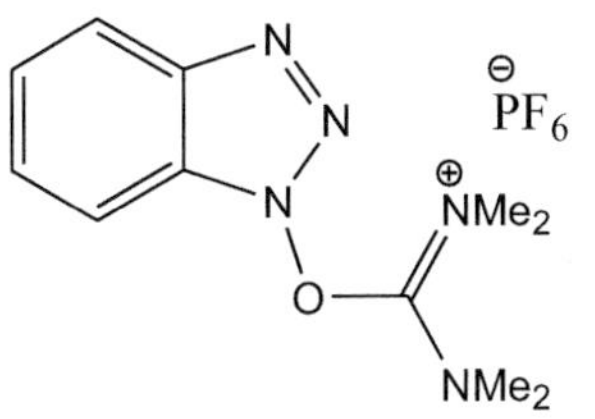

MeO—[aryl]—I (catalyst)
B(OH)₂

PF₆⁻ (HBTU)

PF₆⁻ (HATU)

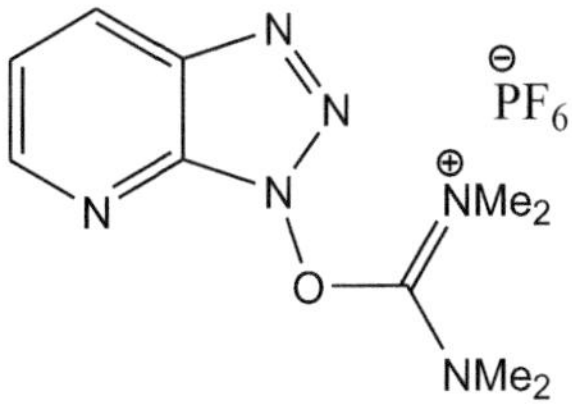

Part 1
Study the network carefully and by inspection single out the most atom economical route to (R)-N-(1-phenylethyl)-benzamide without using any computation. What are its advantages?

Part 2
Enumerate the other routes and give reasons why they are less atom economical.

Part 3
What are the expected by-products of the amidation reaction when each of the coupling agents is used?

PROBLEM 2.17[52]
Propose two synthesis plans for the following target structure using different strategies and compare their expected atom economy performances. Select a strategy by first examining the structural motifs found in the structure and then exploring various ways of parsing the structure from the target synthesis bonds selected.

PROBLEM 2.18[53,54]
2,4,5-Trichlorophenoxyacetic acid (2,4,5-T) is a constituent of the herbicide "Agent Orange." It can be made by the linear route shown in Plan A or by a convergent route shown in Plan B.

Plan A (Linear)

Synthesis Plan Analysis

Plan B (Convergent)

Part 1
For each plan, determine all reaction by-products for each step.

Part 2
Represent each plan as a synthesis tree diagram and determine the respective overall atom economies.

PROBLEM 2.19
A racemic acid (+)**A** is required to be separated into its constituent enantiomers using a chiral amine (+)-**B**. Compare the PMI for the production of the enantiomer that originates from the diastereomeric salt that precipitates out first with the PMI for the production of the enantiomer that originates from the diastereomeric salt that remains dissolved in the mother liquor. Which one is likely to be higher? Explain your answer.

PROBLEM 2.20
A racemic alcohol (+)**A** is required to be separated into its constituent enantiomers using a lipase mediated acetylation reaction. One of the alcohol stereoisomers is selectively acetylated whereas the other is not. Compare the PMI for the production of the unacetylated enantiomer with the PMI for the production of the enantiomer that originates from the acetylation reaction. Which one is likely to be higher? Explain your answer.

PROBLEM 2.21
Papaverine is an alkaloid found in opium that is used as an antispasmodic because of its ability to relax smooth muscle. A synthesis of this compound utilized (3,4-dimethoxy-phenyl)-acetic acid methyl ester as a key hub intermediate. Hub intermediates are used more than once in a synthesis owing to their repeated structural motif

in the final target product. It is imperative that the synthesis sequence that makes such a hub intermediate be optimized as far as possible since that sequence is utilized more than once in the overall plan. If done successfully, this technique results in high overall material efficiency performances for the syntheses of complex structures.

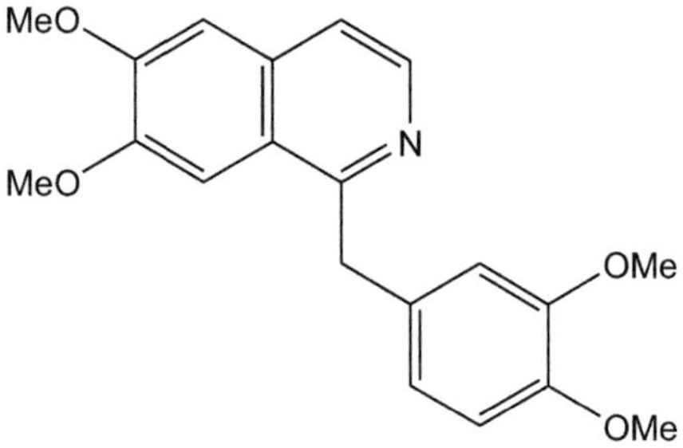

papaverine

Part 1

Find where (3,4-dimethoxy-phenyl)-acetic acid methyl ester maps onto the structure of papaverine.

Part 2

Suggest how (3,4-dimethoxy-phenyl)-acetic acid methyl ester could be made from vanillin. Show balanced chemical equations for each step.

PROBLEM 2.22[55,56]

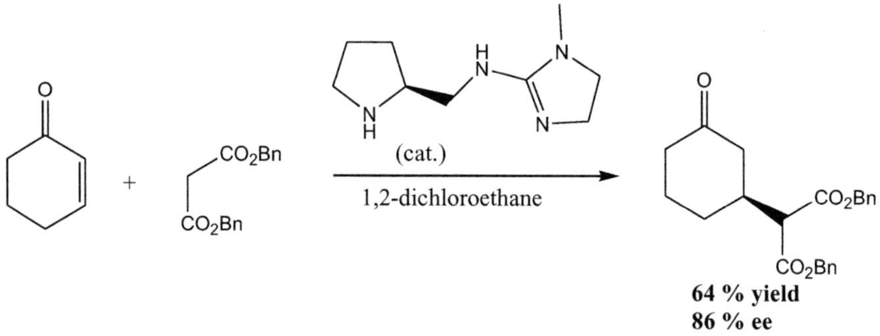

64 % yield
86 % ee

0.5 mmol cyclohexenone
1.2 equivalents malonate
10 mol% proline catalyst
3 mL 1,2-dichloroethane
reaction time: 140 h

Synthesis Plan Analysis

Synthesis of Catalyst:

A solution of N-methylethylenediamine (10.8 g, 146 mmol) in EtOH-H$_2$O (1:1, 90 mL) was preheated to 60°C and CS$_2$ (2.0 mL, 150 mmol) was added dropwise. The resulting mixture was heated at 60°C for 3 h and then conc. HCl (4.7 mL) was slowly added. The temperature was raised to 90°C and stirring was continued for 6 h. After the cooled mixture had been stored at −20°C, it was filtered and the resulting solid dried in vacuo to afford 1-methylimidazolidine-2-thione (8.43 g, 50%).

To a suspension of 1-methylimidazolidine-2-thione (5.17 g, 44.5 mmol) in acetone (50 mL) was added MeI (2.9 mL, 46.6 mmol). The solution was allowed to stir at room temperature for 4 h and the resulting solid was quickly filtered and then dried in vacuo to give 1-methyl-2-(methylthio)-4,5-dihydro-1H-imidazole hydroiodide (8.79 g, 77%) as a beige solid.

To a solution of (S)-N-Boc-2-aminomethyl pyrrolidine (3 g, 15 mmol) in iso-propanol (50 mL) was added 4,5-dihydro-1-methyl-2-(methylthio)-1H-imidazole hydroiodide ((prepared from N-methylethylenediamine by conversion to 1-methyl imidazolidin-2-thione and subsequent reaction with iodomethane), 3.87 g, 15 mmol) at room temperature and the solution was heated to reflux at 95°C for 2 days. The solution was concentrated under reduced pressure to provide 6.13 g (99%) of (S)-tert-butyl 2-((4,5-dihydro-1-methyl-1H-imidazol-2-ylamino)methyl)pyrrolidine-1-carboxylate hydroiodide as a pale yellow solid.

To a stirred solution of the above hydroiodide (1.5 g, 3.61 mmol) in dry CH$_2$Cl$_2$ (5 mL), was added trifluoroacetic acid (5 mL) at 0°C. The solution was brought to room temperature after 30 min., stirred for 3 h, and concentrated under reduced pressure.

The residue was dissolved in ethyl acetate (5 mL) and the solution was extracted with water (2 mL). The aqueous phase was cooled (<5°C), basified with NaOH pellets and the basic solution was extracted with CH$_2$Cl$_2$ (3 × 10 mL). The combined organic layers were dried (Na$_2$SO$_4$) and concentrated under reduced pressure to give 624 mg (93%) of 4,5-dihydro-1-methyl-N-(((S)-pyrrolidin-2-yl)methyl)-1Himidazol-2-amine as a colorless oil.

Part 1

Show a balanced chemical equation for the Michael addition reaction and determine its atom economy.

Part 2

Show a catalytic cycle and mechanism for the reaction.

Part 3

Show a synthesis scheme for the catalyst with balanced chemical equations.

Part 4

Using the *template-REACTION.xls* spreadsheet and the details of the experimental procedure, determine the material efficiency metrics for the reaction excluding the synthesis of the catalyst. State any assumptions made in the calculations. What is the percent increase in the PMI of the Michael reaction if the synthesis of the organocatalyst is included?

PROBLEM 2.23[57,58]

The Craig synthesis plan for (+)-nicotine is outlined below.

Part 1

Determine the overall yield and overall AE for the production of racemic nicotine following the Craig strategy.

Part 2

Draw a synthesis tree diagram for the plan.

Part 3

Based on the tree diagram determine the kernel reaction mass efficiency (RME) for the plan. What is the overall kernel PMI?

Part 4

Track the oxidation state changes of atoms in target synthesis bonds. Highlight the target bonds made throughout the synthesis until the final product is reached.

Synthesis Plan Analysis

PROBLEM 2.24[59,60]
Chemical route

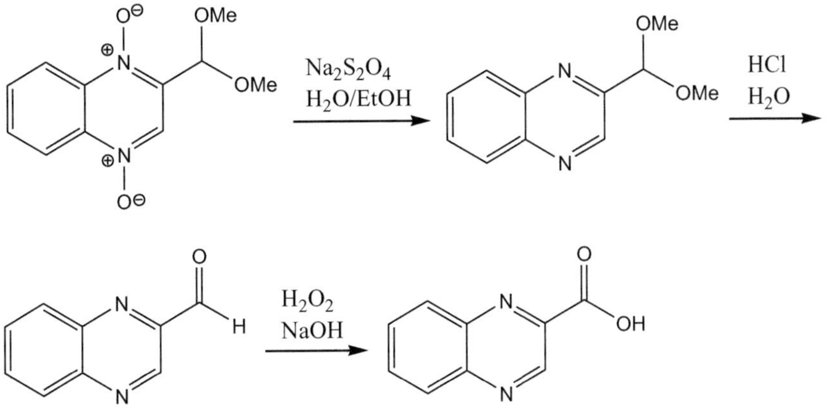

[Scheme: quinoxaline di-N-oxide with CH(OMe)₂ group → Na₂S₂O₄, H₂O/EtOH → quinoxaline with CH(OMe)₂ → HCl, H₂O → quinoxaline-2-carboxaldehyde → H₂O₂, NaOH → quinoxaline-2-carboxylic acid]

Biosynthetic route

[Scheme: 2-methylquinoxaline → Pseudomonas putida monooxygenase → 2-(hydroxymethyl)quinoxaline → benzyl alcohol dehydrogenase → quinoxaline-2-carboxaldehyde → benzaldehyde dehydrogenase → quinoxaline-2-carboxylic acid]

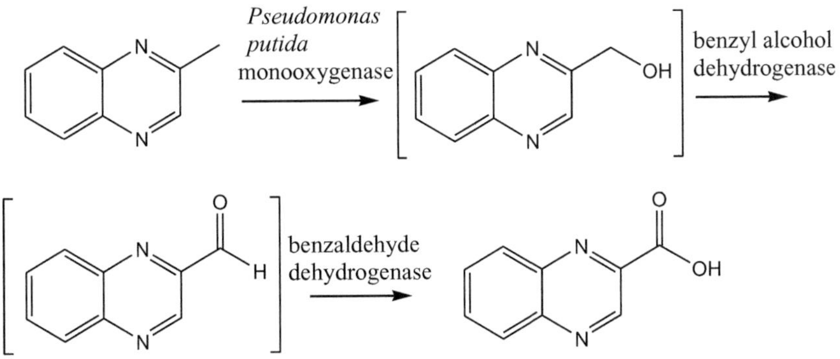

Both routes produce the target product on a 25 kg scale. The overall yield of the chemical route is 35% and that of the biosynthetic route is 86%.

Chemical Route		Biosynthetic Route	
Input material	**Quantity**	**Input material**	**Quantity**
di-N-oxide	3.9 kg	2-Methylquinoxaline	0.97 kg
$Na_2S_2O_4$	5.7 kg	Benzyl Alcohol	2.9 L
35wt% H_2O_2	6.5 L	p-Xylene	0.9 L
4 N HCl	13.6 L	4 N HCl	3.8 L
10wt% NaOH	11.7 L	10wt% NaOH	1.7 L
Chloroform	142 L	Inorganic salts	0.75 kg
N,N-dimethylacetamide	36 L	Trace elements	0.005 kg
EtOH	18 L	Water	79 L

Part 1
Based on these data determine the E-factor and PMI for each process.

Part 2
Verify the yield values.

Part 3
What are the respective atom economies for both routes?

Part 4
Suggest a mechanism for the reduction of the di-N-oxide substrate in the chemical route.

PROBLEM 2.25[61–64]
Adipic acid is a high commodity industrial volume chemical used to make nylon polymer. One traditional way of making it is by oxidizing cyclohexanol in the presence of nitric acid. A new green chemical route and a biosynthetic route have been developed to address the problem of noxious by-products made in the traditional method.

Traditional route:

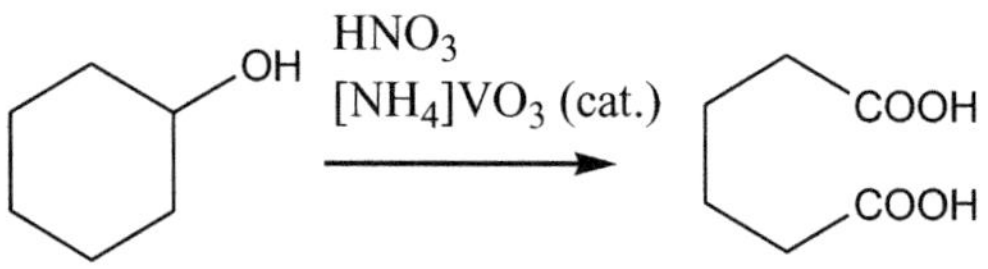

cyclohexanol + HNO_3 / $[NH_4]VO_3$ (cat.) → adipic acid (cyclohexane-1,2-dicarboxylic COOH, COOH structure shown)

Material	Mass (g)
Cyclohexanol	500
Nitric acid (50 wt%)	2100
Ammonium metavanade	1
Water	500
Nitric acid for recrystallization	994
Adipic acid	403.75

Green chemical route:

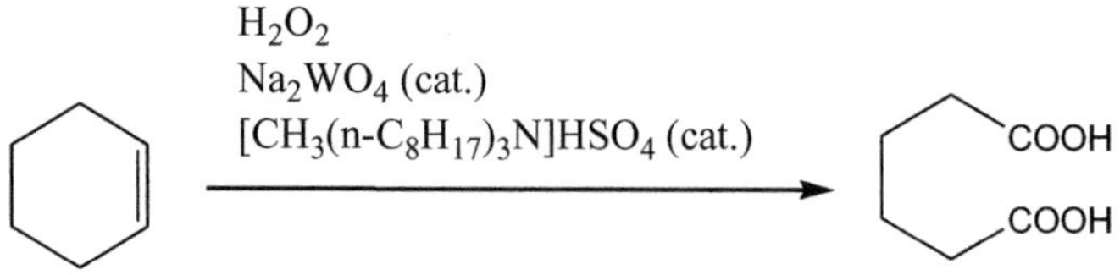

cyclohexene + H_2O_2 / Na_2WO_4 (cat.) / $[CH_3(n-C_8H_{17})_3N]HSO_4$ (cat.) → adipic acid (COOH, COOH)

Synthesis Plan Analysis

Material	Mass (g)
$Na_2WO_4 \cdot 2\ H_2O$	4.1
$[CH_3(n\text{-}C_8H_{17})_3N]HSO_4$	5.67
30wt% H_2O_2	607
Cyclohexene	100
Water	20
Adipic acid	138

Biosynthetic route:

D-glucose →(E. coli, CO_2)→ 3-dehydroshikimic acid →(E. coli, O_2)→ cis-cis-muconic acid →(H_2, Pt (cat.))→ adipic acid

Material	Mass (g)
Bacto tryptone	10
Bacto yeast	5
NaCl	10.5
Na_2HPO_4	6
KH_2PO_4	3
NH_4Cl	1
glucose	10
$MgSO_4$	0.12
Thiamine	0.001
Oxygen (from air)	0.89
Hydrogen	0.22
Adipic acid	2.98

For each route, write out balanced chemical equations and determine the respective yield, AE, and PMI values. For the traditional route, the by-products of nitric acid oxidation are a complex mixture of various nitrogen oxides. For the sake of simplicity in the calculations assume that nitrous acid and hyponitrous acid are the only nitrogen containing by-products. Which route performs better?

PROBLEM 2.26[65,66]
Recently, a procedure was published for the synthesis of methyl nitroacetate that was advertised as "greener" than a previously reported method in *Organic Syntheses*.

Using the *template-REACTION.xls* spreadsheet and the details of the experimental procedures determine the material efficiency metrics for each method. State any assumptions made in the calculations. Are the claims made in the recent report justified? Comment on the quality of the reporting of the experimental procedures.

PROBLEM 2.27[67,68]

Sobetirome is a thyroid hormone agonist belonging to a class of thyromimetics that has been used for treating demyelinating neurological disorders such as X-linked adrenoleukodystrophy and multiple sclerosis.

It was first synthesized by a convergent strategy as shown in the scheme below (Plan A).

A problem was encountered in the final step with a low yield of the intended product since the two phenol groups could not be differentiated with respect to reactivity with pivalyl chloroacetate.

Plan A

Synthesis Plan Analysis

sobetirome

In order to circumvent the problem encountered in the final step of Plan A, a second convergent strategy using protection group chemistry was devised as shown in the scheme below (Plan B).

Plan B

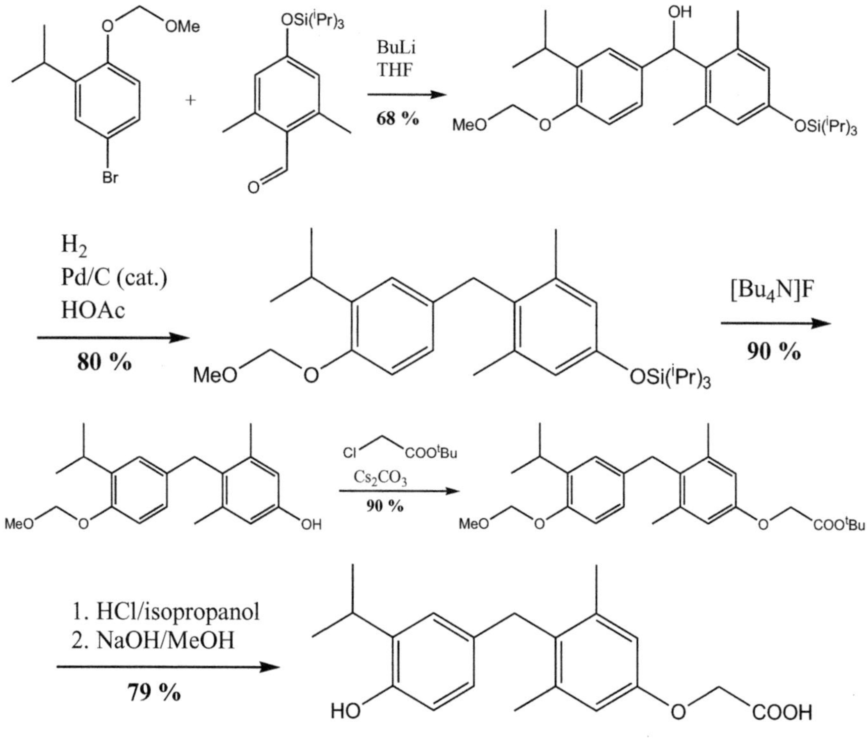

Part 1

From the information given, determine the overall atom economy and overall kernel process mass efficiency for each plan.

Part 2

Suggest an alternative convergent synthesis that does not use protecting groups as in Plan B and avoids the problem encountered in Plan A. Determine its overall AE. How does it compare with the values found in Part 1? Is there a payoff for avoiding protecting groups?

PROBLEM 2.28[69–71]

Ethyl cinnamate can be made via the two synthesis routes shown below. Plan A follows a convergent path using conventional reactions: esterification, hydration, and aldol condensation. Plan B follows a linear path using the following reactions: hydration, Arens–von Dorp–Isler alkylation, and boronic acid catalyzed Meyer–Schuster rearrangement. For each route, show balanced chemical equations for each step, draw a synthesis tree diagram, and determine its overall atom economy, overall yield, and

Synthesis Plan Analysis

overall kernel RME. Show the target bond mapping in the product structure. Which plan is the prevailing one with respect to kernel RME?

Plan A

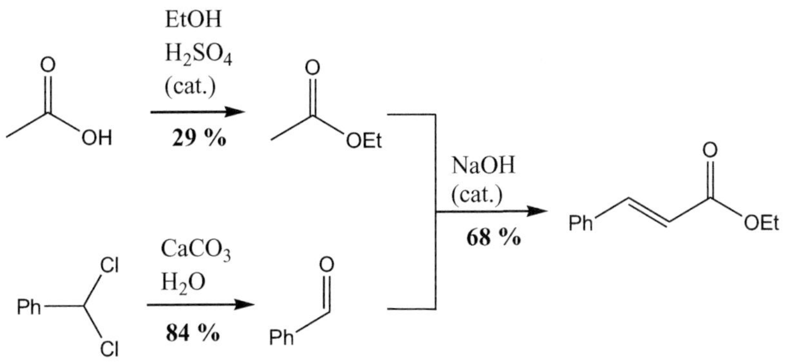

[Scheme: Acetic acid + EtOH, H₂SO₄ (cat.) → ethyl acetate (29%); PhCHCl₂ + CaCO₃, H₂O → PhCHO (84%); combined with NaOH (cat.), 68% → Ph-CH=CH-C(O)-OEt]

Plan B

[Scheme: PhCHCl₂ + CaCO₃, H₂O → PhCHO (84%); CH₂=CHCl + OEt, LiNH₂ → Li-C≡C-OEt, then H₂O → Ph-CH(OH)-C≡C-OEt (90%); C₆F₄H-B(OH)₂ (cat.), toluene, 80% → Ph-CH=CH-C(O)-OEt]

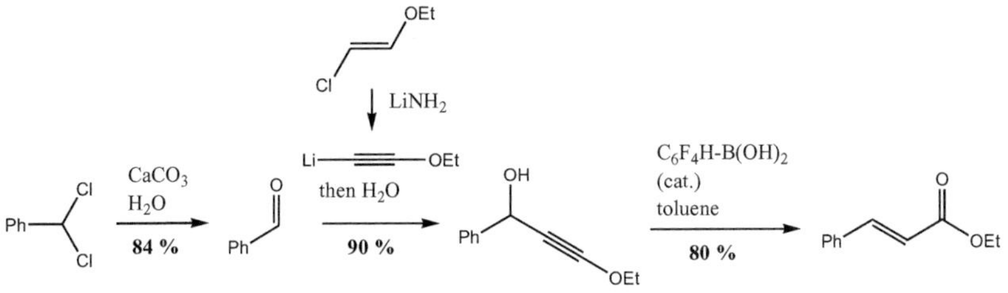

PROBLEM 2.29[72]

A synthesis strategy to prepare a δ-hydroxy ketone is shown below that uses protecting group chemistry. The ketone group is capped as a ketal before reducing the ester, otherwise exposing the ketoester to lithium aluminum hydride will result in both groups being reduced.

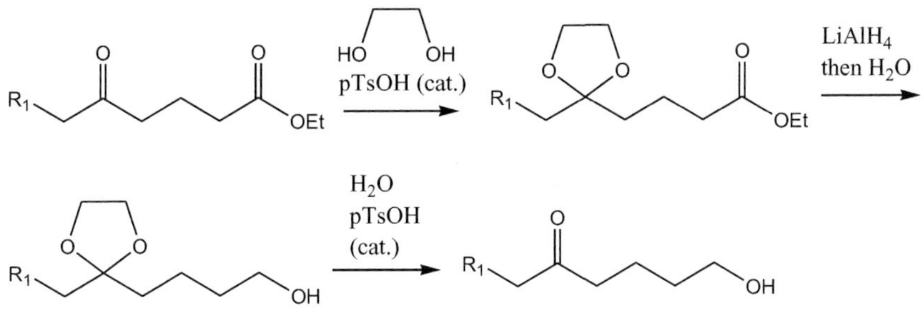

[Scheme: R₁-C(O)-CH₂CH₂CH₂-C(O)-OEt + HO-CH₂CH₂-OH, pTsOH (cat.) → R₁-ketal-CH₂CH₂CH₂-C(O)-OEt; LiAlH₄ then H₂O → R₁-ketal-CH₂CH₂CH₂-CH₂OH; H₂O, pTsOH (cat.) → R₁-C(O)-CH₂CH₂CH₂-CH₂-OH]

Suggest alternative strategies to prepare the target δ-hydroxy ketone that do not use protecting groups.

PROBLEM 2.30[73–76]

Musk ketone (4-*tert*-butyl-2,6-dimethyl-3,5-dinitroacetophenone) is a synthetic musk used in the perfumery industry that mimics the odor of natural musks such as muscone found in musk deer. It is found in Chanel No. 5 perfume at a concentration of 3.5%. In 1891, while doing research on explosives, Albert Baur accidensstally discovered that nitro substituted benzene compounds have an odor that resembles musk. Willem Mallmann synthesized musk ketone according to classical Friedel–Crafts alkylation/acylation and aromatic nitration chemistry as shown in the scheme below.

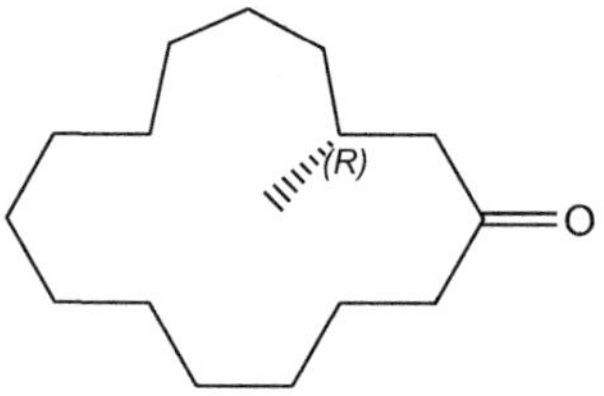

(R)(-)-muscone

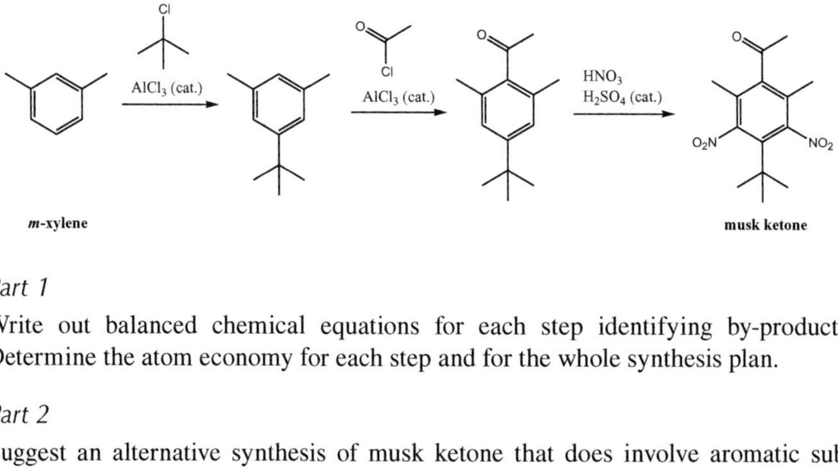

m-xylene

musk ketone

Part 1

Write out balanced chemical equations for each step identifying by-products. Determine the atom economy for each step and for the whole synthesis plan.

Part 2

Suggest an alternative synthesis of musk ketone that does involve aromatic substitution reactions. *Hint*: Exploit the symmetry of the molecule and think of ring construction strategies. For your suggested synthesis, balance all reaction steps and determine the step and overall atom economies.

Part 3

How does your strategy compare in atom economy performance with the classical synthesis? How about environmental impact of by-products and hazards of input materials?

PROBLEM 2.31[77,78]

Two synthesis strategies to prepare the chiral auxiliary (S)-4-(phenylmethyl)-2-oxazolidinone are compared with respect to their material efficiencies. The first

Synthesis Plan Analysis

method is a standard two-step process and the second is a one-pot process that does not isolate the intermediate from the first step.

Two-step process:

L-phenylalanine → (NaBH$_4$, I$_2$/THF) → product 1 → (Cl$_3$C-O-C(=O)-O-CCl$_3$, CH$_2$Cl$_2$) → product 2

Two-step process			
STEP 1		**STEP 2**	
Material	Quantity	Material	Quantity
L-phenylalanine	1.66 g	Product 1	From step 1
THF	8 mL	Dichloromethane	10 mL
NaBH$_4$	0.42 g	Triphosgene solution	1 g in 5 mL CH$_2$Cl$_2$
Iodine solution	5.5 mmol in 20 mL THF	Water	15 mL
Methanol	2 mL	Dichloromethane	16 mL
20wt% NaOH	10 mL		
Water	15 mL		
Dichloromethane	16 mL		
Alumina	1 g		
Product 1	100% yield	Product	92% yield

One-pot process:

L-phenylalanine → (NaBH$_4$, I$_2$/THF) → [intermediate] → (Cl$_3$C-O-C(=O)-O-CCl$_3$, CH$_2$Cl$_2$) → product 2

One-pot process	
Material	**Quantity**
L-phenylalanine	8.3 g
THF	50 mL
NaBH$_4$	3.4 g
I2 solution	12.7 g in 20 mL THF
Methanol	10 mL
20wt% NaOH	50 mL
Dichloromethane	50 mL
Triphosgene solution	5 g in 20 mL CH$_2$Cl$_2$
Isopropyl acetate	25 mL
Heptane	17 mL
Product	6.5 g

Based on these data determine the process mass intensity for each route. Which one is better?

PROBLEM 2.32[160]

(+)-Limonene can be converted to (-)-menthol according to the following sequence:

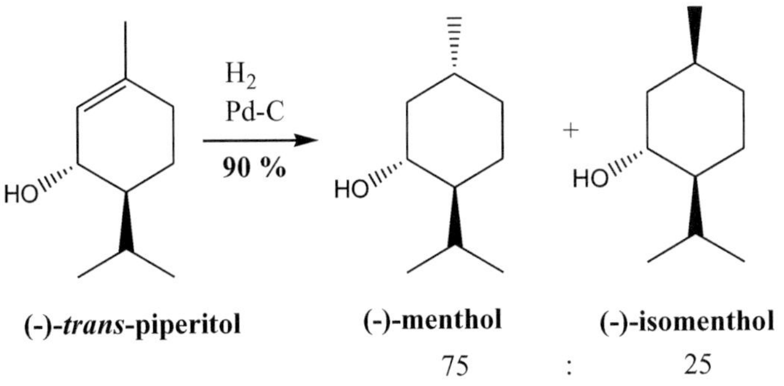

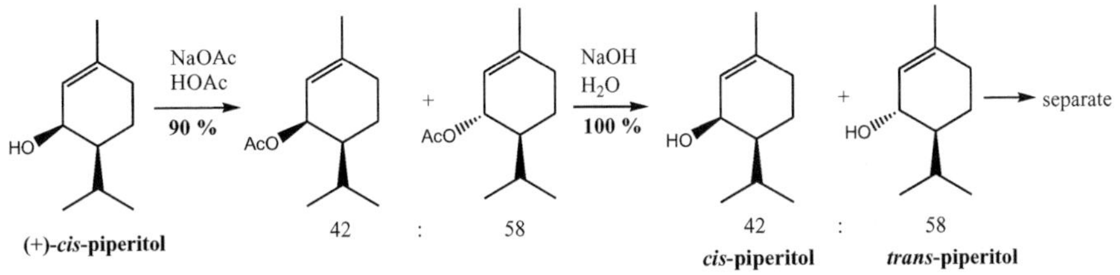

Synthesis Plan Analysis

Part 1
Draw the synthesis tree diagram for the production of (-)-menthol after one pass via (-)-*trans*-piperitol and determine the associated kernel RME.

Part 2
Draw the synthesis tree diagram for the production of (-)-menthol after one pass via (+)-*cis*-piperitol and determine the associated kernel RME.

Part 3
After one pass if one mole of (-)-menthol is collected via (-)-*trans*-piperitol, how many additional moles of (-)-menthol may be collected via the (+)-*cis*-piperitol recycling pathway?

PROBLEM 2.33[79]
An example of the ionic liquid supported synthesis technique is shown below for the preparation of Suzuki coupling products.

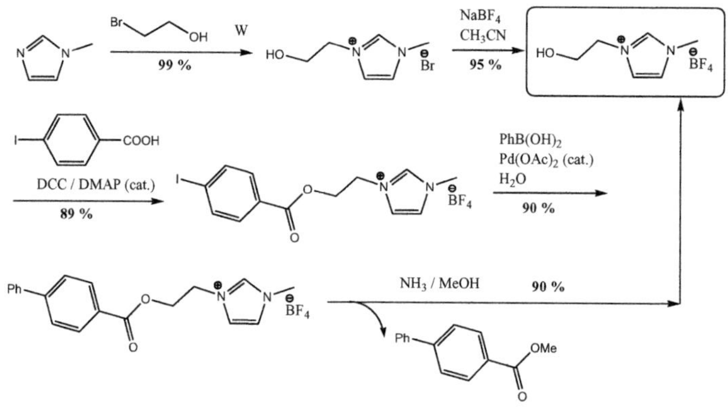

DCC = dicyclohexyldiimide; DMAP = N,N-dimethylaminopyridine
Given the information in the scheme determine

1. the mass of tetrafluoroborate ionic liquid produced after step 2 and the mass of tetrafluoroborate ionic liquid recovered in the last step if 1 kg of 1-methylimidazole is used in step 1,
2. the fractional mass of tetrafluoroborate ionic liquid that is potentially recoverable after one cycle.

PROBLEM 2.34

Four methods are shown below for the synthesis of methyl esters from carboxylic acids or aldehydes.

Fischer esterification

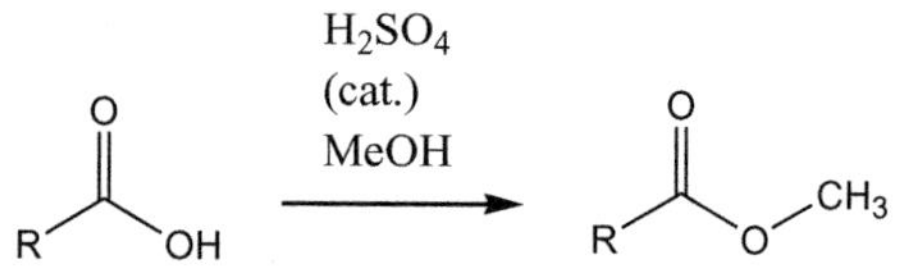

Saponification – substitution sequence

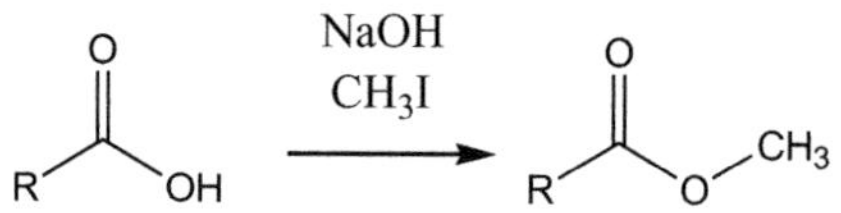

Diazomethane method

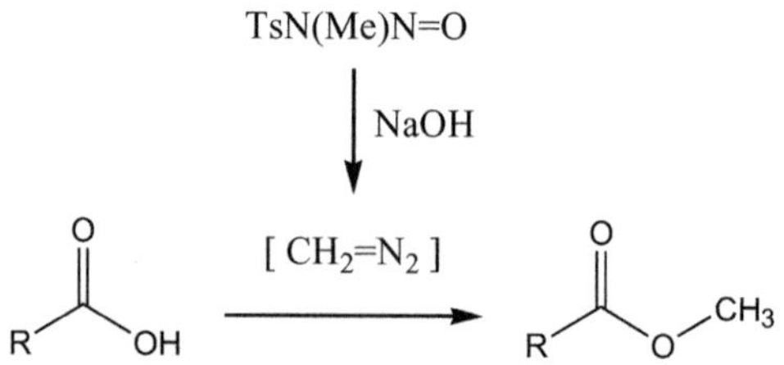

Synthesis Plan Analysis

Cyanohydrin – oxidation – substitution sequence

$$\underset{R}{\overset{O}{\|}}{-}H \xrightarrow[\text{MeOH}]{\text{HCN, MnO}_2} \underset{R}{\overset{O}{\|}}{-}O{-}CH_3$$

For each method provide a balanced chemical equation and determine AE(min) and f(sac)(max). Assume that the smallest R group is a methyl group. Rank the methods according to these metrics from best to worst.

PROBLEM 2.35[80–82]

In the synthesis of mescaline, a hallucinogen alkaloid, a synthesis plan made use of an arene chromium tricarbonyl complex in order to carry out a nucleophilic aromatic substitution reaction in the presence of electron donor groups bonded to the aromatic ring. The chromium tricarbonyl moiety is oxidatively removed from the aromatic ring in the same one-pot procedure.

Part 1

Compare the atom economy performance of the above strategy with a traditional electrophilic aromatic substitution strategy. If the AE of the nucleophilic strategy is to compete with the electrophilic one, what molecular weight should the leaving group X have?

Part 2
Provide mechanisms for each strategy.

Part 3
In step 2 of the nucleophilic aromatic substitution reaction, the authors stated that the chromium tricarbonyl group was removed oxidatively with iodine. However, in the experimental procedure their workup also included washing with an ether solution of sodium dithionite. What was the purpose of sodium dithionite?

Part 4
The chromium tricarbonyl group can also be removed oxidatively with 3,3-dimethyldioxirane according to the reaction shown below. Provide a balanced chemical equation for this transformation. Write out its redox couple. Propose a reasonable mechanism for the reaction.

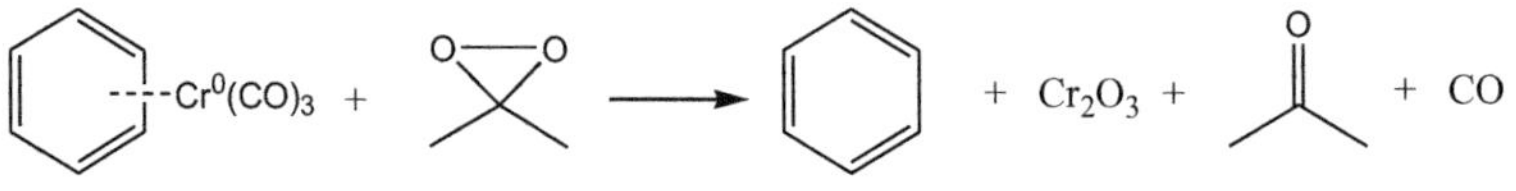

$$\text{Ph-Cr}^0(\text{CO})_3 + \text{dimethyldioxirane} \longrightarrow \text{Ph} + \text{Cr}_2\text{O}_3 + \text{acetone} + \text{CO}$$

PROBLEM 2.36
Four methods to convert keto groups to methylene groups are compared for their material efficiency. For each method write out a balanced chemical equation and from it determine the minimum atom economy and maximum MW fraction of sacrificial reagents. Assume the minimum size group is a methyl group. Rank the methods according to these metrics from best to worst.

Tebbe olefination[83]

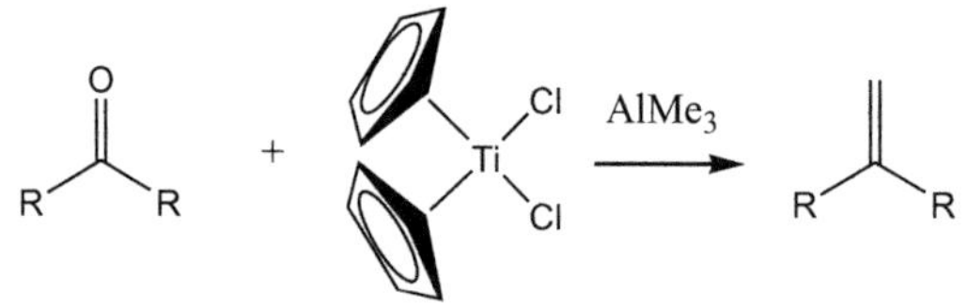

$$R_2C=O + Cp_2TiCl(CH_2AlMe_2Cl) \xrightarrow{AlMe_3} R_2C=CH_2$$

Wittig reaction[84]

$$\text{Br-CH}_3 \xrightarrow{\text{Ph}_3\text{P}} [\text{Ph}_3\text{P}^+\text{-CH}_3 \ \text{Br}^-] \xrightarrow{\text{KO}^t\text{Bu, } R_2C=O} R_2C=CH_2$$

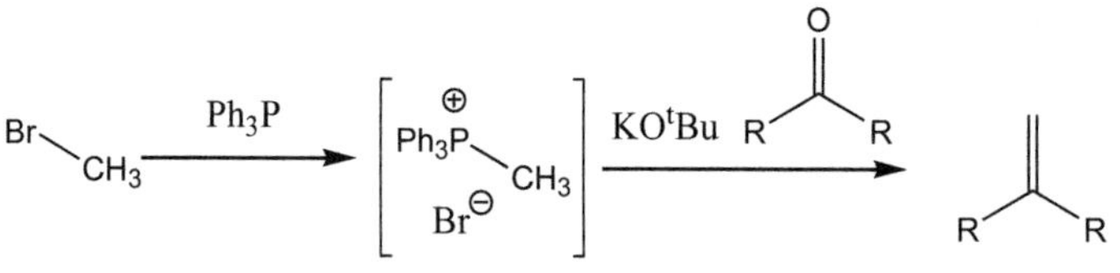

Lombardo olefination[85]

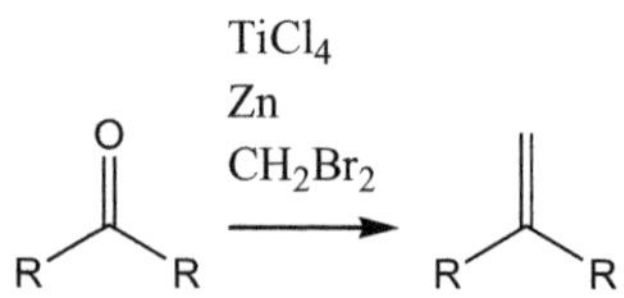

$$R_2C=O \xrightarrow[\text{CH}_2\text{Br}_2]{\text{TiCl}_4, \text{Zn}} R_2C=CH_2$$

Petasis titanocene carbene olefination[86]

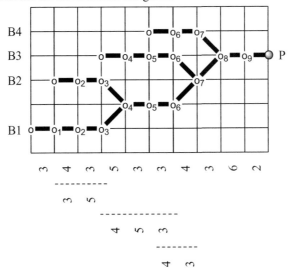

PROBLEM 2.37
A convergent synthesis plan involving four branches is shown below. The operational times in hours for all reactions are given.

Construct a Gantt scheduling diagram that corresponds to the synthesis tree diagram.

Part 1
Determine when branches B2, B3, and B4 should start relative to branch B1.

Part 2
What is the overlap time between branches B3 and B4?
What is the overlap time between branches B2 and B3?
What is the overlap time between branches B1 and B2?

Part 3
For how long will branches B1, B2, and B3 run simultaneously?

Part 4
What fraction of the time taken by branch B1 does not overlap with any of the other three branches?

PROBLEM 2.38

Sarsasapogenin isolation[87]
 Diosgenin isolation[88]
 Marker degradation[89]

Russell E. Marker pioneered the first semi-synthesis of progesterone from starting natural product materials called diosgenin and sarsasapogenin obtained from the rhizomes of *Dioscorea villosa* (or *Trillium erectum*) and Mexican sarsaparilla root, respectively.

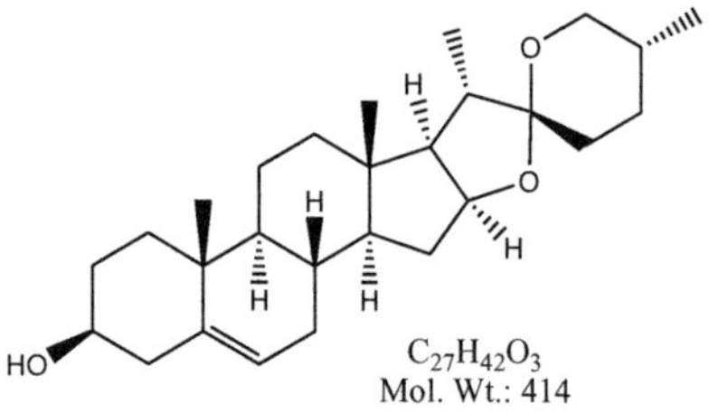

diosgenin $C_{27}H_{42}O_3$ Mol. Wt.: 414

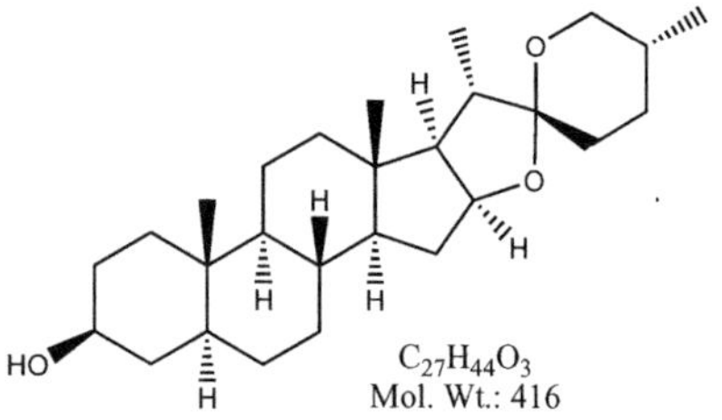

sarsasapogenin $C_{27}H_{44}O_3$ Mol. Wt.: 416

The list of materials used for isolating these starting materials is given below.

Sarsasapogenin extraction materials:

45 kg ground Mexican sarsaparilla root
120 L of 95% ethanol (d = 0.816 g/mL)
ligroin (d = 0.656 g/mL)
diethyl ether (d = 0.708 g/mL)
12 L of 20% ethanol (d = 0.976 g/mL)
2 L of 37 wt% HCl solution (d = 1.1837 g/mL)
10 L 95% ethanol (d = 0.816 g/mL)
600 mL 37 wt% HCl solution (d = 1.1837 g/mL)
3 L water (d = 1 g/mL)
acetic anhydride (d = 1.081 g/mL)
methanol (d = 0.792 g/mL)
2 volumes of ethyl acetate (d = 0.902 g/mL)
Yield of sarsasapogenin acetate: 80 g

Diosgenin extraction materials:

10.5 kg powdered root of *Trillium erectum*
9 gallons of ethanol (d = 0.789 g/mL) (1 US gallon = 3.79 L)
4 L of 85% ethanol (d = 0.850 g/mL)
ligroin (d = 0.656 g/mL)
4 volumes of diethyl ether (d = 0.708 g/mL)
4 L of 20% ethanol (d = 0.976 g/mL)
680 mL of 37 wt% HCl solution (d = 1.1837 g/mL)
ice water (d = 1 g/mL)
1 L of 95% ethanol (d = 0.816 g/mL)

Synthesis Plan Analysis

100 mL of 37 wt% HCl solution (d = 1.1837 g/mL)
50% ethanol (d = 0.934 g/mL)
Yield of diosgenin: 120 g

Part 1

For each case, determine the PMI for the process. State any assumptions made regarding amounts of materials.

Part 2

The Marker degradation sequence from diosgenin to progesterone is shown below.

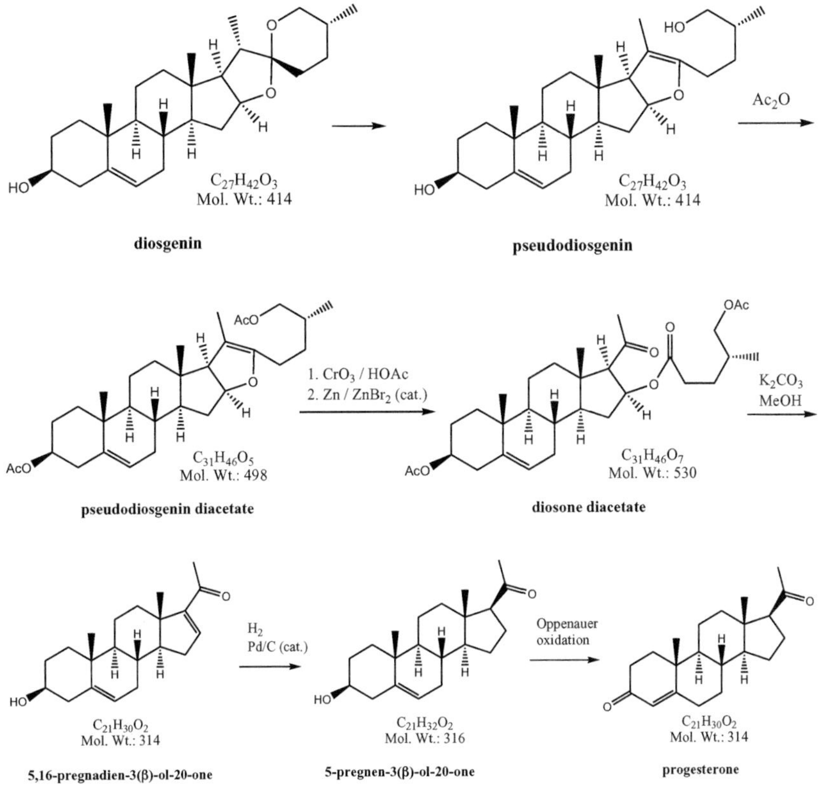

Provide balanced equations for each reaction step and determine the step and overall atom economies.

Part 3

Suggest a mechanism for the third step.

Part 4

What mass of progesterone is expected from 10.5 kg of powdered root of *Trillium erectum*?

Part 5

A competing plan by Parke-Davis[90] produces progesterone from another natural product starting material found in Mexican yams called kappogenin. The one-pot sequence is shown below.

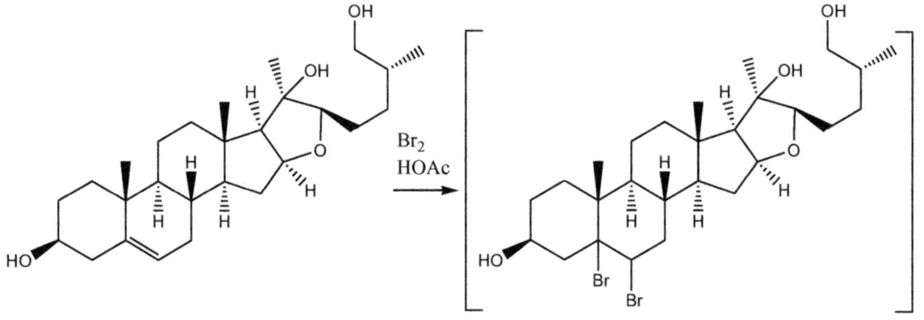

kappogenin **5,6-dibromokappogenin**

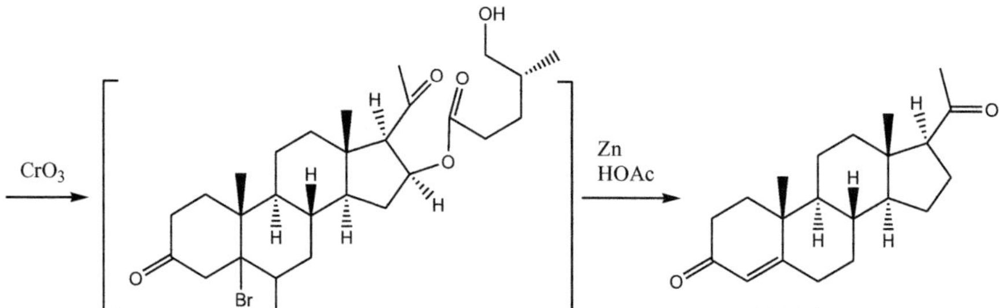

progesterone

Write out the scheme showing balanced chemical equations and determine the overall AE for the one-pot sequence. Compare this value with the one obtained for the diosgenin to progesterone plan.

Part 6

The patent claims were challenged by Marker who showed that kappogenin was actually a mixture of pseudodiosgenin and nologenin[91]. Furthermore, repetition of the patent reaction sequence on the pseudodiosgenin–nologenin mixture led to the isolation of Δ^4-diosgenone-3, not progesterone. Instead, Δ^4-diosgenone-3 was thought to arise from pseudodiosgenin. Further evidence of its structure was also confirmed when Oppenauer oxidation of diosgenin led to the same product, which did not show any melting point depression when a sample of it was mixed with an authentic sample of Δ^4-diosgenone-3. Based on this information, deduce the structure of Δ^4-diosgenone-3 and show how it arose from the sequence of reactions given in the patent. In another experiment, Marker prepared 4,16-pregnadiendione-3,20 from Δ^4-diosgenone-3 according to the following sequence: heating in acetic anhydride to give pseudodiosgenone, chromic acid oxidation, zinc dust reduction in acetic acid, and potassium carbonate hydrolysis in ethanol. Write out balanced chemical equations for the preparation of 4,16-pregnadiendione-3,20.

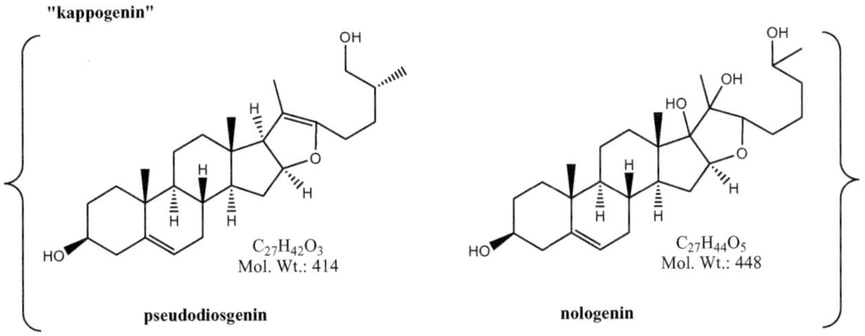

Synthesis Plan Analysis

"kappogenin"

pseudodiosgenin
$C_{27}H_{42}O_3$
Mol. Wt.: 414

nologenin
$C_{27}H_{44}O_5$
Mol. Wt.: 448

PROBLEM 2.39
Diosgenin isolation[88]
  Step 1[92]
  Steps 2 and 3[93]
  Step 4[94]
  Step 5[95]

Progesterone can be made in five steps from the natural product diosgenin found in the rhizomes of *Dioscorea villosa* (or *Trillium erectum*).

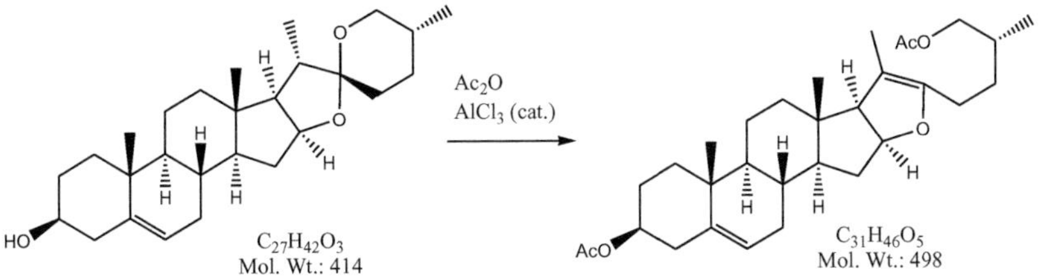

diosgenin
$C_{27}H_{42}O_3$
Mol. Wt.: 414

Ac$_2$O, AlCl$_3$ (cat.)

pseudodiosgenin diacetate
$C_{31}H_{46}O_5$
Mol. Wt.: 498

CrO$_3$
HOAc / H$_2$O

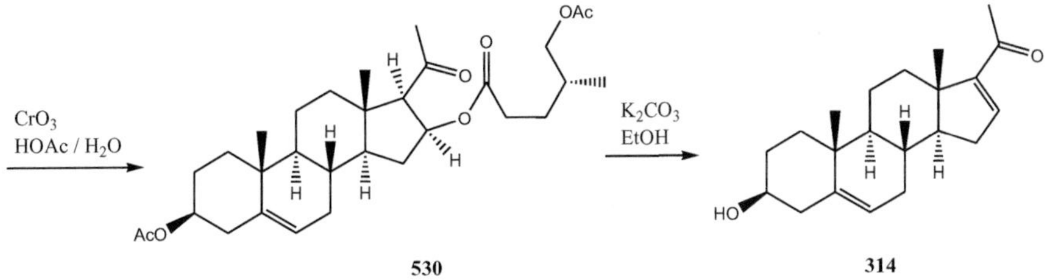

530
diosone diacetate

K$_2$CO$_3$
EtOH

314
5,16-pregnadien-3(β)-ol-20-one

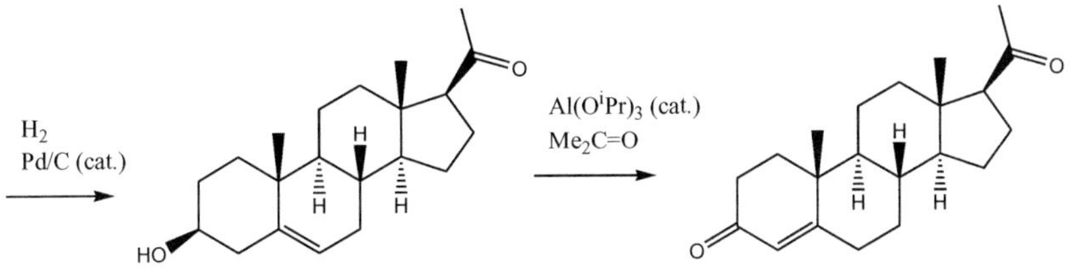

H$_2$
Pd/C (cat.)

316
5-pregnen-3(β)-ol-20-one

Al(O^{i}Pr)$_3$ (cat.)
Me$_2$C=O

314
progesterone

Diosgenin extraction materials:

10.5 kg powdered root of *Trillium erectum*
9 gallons of ethanol (d=0.789 g/mL) (1 US gallon=3.79 L)
4 L of 85% ethanol (d=0.850 g/mL)
ligroin (d=0.656 g/mL)
4 volumes of diethyl ether (d=0.708 g/mL)
4 L of 20% ethanol (d=0.976 g/mL)
680 mL of 37 wt% HCl solution (d=1.1837 g/mL)
ice water (d=1 g/mL)
1 L of 95% ethanol (d=0.816 g/mL)
100 mL of 37 wt% HCl solution (d=1.1837 g/mL)
50% ethanol (d=0.934 g/mL)
Yield of diosgenin: 120 g

Reaction 1:

414 mg diosgenin
20 mL acetic anhydride (d=1.081 g/mL)
67 mg aluminum trichloride
workup: 100 mL ice water
Yield: 488 mg

Reaction 2:

10 g pseudodiosgenin diacetate
7 g chromium trioxide
7 g water
220 mL acetic acid (d=1.0642 g/mL)
workup: 200 mL water (d=1 g/mL); 200 mL diethyl ether (d=0.708 g/mL);
    200 mL saturated sodium bicarbonate solution (d=1.0581 g/mL)
Yield: 4.1 g

Reaction 3:

300 mg diosone diacetate
750 mg potassium carbonate
20 mL ethanol
workup: 20 mL water (d=1 g/mL); 20 mL diethyl ether (d=0.708 g/mL)
Yield: 168 mg

Reaction 4:

150 mg 5,16-pregnadien-3β-ol-20-one
1 g hydrogen gas
50 mL diethyl ether (d=0.708 g/mL)
0.2 g Pd on barium sulfate
Yield: 130 mg

Synthesis Plan Analysis

Reaction 5:

1.6 g 5-pregnen-3β-ol-20-one
29 g acetone
58 g xylene
1.7 g aluminum triisopropoxide
workup: 100 mL 10 wt% HCl solution (d = 1.0474 g/mL)
Yield: 1.2 g progesterone

Part 1
Write out balanced chemical equations for all reaction steps.

Part 2
Draw a synthesis tree diagram for the plan.

Part 3
Using the *template-REACTION.xls* and *template-SYNTHESIS.xls* spreadsheets determine the overall PMI to produce 1 kg of progesterone. Determine the mass of powdered root of *Trillium erectum* that is required. State any assumptions made in the calculations.

PROBLEM 2.40[96]
An iodine-123 labelled compound has recently been developed to facilitate the imaging of cholinesterases found in amyloid plaque deposits in brain tissue afflicted with Alzheimer's disease. Its synthesis plan is shown below.

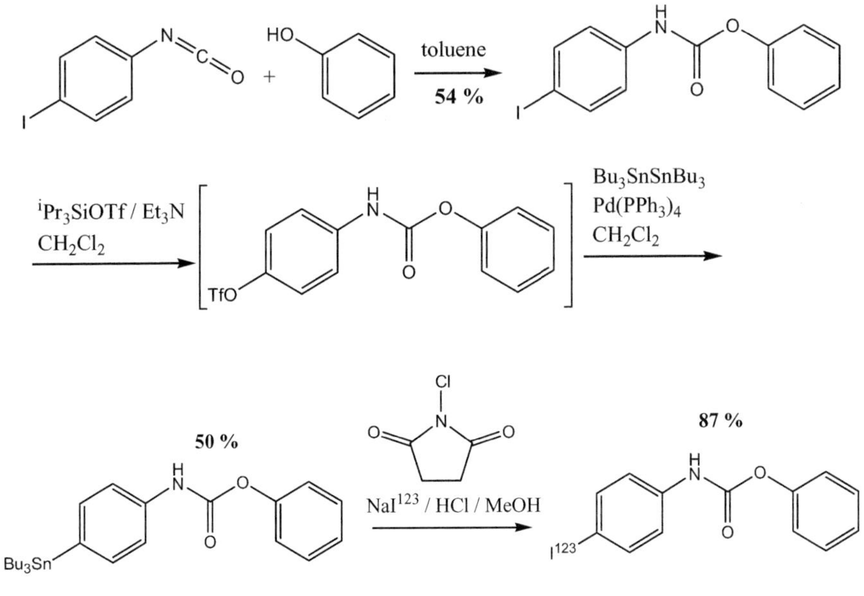

Part 1
Rewrite the scheme showing fully balanced chemical equations for each step.

Plan 2
Determine the overall atom economy of the plan and the MW fraction of sacrificial reagents.

Plan 3
Suggest an alternative synthesis plan for the tracer compound using diazotization chemistry and determine its overall atom economy and MW fraction of sacrificial reagents. How do these metrics compare with the values obtained for the literature synthesis plan?

PROBLEM 2.41
The serotonin re-uptake inhibitor (S)-dapoxetine has been made by three competing plans. Plans A and C are stereoselective syntheses that begin from (R,R)-diethyl tartrate and methyl *trans*-cinnamate, respectively. Plan B utilizes an enzymatic resolution step.

Stereoselective synthesis (Plan A) [97]

Synthesis Plan Analysis

*assumed reaction yield

Enzymatic resolution synthesis (Plan B)[98]

*assumed reaction yield

Stereoselective synthesis (Plan C) [99]

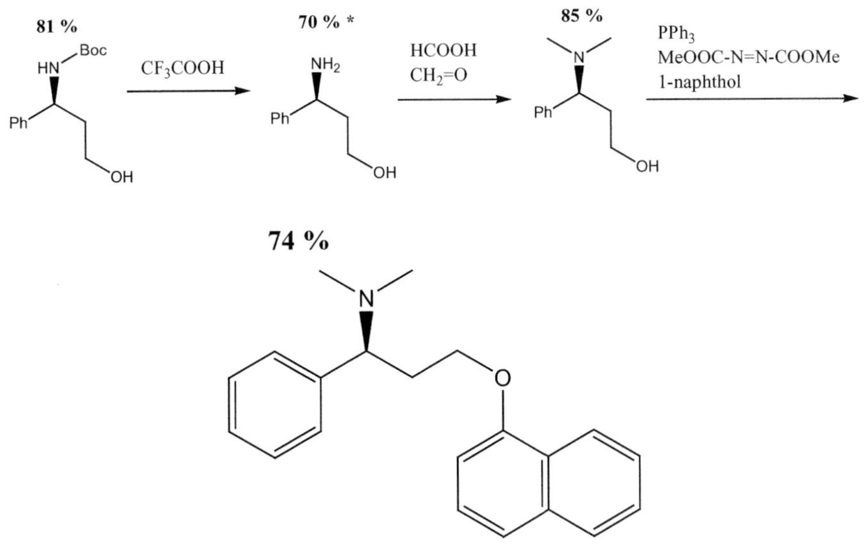

Synthesis Plan Analysis

[Scheme showing synthesis of (S)-dapoxetine:
- Starting material: HN(Boc)-CH(Ph)-CH₂-CH₂-OH, 81%
- CF₃COOH → NH₂-CH(Ph)-CH₂-CH₂-OH, 70% *
- HCOOH, CH₂=O → N(Me)₂-CH(Ph)-CH₂-CH₂-OH, 85%
- PPh₃, MeOOC-N=N-COOMe, 1-naphthol
- Final product: (S)-dapoxetine with N(Me)₂ group, 74%]

*assumed reaction yield

Part 1
Balance all reactions in each plan.

Part 2
Draw synthesis tree diagrams for each plan.

Part 3
Determine the overall yield, overall AE, overall kernel RME, and overall MW fraction of sacrificial reagents for each plan. Rank the plans according to these materials efficiency metrics. Which plan has the highest overall ranking?

Part 4
Suggest a synthesis plan that utilizes a stereoselective imine catalytic hydrogenation strategy to make (S)-dapoxetine directly. Discuss any advantages such a strategy has with respect to green chemistry principles. Determine overall AE and overall MW fraction of sacrificial reagents for such a plan. Based on yield performances for reactions in Plans A, B, and C estimate what the overall kernel RME would be for the conjectured plan.

PROBLEM 2.42[100]
The following reaction procedure yields 705 g epichlorohydrin for a reaction yield of 76%.

In a 5 L flask equipped with a mechanical stirrer are added 3 L anhydrous diethyl ether and 1290 g glycerol α,γ-dichlorohydrin. The flask is surrounded by a cold water bath and 440 g of sodium hydroxide is added in small portions so that the temperature of the reaction is kept at 25–30°C. After addition is complete (about 20 min) the cold water bath is replaced with a warm water bath at 40°C and the mixture is gently boiled for 4 hours. After cooling the ether is decanted from the solid residue, which is rinsed with 2×250 mL diethyl ether. The combined ethereal solutions are fractionally distilled. The portion boiling at 115–117°C is pure epichlorohydrin.

Part 1

It is desired to make 1000 kg of epichlorohydrin by this procedure. One option is to repeat the reaction 1,000,000/705 = 1418 times at the scale of the procedure. This is a so-called scale-out method. A second option is to carry out the reaction once but scaled up appropriately for a target mass of product of 1000 kg. However, for this second option, the reaction yield drops to 60%. Compare and contrast the material efficiency metrics for both options assuming that the necessary equipment and quantities of all materials are available.

Provide some advantages and disadvantages when the first option may be the desired choice.

Provide some advantages and disadvantages when the second option may be the desired choice.

Part 2

The current prices [1] (in CAD) of the materials listed in the procedure above are:
glycerol α,γ-dichlorohydrin: $66.10 for 100 g pack and $98.20 for 500 g pack
diethyl ether (anhydrous grade): $103.50 for 1L and $540 for 18 L
sodium hydroxide: $101.50 for 500 g pack and $181.50 for 1 kg pack
diethyl ether: $94.60 for 1L and $435.50 for 6 L

1. Based on these quotations, calculate the cheapest options for both approaches discussed above for the synthesis of epichlorohydrin.
2. Suppose that workers are hired at $14/hr (CAD) to complete both approaches. Based on the times specified in the original procedure only, what would be the cost difference between Approach 1 and 2?

[1] Sigma-Aldrich, www.sigmaaldrich.com/catalog/AdvancedSearchPage.do (Accessed January 18, 2015).

PROBLEM 2.43

Two plans for the synthesis of tropinone are given below. The Robinson plan is the celebrated single step three-component coupling reaction, which has been described as the first synthesis employing a biomimetic strategy. The preceding Willstätter plan begins from cycloheptanone and employs a linear strategy using classical chemistry.

Synthesis Plan Analysis

Robinson plan[101]

Willstätter plan[102–104]

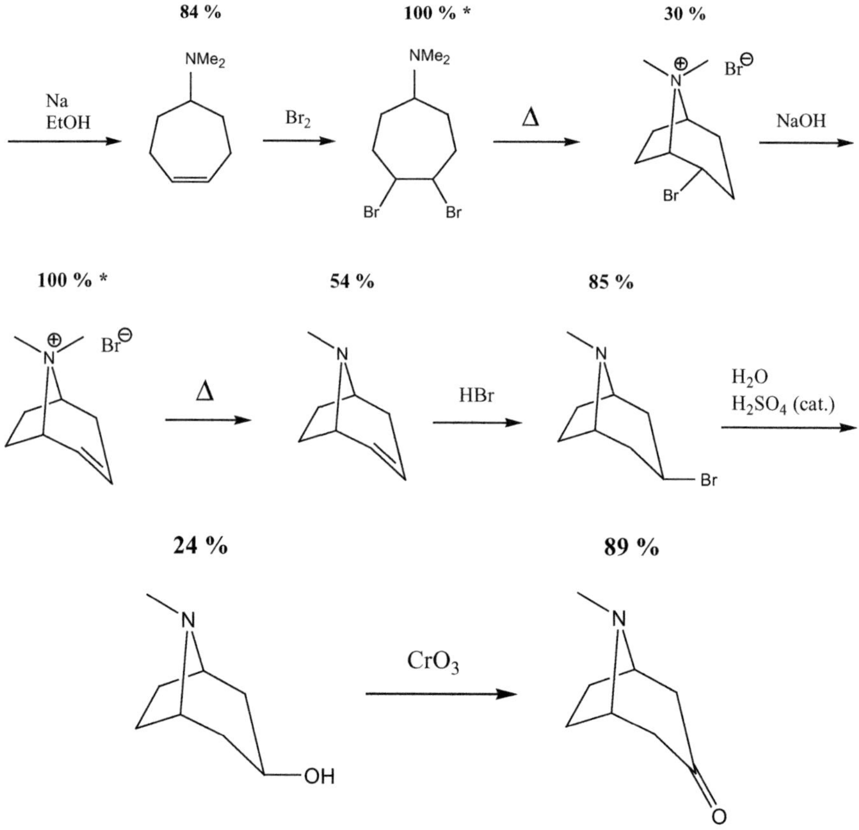

Part 1
Rewrite each plan showing balanced chemical equations for each step.

Part 2
Draw synthesis tree diagrams for each plan.

Part 3
From the tree diagrams determine the overall AE, overall yield, and overall kernel RME.

Part 4
Draw the product structure target bond maps for each plan.

PROBLEM 2.44
Two plans for the synthesis of vitamin C are shown below. Plan A[105] begins from D-glucose and Plan B[106] begins from D-sorbitol.

Step 2 in Plan A is an aerobic oxidation that takes place in a fermenter containing *Acetobacter suboxydans*.

Synthesis Plan Analysis

Step 1 in Plan B is an aerobic oxidation that takes place in a fermenter containing *Gluconobacter suboxydans IFO 3255* and the strain DSM 4025.

Plan A

Plan B

[Scheme: D-sorbitol →(O₂) L-sorbose →(O₂) 2-keto-L-gulonic acid →(NaOH) 2-keto-L-gulonic acid monohydrate monosodium salt (67%) →(2 MeOH, H₂SO₄ cat.) [methyl ester intermediate] → [lactone intermediate] →(NaOMe, heat) L-ascorbic acid monosodium salt (96%)]

Part 1
For each plan, balance all chemical equations.

Part 2
Draw synthesis tree diagrams for each plan and from them determine the overall yields, overall atom economies, and overall kernel reaction mass efficiencies.

Part 3
For each plan, draw the target bond map for vitamin C and itemize all reagents that contribute at least one atom to the final structure. Determine the MW fraction of sacrificial reagents used in each plan. Draw histograms showing the target bond forming reaction profiles for each plan.

Synthesis Plan Analysis

PROBLEM 2.45
Plan A[107,108]
Plan B[109–111]
Plan C[112–116]

Amantadine is a primary amine that has anti-viral activity and has been used to treat influenza. However, it has recently proven ineffective against bird flu and hence the medical profession has counselled physicians not to administer it to patients at risk.

It has been synthesized according to the three routes shown below.

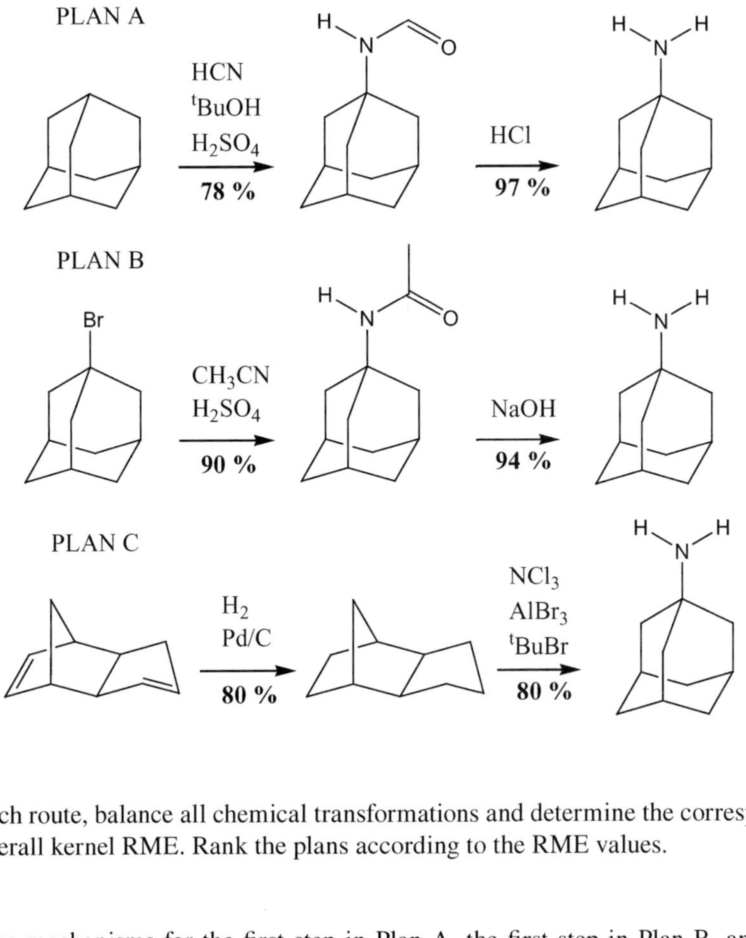

Part 1

For each route, balance all chemical transformations and determine the corresponding overall kernel RME. Rank the plans according to the RME values.

Part 2

Propose mechanisms for the first step in Plan A, the first step in Plan B, and the second step in Plan C.

PROBLEM 2.46[117,118]

The fragrance compound (-)-carvone is chemically synthesized from (+)-limonene according to the following plans. For each plan, draw its synthesis tree diagram and determine its kernel reaction mass efficiency.

Plan A

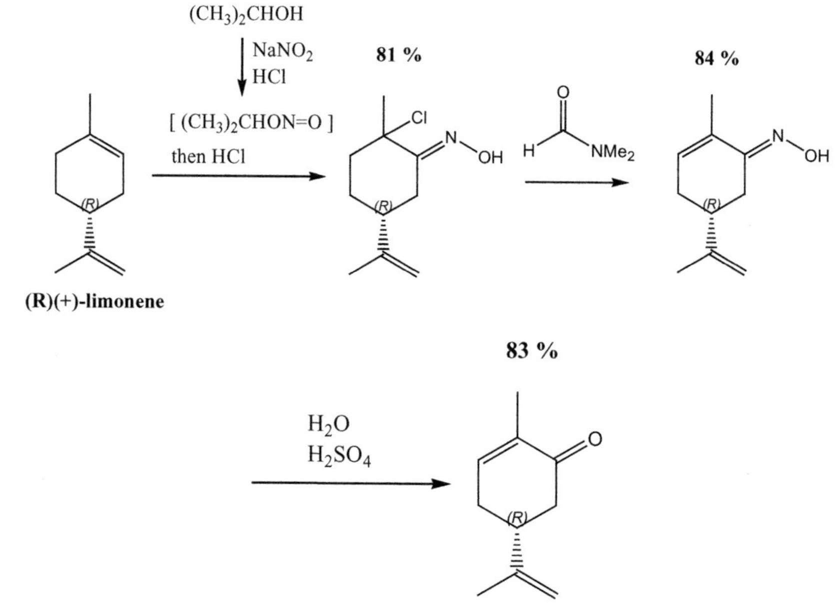

[Reaction scheme: (R)(+)-limonene → (via (CH₃)₂CHOH, NaNO₂, HCl; [(CH₃)₂CHON=O] then HCl) → chloro-nitroso intermediate (81%) → (via H-C(=O)-NMe₂) → oxime intermediate (84%) → (via H₂O, H₂SO₄) → carvone (83%)]

(R)(+)-limonene

Plan B

[Reaction scheme: (R)(+)-limonene → (via CH₃CH₂OH, NaNO₂, H₂SO₄; [CH₃CH₂ON=O] then HCl) → chloro-nitroso intermediate (80%) → (via pyridine) → oxime intermediate (90%) → (via (COOH)₂) → [oxalate ester intermediate] → [rearrangement intermediate] → **(R)(−)-carvone** (78%)]

(R)(+)-limonene

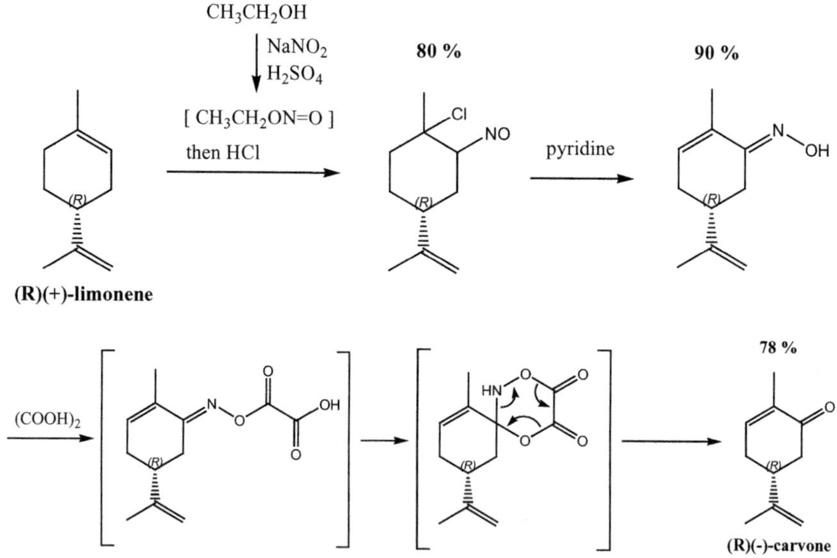

PROBLEM 2.47
Indigo has been made by the following synthetic routes.

Synthesis Plan Analysis

Heumann–Fierz–David synthesis[119–124]

Yamamoto synthesis[125]

Biofeedstock synthesis (using recombinant *Escherichia coli* cell cultures)[126]

Part 1
Rewrite the schemes for the syntheses of indigo showing balanced chemical equations.

Part 2
For each route, draw its synthesis tree diagram.

Part 3
Determine the Curzons reaction mass efficiency for each route.

PROBLEM 2.48[127,128]
A linear and a convergent synthesis for resveratrol have been reported in the literature. Resveratrol is a natural product found in red wines and in the skin of red grapes. It has recently been implicated to have a host of health benefits including anti-cancer, anti-viral, anti-aging, anti-inflammatory, and life-prolonging effects. It is thought to explain the "French paradox" that the incidence of coronary heart disease is relatively low in southern France despite high dietary intakes of saturated fats.

Analyze both synthesis plans fully using synthesis trees. Determine which one has the higher overall reaction mass efficiency.

Plan A (linear)

Synthesis Plan Analysis

Plan B (convergent)

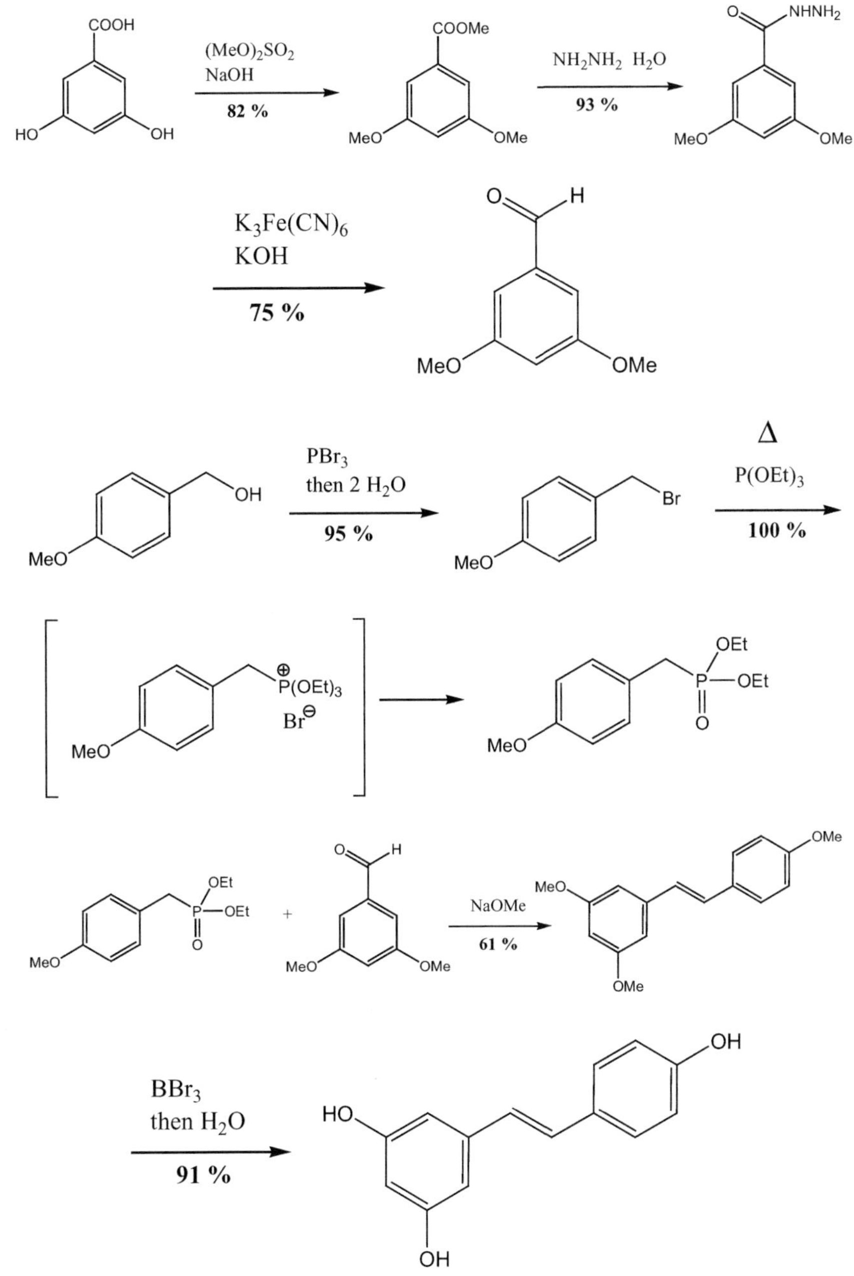

PROBLEM 2.49[129]
A biosynthetic scheme to generate both stereoisomers of a chiral β-lactam using pig liver esterase is shown below.

Determine the overall kernel RME for the synthesis of each β-chiral lactam product from the starting racemic material.

PROBLEM 4.50[130,131]

The Willstätter plan for cyclooctatetrene from pseudo-pelletierine is an example of a synthesis plan that is best described as a degradation sequence rather than as a synthesis sequence.

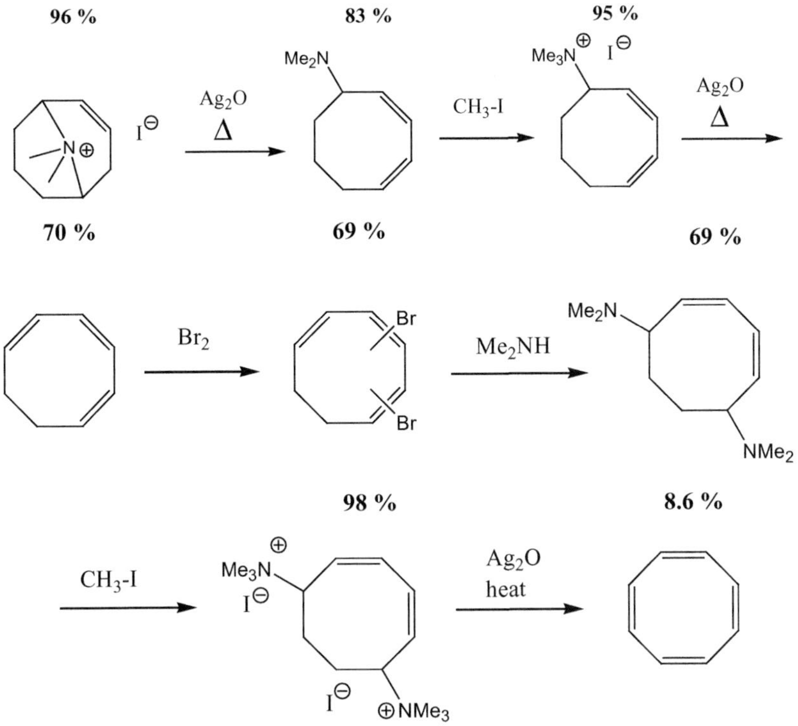

Synthesis Plan Analysis

Part 1
Provide balanced chemical equations for each reaction step.

Part 2
Draw a synthesis tree diagram for the plan and determine the overall atom economy and the molecular weight fraction of sacrificial reagents.

Part 3
Construct a target bond making profile for the plan showing how many steps produce target bonds and how many do not.

PROBLEM 2.51[132]
Alkene hydrogenations typically use palladium metal adsorbed on charcoal as a catalyst. A recent work developed a recyclable catalyst that incorporated two new technologies: ionic liquids and magnetic nanoparticles. The following scheme shows the synthesis of such a catalyst.

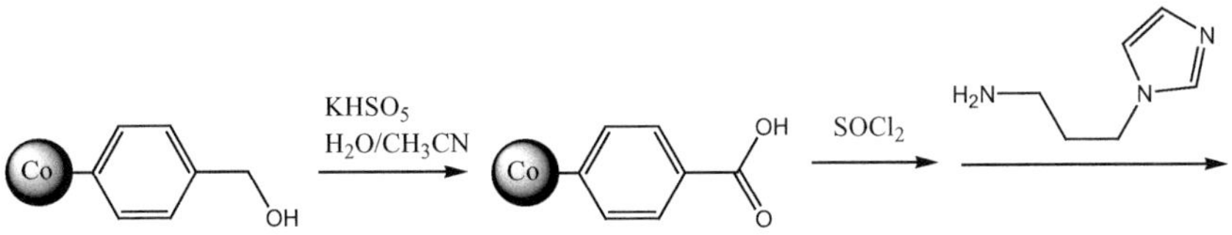

Part 1
Balance all chemical equations in the synthesis plan.

Part 2
Draw a synthesis tree diagram for the synthesis plan.

Part 3
Using the *template-REACTION.xls* and *template-SYNTHESIS.xls* spreadsheets, and the experimental procedure details given in the reference, determine the overall and step material efficiency metrics for the synthesis plan (yield, AE, and PMI). Summarize the results in a table. Show the step and overall E-factor profiles.

Note: The authors did not disclose the loading for PhCOOH@Co/C in mmol per gram for the product nanoparticles for the first step. Assume it is 0.2 mmol per gram.

PROBLEM 2.52[133–135]
Sea-Nine® is a marine antifoulant invented at Rohm & Haas in the late 1970s that is used as a biocide to protect the hulls of ships from unwanted accumulation of barnacles and other organisms that increase the drag force of ships as they travel at sea. This substance is more environmentally friendly compared with tributyltinoxide, which was used previously. Two synthesis plans are shown for making Sea-Nine®.

Plan A

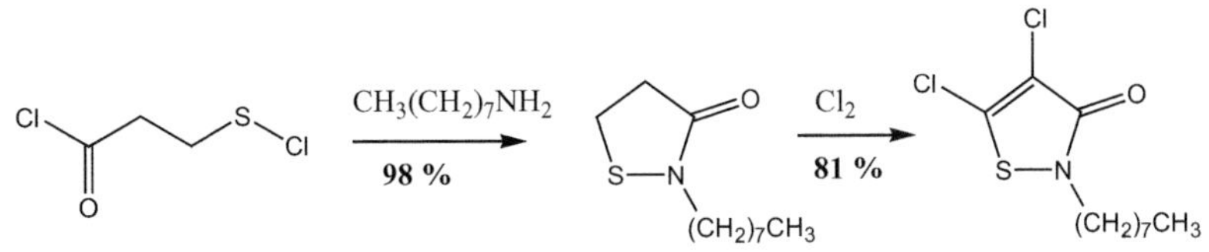

Synthesis Plan Analysis

Plan B

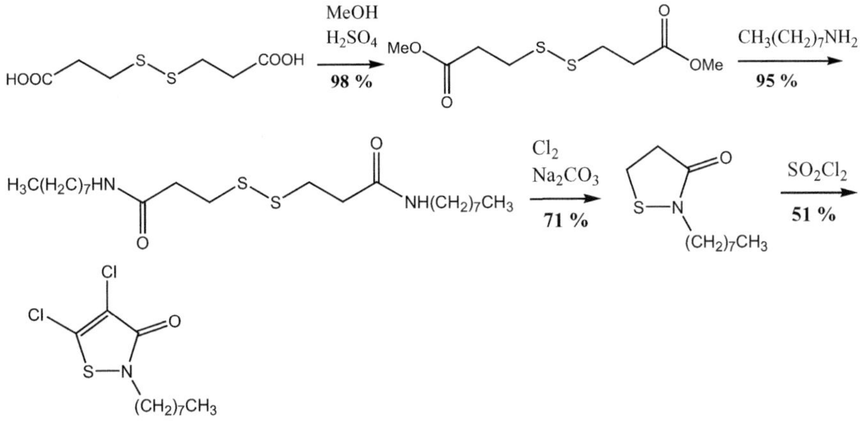

Part 1
Analyze them using synthesis tree diagrams and determine the respective kernel reaction mass efficiencies.

Part 2
The one-step synthesis of tributyltin oxide is shown below. Balance this equation and determine its atom economy. Suggest a mechanism for this reaction.

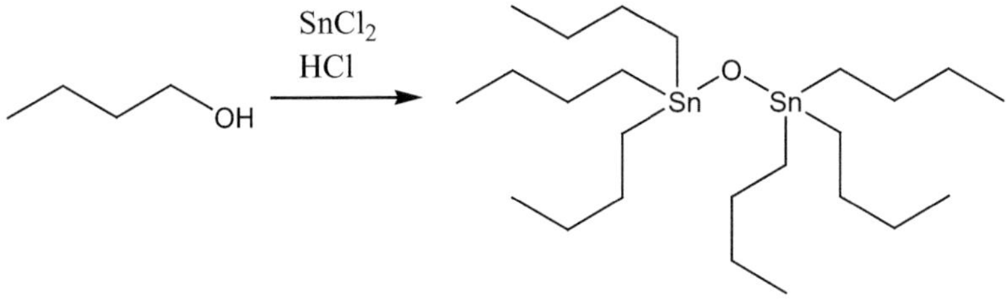

PROBLEM 2.53[136–142]
Salvarsan (arsphenamine) was synthesized by Paul Ehrlich as an anti-syphilis drug in 1910. A synthetic plan for this product is shown below.

Part 1

Show all reagents and reaction by-products and represent it as a synthesis tree.

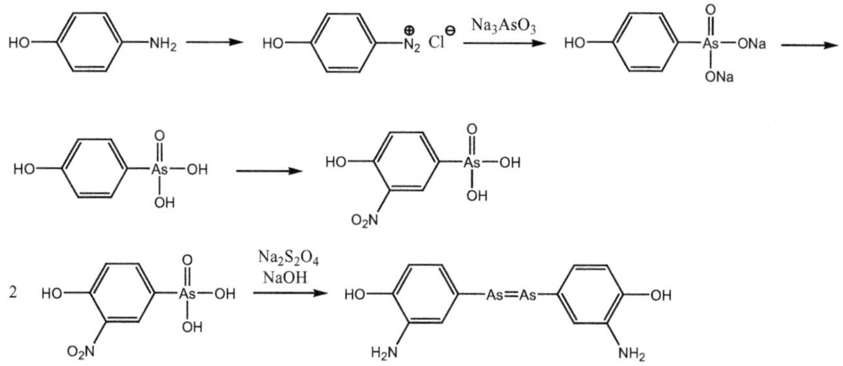

[Reaction scheme: HO-C6H4-NH2 → HO-C6H4-N2+ Cl− → (Na3AsO3) → HO-C6H4-As(=O)(ONa)2 →]

[HO-C6H4-As(=O)(OH)2 → HO-C6H3(O2N)-As(=O)(OH)2]

[2 HO-C6H3(O2N)-As(=O)(OH)2 → (Na2S2O4, NaOH) → HO-C6H3(H2N)-As=As-C6H3(NH2)-OH]

Part 2

The actual structure of salvarsan was determined in 2005 and found to be a mixture of two compounds shown below.[143]

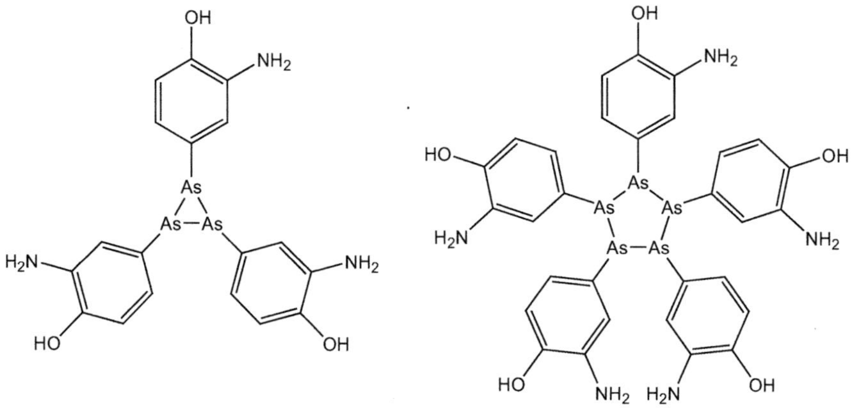

[Structures: a trimer (As)3 and a pentamer (As)5 cyclic arsenic compounds with 3-amino-4-hydroxyphenyl substituents]

Re-evaluate the last redox reaction to each of these product structures. How does this change the synthesis tree diagram in Part 1?

PROBLEM 2.54

The following synthesis tree represents a linear sequence to produce polymers of various length, M1 (1-mer), M2 (2-mer), M3 (3-mer), and so on, of molecular weight m, $2m$, $3m$, and so on. Each polymerization cycle involves two steps using the same set of ingredients A and B of molecular weights a and b, respectively. Assume that the set of reaction yields is the same for each cycle and that no excess reagents are used. Also, assume that the same amount of auxiliary materials (solvents, catalysts, work-up, and purification materials) is used in each cycle, and that all of this material is destined for waste.

Synthesis Plan Analysis

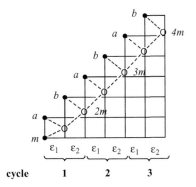

Part 1
Determine the PMI after cycle 1, cycle 2, and cycle 3.

Part 2
Determine the mass of waste after cycle 1, cycle 2, and cycle 3.

Part 3
Generalize the results in Parts 1 and 2 after n cycles.

PROBLEM 2.55[144]
A recent stereoselective synthesis of epinephrine is shown in the scheme below. Represent this as a synthesis tree diagram and determine overall AE and overall kernel RME. Note that intermediate structures in square brackets are not isolated. Suggest a mechanism for the step involving phenyliododiacetate.

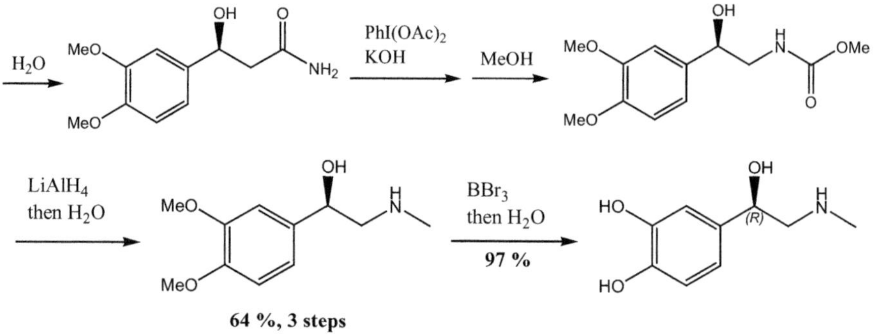

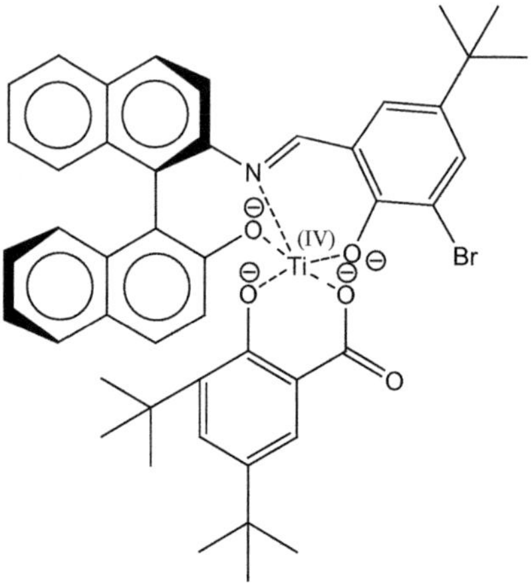

PROBLEM 2.56[145–147]

Carbofuran is a carbamate insecticide belonging to the same class as carbaryl that was manufactured by Union Carbide at Bhopal, India. Three synthetic plans are shown below to manufacture this product industrially. Fill in the missing reagents, balance all chemical equations, and represent the synthetic plans as synthesis trees. From the tree diagrams, determine the overall AE and Curzons RME for each process.

Plan A

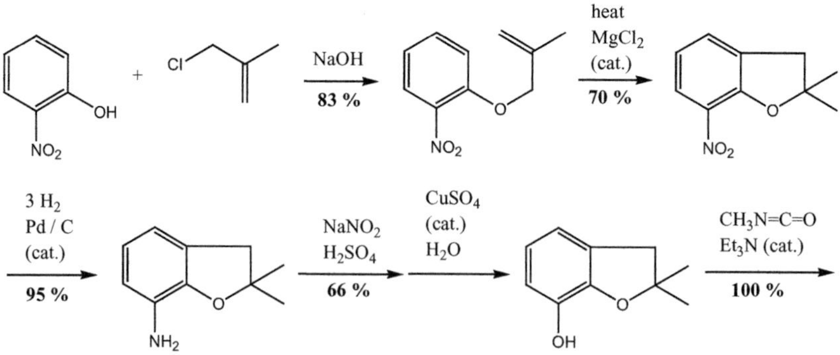

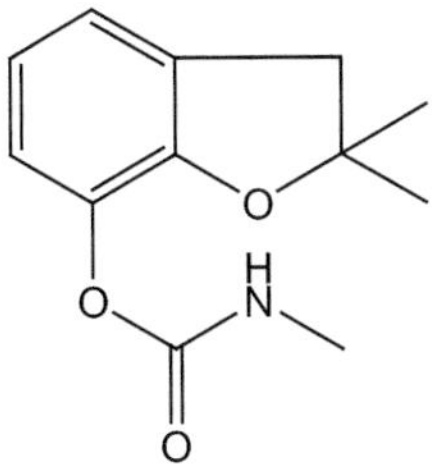

Synthesis Plan Analysis

Plan B

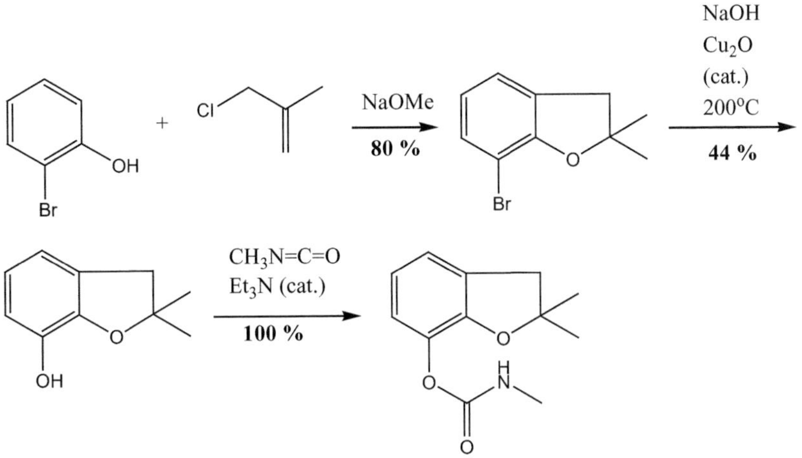

Plan C

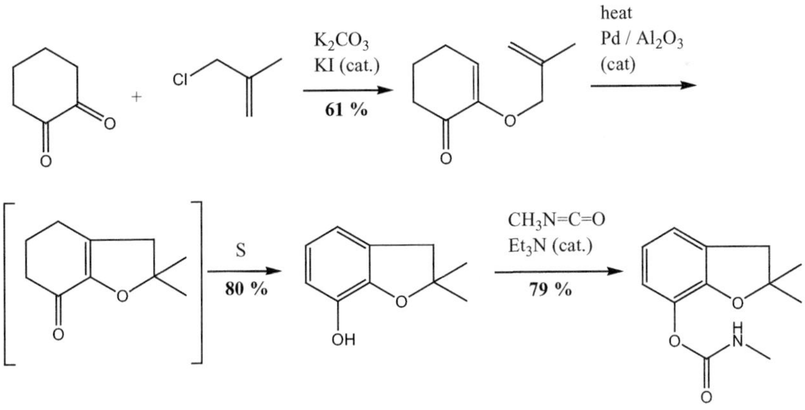

PROBLEM 2.57[148,149]

β-Carotene, or provitamin A, is produced from β-ionone according to the synthesis plan shown. Balance all equations, draw a synthesis tree diagram, and determine the overall atom economy for the plan.

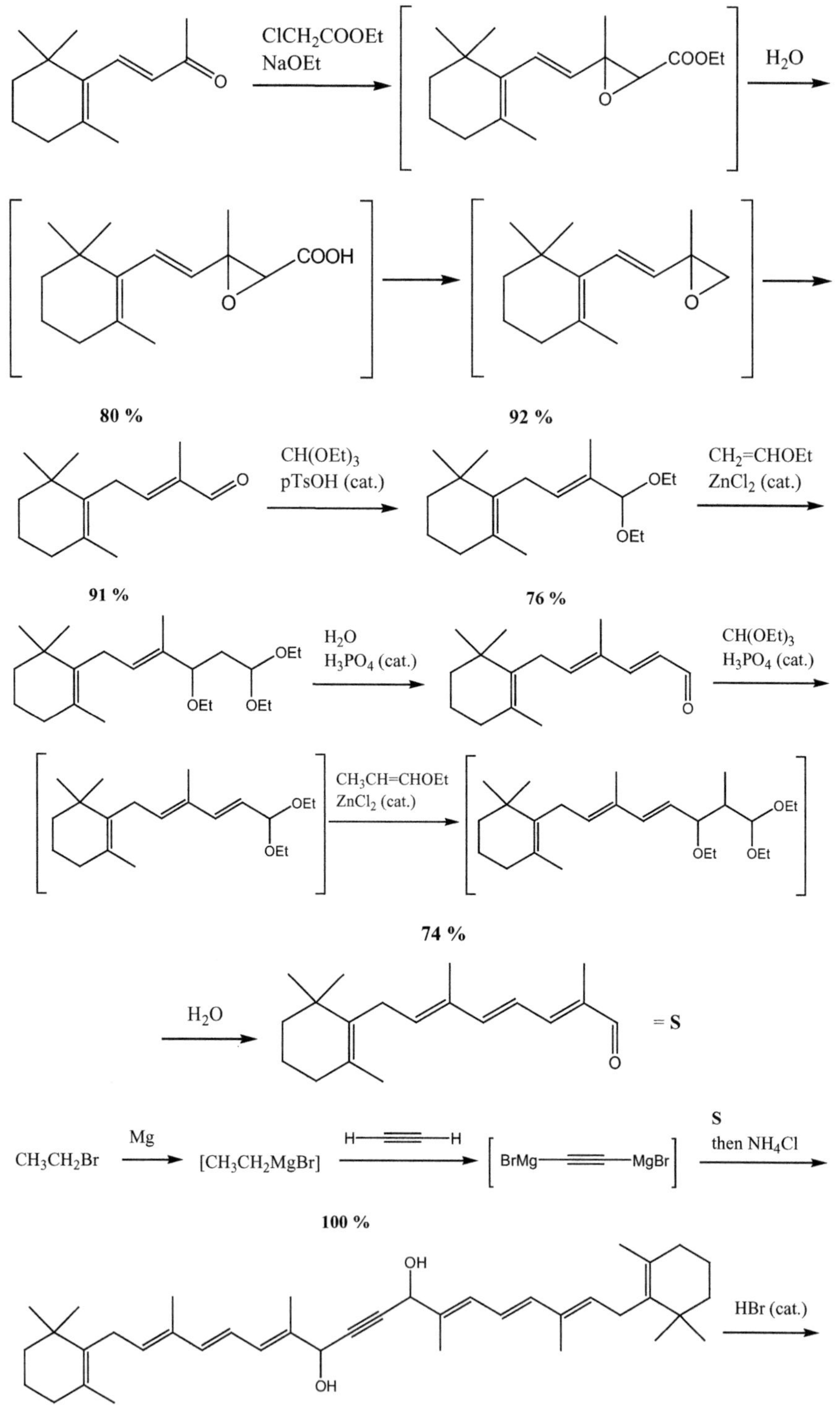
ClCH2COOEt
NaOEt
COOEt
H2O
COOH
80 %
92 %
CH(OEt)3
pTsOH (cat.)
OEt
OEt
CH2=CHOEt
ZnCl2 (cat.)
91 %
76 %
OEt
OEt
OEt
H2O
H3PO4 (cat.)
CH(OEt)3
H3PO4 (cat.)
OEt
OEt
CH3CH=CHOEt
ZnCl2 (cat.)
OEt
OEt
OEt
74 %
H2O
= S
CH3CH2Br
Mg
[CH3CH2MgBr]
H
H
BrMg
MgBr
S
then NH4Cl
100 %
OH
OH
HBr (cat.)

Synthesis Plan Analysis

[Structures showing a synthesis sequence:
- Top structure with 64% yield, treated with H₂ / Lindlar catalyst
- Middle structure with 95% yield, treated with heat
- Bottom structure with 97% yield]

PROBLEM 2.58

(R)(+)-1-Phenylethylamine is a versatile material used in asymmetric synthesis and in classical resolutions of racemic acids. Various methods have been reported for its preparation. Plans A and B both involve classical resolutions using diastereomeric salts, however they differ in that the desired salt precipitates out when resolved with L-(-)-malic acid (Plan A), whereas the desired salt remains in the mother liquor when resolved with L-(+)-tartaric acid (Plan B).

In Plan A, the racemate is prepared from acetophenone and ammonium formate, and then this is resolved using L-malic acid. The crystalline (+)-amine-(-)-malate diastereomeric salt that precipitates is treated with aqueous base to liberate the (+)-enantiomer. Reaction yields shown are those reported by the author.

In Plan B, the racemate is prepared again from acetophenone via a combined imination-hydrogen sequence, and then this is resolved using (+)-tartaric acid. The unwanted (-)-amine-(+)-tartrate salt precipitates while the mother liquor contains the desired (+)-amine-(+)-tartrate salt. After evaporation of the solvent, this salt is treated with sodium hydroxide solution to liberate the crude (+)-amine. The desired pure (+)-amine is obtained by preparing the (+)-amine sulfate salt, recrystallizing it from hot water and acetone, and then treating it with aqueous sodium hydroxide solution.

Plans C, D, and E are direct syntheses of the (+)-amine that involve stereoselective hydrogenation of acetophenone or its imine or hydrazone derivative. In Plan C, the chiral alcohol obtained from hydrogenation is subjected to tandem substitutions via the chloride and azide, and a final reduction. In Plans D and E, an imine or hydrazone derivative of acetophenone is prepared either having a chiral directing group attached (Plan E) or achiral group (Plan D). In Plan E, the subsequent reduction is a rhuthenium catalyzed transfer hydrogenation, which is followed by methanolysis of the chiral directing group. In Plan D, the hydrogenation step requires a chiral ligand attached to the rhodium catalyst which is then followed by reduction of the hydrazone with samarium diiodide.

Plan A (classical resolution with L-(-)-malic acid)[150,151]

Plan B (classical resolution with L-(+)-tartaric acid)[152,153]

Plan C (stereoselective)[154,155]

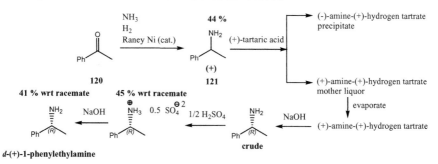

Synthesis Plan Analysis

Plan D (stereoselective)[156]

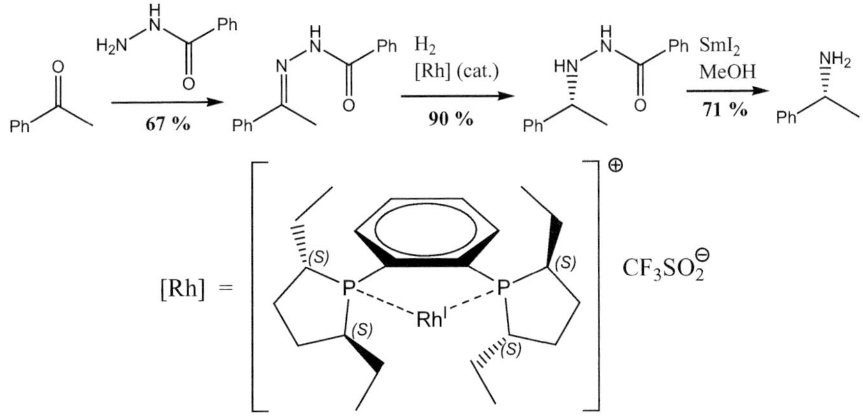

Plan E (stereoselective)[157]

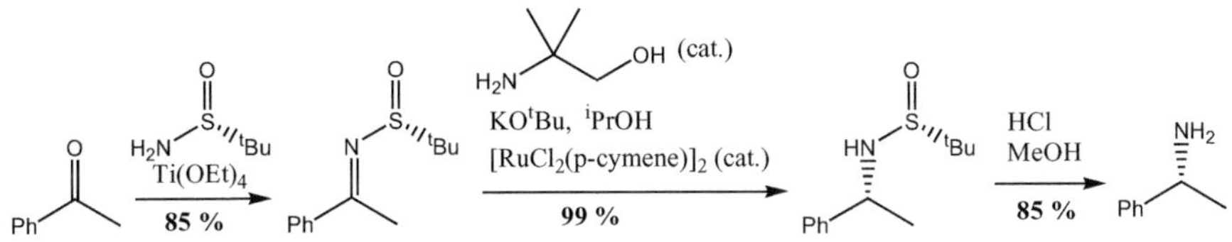

Part 1

For each plan, rewrite the corresponding schemes showing balanced chemical equations.

Part 2

Give a mechanism for step 1 of Plan A for the reaction of ammonium formate and acetophenone.

Part 3

Comment on the reported yield of 63% by the authors of Plan A for the preparation of the (+)-amine-(−)-malate salt from the racemic amine.

Part 4

Explain why the second step of Plan C results in retention of configuration at the asymmetric centre. Support your answer with a mechanism.

Part 5

Using the *template-REACTION.xls* and *template-SYNTHESIS.xls* spreadsheets determine the material efficiency metrics for producing (+)-1-phenylethylamine via each plan. State any assumptions made in the calculations. Which plan is the most material efficient to make this chiral amine?

PROBLEM 2.59[158]

A sixth-generation dendrimer was synthesized from a tri-substituted triazine core and two monomer units according to the following modular template. Each generation is created via tandem click thiol–ene and azide–alkyne reactions.

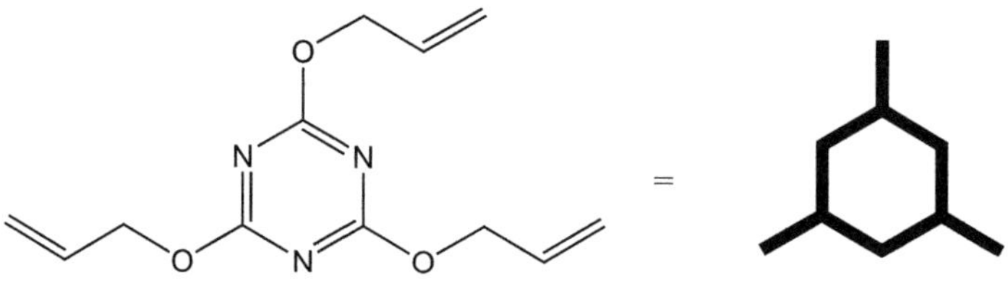

triazine core

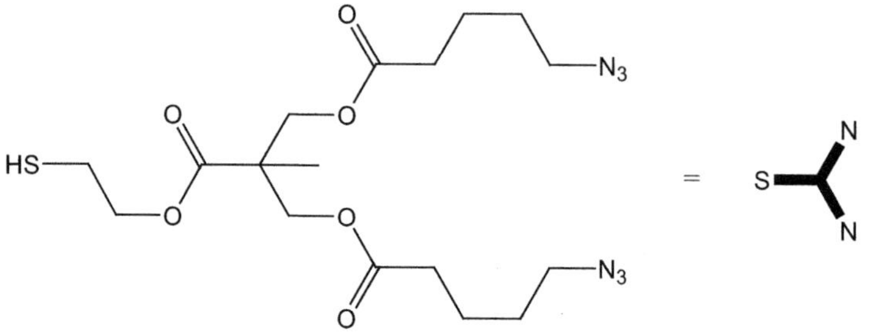

azide-thiol monomer

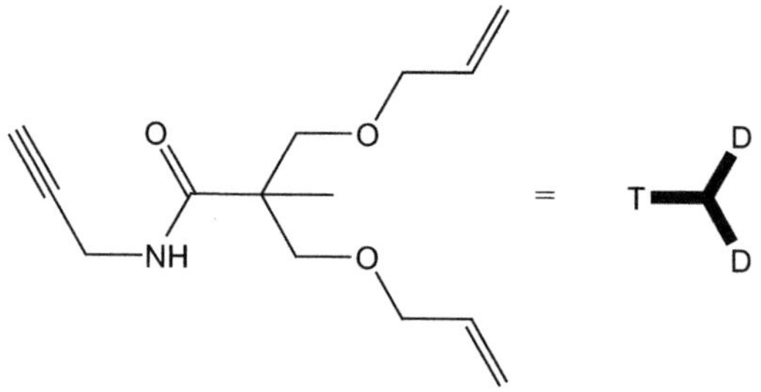

ene-yne monomer

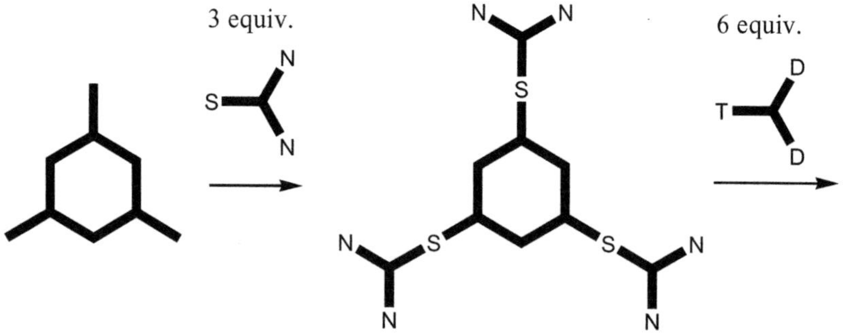

G1 [N]$_6$

Synthesis Plan Analysis

The first generation has 6 azide termini, the second generation has 12 olefin termini, the third generation has 24 azide termini, and the fourth generation has 48 olefin termini.

Part 1
Complete the data in the table below.

Generation no.	No. equiv. AT monomer	No. equiv. TD monomer	No. azide termini	No. olefin termini
1	3	0	6	0
2	0	6	0	12
3	12	0	24	0
4	0	24	0	48
5				
6				

Generalize the results after n generations.

Part 2
The synthesis of the azide–thiol monomer is as follows.

The synthesis of the ene–yne monomer is as follows.

Show balanced equations for each monomer synthesis plan. Draw synthesis tree diagrams and determine the respective overall AE, overall yield, and overall kernel RME.

Part 3

The yield performance for the synthesis of the G6 dendrimer is as follows.

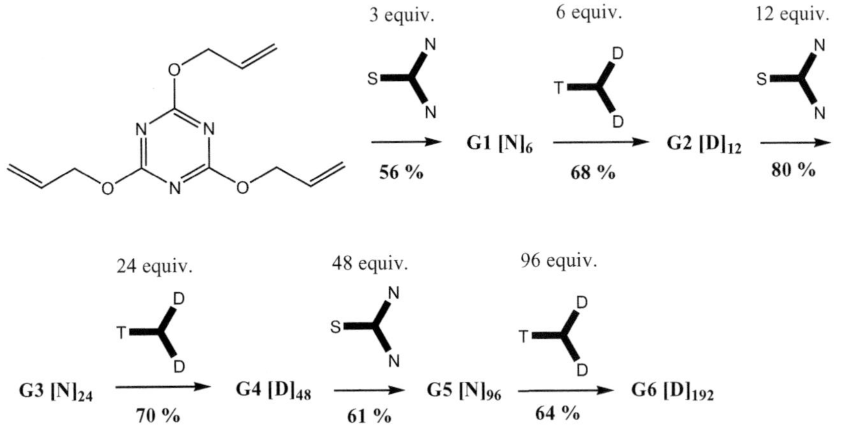

Synthesis Plan Analysis

Show balanced equations for the synthesis plan. Draw a synthesis tree diagram and determine the respective overall AE, overall yield, and overall kernel RME.

Part 4
Repeat the analysis of Part 3 for the synthesis of the G6 dendrimer incorporating the syntheses of the two monomers.

PROBLEM 2.60[159]
A one-pot iterative assembly line strategy was used to synthesize hydrocarbon chains with successively substituted methyl groups via stereochemically controlled homologation of boronic esters. An example is shown below where nine methyl groups were introduced sequentially.

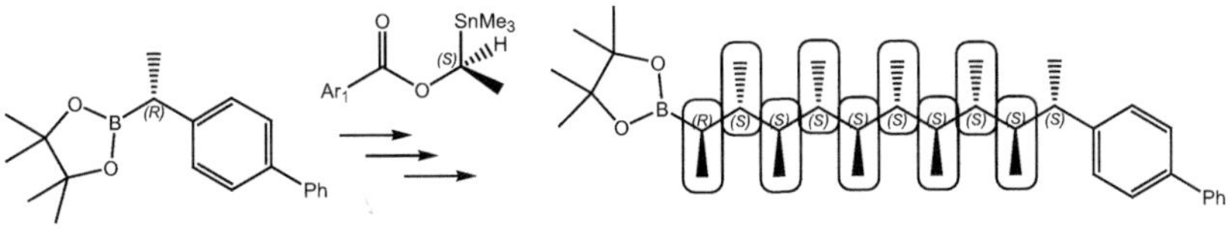

Ar_1 = triisopropylbenzene
The synthesis plan for a single homologation is shown below.

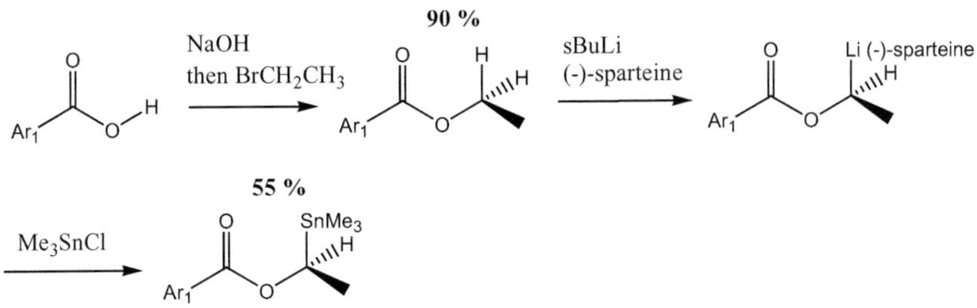

Part 1
Show the complete synthesis plan with balanced chemical equations. Show the target bonds made in the final product structure.

Part 2
Draw a synthesis tree for the plan and determine its kernel material efficiency metrics.

Part 3

What would be the stereochemical sequence of tin reagents required to make the following analogous target molecules? Reaction yields for the assembly line reaction are shown.

helical conformation 44 %

linear conformation 45 %

Part 4

Draw the synthesis trees for the target products in Part 3 and determine the corresponding kernel overall material efficiency metrics.

REFERENCES

1. Andraos, J. *Org. Process Res. Dev.* 2006, *10*, 212.
2. Jaffe, G.M.; Rehl, W.R. US 3714242 (Hoffman-La Roche, 1973).
3. Takabe, K.; Katagiri, T.; Tanaka, J.; Fujita, T.; Watanabe, S.; Suga, K. *Org. Synth.* 1989, *67*, 44.
4. Tani, K.; Yamagata, T.; Otsuka, S.; Kumobayashi, H.; Akutagawa, S. *Org. Synth.* 1989, *67*, 33.
5. Nakatani, Y.; Kawashima, K. *Synlett.* 1978, 147.
6. Ravasio, N.; Poli, N.; Psaro, R.; Saba, M.; Zaccheria, F. *Top. Catal.* 2000, *13*, 195.
7. Pummerer, R.; Prell, E.; Rieche, A. *Chem. Ber.* 1926, *59*, 2159.
8. Takaya, H.; Akutagawa, S.; Noyori, R. *Org. Synth.* 1989, *67*, 20.
9. (R)-fluoxetine: Bracher, F.; Litz, T. *Bioorg. Med. Chem.* 1996, *4*, 877.
10. (S)-dapoxetine: Torre, O.; Gotor-Fernandez, V.; Gotor, V. *Tetrahedron Asymm.* 2006, *17*, 860.
11. GB 3131 (1863).
12. GB 1525 (1872).
13. GB 2143 (1876).
14. GB 999 (1904).
15. Chenier, P. J. *Survey of Industrial Chemistry*, 3rd ed., Kluwer Academic/Plenum Publishers: New York, 2002, p. 69.
16. Jimenez-Gonzalez, C.; Constable, D.J.C. *Green Chemistry and Engineering: A Practical Design Approach*, Wiley: Hoboken, 2011, pp. 122–123.
17. Thalhammer, F.; Gastner, T. US2009163739 (2009).

18. Blicke, F.F.; Norris, P.E. *J. Am. Chem. Soc.* 1954, *76*, 3213.
19. Winterstein, E.; Küng, A. *Z. Physiol. Chem.* 1909, *59*, 141.
20. Vogel, A. *Textbook of Practical Organic Chemistry*, Longman: London, 1948, p. 599.
21. Faith, W.L.; Keyes, D.B.; Clark, R.L. *Industrial Chemicals,* 3rd ed., Wiley: Hoboken, 1966, pp. 101, 541.
22. Shreve, R.N. *Chemical Process Industries,* 3rd ed., McGraw-Hill Book Co.: New York, 1967, p. 812.
23. Burns, N.Z.; Baran, P.S.; Hoffmann, R.W. *Angew Chem. Int. Ed.* 2009, *48*, 2854.
24. Newhouse, T.; Baran, P.S.; Hoffmann, R.W. *Chem Soc Rev.* 2009, *38*, 3010.
25. Sheldon, R.A. *ACS Sust. Chem. Eng.* 2018, *6*, 32.
26. Sauvageau, G.; Gareau, Y. US2015011543 (U. Montreal, 2015).
27. Shapiro, R.; Rossano, L.T.; Mudryk, B.M.; Cuniere, N.; Obelholzer, M.; Zhang, H.; Chen, B.C. Process for Preparing 4,5-Dihydro-pyrazolo-[3,4c]-pyrid-2-ones. US2006069258 (Bristol-Myers Squibb Company, 2006).
28. Gantt, H.L. *Trans. Am. Soc. Mechanical Eng.* 1903, *24*, 1322.
29. Gantt, H.L. *Organizing for Work*, Harcourt, Brace, and Howe: New York, 1919.
30. Ku, H.; Rajagopalan, D.; Karimi, I. *Chem. Eng. Progress* 1987, *83(8)*, 35.
31. Musier, R.F.H.; Evans, L.B. *Chem. Eng. Progress* 1990, *86(6)*, 66.
32. Andraos, J.; Hent, A. *J. Chem. Educ.* 2015, *92*, 1831.
33. Andraos, J. *Org. Process Res. Dev.* 2005, *9*, 149.
34. Chapman, O.L.; Engel, M.R.; Springer, J.P.; Clardy, J.C. *J. Am. Chem. Soc.* 1971, *93*, 6696.
35. Beroza, M. *J. Agr. Food Chem.* 1956, *4*, 49.
36. Alexander, B.H.; Gertler, S.I.; Brown, R.T.; Oda, T.A.; Beroza, M. *J. Org. Chem.* 1959, *24*, 1504.
37. Gaich, T.; Baran, P.S. *J. Org. Chem.* 2010, *75*, 4657.
38. Andraos, J. *Org. Process Res. Dev.* 2009, *13*, 161.
39. Simmons, B.; Walji, A.B.; MacMillan, D.W.C. *Angew. Chem. Int. Ed.* 2009, *48*, 4349.
40. Wender, P.A.; Croatt, M.P.; Witulski, B. *Tetrahedron* 2006, *62*, 7505.
41. Wender, P.A.; Verma, V.A.; Paxton, T.J.; Pillow, T.H. *Acc. Chem. Res.* 2008, *41*, 40.
42. Wender, P.A.; Miller, B.L. *Nature* 2009, *460*, 197.
43. Wender, P.A. *Nat. Prod. Rep.* 2014, *31*, 433.
44. Jimenez-Gonzalez, C.; Constable, D.J.C. *Green Chemistry and Engineering: A Practical Design Approach*, Wiley: Hoboken, 2011, p. 120.
45. Baxendale, I.R.; Ley, S.V.; Nessi, M.; Piutti, C. *Tetrahedron* 2002, *58*, 6285.
46. Baxendale, I.R.; Ley, S.V.; Piutti, C. *Angew. Chem. Int. Ed.* 2002, *41*, 2194.
47. Step 1: Simeonov, S.P.; Coelho, J.A.S.; Afonso, C.A.M. *Org. Synth.* 2016, *93*, 29.
48. Step 2: Sutton, A.D.; Waldie, F.D.; Wu, R.; Schlaf, M.; Silks, L.A.; Gordon, J.C. *Nat. Chem.* 2013, 5, 428.
49. Steps 3 and 4: Silks, L.A.; Gordon, J.C.; Wu, R.; Hanson, S.K. US 2011040110 (Los Alamos National Laboratory, 2011).
50. Sessler, J.L.; Mozaffari, A.; Johnson, M.R. *Org. Synth. Coll. Vol.* 1998, *9*, 242.
51. Hartman, G.D.; Weinstock, L.M. *Org. Synth. Coll. Vol.* 1998, *6*, 620.
52. Jimenez-Gonzalez, C.; Constable, D.J.C. *Green Chemistry and Engineering: A Practical Design Approach*, Wiley: Hoboken, 2011, pp. 113–114.
53. Galat, A. *J. Am. Chem. Soc.* 1952, *74*, 3890.
54. Pokorny, R. *J. Am. Chem. Soc.* 1941, *63*, 1768.
55. Pansare, S.V.; Lingampally, R. *Org. Biol. Chem.* 2009, *7*, 319.
56. Bachand, C.; Belema, M.; Deon, D.H.; Good, A.C.; Goodrich, J.; James, C.A.; Lavoie, R. et al. US2008044380 (Bristol Myers Squibb, 2008).
57. Craig, L.C. *J. Am. Chem. Soc.* 1933, *55*, 2854.

58. Jimenez-Gonzalez, C.; Constable, D.J.C. *Green Chemistry and Engineering: A Practical Design Approach*, Wiley: Hoboken, 2011, p. 129.
59. Jimenez-Gonzalez, C.; Constable, D.J.C. *Green Chemistry and Engineering: A Practical Design Approach*, Wiley: Hoboken, 2011, p. 167.
60. Wong, J.W.; Watson, H.A.; Bouressa, J.F.; Burns, M.P.; Cawley, J.J.; Doro, A.E.; Guzek, D.B. et al. *Org. Process Res. Dev.* 2002, *6*, 477.
61. Jimenez-Gonzalez, C.; Constable, D.J.C. *Green Chemistry and Engineering: A Practical Design Approach*, Wiley: Hoboken, 2011, pp. 226–227.
62. Ellis, B.A. *Org. Synth. Coll.* 1941, *1*, 18
63. Kazuhiko, S.; Masao, A.; Ryoji, N.A. *Science* 1998, *281*, 1646.
64. Draths, K.M.; Frost, J.W. *J. Am. Chem. Soc.* 1994, *116*, 399.
65. Zen, S.; Koyama, M.; Koto, S. *Org. Synth. Coll. Vol.* 1988, *6*, 797.
66. Johnson, E.C.; Guzman, P.E.; Wingard, L.A.; Sabatini, J.J.; Pesce-Rodriguez, R.A. *Org. Process Res. Dev.* 2017, *21*, 1088.
67. Plan A: Chiellini, G.; Aprilett, J.W.; Yoshihara, H.A.I.; Baxter, J.D.; Ribeiro, R.C.J.; Scanlan, T.S. *Chemistry & Biology* 1998, *5*, 299.
68. Plan B: Chiellini, G.; Nguyen, N.H.; Yoshihara, H.A.I.; Scanlan, T.S. *Bioorg. Med. Chem. Lett.* 2000, *10*, 2607.
69. Plan A: Vogel, A.I. *A Textbook of Practical Organic Chemistry*, 3rd ed., Longman: London, 1956, pp. 383, 693.
70. Plan A: Marvel, C.S.; King, W.B. *Org. Synth. Coll.* 1941, *1*, 252.
71. Plan B: Zheng, H.; Lejkowski, M.; Hall, D.G. *Chem. Sci.* 2011, *2*, 1305.
72. Jimenez-Gonzalez, C.; Constable, D.J.C. *Green Chemistry and Engineering: A Practical Design Approach*, Wiley: Hoboken, 2011, p. 125.
73. Baur, A. *Chem. Ber.* 1891, *24*, 2832.
74. Mallmann, W. FR 243951 (1893).
75. Nash, E.G.; Nienhouse, E.J.; Silhavy, T.A.; Humbert, D.E.; Mish, M.J. *J. Chem. Educ.* 1970, *47*, 705.
76. David, O.R.P. *Eur. J. Org. Chem.* 2017, 4.
77. Jimenez-Gonzalez, C.; Constable, D.J.C. *Green Chemistry and Engineering: A Practical Design Approach*, Wiley: Hoboken, 2011, pp. 196–198.
78. Vo, L.; Ciula, J.; Gooding, O. W. *Org. Process Res. Dev.* 2003, *7*, 514.
79. Miao, W.; Chan, T.H. *Org. Lett.* 2003, *5*, 5003.
80. Rose-Munch, F.; Chavignon, R.; Tranchier, J.P.; Gagliardini, V.; Rose, E. *Inorg. Chim. Acta* 2000, *300–302*, 693.
81. Lluch, A.M.; Sanchez-Baeza, F.; Camps, F.; Messeguer, A. *Tetrahedron Lett.* 1991, *32*, 5629.
82. Gilbert, M.; Ferrer, M.; Lluch, A.M.; Sanchez-Baeza, F.; Messeguer, A. *J. Org. Chem.* 1999, *64*, 1591.
83. Tebbe, F.N.; Parshell, G.W.; Reddy, G.S. *J. Am. Chem. Soc.* 1978, *100*, 3611.
84. Wittig, G.; Schöllkopf, U. *Chem. Ber.* 1954, *87*, 1318.
85. Lombardo, L. *Tetrahedron Lett.* 1981, *21*, 4293.
86. Petasis, N.A.; Bzowej, E.I. *J. Am. Chem. Soc.* 1990, *112*, 6392.
87. Marker, R.E.; Rohrmann, E. *J. Am. Chem. Soc.* 1939, *61*, 846.
88. Marker, R.E.; Turner, D.L.; Ulshafer, P.R. *J. Am. Chem. Soc.* 1940, *62*, 2542.
89. Marker, R.E.; Wagner, R.B.; Ulshafer, P.R.; Wittbecker, E.L.; Goldsmith, D.P.J.; Ruof, C.H. *J. Am. Chem. Soc.* 1947, *69*, 2167.
90. Crooks, H.M. Jr.; Jones, E.M. US 2383472.
91. Marker, R.E.; Lopez, J. *J. Am. Chem. Soc.* 1947, *69*, 2380.
92. Baruah, D.; Das, R.N.; Konwar, D. *Synth. Commun.* 2016, *46*, 79.
93. Marker, R.E.; Turner, D.L.; Wagner, R.B.; Ulshafer, P.R.; Crooks, H.M. Jr.; Wittle, E.L. *J. Am. Chem. Soc.* 1941, *63*, 774.

94. Marker, R.E.; Krueger, J. *J. Am. Chem. Soc.* 1940, *62*, 3349.
95. Serini, A.; Koester, H.; Strassberger, L. US 2379832 (Schering Corp., 1945).
96. Macdonald, I.R.; Reid, G.A.; Pottie, I.R.; Martin, E.; Darvesh, S. *J. Nucl. Med.* 2016, *57*, 297.
97. Chincholkar, P.M.; Kale, A.S.; Gumaste, V.K.; Deshmukh, A.R.A.S. *Tetrahedron* 2009, *65*, 2605.
98. Torre, O.; Gotor-Fernandez, V.; Gotor, V. *Tetrahedron Asymm.* 2006, *17*, 860.
99. Siddiqui, S.A.; Srinivasan, K.V. *Tetrahedron Asymm.* 2007, *18*, 2099.
100. Clarke, H.T.; Hartmann, W.W. *Org. Synth. Coll. Vol. 1*, 233 (1941).
101. Robinson, R. *J. Chem. Soc.* 1917, 762.
102. Willstätter, R. *Chem. Ber.* 1901, *34*, 129.
103. Willstätter, R. *Chem. Ber.* 1901, *34*, 3163.
104. Willstätter, R. *Chem. Ber.* 1896, *29*, 936.
105. Reichstein, T.; Grüssner, A. *Helv. Chim. Acta* 1934, *17*, 311.
106. Hoshino, T.; Ojima, S.; Sugisawa, T. EP 518136 (Hoffmann-La Roche, 1992).
107. Haaf, W. US 3152180 (Studlengesellschaft Kohle, 1964).
108. Haaf, W. *Chem. Ber.* 1964, *97*, 3234.
109. Paulshock, M.; Watts, J.C. US 3310469 (Du Pont, 1967).
110. Gerzon, K.; Krumkalns, E.V.; Brindle, R.L.; Marshall, F.J.; Root, M.A. *J. Med. Chem.* 1963, *6*, 760.
111. Stetter, H.; Mayer, J.; Schwarz, M.; Wulff, K. *Chem. Ber.* 1960, *93*, 226.
112. Schleyer, P.R. *J. Am. Chem. Soc.* 1957, *79*, 3292.
113. Schleyer, P.R.; Donaldson, M.M. *J. Am. Chem. Soc.* 1960, *82*, 4645.
114. Engler, E.M.; Farcasiu, M.; Sevin, A.; Cense, J.M.; Schleyer, P.R. *J. Am. Chem. Soc.* 1973, *95*, 5769.
115. Schleyer, P.R.; Donaldson, M.M.; Nicholas, R.D.; Cupas, C. *Org. Synth. Coll.* 1973, *5*, 16.
116. Kovacic, P.; Roskos, P.D. *J. Am. Chem. Soc.* 1969, *91*, 6457.
117. Plan A: Reitsema, R.H. *J. Org. Chem.* 1958, *23*, 2038.
118. Plan B: Royals, E.E.; Horne, S.E. Jr. *J. Am. Chem. Soc.* 1951, *73*, 5856.
119. Heumann, K. *Chem. Ber.* 1890, *23*, 3437.
120. DE 56273 (BASF, 1891).
121. DE 127178 (BASF, 1901).
122. Fierz-David, H.E.; Blangey, L. *The Fundamental Processes of Dye Chemistry*, Interscience Publishing: New York, 1949, pp. 174, 323.
123. For step 1, see: Faith, W.L.; Keyes, D.B.; Clark, R.L. *Industrial Chemicals*, 3rd ed., Wiley: Hoboken, 1966, p. 617.
124. For step 2, see: Noyes, W.A.; Porter, P.K. *Org. Synth. Coll. Vol. 1*, 457 (1941).
125. Yamamoto, Y.; Inoue, Y.; Takaki, U.; Suzuki, H. *Bull. Chem. Soc. Jpn.* 2011, *84*, 82.
126. Berry, A.; Dodge, T.C.; Pepsin, M.; Weyler, W. *J. Ind. Microbiol. Biotech.* 2002, *28*, 127.
127. Plan A: Guiso, M.; Marra, C.; Farina, A. *Tetrahedron Lett.* 2002, *43*, 597.
128. Plan B: Shen, Z.L.; Zhou, G.L.; Jiang, X.Z. *Indian J. Chem. B* 2002, *41B*, 2395.
129. Ohno, M.; Kobayashi, S.; Iimori, T.; Wang, Y.F.; Izawa, T. *J. Am. Chem. Soc.* 1981, *103*, 2405.
130. Willstätter, R.; Waser, E. *Chem. Ber.* 1911, *44*, 3423.
131. Willstätter, R.; Heidelberger, M. *Chem. Ber.* 1913, *46*, 517.
132. Linhardt, R.; Kainz, Q.M.; Grass, R.N.; Stark, W.J.; Reiser, O. *RSC Adv.* 2014, *4*, 8541.
133. Weiler, E.D.; Petigara, R.B.; Wolfersberger, M.H.; Miller, G.A. *J. Heterocyclic Chem.* 1977, *14*, 627.
134. Lewis, S.N.; Miller, G.A.; Hausman, M.; Szamborski, E.C. *J. Heterocyclic Chem.* 1971, *8*, 571.
135. Hahn, S. J.; Kim, J. M., Park, Y. WO 9220664. Sunkyong Industries Co., Ltd., 1992.
136. Ehrlich, P.; Bertheim, A. *Chem. Ber.* 1907, *40*, 3292.

137. Ehrlich, P.; Bertheim, A. *Chem. Ber.* 1910, *43*, 917.
138. Ehrlich, P. *Verhl. 27. Kongr. Innere Med., Wiesbaden* 1910, *27*, 226.
139. Ehrlich, P. *Deutsch. Med. Wschr.* 1910, *36*, 1893.
140. Ehrlich, P. *Munch. Med. Wschr.* 1911, *58*, 1.
141. Ehrlich, P. *Munch. Med. Wschr.* 1911, *58*, 2481.
142. Ehrlich, P.; Bertheim, A. *Chem. Ber.* 1912, *45*, 756.
143. Lloyd, N.C.; Morgan, H.W.; Nicholson, B.K.; Ronimus, R.S. *Angew. Chem. Int. Ed.* 2005, *44*, 941.
144. Singer, R.A.; Carreira, E.M. *Tetrahedron Lett.* 1997, *38*, 927.
145. Plan A: Koerts, A.; de Wit, G.F.; Leopold, R. NL 6602601 (FMC Corp., 1966).
146. Plan B: Orwoll, E.F. US 3356690 (FMC Corp., 1967).
147. Plan C: Hunt, D.A. US 4463184 (Union Carbide, 1984); Scharpf, W.G. US 3474171 (FMC Corp, 1969).
148. Isler, O.; Huber, W.; Ronco, A.; Kofler, M. *Helv. Chim. Acta* 1947, *30*, 1911.
149. Isler, O.; Lindlar, H.; Montavon, M.; Ruegg, R.; Zeller, P. *Helv. Chim. Acta* 1956, *39*, 249.
150. Ingersoll, A.W. *Org. Synth. Coll.* 1943, *2*, 503.
151. Ingersoll, A.W. *Org. Synth. Coll.* 1943, *2*, 506.
152. Robinson, J.C. Jr.; Snyder, H.R. *Org. Synth. Coll.* 1955, *3*, 717.
153. Ault, A. *Org. Synth. Coll.* 1973, *5*, 932.
154. Xie, J.H.; Liu, X.Y.; Xie, J.B.; Wang, L.X.; Zhou, Q.L. *Angew. Chem. Int. Ed.* 2011, *50*, 7329.
155. Levene, P.A.; Rothen, A.; Kuna, M. *J. Biol. Chem.* 1937, *120*, 777.
156. Burk, M.J.; Martinez, J.P.; Feaster, J.E.; Cosford, N. *Tetrahedron* 1994, *50*, 4399.
157. Guijarro, D.; Pablo, O.; Yus, M. *Org. Synth.* 2013, *90*, 338.
158. Antoni, P.; Robb, M.J.; Campos, L.; Montanez, M.; Hult, A.; Malmstrom, E.; Malkoch, M.; Hawker, C.J. *Macromolecules* 2010, *43*, 6625.
159. Burns, M.; Essafi, S.; Bame, J.R.; Bull, S.P.; Webster, M.P.; Balieu, S.; Dale, J.W.; Butts, C.P.; Harvey, J.N.; Aggarwal, V.K. *Nature* 2014, *513*, 183.
160. Seidel, A. (ed.) *Kirk-Othmer Chemical Technology of Cosmetics*, Wiley: Hoboken, 2013, pp. 298–299.

3 Environmental Impact Metrics

3.1 LIFE CYCLE ASSESSMENT

In this chapter, we introduce methods to evaluate the environmental impact of conducting a chemical reaction or synthesis plan. The usual terminology to describe this kind of evaluation is life cycle assessment (LCA), which can take on various degrees of complexity depending on how far back one traces the origin of starting materials or how far forward one traces the desired product. Figure 3.1 shows the most common types of LCA analyses that are typically carried out along with a simple example industrial process to make p-cymene from benzene. The most important take home message from Figure 3.1 is the identification of starting and end points of analysis pertaining to each kind of LCA. The "cradle" starting point refers to virgin starting materials obtained from crude oil, biofeedstocks, ores, natural products, and the atmosphere—essentially what planet Earth provides as a resource. The "grave" end point refers to the final fate of the desired product after disposal and degradation. The "gate" terminology refers to any point in between "cradle" and "grave."

The following list of problem issues arise in conducting any kind of LCA:

1. They involve complex reaction networks tracing the synthetic origin of a given target compound. Currently, there is no consensus or standard on how far back that tracing should go. "Cradle" LCAs typically should trace all the way back to virgin resources found on Earth.
2. LCAs are necessarily broad and multi-layered.
3. LCAs are characterized by variables having no available data or unreliable data.
4. In order to bridge the gap of missing data, computational algorithms are used to estimate key production parameters directly from molecular structure features, for example, QSAR (quantitative structure-activity reactivity) correlations or Hansch's procedure.
5. Functional group contributions obtained from correlations of training data sets are used to estimate missing variables.
6. Most source data needed for environmental impact assessments are restricted to hydrocarbons (aliphatic and aromatic), halogenated hydrocarbons (CFCs, HFCs, and PCBs), pesticides, simple alcohols and amines, simple inorganics (acids and bases), and gases.
7. Fine chemicals used by the pharmaceutical industry would be considered too exotic to apply a thorough analysis based on real experimental data.

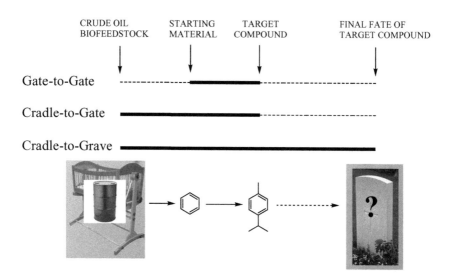

FIGURE 3.1 Common types of LCA analyses.

8. 58,000 industrial chemicals are routinely used with only about 5,000 having some kind of experimental data sets for environmental fate constants.
9. It is generally acknowledged that there is an urgent need to speed up the process of assigning values in order for any kind of environmental assessment to be implemented at low cost. Current research in environmental science is heavily favored toward QSAR modeling to try to meet this challenge.

Computational toxicology has the following ongoing and unresolved issues:

1. Typically, small-sized training sets based on limited experimental data are used to generate correlation equations between parameters having the form of Collander log-log plots. These plots are then used to obtain environmental parameters by extrapolation for new or unknown compounds having no experimental data.
2. Extrapolations are bound to fail the more exotic the structures of compounds are from the training set. This is reminiscent of the problem encountered with linear free energy relationships, for example, Hammett plots.
3. The real drawback in obtaining experimental data is that it is tedious, mundane, and repetitive, and is not compatible with funding agencies that seek to support innovative cutting-edge research.
4. The key variables plagued with these problems are: pK_a, $\log K_{ow}$, Henry law constants, water solubility, vapor pressure, densities, any kind of toxicity data (particularly endocrine disruption), and persistence data (half-lives of chemicals in environment).

Environmental Impact Metrics

The immediate effects on LCA analyses include:

1. Missing or unreliable data pose challenges in the interpretation of LCAs.
2. Missing or unreliable data severely restrict the reliability of conclusions drawn, that is, making decisions as to which reaction or plan is "greener" than another.
3. Limited databases exist containing all of the required impact potentials: hydrocarbons (aliphatic and aromatic), halogenated hydrocarbons (CFCs, HFCs, and PCBs), pesticides, simple alcohols and amines, simple inorganics (acids and bases), and gases.

As a preview to the topics covered in this chapter, Table 3.1 summarizes the list of common environmental impact potentials introduced along with their equations and reference states. Full discussions of these are given in Section 3.3.

3.1.1 Terms, Definitions, and Examples

Cradle-to-Cradle LCA

This conceptual term refers to a type of LCA that tracks the environmental waste and energy impacts of a process from first-generation feedstock starting materials (resource extraction) to final desired products including their use phase. Instead of analyzing their disposal phase, a recycling process is also examined that converts them back to the same starting input materials, if possible, at the end of life of the desired products. This kind of LCA typically has at least one closed loop in its graphical network where an endpoint corresponding to a desired product is connected via a reverse recycling path back to the same starting point corresponding to the source materials from which that desired product was made. It also includes waste streams and may include other network loops in its graphical network. Among all possible LCA analyses this type is the most difficult to carry out, is the most time consuming, and is the least feasible. See *life cycle assessment*.

Cradle-to-Gate LCA

This conceptual term refers to a type of LCA that tracks the environmental waste and energy impacts of a process from first-generation feedstock starting materials (resource extraction) to final desired products that are shipped out from a factory gate. The use phase and disposal phase are not counted. This kind of LCA is by its name a linear analysis, but also includes waste streams and may include other network loops in its graphical network. See *life cycle assessment*.

Cradle-to-Grave LCA

This conceptual term refers to a type of LCA that tracks the environmental waste and energy impacts of a process from first-generation feedstock starting materials (resource extraction) to final desired products including use and disposal of the products after their use. This kind of LCA is by its name a linear analysis, but also includes waste streams and may include other network loops in its graphical network. See *life cycle assessment*.

TABLE 3.1
Summary of Common Potentials Associated With Environmental Impact

Potential	Equation		Reference
Abiotic resource depletion	$(ADP)_j = \sum\limits_{element,i} n_i (ADP)_i$	(3.13)	Antimony (Sb)
Acidification	$(AP)_j = \dfrac{\alpha_j/(MW)_j}{\alpha_{SO_2}/(MW)_{SO_2}}$ $= \dfrac{32\alpha_j}{(MW)_j}$	(3.14)	Sulfur dioxide (SO_2)
Acidification–basification	$ABP = 7 + \log\left\|10^{BP-7} - 10^{AP-7}\right\|$	(3.25)	Water
	$AP = 7 + \log\left[\sum\limits_j X_j \left(10^{(AP)_j - 7}\right)\right]$	(3.26)	
	$BP = 7 + \log\left[\sum\limits_j X_j \left(10^{(BP)_j - 7}\right)\right]$	(3.27)	
	$ABP = 7 + \log\left\|\sum\limits_j X_j \left(10^{(BP)_j - 7}\right)\right.$ $\left. - \sum\limits_j Y_j \left(10^{(AP)_j - 7}\right)\right\|$	(3.28)	
	where X_j is mole ratio of jth acid with respect to all acids present, and Y_j is mole ratio of jth base with respect to all bases present.		
	$ABP_{salt} = 7 + \log\left\|\left(\dfrac{m}{n+m}\right)10^{BP-7}\right.$ $\left. -\left(\dfrac{n}{n+m}\right)10^{AP-7}\right\|$		
	where n = number of cations and m = number of anions in salt		
Bioaccumulation	$BAP = 10^{\log K_{ow,j} - 2.73}$	(3.29)	Toluene
Bioconcentration	$(BCP)_j = 10^{0.76(\log K_{ow,j} - 2.73)}$	(3.30)	Toluene
Global warming	$(GWP)_j = \dfrac{(NC)_j/(MW)_j}{(NC)_{CO_2}/(MW)_{CO_2}}$ $= \dfrac{44(NC)_j}{(MW)_j}$	(3.34)	Carbon dioxide (CO_2)

(Continued)

TABLE 3.1 (CONTINUED)
Summary of Common Potentials Associated With Environmental Impact

Potential	Equation		Reference
Ingestion toxicity	$(INGTP)_j = \dfrac{C_{w,j}/LD_{50,j}}{C_{w,tol}/LD_{50,tol}}$	(3.37)	Toluene
Inhalation toxicity	$(INHTP)_j = \dfrac{C_{a,j}/LC_{50,j}}{C_{a,tol}/LC_{50,tol}}$	(3.40)	Toluene
Ozone depletion	$(ODP)_j = \dfrac{\delta[O_3]_j}{\delta[O_3]_{CFC-11}}$	(3.41)	CFC-11 (CCl_3F)
Smog formation	$(SFP)_j = \dfrac{(MIR)_j}{(MIR)_{ROG}} = \dfrac{(MIR)_j}{3.1}$	(3.45)	Standard mixture of reactive organic gases (ROG)

Curtate LCA[1]

A curtate LCA is any life cycle assessment that ignores upstream and downstream processes. An example of a curtate LCA is a gate-to-gate LCA. The word "curtate" comes from the Latin *curtatus* meaning "shortened" or "truncated."

See *gate-to-gate LCA*.

Environmental Impact

This is a general term used in life cycle assessment literature that estimates the damage done by input materials emitted and waste materials produced in a chemical reaction on the four compartments of the environment (air, water, soil, and sediment) according to several criteria called potentials. Environmental impact is quantified according to the expression given below in Equation (3.1).

$$EI = \frac{\sum_j f_j \Omega_j}{\sum_j \Omega_j} \qquad (3.1)$$

where:
  j refers to the jth compound
  Ω_j is the sum of all environmental potentials
  $f_j =$ (mass of compound j)/(total mass of compounds of type j involved in reaction)

A value of 0 corresponds to complete greenness with respect to the selected environmental impact potentials and a value of 1 corresponds to complete anti-greenness. See *benign index, potential*.

Environmental Index[2,3]

A metric applied to output materials that is the multiplicative product of the E-factor and the environmental quotient factor, Q. This metric is identical to the environmental quotient (EQ). It corrects E-factors for the "unfriendliness" characteristics of contributing waste materials beyond accounting for their mass. The Q-factors estimate qualitative aspects such as raw material availability, environmental pollution, and chemical toxicity. See *environmental quotient*.

Environmental Quotient (EQ)[4]

For a given chemical reaction, the mathematical expression of the environmental quotient is given by Equation (3.2)

$$EQ = \frac{w_1 Q_1 + w_2 Q_2 + \ldots}{m_{\text{product}}} \quad (3.2)$$

where:
- w_j is the mass of waste chemical j
- Q_j is the "unfriendliness" quotient associated with waste chemical j
- m_{product} is the mass of target product collected

The Q-factors are arbitrarily chosen factors reflecting the toxicity and hazard risks caused by a given amount of waste chemical. It is a basic way of distinguishing a kilogram of sodium chloride waste from a kilogram of mercury waste, for example. When the mass of a given waste chemical is multiplied by such a Q-factor, it effectively amplifies its contribution to the waste profile. Waste chemicals that pose no hazards or risks are assigned Q values of one.

Gate-to-Gate LCA

This conceptual term refers to a type of LCA that tracks the environmental waste and energy impacts of a process from a given set of advanced starting materials to final desired products. Gate-to-gate is the most minimal LCA that examines only one process from point A to point B that is a subset of a larger production chain where point A is downstream from virgin extractable raw materials and point B is upstream from further processing to more advanced materials. This kind of LCA is by its name a linear analysis, but also includes waste streams and may include other network loops in its graphical network. See *life cycle assessment*.

Gate-to-Grave LCA

This conceptual term refers to a type of LCA that tracks the environmental waste and energy impacts of a process from a given set of advanced starting materials to final desired products including disposal of the products after their use. This kind of LCA is by its name a linear analysis, but also includes waste streams and may include other network loops in its graphical network. See *life cycle assessment*.

TABLE 3.2
Summary of Equations for Henry's Law Constant

Equation	Variable Definitions	Units of HLC
$H_{px} = P_T \dfrac{Y_j}{X_j}$	P_T = total pressure (atm) X_j = mole fraction of compound j in aqueous solution at equilibrium Y_j = mole fraction of compound j in air phase at equilibrium	atm
$H_{cc} = \dfrac{C_{j,air}}{C_{j,water}}$	$C_{j,air}$ = concentration of compound j in air phase (g/m³ or mol/m³) $C_{j,water}$ = concentration of compound j in water phase (g/m³ or mol/m³)	dimensionless
$H_{yx} = \dfrac{Y_j}{X_j} = K_{aw}$	X_j = mole fraction of compound j in aqueous solution at equilibrium Y_j = mole fraction of compound j in air phase at equilibrium	dimensionless
$H_{pc} = P_T \dfrac{Y_j}{C_{j,water}}$	P_T = total pressure (atm) Y_j = mole fraction of compound j in air phase at equilibrium $C_{j,water}$ = concentration of compound j in water phase (g/m³ or mol/m³)	atm m³/mol

Henry's Law Constant

The Henry's law constant is the proportionality constant between the vapor pressure (partial pressure at equilibrium) of a substance and the concentration of that substance in water, assuming an equilibrium is established between the air-water interface and the solution is ideally dilute. The Henry's law constant is essentially a partition coefficient of the substance between air and water.

The mathematical expression is given by Equation (3.3a):

$$H = \frac{\text{vapor pressure}_{chemical,air}}{\text{concentration}_{chemical,water}} \qquad (3.3a)$$

where vapor pressure is in units of atm and concentration is in units of mol m⁻³. Hence, the standard units of Henry's law constant are atm m³ mol⁻¹. Tabulated data of Henry's law constants are recorded at 25°C.

The reader should be aware that the literature contains various units for the Henry's law constant depending on the author or compilation consulted. Table 3.2 summarizes various formalisms of this constant along with their units.

The connecting relationships are given by Equation (3.3b):

$$H_{cc} = H_{yx}\left(\frac{(MW)_{water}}{(MW)_{air}}\right)\left(\frac{\rho_{air}}{\rho_{water}}\right) = H_{px}\left(\frac{1}{RT}\right)\left(\frac{(MW)_{water}}{\rho_{water}}\right) = H_{pc}\left(\frac{1}{RT}\right) \qquad (3.3b)$$

where:
 (MW)water = 18 g/mol
 (MW)air = 29 g/mol
 T = temperature in degrees Kelvin
 ρ_{water} = 1000 kg/m³
 ρ_{air} = calculated using ideal gas law with R = 8.206 × 10⁻⁵ atm m³/mol K given by Equation (3.3c).

$$\rho_{air} = \frac{p_{vap}(MW)_{air}}{RT} = \frac{1*29}{(8.206\times 10^{-5})T} = \frac{353400}{T} \text{ g/m}^3 = 353.4/T \text{ kg/m}^3 \quad (3.3c)$$

Assuming total pressure is 1 atm, $H_{yx} = H_{px}$.
Therefore, Equation (3.3d) applies.

$$H_{cc} = H_{yx}\left(\frac{0.2194}{T}\right) = H_{px}\left(\frac{0.2194}{T}\right) = H_{pc}\left(\frac{12186}{T}\right) \quad (3.3d)$$

At T = 293 K (20°C), $H_{cc} = H_{yx}(7.48\times 10^{-4}) = H_{px}(7.48\times 10^{-4}) = H_{pc}(41.57)$.
At T = 298 K (25°C), $H_{cc} = H_{yx}(7.36\times 10^{-4}) = H_{px}(7.36\times 10^{-4}) = H_{pc}(40.89)$.
Henry's law constant at T = 298 K obtained from ChemDraw Ultra 7.0 is given by Equation (3.3e).

$$H_{pc} = \frac{H_{cc}}{RT} = \frac{10^{-\log(1/H_{cc})}}{RT} = \frac{10^{-CDV}}{RT} = 0.02445*10^{-CDV}, \quad (3.3e)$$

where:
CDV = ChemDraw value
units of H_{pc} are atm m^3 mol^{-1}

Sources to find reliable Henry's law constants can be found in references [5–19].

Example 3.1

Part 1
The ChemDraw value (CDV) for benzene is 0.66. Determine the value of the Henry's law constant in units of atm m³ mol⁻¹.

SOLUTION
ChemDraw value for log (1/H$_{cc}$) = 0.66 implying that H$_{cc}$ = 10^(−0.66) = 0.219 and H$_{pc}$ = 0.219/40.89 = 5.28e−3 atm m³/mol at 25°C.

Part 2
For benzene, ChemDraw gives the following values for vapor pressure and water solubility: 12940 Pa and 1.79 g/L. From these data, determine the Henry's law constant.

SOLUTION
vapor pressure: 12940 Pa = 12940/101325 = 0.128 atm
water solubility: 1.79 g/L = 1.79/78 = 0.0229 mol/L
Therefore, H$_{pc}$ = 0.128/0.0229 = 5.56 atm L/mol or 5.56e−3 atm m³ mol⁻¹, which agrees with the value obtained in Part 1.
The average value of HLC for benzene from handbook tables is 5.55E−3 atm m³ mol⁻¹.

Hodge–Sterner Toxicity Scale[20]

A scale of toxicities based on ranges of LD50 and LC50 values that are assigned qualitative descriptors where increased severity of toxicity is inversely related to the value of the LD50 or LC50. Hence, substances with low LD50 or LC50 values are assigned the highest scale of toxicity. A summary of the scale is given in Table 3.3.

Impact

The negative effect of chemicals on each of the environmental compartments: air, water, soil, and sediment, when they are released into the environment. This is called environmental impact.

The negative effect of exposure to chemicals occupational workers may experience when they carry out chemical reactions. This is called safety-hazard impact (see Chapter 4).

LC50 (inhalation)

LC50(inhal) is the lethal concentration to kill 50% of a population upon inhalation exposure to a chemical compound for 4 hours expressed in units of g m^{-3}. The reference animal is the rat *Rattus norvegicus* (Wistar). This is a measure of acute toxicity on the basis of a single dose exposure.

Vapors, mists, and dust aerosols are assumed to behave as ideal gases at standard temperature and pressure (25°C, 1 atm).

If LC50(inhal) data are expressed in units of ppm, then they may be converted to units of g m^{-3} according to Equation (3.4).

$$g/m^3 = \frac{(ppm)(MW)}{24450}. \tag{3.4}$$

If LC50(inhal) data are determined in a time exposure of x hours ($x \neq 4$), then they may be converted to the standard unit according to Equation (3.5).

$$LC50_{inhal}^{4h} = LC50_{inhal}^{xh}\left(\frac{x}{4}\right). \tag{3.5}$$

TABLE 3.3
Summary of Hodge–Sterner Toxicity Scales

Range of LD50 (oral, rat) Values (mg/kg or ppm)	Range of LC50 (inhalation, rat) (ppm)	LD50 (dermal, rabbit)(mg/kg or ppm)	Descriptor
LD50 < 1	LC50 < 10	LD50 < 5	Extremely toxic
1 < LD50 < 50	10 < LC50 < 100	5 < LD50 < 43	Highly toxic
50 < LD50 < 500	100 < LC50 < 1000	44 < LD50 < 340	Moderately toxic
500 < LD50 < 5000	1000 < LC50 < 10000	350 < LD50 < 2810	Slightly toxic
5000 < LD50 < 15000	10000 < LC50 < 100000	282 < LD50 < 22590	Practically non-toxic
LD50 > 15000	LC50 > 100000	LD50 > 22600	Relatively harmless

Reference compound: toluene: 49 g m^{-3}, 4 h
 Sources to find reliable LC50(inhal) values can be found in references [6, 21–26]. See *potential: inhalation toxicity*.

Example 3.2

The reported LC50 value for benzene is 10000 ppm for 7 h. Convert this value to g m^{-3} for 4 h.

SOLUTION

For 4 h the LC50 value is 10000*(7/4) = 17500 ppm. Since the molecular weight of benzene is 78 g/mol, the LC50 value becomes 17500*78/24450 = 55.83 g m^{-3}, 4 h.

Example 3.3

The reported LC50 value for toluene is 49 g m^{-3}, 4 h. Convert this value to ppm.

SOLUTION

Since the molecular weight of toluene is 92 g/mol, the LC50 value is (49*24450)/92 = 13022 ppm for 4 h.

Example 3.4

Explain how the number 24450 arises in the formula to convert ppm to g m^{-3} units.

SOLUTION

The underlying assumption is that vapors, mists, and dust aerosols behave as ideal gases at standard temperature and pressure (25 C, 1 atm).
Recall: Ideal gas law $PV = nRT$, $R = 0.08207$ L atm/(deg K mol)
The volume occupied by 1 mole of ideal gas at T = 25°C (298 K) and 1 atm is therefore $V = 1*0.08207*298/1 = 24.45$ L.
Since 1 L = 0.001 m^{-3}, the volume is 0.02445 m^{-3}. When dealing with units of ppm (parts per million), there is a multiplier factor of 1000000, hence 0.02445*1000000 = 24450.

LD50(oral)

LD50(oral) is the lethal dose to kill 50% of population upon oral exposure to compound (ingestion) expressed in units of mg kg^{-1} or ppm. The reference animal is the rat *Rattus norvegicus* (Wistar). This is a measure of acute toxicity on the basis of a single dose exposure.
 Reference compound: toluene; LD50(oral) = 636 mg/kg.
 The main problems with LD50 estimates include:

 1. Variability with sex, age, diet, season, genetic factors, social factors, temperature, and litter size within single species used as a reference population (e.g., mice).

Environmental Impact Metrics

2. Variability between different species used as reference populations (e.g., mice vs. humans). This may improve with developments in genomic mapping of organisms.
3. Since LD50 estimates focus on acute effects only, the effect of a chemical after contact to repeated, small, non-lethal doses causing cumulative damage over long periods of time is ignored.

The overall conclusion is that there is no way of logically standardizing the estimate without introducing some kind of statistical biases. This situation creates a large legal loophole for regulations to be implemented and enforced.

Sources to find reliable LD50(oral) can be found in references [6, 21, 22, 23, 24, 26]. See *potential: ingestion toxicity*.

Life Cycle
The term *life cycle* refers to the notion that a fair, holistic assessment requires the assessment of raw-material production, manufacture, distribution, use, and disposal including all intervening transportation steps necessary or caused by the product's existence.

Life Cycle Assessment or Analysis
Life cycle assessment or analysis is a technique to assess mass flows, energy flows, and environmental impacts associated with all the stages of a product's life from raw material extraction through materials processing, manufacture, distribution, use, repair and maintenance, and disposal or recycling.

Life Cycle Costing
In the context of comparing several chemical reactions or synthesis plans to the same target molecule, life-cycle costing or cost analysis (LCCA) is used to determine the most cost-effective option among the different competing processes to manufacture the chemical, to dispose of it, or to recycle it back to a starting material.

Life Cycle Inventory
Life cycle inventory involves creating flowcharts that map the mass and energy flows from input materials to all output materials in a process. The flows must be mass and energy balanced, that is, the input flows must match the output flows. For example, the total mass of input materials must equal the mass of target product plus the mass of all waste materials and residuals. Similarly, the total energy consumption (input) must equal the sum of all energies released plus all other energy losses (output). The creation of mass and energy inventories is a first step in counting from a given start point to a given end point. The start point can be either a cradle or gate and the end point can be either a cradle, gate, or grave. Hence, there are six possible trajectories of flows: cradle-to-cradle (circular); cradle-to-gate (linear); cradle-to-grave (linear); gate-to-cradle (linear-recycling); gate-to-gate (linear); and gate-to-grave (linear). See *cradle-to-cradle LCA; cradle-to-gate LCA; cradle-to-grave LCA; gate-to-gate LCA; gate-to-grave LCA*.

Materials Safety Data Sheets (MSDS)

These are documents that describe occupational safety and health information, and spill handling procedures for chemical products. MSDS sheets are divided into the following 16 sections:

1. Identification of the substance/mixture and of the chemical supplier
2. Hazards identification
3. Composition/information on ingredients
4. First aid measures
5. Firefighting measures
6. Accidental release measure
7. Handling and storage
8. Exposure controls/personal protection
9. Physical and chemical properties
10. Stability and reactivity
11. Toxicological information
12. Ecological information
13. Disposal considerations
14. Transport information
15. Regulatory information
16. Other information

The following information important for life cycle assessment is found in MSDS sheets:

LD50(oral), LD50(dermal), LC50(inhalation), flash point, lower explosion limit, occupational exposure limits, boiling point, log K_{ow}, water solubility, and risk or hazard phrases.

Octanol–Water Partition Coefficient

The equilibrium ratio of concentrations of a chemical in octanol versus water is given by Equation (3.6).

$$K_{ow} = \frac{[\text{chemical}]_{octanol}}{[\text{chemical}]_{water}} \tag{3.6}$$

This parameter measures the differential solubility of a chemical in a hydrophilic environment (water) versus a hydrophobic environment (octanol). Tabulated data of octanol–water partition coefficients are recorded at 25°C.

Hydrophilic chemicals with low K_{ow} (less than 1) values tend to have high solubilities in water; hence, they have large negative log K_{ow} values. Examples are oxalic acid (−2.22), alanine (−2.94), sodium butyrate (−3.2), and tin(IV)chloride (−3.26). Hydrophobic chemicals with high K_{ow} (greater than 1) values tend to have low solubilities in water; hence, they have large positive log K_{ow} values. Examples are hexachlorobenzene (5.73), DDT (6.53), dioctyl phthalate (7.60), and 3-pentadecylphenol (3.11).

Reference compound: toluene; log K_{ow} = 2.73

Environmental Impact Metrics

Sources for finding reliable log K_{ow} values can be found in references [6, 7, 10, 19, 27–44].

Example 3.5

Verify that log $K_{ow} = -1.38$ for water.

SOLUTION
density of octanol at 25°C = 0.827 g/mL = 827 g/L
solubility of water in octanol at 25°C = 50.4 g water/kg octanol
MW(water) = 18 g/mol
density of water at 25°C = 1 g/mL = 1000 g/L

Therefore, $K_{ow} = \dfrac{\dfrac{50.4/18}{1000/827}}{1000/18} = \dfrac{2.316}{55.56} = 0.0417$ and $\log K_{ow} = \log(0.0417) = -1.38$.

Persistence
The length of time a given chemical remains structurally intact in each of the environmental compartments: air, water, soil, and sediment. The usual timeframes are hours, weeks, months, and "very long."

Risk
In the context of life cycle assessment, risk is the potential or probability of an adverse event or outcome occurring. If a chemical has a hazardous property, then it has the potential to do harm to occupational workers and others exposed to it or to do harm to the environment if it is released in one of the four compartments: air, water, soil, and sediment. The likelihood of an adverse effect is reduced considerably if good laboratory practices, appropriate personnel training, good operating equipment, and appropriate safety procedures are followed and used in the handling of the hazardous chemical.

Risk Index
Risk index for a particular chemical is the product of its potential to do a particular harm and its mass. See *potential*.

TD50
TD50 is the chronic dose-rate in mg/kg body wt/day for a given chemical that would induce cancerous tumors in half the test animals at the end of a standard lifespan for the species. The reference animal is the rat *Rattus norvegicus* (Wistar).
   TD50 units: mg/kg/day
   Reference compound: toluene (3060 mg/kg/day)
   Sources for finding reliable TD50 data can be found in references [45–50].

Toxicity
Toxicity is the degree to which a chemical substance can damage an organism. There are two types of toxicity to consider: acute and chronic.

Acute toxicity looks at lethal effects following oral, dermal, or inhalation exposure to a chemical substance upon a single dose. The usual quantitative measures of acute toxicity are LD50(oral), LD50(dermal), and LC50(inhal).

Chronic toxicity looks at cumulative adverse effects following continuous exposure to a chemical substance over an extended period of time, often measured in months or years. Quantitative measures of chronic toxicity are: (a) no observed effects concentration (NOEC); (b) lowest observed effects concentration (LOEC); (c) maximum acceptable toxicant concentration (MATC); (d) predicted no effects concentration (PNEC); and (e) acute to chronic ratio (AR), which is determined as AR=LC50(inhal)/MATC. See *LC50(inhal); LD50(dermal)* (Chapter 11); *LD50(oral)*.

VOC Emission

Volatile organic compound emission is the ratio of mass of recycled solvent to total mass of solvent used in a process.

3.2 MULTI-COMPARTMENT MODEL

3.2.1 TERMS, DEFINITIONS, AND EXAMPLES

Multi-Compartment Model[6,51–55]

The multi-compartment model (Level I) was developed by Donald Mackay at the University of Toronto and Trent University. This model determines the fate concentrations of a given mass of a chemical released into four environmental compartments: air, water, soil, and sediment.

The multi-compartment model (MCM) calculation determines the mass partitioning of a given amount of chemical into the four environmental compartments (air, water, soil, and sediment) if it were to be released into the environment at 300 K.

The set of equations involved in the MCM calculation are as follows:

Basis mass of chemical released into the 4 compartments: air, water, soil, sediment = 1000 kg

Number of moles of compound j released corresponding to 1000 kg is given by Equation (3.7).

$$n_j = \frac{1000 * 1000}{(MW)_j} \quad (3.7)$$

where $(MW)_j$ is molecular weight in g/mol

Approximate relation between octanol–organic carbon partition coefficient, K_{oc}, and octanol–water partition coefficient, K_{ow} is given by Equation (3.8).

$$K_{oc} = 0.41 K_{ow} \quad (3.8)$$

Z parameters (mol/atm m³) are given by Equations (3.9a) to (3.9d).

Environmental Impact Metrics

$$Z_{water} = \frac{1}{H}$$

$$Z_{air} = \frac{1}{RT}$$

$$Z_{soil} = \frac{K_{oc} 0.02 * 2.4}{H}$$

$$Z_{sed} = \frac{K_{oc} 0.04 * 2.4}{H}$$

(3.9a–d)

where:
- H = Henry's law constant (atm m³ per mol)
- R = gas constant (8.21E–5 atm m³/mol K)
- T = temperature (degrees K)

Fugacity of compound j (atm) is given by Equation (3.10)

$$f_j = \frac{n_j}{Z_{air}V_{air} + Z_{water}V_{water} + Z_{soil}V_{soil} + Z_{sed}V_{sed}} \qquad (3.10)$$

where:
- V(air)   = 1 × 10¹⁴
- V(water) = 2 × 10¹¹
- V(soil) = 9 × 10⁹
- V(sed)   = 1 × 10⁸

Concentrations of compound j in air, water, soil, and sediment compartments (mol/m³) are given by Equations (3.11a) to (3.11d).

$$C_{air,j} = f_j Z_{air} = \frac{f_j}{RT} \qquad (3.11a)$$

$$C_{water,j} = f_j Z_{water} = \frac{f_j}{H} \qquad (3.11b)$$

$$C_{soil,j} = f_j Z_{soil} = \frac{f_j K_{oc} 0.02 * 2.4}{H} \qquad (3.11c)$$

$$C_{sed,j} = f_j Z_{sed} = \frac{f_j K_{oc} 0.04 * 2.4}{H} \qquad (3.11d)$$

Masses of compound j in air, water, soil, and sediment compartments (g) are given by Equations (3.12a) to (3.12d).

$$m_{air,j} = C_{air,j} V_{air} (MW)_j \qquad (3.12a)$$

$$m_{water,j} = C_{water,j} V_{water} (MW)_j \qquad (3.12b)$$

$$m_{soil,j} = C_{soil,j} V_{soil} (MW)_j \qquad (3.12c)$$

$$m_{sed,j} = C_{sed,j} V_{sed} (MW)_j \qquad (3.12d)$$

where $V_{air} = 1 \times 10^{14} \, m^3, V_{water} = 2 \times 10^{11} \, m^3, V_{soil} = 9 \times 10^9 \, m^3, V_{sed} = 1 \times 10^8 \, m^3$.

Example 3.6

Part 1

One tonne of toluene is released into the environment. Based on the MCM estimate the mass partitioning of toluene in air, water, soil, and sediment given the following data. Use the Excel *template-multi-compartment-model.xls* spreadsheet to facilitate the computation.
$H = 6.63E–3$ atm m³/mol
log $K_{ow} = 2.73$

Solution

Mass of toluene in air	9.9E+05 g
Mass of toluene in water	7.3E+03 g
Mass of toluene in soil	3.5E+03 g
Mass of toluene in sediment	7.8E+01 g
Total	1000000 g

98.9% of toluene ends up in the air compartment.

Part 2

A chemical railway car carrying 90 tons of chlorine is punctured after a train derailment. Assuming that all of the chlorine is released to the environment, determine how much ends up in each of the four environmental compartments. Use the Excel *template-multi-compartment-model.xls* spreadsheet to facilitate the computation.
 $H = 1.22E–2$ atm m³/mol
 log $K_{ow} = 0.85$ (estimated)

Solution

1 ton = 907.2 kg
90 tons = 90*907.200 = 81,648 kg

Environmental Impact Metrics

Mass of compound in air	8.13E+07 g
Mass of compound in water	3.28E+05 g
Mass of compound in soil	2.06E+03 g
Mass of compound in sediment	4.57E+01 g
Total	81648000 g

99.6% of chlorine ends up in the air compartment.

Part 3
One tonne of 2,3,7,8-tetrachlorodibenzo-p-dioxin is released into the environment. Based on the MCM estimate how much ends up in each of the four environmental compartments. Use the Excel *template-multi-compartment-model.xls* spreadsheet to facilitate the computation.
$H = 5.00\text{E}{-}5$ atm m³/mol
$\log K_{ow} = 6.80$

SOLUTION

mass of compound in air	1.78E+02 g
mass of compound in water	1.75E+02 g
mass of compound in soil	9.78E+05 g
mass of compound in sediment	2.17E+04 g
TOTAL	1000000 g

97.8% of 2,3,7,8-tetrachlorodibenzo-*p*-dioxin ends up in the soil compartment.

3.3 ENVIRONMENTAL IMPACT POTENTIALS

3.3.1 TERMS, DEFINITIONS, AND EXAMPLES

Potential
An environmental impact variable or safety-hazard impact variable that is capable of developing into an actual impact.
See the various types of potentials denoted by *potential: variable name*.

Potential: Abiotic Resource Depletion[56]
The abiotic resource depletion potential for compound *j* is given by Equation (3.13)

$$(\text{ADP})_j = \sum_{\text{element},i} n_i (\text{ADP})_i \qquad (3.13)$$

where:
$\quad n_i \quad$ is the number of elements of type *i* in compound *j* (ADP)$_i$
$\qquad$ is the abiotic depletion potential of element *i* relative to antimony (Sb)

Reference element: antimony (Sb)

Example 3.7

Determine the ARDP for potassium dichromate.

Solution

ARDP for potassium dichromate ($K_2Cr_2O_7$):

ADP (K) = 3.13E−8
ADP (Cr) = 8.58E−4
ADP (O) = 0
ARDP ($K_2Cr_2O_7$) = 2*(3.13E−8) + 2*(8.58E−4) + 7*0 = 1.72E−3

Potential: Acidification[57,58]

The acidification potential for compound j is given by Equation (3.14).

$$(AP)_j = \frac{\alpha_j / (MW)_j}{\alpha_{SO_2} / (MW)_{SO_2}} = \frac{\alpha_j / (MW)_j}{2/64} = \frac{32\alpha_j}{(MW)_j} \text{ for compound } j \quad (3.14)$$

where:
α_j is the number of dissociable protons in acid if compound j is an acid, or the number of dissociable protons in conjugate acid if compound j is a base
$(MW)_j$ is the molecular weight of compound j (g/mol)

Reference compound: sulfur dioxide (SO_2)

Example 3.8

Determine the acidification potential for the following compounds. (a) sulfuric acid; (b) ammonia; (c) acetic acid; (d) triethylamine; (e) tartaric acid; (f) phenol

Solution

Compound	MW (g/mol)	α	AP
H_2SO_4	98	2	32*2/98 = 0.653
NH_3	17	4	32*4/17 = 7.529
CH_3COOH	60	1	32*1/60 = 0.533
Et_3N	101	1	32*1/101 = 0.317
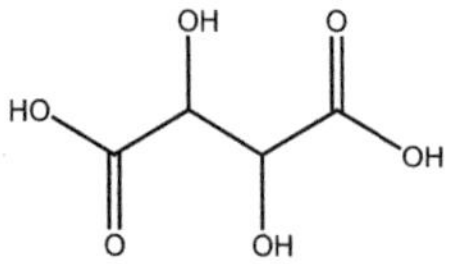 tartaric acid	150	2	32*2/150 = 0.427
Phenol	94	1	32*1/94 = 0.340

Potential: Acidification–Basification[59]

For monoprotic acids, the acidification potential with respect to neutrality is given by either Equations (3.15) or (3.16).

$$AP = \log\left(\frac{(q/p)K_a}{(1/2)K_w}\right) = \log\left(\frac{2(q/p)K_a}{K_w}\right) = \log\left(\frac{2q}{p}\right) + \log K_a - \log K_w \quad (3.15)$$

$$AP = \log\left(\frac{2q}{p}\right) - pK_a + pK_w \quad (3.16)$$

where:
- p = number of equivalent acidic sites on acid
- q = number of equivalent basic sites on conjugate acid
- K_a = acid dissociation constant at 25°C
- pK_w = 14 at 25°C.

For water, $p = 2$ and $q = 1$.

Example 3.9

Determine the acidification potential for the following monoprotic acids.
  (a) acetic acid ($pK_a = 4.76$); (b) phenol ($pK_a = 9.89$); (c) hydrochloric acid ($pK_a = -6.55$); (d) nitric acid ($pK_a = -1.48$).
  Show the ionization equilibrium for each case and determine the p and q values.

Solution

[Acetic acid: $p = 1$ ⇌ $q = 2$ + H⁺]

[Phenol: $p = 1$ ⇌ $q = 1$ + H⁺]

[H—Cl ⇌ Cl⁻ + H⁺; $p = 1$, $q = 1$]

[Nitric acid: $p = 1$ ⇌ $q = 3$ + H⁺]

Acid	pK$_a$	p	q	AP
Acetic acid	4.76	1	2	14−4.76+log(2*2/1)=9.84
Phenol	9.89	1	1	14−9.89+log(2*1/1)=4.41
Hydrochloric acid	−6.55	1	1	14+6.55+log(2*1/1)=20.85
Nitric acid	−1.48	1	3	14+1.48+log(2*3/1)=16.26

For diprotic acids, the acidification potential with respect to neutrality is given by either Equations (3.17) or (3.18).

$$AP = \log\left(\frac{K_{a,avg}}{(1/2)K_w}\right) = \log 2 + \log K_{a,avg} - \log K_w \qquad (3.17)$$

$$AP = \log 2 - pK_{a,avg} + pK_w \qquad (3.18)$$

where $K_{a,avg}$ is given by Equation (3.19).

$$K_{a,avg} = \frac{(q_1/p_1)^2 (K_{a1})^2 + (q_2/p_2)^2 (K_{a2})^2}{(q_1/p_1)(K_{a1}) + (q_2/p_2)(K_{a2})} \qquad (3.19)$$

Example 3.10

Determine the acidification potential for the following diprotic acids.

1. sulfuric acid (pK$_{a,1}$=0.4; pK$_{a,2}$=1.92);
2. salicylic acid (pK$_{a,1}$=2.97; pK$_{a,2}$=13.44);
3. carbonic acid (pK$_{a,1}$=6.37; pK$_{a,2}$=10.25).

Show the ionization equilibria for each case and determine the p and q values.

SOLUTION

Environmental Impact Metrics

[Reaction scheme: carbonic acid dissociation steps]

HO—C(=O)—OH ⇌ [HO—C(=O)—O]⁻ + H⁺   [HO—C(=O)—O]⁻ ⇌ [O=C=O]²⁻ + H⁺

$p = 2$ $q = 2$ $p = 1$ $q = 3$

Acids	pK_a	K_a	q	p	$(q/p)*K_a$	$[(q/p)*K_a]^2$	AP
H_2SO_4 (step 1)	0.4	3.98E−01	3	2	5.97E−01	3.57E−01	
H_2SO_4 (step 2)	1.92	1.20E−02	4	1	4.81E−02	2.31E−03	
				sum	6.45E−01	3.59E−01	14.05
salicylic acid (step 1)	2.97	1.07E−03	2	1	2.14E−03	4.59E−06	
salicylic acid (step 2)	13.44	3.63E−14	1	1	3.63E−14	1.32E−27	
				sum	2.14E−03	4.59E−06	11.63
H_2CO_3 (step 1)	6.37	4.27E−07	2	2	4.27E−07	1.82E−13	
H_2CO_3 (step 2)	10.25	5.62E−11	3	1	1.69E−10	2.85E−20	
				sum	4.27E−07	1.82E−13	7.93

Example 3.11

Determine the AP for phosphoric acid where $pK_{a,1} = 2.12$; $pK_{a,2} = 7.21$; and $pK_{a,3} = 12.32$.

Solution

[Reaction schemes for phosphoric acid dissociation]

HO—P(=O)(OH)—OH ⇌ [HO—P(=O)(OH)—O]⁻ + H⁺   [HO—P(=O)(OH)—O]⁻ ⇌ [HO—P(=O)(O)—O]²⁻ + H⁺

$p = 3$ $q = 2$ $p = 2$ $q = 3$

[HO—P(=O)(O)—O]²⁻ ⇌ [O=P(O)(O)=O]³⁻ + H⁺

$p = 1$ $q = 4$

Acids	pK_a	K_a	q	p	$(q/p)*K_a$	$[(q/p)*K_a]^2$	AP
H_3PO_4 (step 1)	2.12	7.59E−03	2	3	5.06E−03	2.56E−05	
H_3PO_4 (step 2)	7.21	6.17E−08	3	2	9.25E−08	8.55E−15	
H_3PO_4 (step 3)	12.32	4.79E−13	4	1	1.91E−12	3.67E−24	
				sum	5.06E−03	2.56E−05	12.00

For general acids, the acidification potential with respect to neutrality is given by Equation (3.20).

$$AP = \log 2 + pK_w + \log\left[\frac{\sum_j (q_j/p_j)^2 (K_{aj})^2}{\sum_j (q_j/p_j)(K_{aj})}\right]$$

$$= 14.301 + \log\left[\frac{\sum_j (q_j/p_j)^2 (K_{aj})^2}{\sum_j (q_j/p_j)(K_{aj})}\right] \quad (3.20)$$

For monobasic bases, the basification potential with respect to neutrality is given by either Equation (3.21a) or (3.21b).

$$BP = \log\left(\frac{(p/q)K_b}{(1/2)K_w}\right) = \log\left(\frac{2(p/q)K_b}{K_w}\right)$$

$$= \log\left(\frac{2p}{q}\right) + \log K_b - \log K_w \quad (3.21a)$$

$$BP = \log\left(\frac{2p}{q}\right) - pK_b + pK_w = \log\left(\frac{2p}{q}\right) + pK_a \quad (3.21b)$$

where:
- p = number of equivalent acidic sites on acid
- q = number of equivalent basic sites on conjugate acid
- pK_w = 14 at 25°C

For water, $p = 2$ and $q = 1$.

Example 3.12

Determine the basification potential (BP) for the following monobasic bases.
(a) aniline ($pK_a = 4.63$); (b) triethylamine ($pK_a = 10.75$); (c) ammonia ($pK_a = 9.25$)
Show the ionization equilibrium for each case and determine the p and q values.

PhNH$_2$ + H$_2$O ⇌ PhNH$_3^+$ + HO$^-$
$q = 1$ $p = 3$

Et$_3$N + H$_2$O ⇌ Et$_3$NH$^+$ + HO$^-$
$q = 1$ $p = 1$

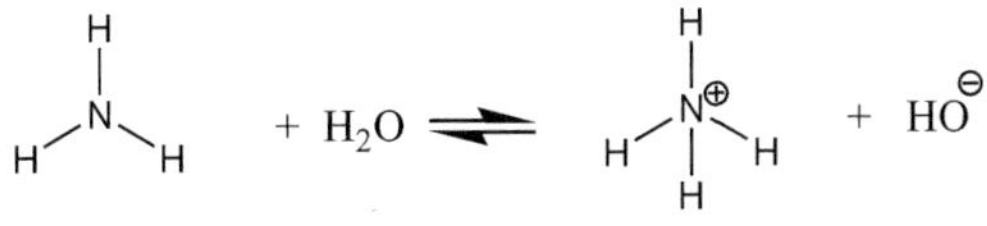

Environmental Impact Metrics

$$H_3N + H_2O \rightleftharpoons H_4N^{\oplus} + HO^{\ominus}$$

q = 1   p = 4

Base	pK$_a$	p	q	BP
Aniline	4.63	3	1	4.63 + log(2*3/1) = 5.41
Triethylamine	10.75	1	1	10.75 + log(2*1/1) = 11.05
Ammonia	9.25	4	1	9.25 + log(2*4/1) = 10.15

For dibasic bases, the basification potential with respect to neutrality is given by either Equations (3.22) or (3.23).

$$BP = \log\left(\frac{K_{b,\text{avg}}}{(1/2)K_w}\right) = \log 2 + \log K_{b,\text{avg}} - \log K_w \quad (3.22)$$

$$BP = \log 2 - pK_{b,\text{avg}} + pK_w \quad (3.23)$$

where $K_{b,\text{avg}}$ is given by Equation (3.24).

$$K_{b,\text{avg}} = \frac{(p_1/q_1)^2 (K_{b1})^2 + (p_2/q_2)^2 (K_{b2})^2}{(p_1/q_1)(K_{b1}) + (p_2/q_2)(K_{b2})} \quad (3.24)$$

Example 3.13

Determine the basification potential for the following dibasic bases.

1. benzidine (pK$_{a,1}$ = 3.75; pK$_{a,2}$ = 4.97);
2. piperazine (pK$_{a,1}$ = 5.57; pK$_{a,2}$ = 9.81);
3. nicotine (pK$_{a,1}$ = 3.15; pK$_{a,2}$ = 7.85).

Show the ionization equilibria for each case and determine the p and q values.

benzidine

$H_2N-C_6H_4-C_6H_4-NH_2 + H_2O \rightleftharpoons H_2N-C_6H_4-C_6H_4-NH_3^{\oplus} + HO^{\ominus}$

q = 2   p = 3

$H_2N-C_6H_4-C_6H_4-NH_3^{\oplus} + H_2O \rightleftharpoons H_3N^{\oplus}-C_6H_4-C_6H_4-NH_3^{\oplus} + HO^{\ominus}$

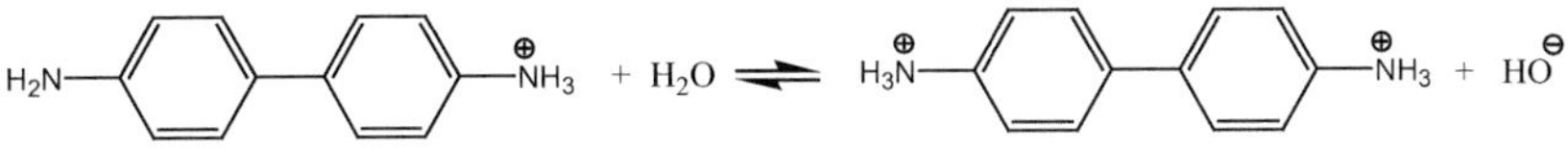

q = 1   p = 6

piperazine

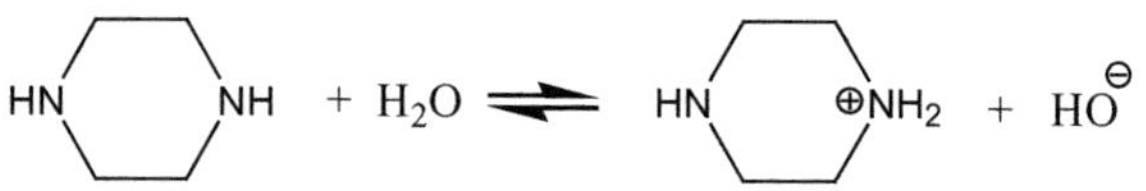

HN⌐NH + H₂O ⇌ HN⌐⊕NH₂ + HO⁻

$q = 2$ ... $p = 2$

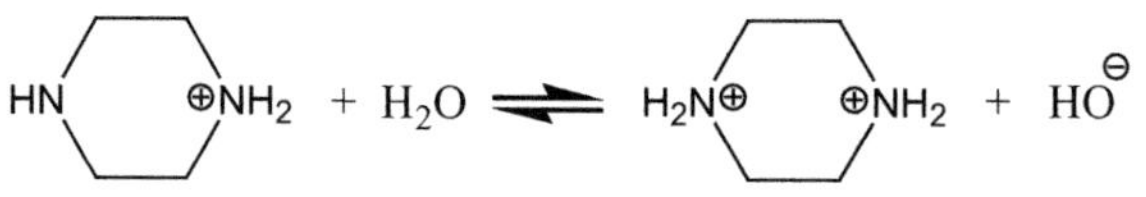

HN⌐⊕NH₂ + H₂O ⇌ H₂N⊕⌐⊕NH₂ + HO⁻

$q = 1$ ... $p = 4$

nicotine

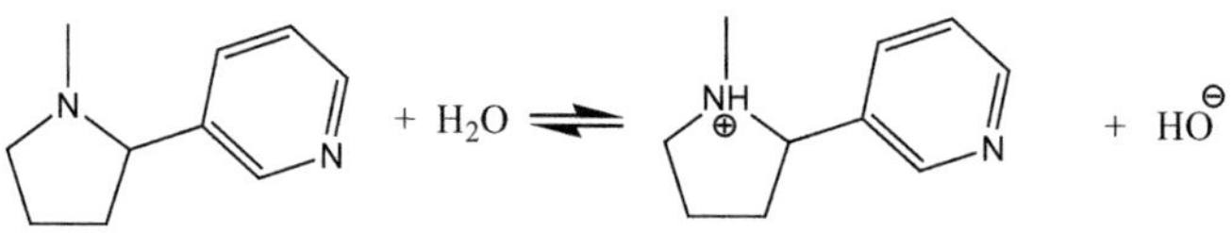

[structure] + H₂O ⇌ [structure] + HO⁻

$q = 1$ ... $p = 1$

[structure] + H₂O ⇌ [structure] + HO⁻

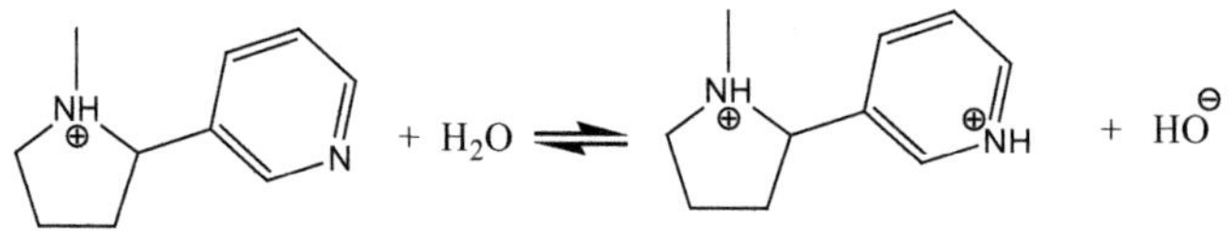

$q = 1$ ... $p = 1$

pK_b values: $pK_b = 14 - pK_a$

1. benzidine ($pK_{b,2} = 10.25$; $pK_{b,1} = 9.03$);
2. piperazine ($pK_{b,2} = 8.43$; $pK_{b,1} = 4.19$);
3. nicotine ($pK_{b,2} = 10.85$; $pK_{b,1} = 6.15$).

Bases	pK_b	K_b	p	q	$(p/q)*K_b$	$[(p/q)*K_b]^2$	BP
Benzidine (step 1)	9.03	9.33E−10	3	2	1.40E−09	1.96E−18	
Benzidine (step 2)	10.25	5.62E−11	6	1	3.37E−10	1.14E−19	
				sum	1.74E−09	2.07E−18	5.38E+00
Piperazine (step 1)	4.19	6.46E−05	2	2	6.46E−05	4.17E−09	
Piperazine (step 2)	8.43	3.72E−09	4	1	1.49E−08	2.21E−16	
				sum	6.46E−05	4.17E−09	1.01E+01
Nicotine (step 1)	6.15	7.08E−07	1	1	7.08E−07	5.01E−13	
Nicotine (step 2)	10.85	1.41E−11	1	1	1.41E−11	2.00E−22	
				sum	7.08E−07	5.01E−13	8.15E+00

Environmental Impact Metrics

For general bases, the basification potential with respect to neutrality is given by Equation (3.25).

$$BP = \log 2 + pK_w + \log\left[\frac{\sum_j (p_j/q_j)^2 (K_{bj})^2}{\sum_j (p_j/q_j)(K_{bj})}\right]$$

$$= 14.301 + \log\left[\frac{\sum_j (p_j/q_j)^2 (K_{bj})^2}{\sum_j (p_j/q_j)(K_{bj})}\right] \quad (3.25)$$

Overall acidification–basification potential with respect to neutrality is given by Equation (3.26).

$$ABP = \log\left(\frac{|10^{BP-7} - 10^{AP-7}|}{10^{-7}}\right) = 7 + \log|10^{BP-7} - 10^{AP-7}| \quad (3.26)$$

For combinations of several acids, the acidification potential with respect to neutrality is given by Equation (3.27).

$$AP = 7 + \log\left[\sum_j X_j \left(10^{(AP)_j - 7}\right)\right] \quad (3.27)$$

Where X_j is the mole fraction of jth acid.

For combinations of several bases, the basification potential with respect to neutrality is given by Equation (3.28).

$$BP = 7 + \log\left[\sum_j X_j \left(10^{(BP)_j - 7}\right)\right] \quad (3.28)$$

where X_j is the mole fraction of jth base.

For combinations of mixed acids and bases, the acidification–basification potential with respect to neutrality is given by Equation (3.29).

$$ABP = 7 + \log\left|\sum_j X_j \left(10^{(BP)_j - 7}\right) - \sum_j Y_j \left(10^{(AP)_j - 7}\right)\right| \quad (3.29)$$

where:
X_j is mole ratio of jth acid with respect to all acids present
Y_j is mole ratio of jth base with respect to all bases present

Sources for reliable pK_a data can be found in references [60–69].

Example 3.14

Part 1

Determine the acids produced when the following compounds decompose in water.

Acetic anhydride
Chlorine
Methyl chloroformate
Carbon dioxide
Diphosgene
Phosphorus oxychloride
Phosgene
Sulfur dioxide
Sulfur trioxide
Triphosgene

SOLUTION

Compounds that Decompose in Water	Equivalent Acids
Acetic anhydride	2 HOAc
Chlorine	HOCl + HCl
ClCOOMe	MeOH + HCl + H_2CO_3
CO_2	H_2CO_3
Diphosgene	4 HCl + 2 H_2CO_3
$O=PCl_3$	3 HCl + H_3PO_4
phosgene	2 HCl + H_2CO_3
SO_2	H_2SO_3
SO_3	H_2SO_4
Triphosgene	6 HCl + 3 H_2CO_3

Part 2

For each case in Part 1, determine the corresponding ABP parameter for the combined acids generated.

SOLUTION

Monoprotic acids:

Acid	pK_a	K_a	q	p	$2(q/p)K_a/K_w$	AP
HOAc	4.76	1.74E−05	2	1	6.95E+09	9.842
HOCl	7.46	3.47E−08	1	1	6.93E+06	6.841
HCl	−6.55	3.55E+06	1	1	7.10E+20	20.851
MeOH	15.5	3.16E−16	1	1	6.32E−02	−1.199

Environmental Impact Metrics

Diprotic acids:

Acid	pK$_a$	K$_a$	q	p	(q/p)K$_a$	[(q/p)K$_a$]^2	2 K(avg)/K$_w$	AP
H$_2$CO$_3$ (step 1)	6.35	4.47E−07	2	2	4.47E−07	1.995E−13		
H$_2$CO$_3$ (step 2)	10.33	4.68E−11	3	1	1.40E−10	1.97E−20		
				sum	4.47E−07	2.00E−13	8.93E+07	7.951
H$_3$PO$_4$ (step 1)	2.16	6.92E−03	2	3	4.61E−03	2.127E−05		
H$_3$PO$_4$ (step 2)	7.21	6.17E−08	3	2	9.25E−08	8.554E−15		
H$_3$PO$_4$ (step 3)	12.32	4.79E−13	4	1	1.91E−12	3.665E−24		
				sum	4.61E−03	2.13E−05	9.22E+11	11.965
H$_2$SO$_3$ (step 1)	1.76	1.74E−02	2	2	1.74E−02	0.000302		
H$_2$SO$_3$ (step 2)	7.2	6.31E−08	3	1	1.89E−07	3.583E−14		
				sum	1.74E−02	3.02E−04	3.48E+12	12.541
H$_2$SO$_4$ (step 1)	0.4	3.98E−01	3	2	5.97E−01	0.356601		
H$_2$SO$_4$ (step 2)	1.92	1.20E−02	4	1	4.81E−02	0.0023127		
				sum	6.45E−01	3.59E−01	1.11E+14	14.046

	Acid 1	X1	AP (acid 1)	Acid 2	X2	AP (acid 2)	Acid 3	X3	AP (acid 3)	ABP
Acetic anhydride	HOAc	0.5	9.842	HOAc	0.5	9.842				9.842
Chlorine	HOCl	0.5	6.841	HCl	0.5	20.851				20.550
ClCOOMe	MeOH	0.333	−1.199	HCl	0.333	20.851	H$_2$CO$_3$	0.333	7.951	20.374
CO$_2$	H$_2$CO$_3$	1	7.951							7.951
Diphosgene	HCl	0.667	20.851	H$_2$CO$_3$	0.333	7.951				20.675
O=PCl$_3$	HCl	0.75	20.851	H$_3$PO$_4$	0.25	11.965				20.726
Phosgene	HCl	0.667	20.851	H$_2$CO$_3$	0.333					20.675
SO$_2$	H$_2$SO$_3$	1	12.541							12.541
SO$_3$	H$_2$SO$_4$	1	14.046							14.046
Triphosgene	HCl	0.667	20.851	H$_2$CO$_3$	0.333	7.951				20.675

Example 3.15

For the following salts dissolved in water, determine the acidification–basification potential (ABP) for the solution. Assume that all salts are soluble in water.

1. Calcium phosphate
2. Iron(III)chloride
3. Potassium hydroxide

SOLUTION

1. Calcium phosphate

$$Ca_3(PO_4)_2 = 3Ca^{+2} + 2PO_4^{-3}$$

number of cations, $n = 3$
number of anions, $m = 2$
number of ionisable protons or charge of cation, $p = 2$
number of basic sites on conjugate base of cation, $q = 1$
number of ionisable protons on conjugate acid of anion, $p = 1$
number of basic sites on anion, $q = 4$
pK_a of cation $= 1.37$
pK_b of anion $= 1.68$
AP contribution from cation:

$$AP_{cation} = \log 2 + \log\left(\frac{q}{p}\right) - pK_a + 14$$

Therefore, AP(cation) = log(2) + log(1/2) − 1.37 + 14 = 12.63
BP contribution from anion:

$$BP_{anion} = \log 2 + \log\left(\frac{p}{q}\right) - pK_b + 14$$

Therefore, BP(anion) = log(2) + log (1/4) − 1.68 + 14 = 12.02

$$ABP_{salt} = 7 + \log\left|\left(\frac{m}{n+m}\right)10^{BP-7} - \left(\frac{n}{n+m}\right)10^{AP-7}\right|$$

Therefore, ABP(salt) = 7 + log(abs((2/5)*10^(12.02−7)−(3/5)*10^(12.63−7))) = 12.33

2. Iron(III)chloride

$$FeCl_3 = Fe^{+3} + 3Cl^-$$

pK_a of cation $= 2.47$
pK_b of anion $= 20.55$

$$AP_{cation} = \log 2 + \log\left(\frac{q}{p}\right) - pK_a + 14$$

Therefore, AP(cation) = log(2) + log(1/3) − 2.47 + 14 = 11.35

$$BP_{anion} = \log 2 + \log\left(\frac{p}{q}\right) - pK_b + 14$$

Therefore, BP(anion) = log(2) + log (1/1) − 20.55 + 14 = −6.25

$$ABP_{salt} = 7 + \log\left|\left(\frac{m}{n+m}\right)10^{BP-7} - \left(\frac{n}{n+m}\right)10^{AP-7}\right|$$

Therefore, ABP(salt) = 7 + log(abs((3/4)*10^(−6.25−7)−(1/4)*10^(11.35−7))) = 10.75
potassium hydroxide

$$KOH = K^+ + OH^-$$

Environmental Impact Metrics

pK_a of cation $= -2.25$
pK_b of anion $= 0$

$$AP_{cation} = \log 2 + \log\left(\frac{q}{p}\right) - pK_a + 14$$

Therefore, $AP(cation) = \log(2) + \log(1/1) + 2.25 + 14 = 16.55$

$$BP_{anion} = \log 2 + \log\left(\frac{p}{q}\right) - pK_b + 14$$

Therefore, $BP(anion) = \log(2) + \log(2/1) - 0 + 14 = 14.60$

$$ABP_{salt} = 7 + \log\left|\left(\frac{m}{n+m}\right)10^{BP-7} - \left(\frac{n}{n+m}\right)10^{AP-7}\right|$$

Therefore, $ABP(salt) = 7 + \log(abs((1/2)*10^{\wedge}(14.60-7) - (1/2)*10^{\wedge}(16.55-7))) = 16.24$

Potential: Bioaccumulation Potential[6,7,10,28,29,70,71]

The bioaccumulation potential for compound j is given by Equation (3.30).

$$BAP = 10^{\log K_{ow,j} - \log K_{ow,toluene}} = 10^{\log K_{ow,j} - 2.73} \qquad (3.30)$$

where:
$K_{ow,j}$ is the octanol–water partition coefficient for compound j
Reference compound: toluene
$\log K_{ow}$ (toluene) $= 2.73$

Hydrophilic chemicals with low K_{ow} (less than 1) values tend to have high solubilities in water. Hence, they have log K_{ow} values that are large negative numbers. This translates to small values for BAP.

Hydrophobic chemicals with large K_{ow} (larger than 1) values tend to have low solubilities in water. Hence, they have log K_{ow} values that are large positive numbers. This translates to large values for BAP. In this case the risk involves potential accumulation of the offending chemical in fat tissue in fish and mammals.

Example 3.16

Determine the BAP for aniline given that log K_{ow} (aniline) $= 0.9$.

SOLUTION

$BAP = 10^{\wedge}(0.9 - 2.73) = 10^{\wedge}(-1.83) = 0.0148$

Potential: Bioconcentration Potential[72–78]

The bioconcentration potential for compound j is given by Equation (3.31).

$$(BCP)_j = \frac{(BCF)_j}{(BCF)_{toluene}} = 10^{0.76(\log K_{ow,j} - 2.73)} \qquad (3.31)$$

where:

$K_{ow,j}$ is the octanol–water partition coefficient for compound j
Reference compound: toluene
log K_{ow} (toluene) = 2.73

Example 3.17

Determine the BCP for aniline given that log K_{ow} = 0.9.

SOLUTION

$$BCP = 10^{(0.76*(0.9-2.73))} = 10^{(1.391)} = 0.0407$$

Potential: Cancer Potency[45–50]

The cancer potency potential for compound j is given by Equation (3.32).

$$(CPP)_j = \frac{(C_{w,j} + C_{a,j})/TD_{50,j}}{(C_{w,tol} + C_{a,tol})/TD_{50,tol}} \qquad (3.32)$$

Where:
- $C(w,j)$ is the steady-state concentration of compound j in water after release of 1000 kg into water compartment as predicted by MCM (mol/m³)
- $C(a,j)$ is the steady-state concentration of compound j in air after release of 1000 kg into air compartment as predicted by MCM (mol/m³)
- $TD50(j)$ is the chronic dose-rate in mg/kg body wt/day for compound j that would induce tumors in half the test animals at the end of a standard lifespan for the species

Reference compound: toluene; TD50 (toluene) = 3060 mg/kg/day
See TD50.

Potential: Endocrine Disruption[79,80]

The endocrine disruption potential for compound j is given by Equation (3.33).

$$(EDP)_j = 10^{\log(RBA)_j - \log(RBA)_{BPA}} = 10^{\log(RBA)_j + 2.11} \qquad (3.33)$$

where:
- $(RBA)_j$ is the relative binding affinity of compound j to the estrogen receptor (see Equation (3.34))
- $(RBA)_{BPA}$ is the relative binding affinity of bisphenol A to the estrogen receptor

$$\text{compound J + Estrogen Receptor (ER)} \underset{}{\overset{(RBA)j}{\rightleftarrows}} \text{J-ER (bound)}$$

$$(RBA)_j = \frac{[J\text{-}ER]_{bound}}{[J]_{unbound}[ER]} \qquad (3.34)$$

Reference compound: bisphenol A (BPA)

$$\log(RBA)_{BPA} = -2.11$$

Example 3.18

Determine the EDP for estrone given that log RBA (estrone) = 0.86.

SOLUTION

$$EDP = 10^{(0.86+2.11)} = 10^{2.97} = 933.3$$

Potential: Global Warming[81]

The global warming potential for compound j is given by Equation (3.35).

$$(GWP)_j = \frac{(NC)_j/(MW)_j}{(NC)_{CO_2}/(MW)_{CO_2}} = \frac{(NC)_j/(MW)_j}{1/44} = \frac{44(NC)_j}{(MW)_j} \qquad (3.35)$$

where NC is the number of carbon atoms and MW is molecular weight (g/mol). Reference compound: carbon dioxide (CO_2)

Example 3.19

Determine the GWP for the following substances. acetone, carbon tetrachloride, diethyl ether, p-xylene

SOLUTION

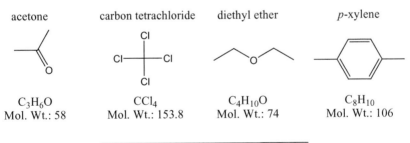

acetone — C_3H_6O Mol. Wt.: 58

carbon tetrachloride — CCl_4 Mol. Wt.: 153.8

diethyl ether — $C_4H_{10}O$ Mol. Wt.: 74

p-xylene — C_8H_{10} Mol. Wt.: 106

Substance	NC	MW	GWP
Acetone	3	58	2.276
Diethyl ether	4	74	2.378
Carbon tetrachloride	1	153.8	0.286
p-Xylene	8	106	3.321

Potential: Ingestion Carcinogenicity[82–85]

The ingestion carcinogencity potential for compound j is given by either Equations (3.36) or (3.37).

$$(INGCP)_j = \frac{C_{w,j}(OSF)_j}{C_{w,benzene}(OSF)_{benzene}} \quad (3.36)$$

$$(INGCP)_j = \frac{C_{w,j}(HV)_j}{C_{w,benzene}(HV)_{benzene}} \quad \text{(when OSF data are not available)} \quad (3.37)$$

where:
- C(w,j)   is the steady-state concentration of compound j in water after release of 1000 kg into water compartment as predicted by MCM (mol/m³)
- (OSF)j   is the oral slope factor for compound j
- (HV)j   is the hazard value for compound j

Reference compound: benzene
C(w,benzene) = 5.755e−10 mol/m³
(OSF)benzene = 0.015
(HV)benzene = unknown

Potential: Ingestion Toxicity

The ingestion toxicity potential for compound j is given by Equation (3.38).

$$(INGTP)_j = \frac{C_{w,j}/LD_{50,j}}{C_{w,tol}/LD_{50,tol}} \quad (3.38)$$

where:
- C(w,j)   is the steady-state concentration of compound j in water after release of 1000 kg into water compartment as predicted by MCM (mol/m³)
- LD50(j)   is the lethal dose to kill 50% of population upon exposure to compound j

Reference compound: toluene; LD50,tol = 636 (oral rat, mg/kg)
See *LD50(oral)*.

Example 3.20

Determine the INGTP for benzene given that LD50 = 930 mg/kg.

SOLUTION

Multi-compartment model calculation for release of 1000 kg benzene

$$C(\text{water, benzene}) = 5.63E-10 \text{ mol}/m^3$$

$$INGTP = (5.63E-10 / 4.062E-10) * (636/930) = 0.948$$

Environmental Impact Metrics

Potential: Inhalation Carcinogenicity[82,86]

The inhalation carcinogenicity potential for compound j is given by either Equations (3.39) or (3.40).

$$(INHCP)_j = \frac{C_{a,j}(ISF)_j}{C_{a,benzene}(ISF)_{benzene}} \qquad (3.39)$$

$$(INHCP)_j = \frac{C_{a,j}(HV)_j}{C_{a,benzene}(HV)_{benzene}} \text{ (used when ISF data are not available)} \qquad (3.40)$$

where:
- C(a,j) is the steady-state concentration of compound j in air after release of 1000 kg into air compartment as predicted by MCM (mol/m³)
- (ISF)j is the inhalation slope factor for compound j
- (HV)j is the hazard value for compound j

Reference compound: benzene
C(a,benzene) = 1.269E−10 mol/m³

Potential: Inhalation Toxicity

The inhalation toxicity potential for compound j is given by Equation (3.41).

$$(INHTP)_j = \frac{C_{a,j}/LC_{50,j}}{C_{a,tol}/LC_{50,tol}} \qquad (3.41)$$

where:
- C(a,j) is the steady-state concentration of compound j in air after release of 1000 kg into air compartment as predicted by MCM (mol/m³)
- LC50(j) is the lethal concentration to kill 50% of population upon exposure to compound j

Reference compound: toluene; LC50,tol = 49 (inhalation rat, g/m³, 4 h)
See LC50(inhalation).

Example 3.21

Part 1
Determine the INHTP for benzene given that LC50 = 31.9 g/m³, 4 h

Solution
Multi-compartment model calculation for release of 1000 kg benzene

$$C(air, benzene) = 1.27E-10 \text{ mol/m}^3$$

$$INHTP = (1.27E-10/1.075E-10)*(49/31.9) = 1.815$$

Part 2
Unit conversions for LC50 values:

$$\text{ppm} = \frac{24.45 * 1000 * (g/m^3)}{(MW)} \text{ (converting } g/m^3 \text{ to ppm)}$$

$$(g/m^3) = \frac{(\text{ppm}) * (MW)}{24.45 * 1000} \text{ (converting ppm to } g/m^3)$$

$LC_{50}(4h) = LC_{50}(xh) * (x/4)$ (converting LC50 at x hours to LC50 at 4 hours)

Convert the following ppm values of LC50 to g/m^3

Chemical Name	LC50 (ppm)	MW (g/mol)
1,1,1-trichloroethane	2000	133.35
1,1,2-trichloroethane	2000	133.35
1,2,4-trichlorobenzene	1100	181.35
1,2,4-trimethylbenzene	3655	120
1,2-dichlorobenzene	1700	147.9
1,2-dichloroethane	2063	98.9
1,2-dichloropropane	5554	112.9
1,3-butadiene	128850	56
1,3-dichloropropene	996	110.9
1,4-dichlorobenzene	1100	146.9
1,4-dioxane	6368	88

SOLUTION

Chemical Name	LC50 (ppm)	MW (g/mol)	LC50 (g/m^3)
1,1,1-trichloroethane	2000	133.35	10.91
1,1,2-trichloroethane	2000	133.35	10.91
1,2,4-trichlorobenzene	1100	181.35	8.16
1,2,4-trimethylbenzene	3655	120	17.94
1,2-dichlorobenzene	1700	147.9	10.28
1,2-dichloroethane	2063	98.9	8.34
1,2-dichloropropane	5554	112.9	25.65
1,3-butadiene	128850	56	295.12
1,3-dichloropropene	996	110.9	4.52
1,4-dichlorobenzene	1100	146.9	6.61
1,4-dioxane	6368	88	22.92

Part 3
Convert the following LC50 values recorded at various times to the standardized 4-hour limit.

Environmental Impact Metrics

Chemical Name	LC50 (g/m³)	Time (h)
Sodium carbonate	2.3	2
Sulfur trioxide	30	6
Sulfuric acid	0.51	2
Tetrahydrofuran	72	2
Triphenoxyphosphine	6.7	1

SOLUTION

Chemical Name	LC50 (g/m³)	Time (h)	LC50 (g/m³, 4 h)
Sodium carbonate	2.3	2	1.15
Sulfur trioxide	30	6	45
Sulfuric acid	0.51	2	0.255
Tetrahydrofuran	72	2	36
Triphenoxyphosphine	6.7	1	1.675

Potential: Ozone Depletion[87–99]

The ozone depletion potential for compound j is given by Equation (3.42).

$$(ODP)_j = \frac{\delta[O_3]_j}{\delta[O_3]_{CFC-11}} \tag{3.42}$$

$\delta[O_3]_j$ is the integrated change in stratospheric ozone due to compound j
Reference compound: CFC-11 (CCl_3F)

Persistence Potential[100–109]

The persistence potential for compound j is given by Equation (3.43).

$$(PER)_j = \frac{r_{tol,air} + r_{tol,w} + r_{tol,soil} + r_{tol,sed}}{r_{j,air} + r_{j,w} + r_{j,soil} + r_{j,sed}}$$

$$= \frac{\dfrac{C_{tol,air}}{\tau_{tol,air}} + \dfrac{C_{tol,w}}{\tau_{tol,w}} + \dfrac{C_{tol,soil}}{\tau_{tol,soil}} + \dfrac{C_{tol,sed}}{\tau_{tol,sed}}}{\dfrac{C_{j,air}}{\tau_{j,air}} + \dfrac{C_{j,w}}{\tau_{j,w}} + \dfrac{C_{j,soil}}{\tau_{j,soil}} + \dfrac{C_{j,sed}}{\tau_{j,sed}}} \tag{3.43}$$

where:
r(j,air)   is the initial rate of first-order degradation of compound j in the air compartment
r(j,w)   is the initial rate of first-order degradation of compound j in the water compartment

r(j,soil)   is the initial rate of first-order degradation of compound j in the soil compartment
r(j,sed)   is the initial rate of first-order degradation of compound j in the sediment compartment
C(j,air)   is the steady-state concentration of compound j in air after release of 1000 kg into the air compartment as predicted by MCM (mol/m^3)
C(j,w)   as above for the water compartment (mol/m^3)
C(j,soil) as above for the soil compartment (mol/m^3)
C(j,sed) as above for the sediment compartment (mol/m^3)
τ (j,air)   is the degradation half-life of compound j in the air compartment (h)
τ (j,w)   is the degradation half-life of compound j in the water compartment (h)
τ (j,soil) is the degradation half-life of compound j in the soil compartment (h)
τ (j,sed) is the degradation half-life of compound j in the sediment compartment (h)

Reference compound: toluene

C(air)	1.07E–10 mol/m^3
C(water)	4.06E–10 mol/m^3
C(soil)	4.29E–09 mol/m^3
C(sediment)	8.59E–09 mol/m^3

Tau(tol,air) = 17 h
Tau(tol,water) = 550 h
Tau(tol,soil) = 1700 h
Tau(tol,sediment) = 5500 h

The rate of degradation of toluene = (1.075E–10/17) + (4.062E–10/550) + (4.29E–9/1700) + (8.59E–9/5500)
= 1.107E–11 mol/m^3 h

Therefore, the persistence potential for compound j relative to toluene is given by Equation (3.44)

$$(\text{PER})_j = \frac{1.107 \times 10^{-11}}{\dfrac{C_{j,\text{air}}}{\tau_{j,\text{air}}} + \dfrac{C_{j,w}}{\tau_{j,w}} + \dfrac{C_{j,\text{soil}}}{\tau_{j,\text{soil}}} + \dfrac{C_{j,\text{sed}}}{\tau_{j,\text{sed}}}} \quad (3.44)$$

If the rate of degradation of compound j is larger than that of toluene, then compound j persists less in the environment than toluene and PER < 1. If the rate of degradation of compound j is less than that of toluene, then compound j persists more in the environment than toluene and PER > 1.

See references under MCM model.

Environmental Impact Metrics

Example 3.22

Using the multi-compartment model, determine the persistence potential for release of 1000 kg hexane into the environment using the following data. Use the Excel *template-persistence-potential.xls* spreadsheet to facilitate the computation.

τ (hexane, air) = 17 h
τ (hexane, water) = 550 h
τ (hexane, soil) = 1700 h
τ (hexane, sediment) = 5500 h
log K_{ow} = 3.9
H = 1.80 atm m³/mol

Solution

Result of multi-compartment model for release of 1000 kg hexane into the environment:

C(air) = 1.16E–10 mol/m³
C(water) = 1.59E–12 mol/m³
C(soil) = 2.49E–10 mol/m³
C(sediment) = 4.97E–10 mol/m³
Rate of degradation = 7.078E–12 mol/m³ h
Persistence potential relative to toluene = 1.107E–11/7.078E–12 = 1.56
Therefore, hexane persists in the environment 1.56 times more than toluene.

Potential: Photochemical Ozone Creation[110–114]

The photochemical ozone creation potential for volatile organic compound j is given by Equation (3.45).

$$(POCP)_j = \frac{m_{\text{ethylene eq.}}}{m_j} = \frac{a_j/b_j}{a_{\text{ethylene}}/b_{\text{ethylene}}} \qquad (3.45)$$

where:
- m(ethylene eq.) is the equivalent mass of ethylene for compound j
- m_j is the mass of compound j
- a_j is the change in ozone concentration due to a change in emission of compound j over a time t
- b_j is the integrated emission of compound j over that time reference compound: ethylene

Potential: Smog Formation[115–117]

The smog formation potential for compound j is given by Equation (3.46).

$$(SFP)_j = \frac{(MIR)_j}{(MIR)_{ROG}} = \frac{(MIR)_j}{3.1} \qquad (3.46)$$

where $(MIR)_j$ is the maximum incremental reactivity of compound j
Reference compound: standard mixture of reactive organic gases (ROG)
$(MIR)_{ROG}$ = 3.1

Example 3.23

Determine the SFP values for the following substances.

Chemical Name	MIR
1,2,3-trimethylbenzene	8.9
1,2,4-trimethylbenzene	8.8
1,3,5-trimethylbenzene	10.1
1,3,5-trimethylcyclohexane	1.7
1,3-butadiene	10.9
1,3-diethyl-4-methylcyclohexane	1.9
1,3-diethylcyclohexane	1.8
1,3-dimethylcyclohexane	2.5
1-butene	8.9
1-ethyl-4-methylcyclohexane	2.3
1-heptene	3.5
1-hexene	4.4
1-nonene	2.2
1-octene	2.7
1-pentene	6.2

Solution

Chemical Name	MIR	SFP
1,2,3-trimethylbenzene	8.9	2.87
1,2,4-trimethylbenzene	8.8	2.84
1,3,5-trimethylbenzene	10.1	3.26
1,3,5-trimethylcyclohexane	1.7	0.55
1,3-butadiene	10.9	3.52
1,3-diethyl-4-methylcyclohexane	1.9	0.61
1,3-diethylcyclohexane	1.8	0.58
1,3-dimethylcyclohexane	2.5	0.81
1-butene	8.9	2.87
1-ethyl-4-methylcyclohexane	2.3	0.74
1-heptene	3.5	1.13
1-hexene	4.4	1.42
1-nonene	2.2	0.71
1-octene	2.7	0.87
1-pentene	6.2	2.00

Environmental Impact Metrics

3.4 BENIGN INDEX

3.4.1 TERMS, DEFINITIONS, AND EXAMPLES

Benign Index (BI)[59]

The benign index is a measure of the degree of greenness of a chemical reaction or synthesis with respect to a set of environmental impact potentials. It can be applied to input materials or to waste materials. The mathematical definition is given by Equation (3.47).

$$BI = 1 - EI = 1 - \frac{\sum_j f_j \Omega_j}{\sum_j \Omega_j} \qquad (3.47)$$

where:
  EI is the environmental impact
  j refers to the jth compound
  Ω_j is the sum of all environmental potentials
  f_j = (mass of compound j)/(total mass of compounds of type j involved in reaction)

A value of 1 corresponds to complete greenness with respect to the selected environmental impact potentials and a value of 0 corresponds to complete anti-greenness. See *environmental impact*; *potential*.

Example 3.24

For a general chemical reaction given by
  Input materials → Desired product + Waste materials
the benign indices (BI) for input materials and waste materials is defined according to Equations (3.48) and (3.49).

$$BI(\text{in}) = 1 - EI(\text{in}) = 1 - \frac{\sum_j f(\text{input})_j \Omega_j}{\sum_j \Omega_j} \qquad (3.48)$$

and

$$BI(\text{waste}) = 1 - EI(\text{waste}) = 1 - \frac{\sum_j f(\text{waste})_j \Omega_j}{\sum_j \Omega_j} \qquad (3.49)$$

where EI(in) and EI(waste) are environmental impacts of input and waste materials given by Equations (3.50) and (3.51).

$$f(\text{input})_j = \frac{\text{mass of input compound } j}{\text{total mass of input compounds used in reaction}} \qquad (3.50)$$

$$f(\text{waste})_j = \frac{\text{mass of waste compound } j}{\text{total mass of waste compounds produced in reaction}} \quad (3.51)$$

and Ω_j is the sum of all potentials pertaining to compound j given by

$$\Omega_j = (\text{ABP})_j + (\text{ODP})_j + (\text{SFP})_j + (\text{GWP})_j$$
$$+ (\text{INHTP})_j + (\text{INGTP})_j + (\text{BCP})_j + (\text{ARDP})_j$$

ABP = acidification–basification potential
ARDP = abiotic resource depletion potential
BCP = bioconcentration potential
GWP = global warming potential
INGTP = ingestion toxicity potential
INHTP = inhalation toxicity potential
ODP = ozone depletion potential
SFP = smog formation potential

An industrial synthesis for the production of aniline by catalytic hydrogenation of nitrobenzene is given in the scheme below.[118]

600 kg/h nitrobenzene
6000 cubic meters per hour of hydrogen gas at T = 200°C and p = 5 psig
3500 kg aqueous copper carbonate solution (0.07 kg Cu per 100 kg aniline produced)
workup: 2500 kg 50 wt% sodium silicate aqueous solution
Yield of aniline = 98%

Using the REACTION template spreadsheet (*template-REACTION.xls*), the benign index calculator (*calculator-benign-index.xls*), and the data given below, determine BI(in) and BI(waste) for the reaction. Comment on which chemicals have the most impact.

Substance	ABP	ODP	SFP	LD50 (mg/kg)	LC50 (g/m³, 4 h)	H (atm m³/mol)	log Kow	ARDP
Nitrobenzene	0	0	0.02	682	2.8	2.40E−05	1.85	0
Hydrogen	0	0	0	NA	1.28	0.45	0	
Water	0	0	0	90000		8.48E−09	−1.38	0
CuCO₃	9.82	0	0	320	NA	1.0E−100	−1.0E+100	0.00194
Na₄SiO₄	10.83	0	0	1153	NA	1.0E−100	−1.0E+100	3.60E−10

NA = not available

SOLUTION
Balanced chemical equation:

$$C_6H_5NO_2 \xrightarrow[\text{- 2 H}_2\text{O}]{\substack{3\ H_2 \\ CuCO_3\ (\text{cat.})}} C_6H_5NH_2$$

123 → 93

Reactants	MW(g/mol)	Stoich. Coeff. (SC)	Mass Used (g)	Moles
Nitrobenzene	123	1	600000	4878
Dihydrogen	2	3	105174	52587

We assume hydrogen gas obeys the ideal gas law.

Number of moles of hydrogen = $(P\ V)/(R\ T)$
$p = 5$ psig = 0.3402 atm
$V = 6000$ cu m = 6000000 L
$T = 200 + 273 = 473$ K

Therefore, $n = (0.3402*6000000)/(0.08207*473) = 52587$ moles.

The limiting reagent is nitrobenzene (4878 moles).
Moles of aniline produced = $0.98*4878 = 4780.5$
Mass of aniline produced = $4780.5*93 = 444585$ g
Mass of water produced = $2*4780.5*18 = 172098$ g
Mass of unreacted nitrobenzene = $(4878-4780.5)*123 = 11992.5$ g
Mass of unreacted hydrogen = $(52587-3*4780.5)*2 = 76491$ g

Check mass balance:

Input Materials	Mass (g)	Output Materials	Mass (g)
Nitrobenzene	600000	Aniline	444585
Hydrogen	105174	Water	172098
$CuCO_3$ (aq)	3500000	Nitrobenzene (unreacted)	11992.5
Na_4SiO_4 (aq)	2500000	Hydrogen (unreacted)	76491
SUM	6705174	$CuCO_3$ (aq)	3500000
		Na_4SiO_4 (aq)	2500000
		SUM	6705167

Input Materials

Substance	Mass (g)	Proportion	ABP	ODP	SFP	GWP	INHTP	INGTP	BCP	ARDP	Total	EI Contribution
Nitrobenzene	600000	0.0895	0	0	0.02	2.146341	4.218157	60.98576	0.214388	0.000000	67.58465	6.0477
Hydrogen	105174	0.0157	0	0	0	0	0	0	0.018501	0.000000	0.018501	0.0003
Water	4749184	0.7083	0	0	0	0	0	4.822061	0.000752	0.000000	4.822813	3.4159
CuCO$_3$	816.54	0.0001	9.82	0	0	0	0	197.6241	0.000000	0.001940	207.4461	0.0253
Na$_4$SiO$_4$	1250000	0.1864	10.83	0	0	0	0	36.82863	0.000000	0.000000	47.65863	8.8847
Total	6705174	1								Total	327.5307	18.3738
										Overall EI		0.0561
										BI (input)		0.9439

Waste Materials

Substance	Mass (g)	Proportion	ABP	ODP	SFP	GWP	INHTP	INGTP	BCP	ARDP	Total	EI Contribution
Water (by-product)	172098	0.0275	0	0	0	0	0	4.822061	0.000752	0.000000	4.822813	0.1326
Nitrobenzene (unreacted)	11992.5	0.0019	0	0	0.02	2.146341	4.218157	60.98576	0.214388	0.000000	67.58465	0.1295
Hydrogen (unreacted)	76491	0.0122	0	0	0	0	0	0	0.018501	0.000000	0.018501	0.0002
Water	4749184	0.7586	0	0	0	0	0	4.822061	0.000752	0.000000	4.822813	3.6585
$CuCO_3$	816.54	0.0001	9.82	0	0	0.356131	0	197.6241	0.000000	0.001940	207.8022	0.0271
Na_4SiO_4	1250000	0.1997	10.83	0	0	0	0	36.82863	0.000000	0.000000	47.65863	9.5156
Total	6260582	1								Total	332.7096	13.4635
										Overall EI		0.0405
										BI (waste)		0.9595

BI(waste) exceeds BI(input) suggesting that the waste produced in the reaction is more benign than the input materials used at the outset.

The ingestion toxicity potential contributes the highest proportion of all potentials for both input materials and waste materials at 86% and 84%, respectively. Nitrobenzene and sodium silicate contribute most to INGTP as input materials; whereas, sodium silicate contributes the most to INGTP as a waste material. Nitrobenzene and sodium silicate have the highest environmental impact as input materials. Sodium silicate has the highest environmental impact as a waste material.

Uncertainty in BI[119]

This is a measure of error in the benign index (BI) determination when data are not available or are unreliable. The mathematical expression for uncertainty is given by Equation (3.52).

$$\%\text{uncertainty} = \left(\frac{x}{nC}\right)100 \tag{3.52}$$

where x is the number of missing parameters for all the substances used in a given reaction, n is the number of parameters needed to estimate BI ($n = 8$) for each substance, and C is the total number of chemicals required for carrying out a given reaction including reagents, catalysts, additives, ligands, reaction solvents, workup materials, purification materials, and its associated by-products. See *benign index*.

Example 3.25

In Example 3.24, determine the % uncertainties in BI(input) and BI(waste).

SOLUTION

For the input materials, we have $n = 8$, $C = 5$, and $x = 3$. Hence the % uncertainty in BI(input) is 100*(3/(8*5)) = 7.5%.

For the waste materials, we have $n = 8$, $C = 6$, and $x = 3$. Hence the % uncertainty in BI(waste) is 100*(3/(8*6)) = 6.25%.

3.5   EXCEL SPREADSHEET TOOLS: PART 2

3.5.1   DESCRIPTIONS OF SPREADSHEETS

The following Excel spreadsheet tools are available in the accompanying CD of this book, which facilitate computations that are typically tedious tasks if done otherwise.

1. *benign-index-template.xls*
2. *calculator-benign-index.xls*
3. *database-industrial-chemicals-LCA-parameters.xls*
4. *database-LCA-parameters.xls*
5. *synthesis-plans-LCA-waste-template.xls*
6. *synthesis-plans-LCA-inputs-template.xls*

7. template-LCA.xls
8. template-multi-compartment-model.xls
9. template-persistence-potential.xls

For each spreadsheet, we briefly describe the kind of inputs required and the results obtained.

The *benign-index-template.xls* spreadsheet is composed of three sheets: a materials sheet, a reaction sheet, and a synthesis sheet. The materials inventory sheet requires the input of the following parameters for a chemical reaction: (a) chemical name of each input material, by-product, and target product; (b) chemical abstracts number (CAS) for each chemical; (c) molecular weight; (d) LD50(oral) in ppm units; (e) LD50(dermal) in ppm units; (f) LC50 inhalation in ppm units; (g) occupational exposure limit (OEL) in ppm units; (h) log K_{ow}; (i) number of carbon atoms; (j) number of acidic hydrogen atoms with $pK_a < 7$; (k) lower explosion limit (LEL); (l) flash point (deg C); (m) boiling point (deg C); (n) whether the chemical is flammable, corrosive, explosive, pyrophoric, a strong oxidizer, or reacts violently with water; (o) acidification–basification potential (ABP); (p) ozone depletion potential (ODP); (q) smog formation potential (SFP); and (r) Henry's law constant in atm m^3 mol^{-1} units. Once the balanced chemical equation is established all of the relevant information in the materials inventory sheet is fed into the reaction sheet. In addition, the masses of all materials are entered as appropriate depending on whether the chemical is a reagent, catalyst, reaction solvent, workup material, purification material, or target product. Next, the number of LCA parameters that are not known is also entered. The spreadsheet will automatically calculate the following material efficiency metrics: yield, atom economy (AE); stoichiometric factor (SF); E-factor and all its contributors; process mass intensity (PMI); and reaction mass efficiency (RME). If the unit cost of each material is known, then the overall material cost of carrying out the reaction is determined. The following environmental impact metrics are calculated: benign indices (BIs) for input, output, and waste; and the percent uncertainty in the BI determinations. This reaction sheet is repeated for every reaction in a synthesis plan as appropriate. The synthesis sheet determines the overall E-factor and overall BI(waste) metrics once the scaling factors are determined for each reaction in a synthesis plan. The input information required in the synthesis sheet includes: the list of substances with their descriptors; molecular weights; masses from the reaction sheets; scaling factors; and numerical values of potentials AP, ODP, SFP, GWP, INHTP, INGTP, BP, and ADP for each substance. All calculations are worked out using a 1000 g basis for the target product of a synthesis plan.

The *calculator-benign-index.xls* spreadsheet determines the value of the BI index for input materials (BI(input)), or waste materials (BI(waste)) for any chemical reaction. This calculator also determines the fractional contribution of each potential for each chemical substance so the most offending chemicals can be identified as well as the corresponding potentials that are responsible for those offences. The input information is the same as in the *benign-index-template.xls* spreadsheet.

The *database-industrial chemicals-LCA-parameters.xls* and *database-LCA-parameters.xls* spreadsheets are a repository of relevant parameters for many thousands of chemicals that are available in a single resource. These database spreadsheets greatly increase the speed in carrying out determinations of the BI indices.

The *synthesis-plans-LCA-waste-template.xls* spreadsheet is identical to the synthesis sheet in the *benign-index-template* spreadsheet.

The *synthesis-plans-LCA-input-template.xls* spreadsheet is similar to the *synthesis-plans-LCA-waste-template.xls* spreadsheet except that it deals with the masses of input materials as opposed to unreacted input materials. Also, it does not include by-products since they are not input materials.

The *template-LCA.xls* spreadsheet is identical to the reaction sheet in the *benign-index-template.xls* spreadsheet.

The *template-multi-compartment-model.xls* spreadsheet calculates the masses of a given chemical if 1000 kg of it were to be released into the 4 compartments of the environment: air; water; soil; and sediment. It is based on the Mackay model discussed in Section 3.1.1, which uses the following volumes in cubic meters of each compartment: 1E14 (air); 2E11 (water); 9E9 (soil); and 1E8 (sediment). The input parameters include: the name of the substance; molecular weight; Henry's law constant; and log K_{ow}.

The *template-persistence-potential.xls* spreadsheet is an extension of the *template-multi-compartment-model.xls* spreadsheet. Based on the half-life inputs in hours of a chemical in each of the four environmental compartments, it calculates the corresponding rate of degradation of that chemical in mol m^{-3} h^{-1} units. In addition, it calculates the persistence potential relative to toluene as a reference compound.

3.5.2 Assumptions and Dealing with Missing Data

When K_{ow} is not known for a substance assume that for substances that are expected to be water soluble, assume $K_{ow} \to 0$ and log $K_{ow} \to$ minus infinity (enter – 1e100 in spreadsheet so that BAP=0).

For aqueous wash solutions (acidic or basic) assume that log K_{ow} = –1.38 (value for water).

If a substance is insoluble in either water or 1-octanol, enter "not applicable" in the appropriate cell.

For gases, LD50(oral) does not apply.

LC50(inhal) typically applies to gases and volatile liquids and solids having appreciable vapor pressures. It is not applicable for non-volatile substances.

For aqueous acidic wash solutions, AP does not apply since the acidic compounds are diluted in water (less volatile than in pure concentrated form) and MW is not applicable (due to mixture).

For non-volatile solids the Henry's law constant is set to 1E–100 (close to 0) so as to avoid a division by zero error in the computations in the spreadsheets.

Below are listed references for estimating missing data parameters needed for an LCA.

Estimating bioconcentration factors:

From log K_{ow}: references [120–125].
From liposome-water or lipid-water partitioning: references [126, 127].
From aqueous solubility: reference [128].

Environmental Impact Metrics

Estimating log K_{ow}:

From aqueous solubility data: references [32, 128, 129–131].
From atom/fragment contributions: references [132, 133].
From liquid-solute water solubilities and molar volumes: reference [134].
From GLC (gas liquid chromatography) retention indices: reference [135].
From group contribution solvation model: reference [136].
From PHYSPROP database using artificial neural networks and electrotopological state indices: references [137, 138].
From separation methods: reference [139].
From on-line ALOGSP 2.1 program: references [140–143].
From mobile order and disorder thermodynamics: reference [144].
From specific high log K_{ow} values: reference [145].
Using XLOGP3 software: reference [146].
From combined additive modeling: reference [147].
From quantitative structure property relationships (QSPR): reference [148].
From quantum mechanical calculations: reference [149].
From molecular surface area and electrostatic potential descriptors: reference [150].
From UNIFAC method: reference [151].
From QSAR: references [152–154].
From Monte Carlo methods: reference [155].

Estimating log K_{oc}:

From aqueous solubility and log K_{ow}: references [28, 156–160].
From log K_{ow}: reference [161].
From molecular topology/fragment contribution method: reference [162].
From linear free energy relationship: reference [163].
From high performance reverse phase liquid chromatography: reference [164].

Estimating Henry's law constants:

From bond contribution method: reference [165].
From aqueous solubilities and vapor pressures: reference [166].
From molecular structure: references [167–170].
Estimating vapor pressures: references [166, 171–173].
Estimating aqueous solubilities: references [138, 166, 174–176].
Below are listed references pertaining to the reliability of LCA data.
Reliability of Kow data: references [177–186].
Reliability of Koc data: reference [187].
Reliability of QSAR modeling: references [188–191].
Reliability of literature data: references [192–198].
Reliability of computational toxicology: reference [199].

3.6 PROBLEMS

PROBLEM 3.1
Using the data given below verify that log $K_{ow} = 3.25$ for octanol.
density of water at 25 C = 1 g/mL = 1000 g/L
solubility of octanol in water at 25°C = 0.46 g octanol/kg water
MW(octanol) = 130 g/mol
density of octanol at 25°C = 0.827 g/mL = 827 g/L

PROBLEM 3.2

If $AP = \log\left(\dfrac{2q}{p}\right) - pK_a + pK_w$ for monoprotic acids and $BP = \log\left(\dfrac{2p}{q}\right) - pK_b +$

$pK_w = \log\left(\dfrac{2p}{q}\right) + pK_a$ for monobasic bases, what is the connecting relationship between AP and BP?

PROBLEM 3.3
If $AP = \log 2 - pK_{a,avg} + pK_w$ for diprotic acids where

$$K_{a,avg} = \dfrac{(q_1/p_1)^2 (K_{a1})^2 + (q_2/p_2)^2 (K_{a2})^2}{(q_1/p_1)(K_{a1}) + (q_2/p_2)(K_{a2})}$$

and $BP = \log 2 - pK_{b,avg} + pK_w$ for dibasic bases where

$$K_{b,avg} = \dfrac{(p_1/q_1)^2 (K_{b1})^2 + (p_2/q_2)^2 (K_{b2})^2}{(p_1/q_1)(K_{b1}) + (p_2/q_2)(K_{b2})},$$

what is the connecting relationship between AP and BP?

PROBLEM 3.4
Based on the results of Problem 3.3, determine the acidification potentials (AP) of the following dibasic bases.

1. benzidine ($pK_{a,1} = 3.75$; $pK_{a,2} = 4.97$)
2. piperazine ($pK_{a,1} = 5.57$; $pK_{a,2} = 9.81$)
3. nicotine ($pK_{a,1} = 3.15$; $pK_{a,2} = 7.85$).

PROBLEM 3.5
For the following salts dissolved in water, determine the acidification–basification potential (ABP) for the solution. Assume that all salts are soluble in water.

1. Sodium carbonate
2. Sodium chloride
3. Magnetite

Environmental Impact Metrics

PROBLEM 3.6
One tonne of acetic acid is released into the environment. Based on the MCM estimate, how much ends up in each of the four environmental compartments. Use the Excel *template-multi-compartment-model.xls* spreadsheet to facilitate the computation.

$H = 1E-7$ atm m^3/mol
$\log K_{ow} = -0.17$

PROBLEM 3.7
One tonne of phosgene is released into the environment. Based on the MCM estimate, how much ends up in each of the four environmental compartments. Use the Excel *template-multi-compartment-model.xls* spreadsheet to facilitate the computation.

$H = 8.92E-3$ atm m^3/mol
$\log K_{ow} = -0.71$

PROBLEM 3.8
In Bhopal, India, the largest industrial chemical release of methyl isocyanate occurred in December 1984. An estimated 40 tonnes of this substance was released into the environment. Based on the MCM estimate, how much ends up in each of the four environmental compartments. Use the Excel *template-multi-compartment-model.xls* spreadsheet to facilitate the computation.

$H = 1.13E-3$ atm m^3/mol
$\log K_{ow} = 0.79$

PROBLEM 3.9
Using the multi-compartment model, determine the persistence potential for release of 1000 kg 2,3,7,8-tetrachlorodibenzo-*p*-dioxin into the environment using the following data. Use the Excel *template-persistence-potential.xls* spreadsheet to facilitate the computation.

τ (hexane,air) = 30 d
τ (hexane,water) = 430 d
τ (hexane,soil) = 6700 d
τ (hexane,sediment) = 2100 d
$\log K_{ow} = 6.8$
$H = 5.00E-5$ atm m^3/mol

PROBLEM 3.10[200]
An industrial synthesis for the production of 2000 lbs of aniline by reducing nitrobenzene with iron filings is given in the scheme below. The authors stated that the by-product of the reaction is Fe_3O_4.

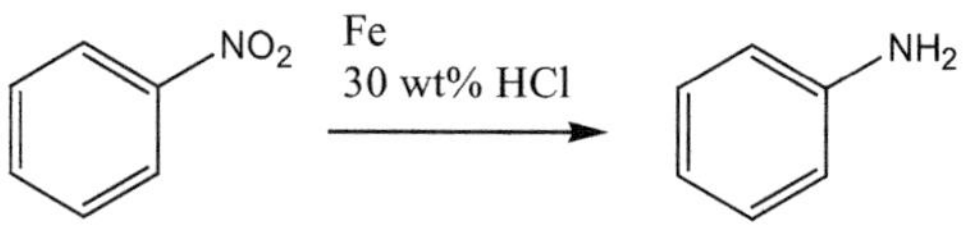

$$\text{C}_6\text{H}_5\text{NO}_2 \xrightarrow[\text{30 wt\% HCl}]{\text{Fe}} \text{C}_6\text{H}_5\text{NH}_2$$

2780 lbs nitrobenzene
3200 lbs iron filings
2500 lbs 30 wt% HCl

Using the REACTION template spreadsheet (*template-REACTION.xls*), the benign index calculator (*calculator-benign-index.xls*), and the data given below, determine BI(in) and BI(waste) for the reaction. Comment on which chemicals have the most impact.

Substance	ABP	ODP	SFP	LD50 (mg/kg)	LC50 (g/m³, 4 h)	H (atm m³/mol)	log Kow	ARDP
Nitrobenzene	0	0	0.02	682	2.8	2.40E–05	1.85	0
Iron	0	0	0	750		1.0E–100	–1.0E+100	8.43E–08
Water	0	0	0	90000		8.48E–09	–1.38	0
HCl	20.85	0	0	900	1.16	5.26E–05	1.1	4.86E–08
Fe₃O₄	24.06	0	0	NA		1.0E–100	–1.0E+100	2.53E–07

NA = not available

Determine the % uncertainties in the BI(input) and BI(waste) estimates.

PROBLEM 3.11[201]
The cradle-to-gate life cycle input material inventory for the production of 1000 kg of acetaldehyde are shown below.

Input Materials	Mass (kg)
Air	618
Alum (AlK(SO$_4$)$_2$ 12 H$_2$O)	0.0374
Barium carbonate	0.0714
Chlorine	4.38
Crude oil	859
Ethylene	741
Hydrofluorosilicic acid (H$_2$SiF$_6$)	0.00531
Hydrogen chloride	17.4
Lime	0.0243
Sodium carbonate	0.0241
Naphtha	837
Oxygen	448
Salt rock	6.04
Sodium chloride	9.66
Sodium hydroxide	18.1
Water for reaction	2.99
Water (total)	1330
Total	**4891.733**

Environmental Impact Metrics

The energy input, air emissions, water emissions, and solid waste inventories are given below.

Energy Input	MJ
Coal	125
Cooling water	21
Diesel	1350
Electricity	2490
Heating fuel	14.6
Heavy oil	1180
Hydro power	8.37
Natural gas	2800
Nuclear power	8.37
Refrigeration	563
Steam	11800
Potential energy recovery	−4650
Total	**15710.34**

Air Emissions	Mass (kg)	Water Emissions	Mass (kg)
1,2-dichloroethane	0.0955	Acetaldehyde	42.2
Acetaldehyde	42.2	Acetic acid	109
Acetic acid	1.15	Barium sulfate	0.0418
Carbon dioxide	1430	BOD	0.274
Carbon monoxide	19.3	Calcium carbonate	0.0178
Chlorine	0.244	COD	195
Ethane	0.593	Magnesium hydroxide	0.00298
Ethylene	6.72	Mercury	3.12E−06
Ethylene 1,2-dichloride	29.6	Sodium sulfide	6.3
Hydrogen	7.9	TDS	4.73
Hydrogen chloride	0.0986	TOC	66.8
Hydrogen sulfide	25	Waste Water	1310
Hypochlorous acid	0.000173	**Total**	**1734.367**
Methane	6.08		
NMVOC	10.4		
NO_x	7.22		
SO_x	5.59		
Vinyl chloride	0.0193		
Total	**1592.211**		

Solid Waste	Mass (kg)
Barium sulfate	0.0434
Calcium carbonate	0.0193
Magnesium hydroxide	0.000482
Other	28.7
Total	**28.76318**

BOD = biological oxygen demand
COD = chemical oxygen demand
NMVOC = nonmethane volatile organic compounds
TDS = total dissolved solids
TOC = total organic carbon

Part 1

Based on the information given in the input material inventory, write out a synthesis plan for the production of acetaldehyde. Show balanced equations for each step.

Part 2

Determine the process mass intensity and energy intensity for the process.

Part 3

Using the *calculator-benign-index.xls* spreadsheet and the data given below, determine the benign indices for inputs, air emissions, water emissions, and solid waste. Which potentials have the largest impact in each category? Which chemicals have the most impact?

Substance	MW (g/mol)	No. Carbon Atoms	Alpha No. Acidic Hs	ABP	ODP
Air	28.56	0	0		
Alum (AlK(SO$_4$)$_2$ 12 H$_2$O)	478	0	0	15.95	
Barium carbonate	197.33	1	0	9.82	
Chlorine	70.9	0	0	0	
Crude oil					
Ethylene	28	2	0	0	
Hydrofluorosilicic acid (H$_2$SiF$_6$)	144	0	2		
Hydrogen chloride (g)	36.45	0	1	20.851	
Lime (CaO)	56.8	0	0	24	
Sodium carbonate	106	1	0	9.65	
Naphtha				0	
Oxygen	32	0	0	0	
Salt rock					
Sodium chloride	58.45	0	0	−0.8	
Sodium hydroxide	40	0	0	14.3	
Water	18	0	0	0	

Environmental Impact Metrics

Substance	MW (g/mol)	No. Carbon Atoms	Alpha No. Acidic Hs	ABP	ODP
1,2-dichloroethane	98.9	2	0	0	
Acetaldehyde	44	2	0	0	
Acetic acid	60	2	1	9.842	
Carbon dioxide	44	1	0	7.95	
Carbon monoxide	28	1	0	0	
Ethane	30	2	0	0	
Ethane 1,2-dichoro	96.9	2	0	0	
Hydrogen	2	0	0	0	
Hydrogen sulfide	34	0	2	7.057	
Hypochlorous acid	52.45	0	1	6.841	
Methane	16	1	0	0	
Vinyl chloride	62.45	2	0	0	
Barium sulfate	233.33	0	0	1.27	
Calcium carbonate	100.08	1	0	9.82	
Magnesium hydroxide	58.3	0	0	14.43	
Mercury	200.59	0	0	0	
Sodium sulfate	78	0	0	14.74	

Substance	SFP	LD50 (mg/kg)	LC50 (g/m3 4 h)	H (atm m3/mol)	log Kow	ADP
Air						0
alum (AlK(SO$_4$)$_2$ 12 H$_2$O)				1.0E–100	–1E+100	7.16E–04
Barium carbonate		418		1.0E–100	–1E+100	1.06E–10
Chlorine		8910	0.21	1.22E–02	–1.32	9.72E–08
Crude oil						0
Ethylene	2.39		0.05	0.228	1.13	0
Hydrofluorosilicic acid (H$_2$SiF$_6$)		430		1.0E–100	–1E+100	1.78E–05
Hydrogen chloride (g)		900	0.41	5.26E–05	1.1	4.86E–08
Lime (CaO)				1.0E–100	–1E+100	7.08E–10
Sodium carbonate		4090	1.15	1.0E–100	–1E+100	8.24E–11
Naphtha						0
Oxygen				0.769	0.65	0
Salt rock				1.0E–100	–1E+100	0
Sodium chloride		3560		1.0E–100	–2.96	4.87E–08
Sodium hydroxide		40		1.0E–100	–1E+100	8.24E–11
Water		90000		8.48E–09	–1.38	0
1,2-dichloroethane	0.31	500	8.34	9.79E–04	1.47	9.72E08
Acetaldehyde	1.67	661	23.4	6.67E–05	–0.34	0
Acetic acid		3310	39.3	1.00E–07	–0.17	0
Carbon dioxide			105.7	2.88E–02	–1.33	0
Carbon monoxide	0.029		1.9	1.09	–1.0E+100	0
Ethane	0.08			4.91E–01	1.81	0
Ethane 1,2-dichoro		200	25	2.61E–02	2.13	0

Substance	SFP	LD50 (mg/kg)	LC50 (g/m3 4 h)	H (atm m3/mol)	log Kow	ADP
Hydrogen				1.28	0.45	0
Hydrogen sulfide			1	1.05E−02		0
Hypochlorous acid		8910		1.63E−06		4.86E−08
Methane	0.005		163	6.58E−01	1.09	0
Vinyl chloride	1	500	0.26	2.78E−02	1.62	4.86E−08
Barium sulfate				1.0E−100	−1.0E+100	3.58E−04
Calcium carbonate		6450		1.0E−100	−1.0E+100	7.08E−10
Magnesium hydroxide		8500		1.0E−100	−1.0E+100	3.73E−09
Mercury				1.0E−100	−1.0E+100	0.495
Sodium sulfate		208		1.0E−100	−1.0E+100	3.58E−04

PROBLEM 3.12[202]

A chemical plant produces ethylene glycol, diethylene glycol, and triethylene glycol from ethylene oxide and water. Ethylene glycol is produced first, then the higher glycols. Products are separated by distillation. The relevant data about the process are given below.

Input Material	Mass (kg)	Product	Mass (kg)	% purity
Ethylene oxide	799	Ethylene glycol	1000	99.7
Water	310	Ethylene glycol, di-	94.6	89.3
		Ethylene glycol, tri-	10	91.1

Emissions	Mass (kg)	Energy input	MJ
Ethylene oxide	3.97	Electricity	412
Ethylene glycol	0	Heating steam	15100
Ethylene glycol, di-	0	Cooling water	−14500
Ethylene glycol, tri-	0.0893		
Higher glycols	0.434		

Part 1

Write out a synthesis plan that accounts for the formation of the three products. Show balanced equations.

Part 2

Determine the percent reaction yields for the production of ethylene glycol, diethylene glycol, and triethylene glycol with respect to either ethylene oxide or water whichever is the limiting reagent.

Environmental Impact Metrics

Part 3

Determine the percent conversion of ethylene oxide to glycol products and the percent selectivities to each glycol.

Part 4

Track the number of moles of ethylene oxide from the start of the reaction and the number of moles of each glycol product made.

Part 5

Based on the tracking of the number of moles of ethylene oxide in Part 4, determine the percent reaction yields for production of glycol products with respect to ethylene oxide at each stage of the process.

Part 6

Draw synthesis trees for the production of ethylene glycol, diethylene glycol, and triethylene glycol.

Part 7

Determine the PMI values for production of each glycol product.

Part 8

Determine the energy intensity for production of each glycol product.

Part 9

Using the *calculator-benign-index.xls* spreadsheet and the data given below, determine the benign indices for inputs, products, and emissions. Which potentials have the largest impact in each category? Which chemicals have the most impact?

Substance	Mass (g)	MW (g/mol)	No. Carbon Atoms	Alpha No. Acidic Hs	ABP	ODP	SFP
Ethylene oxide	799000	44	2	0	0		
Water	310000	18	0	2	0		
Ethylene glycol	997000	62	2	2	−0.22		2.31
Ethyleneglycol,di-	84480	106	4	2	−0.22		
Ethyleneglycol,tri-	9110	150	6	2	−0.22		

Substance	LD50 (mg/kg)	LC50 (g/m3 4 h)	H (atm m3/mol)	log Kow	ADP
Ethylene oxide	72	1.44	1.48E−04	−0.3	0
Water	90000		8.48E−09	−1.38	0
Ethylene glycol	4700	2.54	6.00E−08	−1.36	0
Ethylene glycol, di-	12565		2.00E−09	−1.47	0
Ethylene glycol, tri-	15000	4.4	2.61E−10	−1.98	0

REFERENCES

1. Schneider, F.; Szuppa, T.; Stolle, A.; Ondruschka, B.; Hopf, H., *Green Chem.*, 2009, 11, 1894.
2. Heinzle, E.; Weirich, D.; Brogli, F.; Hoffmann, V.H.; Koller, G.; Verduyn, M.A.; Hungerbuhler, K., *Ind. Eng. Chem. Res.*, 1998, 37, 3395.
3. Koller, G.; Fischer, U.; Hungerbuhler, K., *Ind. Eng. Chem. Res.*, 2000, 39, 960.
4. Sheldon, R.A., *ChemTech*, 1994, 24 (3), 38.
5. Mackay, D.; Shiu, W.Y., *J. Phys. Chem. Ref. Data*, 1981, 10, 1175.
6. Mackay, D.; Shiu, W.Y.; Ma, K.C.; Lee, S.C., *Handbook of Physical-Chemical Properties and Environmental Fate for Organic Chemicals*, CRC Press: Boca Raton, 2006, Vol. 1–4.
7. Howard, P.H.; Meylan, W.M. (eds.) *Handbook of Physical Properties of Organic Chemicals*, Lewis Publishers: Boca Raton, 1997.
8. Sander, R. Compilation of Henry's Law Constants for Inorganic and Organic Species of Potential Importance in Environmental Chemistry, Version 3.0, 1999. (www.henrys-law.org)
9. Haynes, W.M. (Ed.), *CRC Handbook of Chemistry and Physics*, 91st ed., CRC Press: Boca Raton, 2009, pp. 87–120.
10. Verschueren, K., *Handbook of Environmental Data on Organic Chemicals*, Wiley: Hoboken, 2009.
11. Sangster, J.M., Henry's law constants for compounds stable in water. In: Fogg, P.G.T. and Sangster, J.M. (Eds.), *Chemicals in the Atmosphere – Solubility, Sources and Reactivity*. Wiley: Chichester, 2003, pp. 255–397.
12. Staudinger, J.; Roberts, P.V., *Crit. Rev. Environ. Sci. Technol.*, 1996, 26, 205–297.
13. Staudinger, J.; Roberts, P.V., *Chemosphere*, 2001, 44, 561–576.
14. Ashworth, R.A.; Howe, G.B.; Mullins, M.E.; Rogers, T.N. *J. Haz. Mater.*, 1988, 18, 25.
15. Yaws, C.; Yang, H.C.; Pan, X., *Chem. Eng.*, 1991, 98, 179.
16. Suntio, L.R.; Shiu, W.Y.; Mackay, D.; Seiber, J.N.; Glotfelty, D., *Rev. Env. Contam. Toxicol.*, 1988, 103, 1.
17. EPA/540/R95/128, Soil Screening Guidance: Technical Background Document, 1996, Part 5 Table 36 (www.epa.gov/superfund/health/conmedia/soil/pdfs/part_5.pdf).
18. Yaws, C.L., *Chemical Properties Handbook*, McGraw-Hill: New York, 1999, pp. 403–424.
19. ChemSpider (www.chemspider.com).
20. Hodge, H.C.; Sterner, J.H. Tabulation of toxicity classes, *Am. Industrial Hygiene Assoc. Quart.*, 1949, 10, 93.
21. Registry of Toxic Effects of Chemical Substances (RTECS) (http://ccinfoweb.ccohs.ca/rtecs/search.html).
22. Lewis, R.J. Jr. Sax's Dangerous Properties of Industrial Materials, Wiley: Hoboken, 2004, Vol. 1–3.
23. Allen, D.T.; Shonnard, D.R., *Green Engineering: Environmentally Conscious Design of Chemical Processes*, Prentice Hall: Upper Saddle River, 2002, p. 386.
24. Davis, G.A.; Kincaid, L.; Swanson, M.; Schultz, T.; Bartmess, J.; Griffith, B.; Jones, S. EPA-600-R-94-177 Chemical Hazard Evaluation for Management Strategies: a method for ranking and scoring chemicals by potential human health and environmental impacts, 1994.
25. EPA-540-R-97-036 Health Effects Assessment Summary Tables (HEAST), 1997. Report of the International Workshop on In Vitro Methods for Assessing Acute Systemic Toxicity, Appendix H, National Toxicity Program, 2006. (http://iccvam.niehs.nih.gov/docs/acutetox_docs/brdval031706/rbrd/appH.pdf).
26. Material Safety Data Sheets (MSDS).

27. Haynes, W.M. (Ed.), *CRC Handbook of Chemistry and Physics*, 91st ed., CRC Press: Boca Raton, 2009, Chapter 16, pp. 43–47.
28. Huuskonen, J., *J. Chem. Inf. Comput. Sci.*, 2003, 43, 1457.
29. Sangster, J., *J. Phys. Chem. Ref. Data*, 1989, 18, 1111.
30. Bowman, B.T.; Sans, W.W., *J. Env. Sci. Health Pt. B*, 1983, 18, 667.
31. Hansch, C.; Leo, A.; Hoeckman, D., *Exploring QSAR. Hydrophobic, Electronic, and Steric Constants*, American Chemical Society: Washington, 1995, Vol. 1, 2.
32. Sangster, J., *Octanol–Water Partition Coefficients: Fundamentals and Physical Chemistry*, Wiley: Hoboken, 1997.
33. Leo, A.; Hansch, C.; Elkins, D., *Chem. Rev.*, 1971, 71, 525.
34. Tewari, Y.B.; Miller, M.M.; Wasik, S.P.; Martire, D.E., *J. Chem. Eng. Data*, 1982, 27, 451.
35. Isnard, P.; Lambert, S., *Chemosphere*, 1989, 18, 1837.
36. Hansch, C.; Kim, D.; Leo, A.J.; Novellino, E.; Silipo, C.; Vittoria, A., *Crit. Rev. Toxicol.*, 1989, 19, 185.
37. EPA/540/R95/128, Soil Screening Guidance: Technical Background Document, 1996, Part 5 Table 36, Appendix K. (www.epa.gov/superfund/health/conmedia/soil/pdfs/part_5.pdf) (www.epa.gov/superfund/health/conmedia/soil/pdfs/appd_k.pdf).
38. de Bruijn, J.; Busser, F.; Seinen, W.; Hermens, J., *Env. Tox. Chem.*, 1989, 8, 499.
39. Briggs, G.G., *J. Agric. Food Chem.*, 1981, 29, 1050.
40. Mirrlees, M.S.; Moulton, S.J.; Murphy, C.T.; Taylor, P.J., *J. Med. Chem.*, 1976, 19, 615.
41. Wasik, S.P.; Tewari, Y.B.; Miller, M.M.; Purnell, J.H., *J. Res. Nat. Bur. Stnd.*, 1982, 87, 311.
42. Veith, G.D.; Austin, N.M.; Morris, R.T., *Water Res.*, 1979, 13, 43.
43. Chin, Y.P.; Weber, W.J. Jr.; Voice, T.C., *Water Res.*, 1986, 20, 1443.
44. Veith, G.D.; Morris, R.T. EPA-600-3-78-049, A Rapid Method for Estimating log P for Organic Chemicals, 1978.
45. Gold, L.S.; Zeiger, E., Cancer Potency Database (CPDB) (http://potency.berkeley.edu/pdfs/ChemicalTable.pdf) (http://potency.berkeley.edu/cpdb.html).
46. Peto; R.; Pike, M.C.; Bernstein, L.; Gold, L.S.; Ames, B.N., *Env. Health Persp.*. 1984, 58, 1.
47. Sawyer, C.; Peto, R., Bernstein, L.; Pike, M.C., *Biometrics*, 1984, 40, 27.
48. Gold, L.S.; Slone, T.H.; Bernstein, L., *Environ. Health Perspect.*, 1989, 79, 259.
49. Gold, L. S.; Slone, T. H.; Ames, B. N., *Drug Metabol. Rev.*, 1998, 30, 359.
50. Gold, L.S.; Slone, T.H.; Ames, B.N., Overview of analyses of the carcinogenic potency database. In: Gold, L.S., and Zeiger, E., Eds. *Handbook of Carcinogenic Potency and Genotoxicity Databases*. CRC Press: Boca Raton, FL:, 1997, pp. 661–685.
51. Mackay, D.; Arnot, J. A.; Webster, E.; Reid, L. The Evolution and Future of Environmental Fugacity Models. In: Devillers, J., Ed. *Ecotoxicology Modeling, Emerging Topics in Ecotoxicology: Principles, Approaches, and Perspectives.* Springer Science: New York, 2009, pp. 355–375.
52. Mackay, D.; Arnot, J. A., *J. Chem. Eng. Data*, 2011, 56, 1348.
53. Mackay, D., *Environ. Sci. Technol.*, 1979, 13, 1218.
54. Mackay, D.; Paterson, S., *Environ. Sci. Technol.*, 1981, 15, 1006.
55. Mackay, D.; Paterson, S., *Environ. Sci. Technol.*, 1982, 16, 654A.
56. Guinée, J.B., ed., *Handbook on Life Cycle Assessment*, Kluwer Academic Publishers: Dordrecht, 2002, Part 2b, p. 167.
57. Guinée, J.B., ed., *Handbook on Life Cycle Assessment*, Kluwer Academic Publishers: Dordrecht, 2002, Part 2b, p. 344.
58. Allen, D.T.; Shonnard, D.R., *Green Engineering: Environmentally Conscious Design Of Chemical Processes*, Prentice Hall: Upper Saddle River, 2002, p. 379.
59. Andraos, J., *Org. Process Res. Dev.*, 2012, 16, 1482.
60. Lide, D.R. (Ed.), *CRC Handbook of Chemistry and Physics*, 87th ed., CRC Press: Boca Raton, 2006, pp. 42–51.

61. Lange, N.A., *Handbook of Chemistry*, 10th ed., McGraw-Hill Book Co., Inc.: New York, 1961, pp. 1198–1204.
62. www.zirchrom.com/organic.htm.
63. Perrin, D.D., *Ionization Constants of Inorganic Acids and Bases in Aqueous Solution*, 2nd ed., Pergamon Press: Oxford, 1982.
64. Haynes, W.M., ed., *CRC Handbook of Chemistry and Physics*, 92nd ed., CRC Press: Boca Raton, 2011, pp. 92–93.
65. Sober, H.A., Ed., *CRC Handbook of Biochemistry*, The Chemical Rubber Co.: Cleveland, 1970, pp. J187–J226.
66. Kortüm, G.; Vogel, W.; Andrussov, K., *Dissociation Constants of Organic Acids in Aqueous Solution*, Butterworths: London, 1961.
67. Perrin, D.D. *Dissociation Constants of Organic Bases in Aqueous Solution*, Butterworths: London, 1961.
68. Perrin, D.D., *Dissociation Constants of Organic Bases in Aqueous Solution (Supplement 1972)*, Butterworths: London, 1965.
69. Gordon, A.J.; Ford, R.A., *The Chemist's Companion*, Wiley: Hoboken, 1972, pp. 58–63.
70. Shiu, W.Y.; Mackay, D., *J. Phys. Chem. Ref. Data*, 1986, 15, 911.
71. Haynes, W.M. (Ed.), *CRC Handbook of Chemistry and Physics*, 91st ed., CRC Press: Boca Raton, 2009, pp. 43–47 (given as log P values).
72. Veith, G.D.; DeFoe, D.L.; Bergstedt, B.V., *J. Fisheries Res. Board Canada*, 1979, 36, 1040.
73. Meylan, W.M.; Howard, P.H.; Boethling, R.S.; Aronson, D.; Printup, H.; Gouchie, S., *Env. Toxicol. Chem.*, 1999, 18, 664.
74. Chiou, C.T. *Env. Sci. Tech.*, 1985, 19, 57.
75. Allen, D.T.; Shonnard, D.R., *Green Engineering: Environmentally Conscious Design of Chemical Processes*, Prentice Hall: Upper Saddle River, 2002, p. 108.
76. Veith, G.D.; Kosian, P., Mackay, D., ed., *Physical Behaviour of PCBs in the Great Lakes*, Ann Arbor Science, 1983, pp. 269–282.
77. Ernst, W., *Chemosphere,* 1977, 11, 731.
78. Gobas, F.A.P.C.; Morrison, H.A. Bioconcentration and Biomagnification in the Aquatic Environment. In Boethling, R.S; Mackay, D. (Eds.), *Handbook of Property Estimation Methods for Chemicals: Environmental and Health Sciences*, Lewis Publishers: Boca Raton, 2000, Chapter 9, pp. 189–232.
79. Shi, L.M.; Fang, H.; Tong, W.; Wu, J.; Perkins, R.; Blair, R.M.; Branham, W.S.; Dial, S.L.; Moland, C.L.; Sheehan, D.M., *J. Chem. Inf. Comput. Sci.*, 2001, 41, 186.
80. Whaley, D.A.; Keyes, D.; Khorrami, B., *Drug Chem. Toxicol.*, 2001, 24, 359.
81. Allen, D.T.; Shonnard, D.R., *Green Engineering: Environmentally Conscious Design of Chemical Processes*, Prentice Hall: Upper Saddle River, 2002, p. 377; Appendix D.
82. Allen, D.T.; Shonnard, D.R., *Green Engineering: Environmentally Conscious Design of Chemical Processes*, Prentice Hall: Upper Saddle River, 2002, p. 388.
83. USEPA, 2009. Integrated Risk Information System. Washington, DC: US Environmental Protection Agency, National Center for Environmental Assessment. Available from: www.epa.gov/iris/.
84. Wang, N.C.Y.; Venkatapathy, R.; Bruce, R.M.; Moudgal, C., *Reg. Tox. Pharmacol.*, 2011, 59, 215.
85. Venkatapathy, R.; Wang, C.Y.; Bruce, R.M.; Moudgal, C., *Tox. Appl. Pharmacol.*, 2009, 234, 209.
86. US EPA, Health Effects Assessment Summary Tables (HEAST), EPA-540-R-97-036, 1997.
87. Allen, D.T.; Shonnard, D.R., *Green Engineering: Environmentally Conscious Design of Chemical Processes*, Prentice Hall: Upper Saddle River, 2002, p. 378; Appendix D.

88. Wuebbles, D.J. *Rep. UCID-18924* (Lawrence Livermore National Laboratory, Livermore, 1981).
89. Fisher, D.A.; Hales, C.H.; Filkin, D.L.; Ko, M.K.W.; Sze, N.D.; Connell, P.S.; Wuebbles, D.J.; Isaksen, I.S.A.; Stordal, F., *Nature*, 1990, 344, 508.
90. Solomon, S.; Albritton, D.L., *Nature,* 1992, 357, 33.
91. Guinée, J.B., ed., *Handbook on Life Cycle Assessment*, Kluwer Academic Publishers: Dordrecht, 2002, Part 2b, p. 188.
92. Solomon, S.; Mills, M.; Heidt, L.E.; Pollock, W.H.; Tuck, A.F., *J. Geophys. Res.*, 1992, 97, 825.
93. Nimitz, J.S.; Skaggs, S.R., *Env. Sci. Tech.,* 1992, 26, 739.
94. Japar, S.M.; Wallington, T.J.; Rudy, S.J.; Chang, T.Y. *Env. Sci. Tech.*, 1991, 25, 415.
95. Bufalini, J.J.; Dodge, M.C., *Env. Sci. Tech.*, 1983, 17, 308.
96. Pyle, J.A.; Solomon, S.; Wuebbles, D.; Zvenigorodsky, S. Ozone depletion and chlorine loading potentials. In: *Scientific Assessment of Ozone Depletion: 1991.* World Meteorological Organization Global Ozone Research and Monitoring Project--Report no. Chapter 6, Geneva: World Meteorological Organization, 1992. (www.ciesin.org/docs/011-551/permit.html).
97. US EPA, Ozone Layer Protection. (www.epa.gov/ozone/science/ods/classone.html).
98. US EPA, Ozone Layer Protection. (www.epa.gov/ozone/science/ods/classtwo.html).
99. IPCC Third Assessment Report - Climate Change 2001, Chapter 6. (www.grida.no/climate/ipcc_tar/wg1/248.htm).
100. Briggs, G.G.; Tinker, P.B.; Graham-Bryce, I.J., *Phil. Trans. Biol. Sci.*, 1990, 329, 375.
101. Howard, P.H. Biodegradation. In Boethling, R.S; Mackay, D. (Eds.) *Handbook of Property Estimation Methods for Chemicals: Environmental and Health Sciences*, Lewis Publishers: Boca Raton, 2000, Chapter 12, pp. 281–310.
102. Waskom, R.M., Best Management Practices for Agricultural Pesticide Use, Colorado State University Cooperative Extension 1995. (www.npscolorado.com/xcm177.pdf).
103. Johnson, W.S.; McKie, P.; Moses, C., Pesticide Adsorption and Half-life in Soils, U Nevada Cooperative Extension, 1998. (http://w3.ualg.pt/~lnunes/Textosdeapoio/Half-life of pesticides.pdf).
104. Menzie, C.M., *Ann. Rev. Entomol.*, 1972, 17, 199.
105. Pimentel, D., ed., *CRC Handbook of Pest Management in Agriculture*, Vol. 3, CRC Press: Boca Raton, 1981, p. 149.
106. Knisel, W.G., ed., Groundwater Loading Effects of Agricultural Management System (GLEAMS), Version 2.10, U.S. Dept. Agriculture, 1996. (www.tifton.uga.edu/sewrl/Gleams/glm30pst.pdf).
107. Jury, W.A.; Focht, D.D.; Farmer, W.J., *J. Environ. Qual.*, 1987, 16, 422.
108. Jury, W.A.; Spencer, W.F.; Farmer, W.J., *J. Environ. Qual.*, 1984, 13, 573.
109. Wauchope, R.D.; Buttler, T.M.; Hornsby, A.G.; Beckers, P.W.M.A.; Burt, J.P., *Rev. Environ. Contamin. Toxicol.*, 1992, 123, 1.
110. Guinée, J.B., Ed., *Handbook on Life Cycle Assessment*, Kluwer Academic Publishers: Dordrecht, 2002, Part 2b, pp. 332.
111. Heijungs, R., Ed., *Environmental Life Cycle Assessment of Products*, Centre of Environmental Science: Leiden, 1992, p. 73.
112. Derwent, R.G.; Jenkin, M.E.; Saunders, S.M., *Atmos. Environ.*, 1996, 30, 181.
113. Derwent, R.G.; Jenkin; M.E.; Saunders, S.M.; Pilling, M.J. *Atmos. Environ.,* 1998, 32, 2429.
114. Jenkin, M.E.; Hayman, G.D., *Atmos. Environ.*, 1999, 33, 1275.
115. Guinée, J.B., Ed., *Handbook on Life Cycle Assessment*, Kluwer Academic Publishers: Dordrecht, 2002, Part 2b, p. 335.
116. Allen, D.T.; Shonnard, D.R., *Green Engineering: Environmentally Conscious Design of Chemical Processes*, Prentice Hall: Upper Saddle River, 2002, p. 380.

117. Carter, W.P.L., *J. Air Waste Management Assoc.,* 1994, 44, 881.
118. Groggins, P.H., *Unit Operations in Organic Syntheses,* 5th ed., McGraw-Hill Book Co.: London, pp. 180–182.
119. Andraos, J.; Ballerini, E.; Vaccaro, L. *Green Chem.,* 2015, 17, 913.
120. Banerjee, S.; Baughman, G.L., *Env. Sci. Tech.,* 1991, 25, 536.
121. Dimitrov, S.D.; Dimitrova, N.C.; Walker, J.D.; Veith, G.D.; Mekenyan, O.G., *QSAR Comb. Sci.,* 2003, 22, 58.
122. Dimitrov, S.; Breton, R.; MacDonald, D.; Walker, J.D.; Mekenyan, O., *SAR QSAR Env. Res.,* 2002, 13, 445.
123. Voutsas, E.; Magoulas, K.; Tassios, D., *Chemosphere,* 2002, 48, 645.
124. Meylan, W.M.; Howard, P.H.; Boethling, R.S.; Aronson, D.; Printup, H.; Gouchie, S., *Env. Tox. Chem.,* 1999, 18, 664.
125. Neely, W.B.; Branson, D.R.; Blau, G.E. *Env. Sci. Tech.,* 1974, 8, 1113.
126. van der Heijden, S.A.; Jonker, M.T.O. *Env. Sci. Tech.,* 2009, 43, 8854.
127. Chiou, C.T., *Env. Sci. Tech.,* 1985, 19, 57.
128. Chiou, C.T.; Freed, V.H.; Schmedding, D.W.; Kohnert, R.L., *Env. Sci. Tech.,* 1977, 11, 475.
129. Miller, M.M.; Wasik, S.P.; Huang, G.L.; Shiu, W.Y.; Mackay, D., *Env. Sci. Tech.,* 1985, 19, 522.
130. Niimi, A.J., *Water Res.,* 1991, 25, 1515.
131. Chiou, C.T.; Schmedding, D.W.; Manes, M., *Env. Sci. Tech.,* 1982, 16, 4.
132. Meylan, W.M.; Howard, P.H. *J. Pharm. Sci.,* 1995, 84, 83.
133. Leo, A.J., *Chem. Rev.,* 1993, 93, 1281.
134. Chiou, C.T.; Schmedding, D.W.; Manes, M. *Env. Sci. Tech.,* 2005, 35, 8840.
135. Spafiu, F.; Mischie, A.; Ionita, P.; Beteringhe, A.; Constantinescu, Balaban, A.T. *ARKIVOC* 2009, 10, 174.
136. Lin, S.T.; Sandler, S.I., *Ind. Eng. Chem. Res.,* 1999, 38, 408.
137. Tetko, I.V.; Tanchuk, V.Y.; Villa, A.E.P., *J. Chem. Inf. Comput. Sci.,* 2001, 41, 1407.
138. Tetko, I.V.; Tanchuk, V.Y.; Kasheva, T.N.; Villa, A.E. *J. Chem. Inf. Comput. Sci.,* 2001, 41, 1488.
139. Poole, S.K.; Poole, C.F., *J. Chromatography B,* 2003, 797, 3.
140. On-line Lipophilicity/Aqueous Solubility Calculation Software using PHYSPROP database, Virtual Computational Chemistry Laboratory. (www.vcclab.org/lab/alogps).
141. Tetko, I.V.; Tanchuk, V.Y., *J. Chem. Inf. Comput. Sci.,* 2002, 42, 1136.
142. Tetko, I.V.; Poda, G.I., *J. Med. Chem.,* 2004, 47, 5601.
143. Tetko, I.V.; Bruneau, P. *J Pharm Sci,* 2004, 93, 3103.
144. Ruelle, P., *Chemosphere,* 2000, 40, 457.
145. Brooke, D.N.; Dobbs, A.J.; Williams, N., *Ecotox. Env. Safety,* 1986, 11, 251.
146. Cheng, T.; Zhao, Y.; Li, X.; Lin, F.; Xu, Y.; Zhang, X.; Li, Y.; Wang, R.; Lai, L., *J. Chem. Inf. Model.,* 2007, 47, 2140.
147. Suzuki, T.; Kudo, Y., *J. Computer Aided Mol. Design,* 1990, *4,* 155.
148. Tehrany, E.A.; Fournier, F.; Desobry, S., *J. Food Eng.,* 2004, 64, 315.
149. Eisfeld, W.; Maurer, G. *J. Phys. Chem. B,* 1999, 103, 5716.
150. Haeberlein, M.; Brinck, T., *J. Chem. Soc. Perkin Trans.,* 2 1997, 289.
151. Wienke, G.; Gmehling, J., *Tox. Env. Chem.,* 1998, 65, 57.
152. Moriguchi, I.; Hirono, S.; Liu, Q.; Nakagome, I.; Matsushita, Y., *Chem. Pharm. Bull.,* 1992, 40, 127.
153. Ghose, A.K.; Crippen, G.M., *J. Chem. Inf. Comput. Sci.,* 1987, 27, 21 (used by ChemDraw).
154. Viswanadhan, V.N.; Ghose, A.K.; Revankar, G.R.; Robins, R.K. *J. Chem. Inf. Comput. Sci.,* 1989, 29, 163 (used by ChemDraw).
155. Broto, P.; Moreau, G.; Vandycke, C., *Eur. J. Med. Chem.,* 1984, 19, 71 (used by ChemDraw).

156. Karickhoff, S.W., *Chemosphere*, 1981, 10, 833.
157. Leo, A. Octanol/Water Partition Coefficient. In Boethling, R.S; Mackay, D. (Eds.) *Handbook of Property Estimation Methods for Chemicals: Environmental and Health Sciences*, Lewis Publishers: Boca Raton, 2000, Chapter 5, pp. 89–114.
158. Chiou, C.T.; Porter, P.E.; Schmedding, D.W., *Env. Sci. Tech.*, 1983, 17, 227.
159. Chiou, C.T.; Peters, L.J.; Freed, V.H. *Science*, 1979, 206, 831.
160. Chiou, C.T., *Partition and Adsorption of Organic Contaminants in Environmental Systems*, Wiley: Hoboken, 2002.
161. Seth, R.; Mackay, D.; Munthe, J., *Env. Sci. Tech.*, 1999, 33, 2390.
162. Meylan, W.; Howard, P.H.; Boethling, R.S., *Env. Sci. Tech.*, 1992, 26, 1560.
163. Endo, S.; Grathwohl, P.; Haderlein, S.B.; Schmidt, T.C., *Env. Sci. Tech.*, 2009, 43, 3094.
164. Chin, Y.P.; Peven, C.S.; Weber, W.J. Jr., *Water Res.*, 1988, 22, 873.
165. Meylan, W.M.; Howard, P.H., *Env. Toxicol. Chem.*, 1991, 10, 1283.
166. (a) Sage, M.L.; Sage, G.W. Vapor Pressure. In Boethling, R.S; Mackay, D. (Eds.) *Handbook of Property Estimation Methods for Chemicals: Environmental and Health Sciences*, Lewis Publishers: Boca Raton, 2000, Chapter 3, pp. 53–68; (b) Mackay, D. Solubility in Water. In Boethling, R.S; Mackay, D. (Eds.) *Handbook of Property Estimation Methods for Chemicals: Environmental and Health Sciences*, Lewis Publishers: Boca Raton, 2000, Chapter 7, pp. 125–140.
167. Dearden, J.C.; Schüürmann, G. *Env. Tox. Chem.*, 2003, 22, 1755.
168. Russell, C.J.; Dixon, S.L.; Jurs, P.C. *Anal. Chem.*, 1992, 64, 1350.
169. Suzuki, T.; Ohtaguchi, K.; Koide, K., *Comput. Chem.*, 1992, 16, 41.
170. Nirmalakhandan, N.N.; Speece, R.E., *Env. Sci. Tech.*, 1988, 22, 1349.
171. Daubert, T.E.; Danner, R.P. *Physical and Thermodynamic Properties of Pure Chemicals: Data Compilation*, American Institute of Chemical Engineering: New York, 1989, Vol. 1–4.
172. Boublik, T.; Fried, V.; Hala, E. *The Vapour Pressures of Pure Substances*, Elsevier: Amsterdam, 1984.
173. Burkhard, L.P.; Andren, A.W.; Armstrong, D.E. *Env. Sci. Tech.* 1985, 19, 500.
174. Yalkowsky, S.H.; Banerjee, S. *Aqueous Solubility: Methods of Estimation for Organic Compounds*, Marcel Dekker: New York, 1992.
175. Yalkowsky, S.H.; Valvani, S.C.; Roseman, T.J.*J. Pharm. Sci.*, 1983, 72, 866.
176. Tewari, Y.B.; Miller, M.M.; Wasik, S.P. *J. Res. Nat. Bur. Stnd.* 1982, 87, 155.
177. Pontolillo, J.; Eganhouse, R.P. Water-Resources Investigations Report 01-4201, US Geological Survey, 2001. (http://pubs.water.usgs.gov/wri01-4201/).
178. Kollig H. *Toxicol. Environ. Chem.* 1988, 17, 287.
179. Kollig, H.P.; Kitchens, B.E. *Toxicol. Env. Chem.* 1990, 28, 95.
180. Reichhardt T. *Nature* 2002, 416, 249.
181. Eganhouse, R.P.; Pontolillo, J. *SETAC Globe: Learned Discourse* 2002, *3*, 34.
182. Eganhouse, R.P.; Pontolillo, J. *SETAC Globe: Learned Discourse* (Online Supplement) January-February 2003.
183. Burkhard, L.P.; Mount, D.R. *SETAC Globe: Learned Discourse* 2002, *3*, 40.
184. Heller, S.R.; Bigwood, D.W.; May, W.E. *J. Chem. Int. Comput. Sci.* 1994, 34, 627.
185. Isnard, P.; Lambert, S. *Chemosphere* 1989, 18, 1837.
186. Banerjee, S.; Yalkowsky, S.H.; Valvani, S.C. *Env. Sci. Tech.* 1980, 14, 1227.
187. Grathwohl, P. *Env. Sci. Tech.* 1990, 24, 1687.
188. Tong, W.D.; Hong, H.X.; Xie, Q.; Shi, L.M.; Fang, H.; Perkins, R. *Curr. Computer-Aided Drug Design* 2005, *1*, 195.
189. Kar, S.; Roy, K. *J. Ind. Chem. Soc.* 2010, 87, 1455.
190. Karickhoff, S.W.; McDaniel, V.K.; Melton, C.; Vellino, A.N.; Nute, D.E.; Carreira, L.A. *Env. Tox. Chem.* 1991, 10, 1405.
191. Peijenburg, W.J.G.M. *Pure Appl. Chem.* 1991, 63, 1667.

192. Kollig, H.P. *Toxicol. Env. Chem.* 1988, 17, 287.
193. Bunnett, J.F. *Acc. Chem. Res.* 1985, 18, 291.
194. Sabine, J.R. *BioScience* 1985, 35, 358.
195. Mill, T.; Walton, B.T. *Env. Toxicol. Chem.* 1987, 6, 161.
196. Boethling, R.S.; Howard, P.H.; Meylan, W.M. *Env. Toxicol. Chem.* 2004, 23, 2290.
197. Kollig, H.P. *Toxicol. Env. Chem.* 1990, 25, 171.
198. Kollig, H.P.; Kitchens, B.E. *Toxicol. Env. Chem.* 1990, 28, 95.
199. Greene, N.; Pennie, W. *Toxicol. Res.* 2015, *4*, 1159.
200. Faith, W. L.; Keyes, D. B.; Clark, R. L. *Industrial Chemicals*, 3rd ed., Wiley: Hoboken, 1966, p. 101.
201. Jimenez-Gonzalez, C.; Constable, D.J.C. *Green Chemistry and Engineering: A Practical Design Approach*, Wiley: Hoboken, 2011, pp. 485–487.
202. Jimenez-Gonzalez, C.; Constable, D.J.C. *Green Chemistry and Engineering: A Practical Design Approach*, Wiley: Hoboken, 2011, pp. 487–488.

4 Safety-Hazard Impact Metrics

This chapter introduces safety-hazard impact metrics in a similar fashion as the environmental impact metrics discussed in Chapter 3. Table 4.1 summarizes the list of safety-hazard potentials considered along with their equations and reference states.

4.1 PARAMETERS

4.1.1 Terms, Definitions, and Examples

Flash Point

The flash point of a volatile material is the lowest temperature at which it can vaporize to form an ignitable mixture in air; it requires an ignition source. It is an empirical measurement that is made according to two methods. In the open cup method, a sample is contained in an open cup which is heated and the flame is brought over surface. The measured flash point will actually vary with the height of the flame above the liquid surface. In the closed cup method, cups are sealed with a lid through which the ignition source can be introduced. This method gives lower values for the flash point than open cup (5 to 10°C lower). It gives a better approximation to the temperature at which the vapor pressure reaches the lower flammable limit. For the purposes of determining the flash point, potential data from the closed cup method are used.

See *potential: flash point*.

Hazard

In the context of life cycle assessment or analysis of a given chemical reaction or synthesis plan, a hazard is the property of a chemical to cause harm or damage to the health of occupational workers and others upon exposure to that chemical, or to cause harm or damage to the environment upon release of that chemical into one or more of the four environmental compartments: air, water, soil, and sediment.

Hazard Waste Index[1]

The Gupta–Babu hazard waste index (HWI) follows the National Fire Protection Association (NFPA) scoring scale ranging between 0 and 4 for flammability and reactivity contributors, the Dow Chemical rating for corrosivity, and threshold limit values (TLV) for the toxicity rating. All four components are mass weighted with respect to the mass fraction of the total waste associated with a given chemical. However, the HWI is not treated as a single number derived as a sum or product of contributing factors; rather, it is depicted as an overall code. Since HWI is not a number, it cannot be used for ranking purposes. Furthermore, HWI does not take into account reaction yields, pressure and temperature conditions, and explosiveness potential.

TABLE 4.1
Summary of Potentials Associated with Safety And Hazard

Potential	Equation		Reference
Reaction pressure	$(\text{RPHI})_j = \exp\left[-\dfrac{|P_j - P_{\text{ambient}}|}{P_{\text{ambient}}}\right]$	(4.1)	1 atm
Reaction temperature	$(\text{RTHI})_j = \exp\left[-\dfrac{|T_j - T_{\text{ambient}}|}{T_{\text{ambient}}}\right]$	(4.2)	25°C
Corrosiveness as a gas/vapor	$(\text{CGP})_j = \dfrac{\text{LC}_{50,\text{diethyl-ether}}}{\text{LC}_{50,j}}$ LC50(inhalation)	(4.5)	Diethyl ether
Corrosiveness as a liquid/solid	$(\text{CLP})_j = \dfrac{\text{LD}_{50,\text{toluene}}}{\text{LD}_{50,j}}$ LD50(dermal)	(4.6)	Toluene
Dermal absorption	$(\text{DAP})_j = \dfrac{(\text{Fl}^*)_j}{(\text{Fl}^*)_{\text{toluene}}}$ critical flux for dermal absorption in mg/cm²/h	(4.7)	Toluene
Explosive strength	$(\text{XSP})_j = \dfrac{(\text{LBT})_j}{(\text{LBT})_{\text{RDX}}}$ Trauzl lead block test (volume change caused by an explosion)	(4.8)	RDX
Explosiveness as a vapor	$(\text{XVP})_j = \dfrac{(\text{LEL})_{\text{diethyl-ether}}}{(\text{LEL})_j}$ lower explosion limit	(4.9)	Diethyl ether
Flammability	$(\text{FP})_j = \dfrac{(\text{FLP})_{\text{diethyl-ether}}}{(\text{FLP})_j}$ flashpoint in degrees K	(4.10)	Diethyl ether
Hydrogen generation	$(\text{HGP})_j = \dfrac{M_{H_2}/(\text{MW})_j}{1/23}$ mass of H_2 generated	(4.11)	Sodium metal
Impact sensitivity	$(\text{ISP})_j = \dfrac{(\text{IS})_{\text{RDX}}}{(\text{IS})_j}$ impact sensitivity (minimum force of impact to cause an explosion)	(4.12)	RDX

(Continued)

Safety-Hazard Impact Metrics

TABLE 4.1 (CONTINUED)
Summary of Potentials Associated with Safety And Hazard

Potential	Equation		Reference
Occupational exposure limit (NOISH)	$(\text{OELP})_j = \dfrac{(\text{OEL})_{\text{toluene}}}{(\text{OEL})_j}$ occupational exposure limit in mmol/m³	(4.14)	Toluene
Oxygen balance (combustion or calcination)	$(\text{OBP})_j = \dfrac{(\text{OB})_j}{(\text{OB})_{\text{methane}}}$ $= \dfrac{16\left[n_{\text{substrate}} - n_{\text{products}}\right]/(\text{MW})j}{-4}$ minimum mass of O_2 consumed $n_{\text{substrate}} < n_{\text{products}}$	(4.15a)	Methane
Oxygen balance (oxidation)	$(\text{OBP})_j = \dfrac{(\text{OB})_j}{(\text{OB})_{\text{hydrogen-peroxide}}}$ $= \dfrac{16\left[n_{\text{substrate}} - n_{\text{products}}\right]/(\text{MW})j}{0.471}$ maximum mass of O_2 generated $n_{\text{substrate}} > n_{\text{products}}$	(4.15b)	Hydrogen peroxide
Risk phrase	$(\text{RRP})_j = \dfrac{\log\left[10^n \prod_{k=1}^{n}(Q-\text{Eissen})_{k,j}\right]}{\log[300]}$ $= \dfrac{n + \log\left[\prod_{k=1}^{n}(Q-\text{Eissen})_{k,j}\right]}{2.477}$ n = total number of risk phrases Q-Eissen factors ranging from 0 (no risk) to 5 (maximum risk)	(4.18)	Toluene
Skin dose	$(\text{SDP})_j = \dfrac{(K_p)_j (\text{WS})_j}{(K_p)_{\text{toluene}} (\text{WS})_{\text{toluene}}}$, WS = water solubility (mg/mL) K_p = transdermal permeation coefficient in cm/h	(4.19)	Toluene

Inherent Safety Index (ISI)[2]

The inherent safety index is based on an arbitrary penalty point scoring system with no mass weighting of scores and a limited number of parameters.

It is based on simple chemical scores based on inventory, flash point, boiling point, flammability, upper and lower exposure limits, explosiveness, and toxicity;

and simple process scores based on reaction temperature, reaction pressure, and % yield.

LD50(dermal)

LD50(dermal) is the lethal dose to kill 50% of population upon dermal exposure to a chemical compound (penetration through skin) expressed in units of mg kg^{-1} or ppm. The reference animal is the white rabbit *Oryctolagus cuniculus*. This is a measure of acute toxicity on the basis of a single dose exposure.

Reference compound: toluene; LD(dermal) = 2250 mg/kg.
Sources to find reliable LD50(dermal) values: see references [3 and 4].
See *potential: dermal absorption*.

Lower Exposure Limit (LEL)

The lowest concentration (expressed as a percentage) of a gas or a vapor in air capable of producing a flash of fire in the presence of an ignition source (arc, flame, heat).

Reference compound: diethyl ether; 1.7%
Sources to find LEL values: see references [4–7].
See *potential: explosive vapor*.

Oxygen Balance (OB)[8,9]

Oxygen balance covers two kinds of reactions involving oxygen: (1) combustible chemicals are parametrized by the minimum amount of oxygen consumed in the combustion process (negative oxygen balance); and (2) oxidizing chemicals are parametrized by the maximum amount of oxygen available for a complete oxidation process (positive oxygen balance). Higher positive oxygen balances correlate with more positive reduction potentials and higher oxygen weight percent content suggesting that stronger oxidizing agents generally have larger OB values.

See *potential: oxygen balance*.

Reaction Pressure Hazard Index[9]

The reaction pressure hazard index is given by Equation (4.1).

$$\text{RPHI} = \exp\left[-\frac{|P - P_{\text{ambient}}|}{P_{\text{ambient}}}\right], 0 < \text{RPHI} < 1 \quad (4.1)$$

where $P_{\text{ambient}} = 1$ atm and P is the pressure of the reaction (atm). If $P = 1$ atm, then RPHI = 1, which corresponds to a non-hazardous situation. If RPHI = 0, then the reaction poses a very high pressure hazard. Figure 4.1 illustrates graphically Equation (4.1).

Example 4.1

Determine RPHI for the following reaction pressures: 5 atm, 20 atm, 200 atm, 1000 atm.

Safety-Hazard Impact Metrics

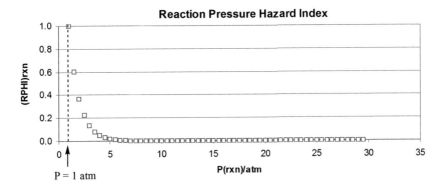

FIGURE 4.1. Graph of Equation (4.1).

SOLUTION

For $P = 5$ atm, RPHI $= \exp(-\mathrm{abs}(5-1)/1) = 0.0183$
For $P = 20$ atm, RPHI $= \exp(-\mathrm{abs}(20-1)/1) = 5.6\mathrm{E}{-9}$
For $P = 200$ atm, RPHI $= \exp(-\mathrm{abs}(200-1)/1) = 3.8\mathrm{E}{-87}$
For $P = 1000$ atm, RPHI $= \exp(-\mathrm{abs}(1000-1)/1) = 0$

Reaction Temperature Hazard Index[9]

The reaction temperature hazard index is given by Equation (4.2).

$$\mathrm{RTHI} = \exp\left[-\frac{|T - T_{\mathrm{ambient}}|}{T_{\mathrm{ambient}}}\right], \quad 0 < \mathrm{RTHI} < 1 \qquad (4.2)$$

where $T_{\mathrm{ambient}} = 298$ K and T is the temperature of the reaction (K). If $T = 298$ K, then RTHI $= 1$, which corresponds to a non-hazardous situation. If RTHI $= 0$ then the reaction poses a very high temperature hazard. Figure 4.2 illustrates graphically Equation (4.2).

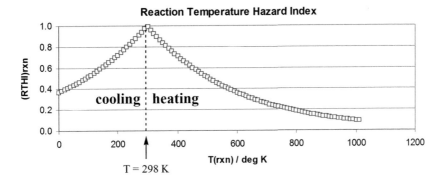

FIGURE 4.2. Graph of Equation (4.2).

Example 4.2

Determine RTHI for the following reaction temperatures: 200 K, 500 K, 800 K, 1000 K, 2000 K, and 5000 K.

SOLUTION

For T = 200 K, RTHI = exp(−abs(200−298)/298) = 0.7197
For T = 500 K, RTHI = exp(−abs(500−298)/298) = 0.5077
For T = 800 K, RTHI = exp(−abs(800−298)/298) = 0.1855
For T = 1000 K, RTHI = exp(−abs(1000−298)/298) = 0.0948
For T = 2000 K, RTHI = exp(−abs(2000−298)/298) = 0.00331
For T = 5000 K, RTHI = exp(−abs(5000−298)/298) = 1.4E−7

Risk

In the context of life cycle assessment, risk is the potential or probability of an adverse event or outcome occurring. If a chemical has a hazardous property, then it has the potential to do harm to occupational workers and others exposed to it or to do harm to the environment if it is released in one of the four compartments: air, water, soil, and sediment. The likelihood of an adverse effect is reduced considerably if good laboratory practices, appropriate personnel training, good operating equipment, and appropriate safety procedures are followed and used in the handling of the hazardous chemical.

Risk Index

Risk index for a particular chemical is the product of its potential to do a particular harm and its mass.
See *potential*.

Threshold Limit Value

The threshold limit value (TLV) of a chemical substance is a level to which it is believed an occupational worker can be exposed day after day for a working lifetime without adverse effects.

TLVs issued by the American Conference of Governmental Industrial Hygienists are the most widely accepted occupational exposure limits both in the United States and most other countries.

The TLV for chemical substances is defined as a concentration in air, typically for inhalation or skin exposure. Its units are in parts per million (ppm) for gases and in milligrams per cubic meter (mg/m^3) for particulates such as dust, smoke, and mist. The basic formula for converting between ppm and mg/m^3 for gases is ppm = (mg/m^3)*24.45/molecular weight. This formula is not applicable to airborne particles.

Three types of TLVs for chemical substances are defined:

1. Threshold limit value. Time weighted average (TLV-TWA): average exposure on the basis of an 8h/day, 40h/week work schedule
2. Threshold limit value. Short-term exposure limit (TLV-STEL): spot exposure for a duration of 15 min that cannot be repeated more than 4 times per day with at least 60 min between exposure periods

Safety-Hazard Impact Metrics

3. Threshold limit value: Ceiling limit (TLV-C): absolute exposure limit that should not be exceeded at any time.

Time Weighted Average
A threshold limit value that corresponds to average exposure to a chemical on the basis of an 8-hour day and 40-hour week work schedule for an occupational worker exposed to that chemical.
See *threshold limit value*.

Water Solubility
The maximum mass of a chemical substance (g) that can dissolve in a given volume of water (1 L) of water. The usual units are g/L, mol/L, mg/mL, mg/cm³.
Sources for obtaining water solubility data: see references [10–16].
In cases when no experimental water solubility data are available they may be estimated
using the Mackay Equation (4.3a).[17]

$$WS = \frac{(MW)10^{3.25-\log K_{ow}}}{1000} \tag{4.3a}$$

where $\log K_{ow}$ is the octanol–water partition coefficient, and WS is in units of mg/mL or, the Suzuki Equations (4.3b) and (4.3c)[18]

$$-\log(WS) = 1.05\log K_{ow} - 0.515 + 0.00956(mp - 25), \tag{4.3b}$$

for substances with melting points higher than 25°C
and

$$-\log(WS) = 1.05\log K_{ow} - 0.515, \tag{4.3c}$$

for substances with melting points less than 25°C
where mp is melting point in degrees celsius; WS is in units of mol/L.
For salts, X_nY_m, their aqueous solubility may be estimated from solubility product constants, K_{sp} values according to Equation (4.4).

$$WS = \left(\frac{K_{sp}}{n^n m^m}\right)^{1/(n+m)} \tag{4.4}$$

where WS is in units of mol/L.
Sources for reliable K_{sp} data: see references [19–24].

Example 4.3[19]

Determine water solubilities in g/100 mL of the following salts based on the given K_{sp} values: (a) aluminum hydroxide (1.90×10^{-33}); (b) zinc sulfide (1.2×10^{-23}); (c) magnesium ammonium phosphate (2.5×10^{-13}). In each case, write out the corresponding equilibrium equation.

SOLUTION

(a)

$$Al(OH)_3 \rightleftharpoons Al^{+3} + 3OH^-$$

$$K = \frac{[Al^{+3}][3OH^-]^3}{[Al(OH)_3]} \Rightarrow K_{sp} = [Al^{+3}][3OH^-]^3$$

$n = 1, m = 3$
From Equation (4.4), we have

$$WS = \left(\frac{1.90 \times 10^{-33}}{1^1 3^3}\right)^{1/(1+3)} = \left(\frac{1.90 \times 10^{-33}}{27}\right)^{1/4} = 2.896 \times 10^{-9} \text{mol/L}$$

Since the molecular weight of aluminum hydroxide is 78 g/mol and 1 L = 1000 mL, the water solubility is then 2.896E−9*78/1000 = 2.259E−10 g/mL = 2.259E−8 g/100 mL.

(b)

$$ZnS \rightleftharpoons Zn^{+2} + S^{-2}$$

$$K = \frac{[Zn^{+2}][S^{-2}]}{[ZnS]} \Rightarrow K_{sp} = [Zn^{+2}][S^{-2}]$$

$n = 1, m = 1$
From Equation (4.4), we have

$$WS = \left(\frac{1.20 \times 10^{-23}}{1^1 1^1}\right)^{1/(1+1)} = \left(\frac{1.20 \times 10^{-23}}{1}\right)^{1/2} = 3.464 \times 10^{-12} \text{ mol/L}$$

Since molecular weight of zinc sulfide is 97.38 g/mol and 1 L = 1000 mL, the water solubility is then 3.464E−12*97.38/1000 = 3.373E−13 g/mL = 3.373E−11 g/100 mL.

(c)

$$Mg(NH_4)(PO_4) \rightleftharpoons Mg^{+2} + NH_4^+ + PO_4^{-3}$$

$$K = \frac{[Mg^{+2}][NH_4^+][PO_4^{-3}]}{[Mg(NH_4)(PO_4)]} \Rightarrow K_{sp} = [Mg^{+2}][NH_4^+][PO_4^{-3}]$$

$n = 1, m = 1, l = 1$
From Equation (4.4), we have

$$WS = \left(\frac{2.50 \times 10^{-13}}{1^1 1^1 1^1}\right)^{1/(1+1+1)} = \left(\frac{2.50 \times 10^{-13}}{1}\right)^{1/3} = 6.30 \times 10^{-5} \text{ g/mol}$$

Since molecular weight of magnesium ammonium phosphate is 137.3 g/mol and 1 L = 1000 mL, the water solubility is then 6.30E−5*137.3/1000 = 8.650E−6 g/mL = 8.650E−4 g/100 mL.

4.2 SAFETY-HAZARD IMPACT POTENTIALS

4.2.1 TERMS, DEFINITIONS, AND EXAMPLES

Potential: Corrosiveness as a Gas[3,25–29]

The corrosiveness as a gas potential for compound j is given by Equation (4.5).

$$(CGP)_j = \frac{LC_{50,\text{diethyl-ether}}}{LC_{50,j}} \qquad (4.5)$$

$(CGP)_j = 0$, if compound j is a non-corrosive gas
where, $LC_{50(j)}$ is the lethal concentration to kill 50% of population upon exposure to compound j
Reference compound: diethyl ether; LC(50)diethyl-ether = 0.52 g/m³, 4 h (mouse)
Human health hazard: pulmonary

Example 4.4

Determine the corrosiveness potential as a gas for the following substances.

Toluene, LC50(toluene) = 49 g/m³, 4 h
Phosgene, LC50(phosgene) = 0.0072 g/m³, 4 h

SOLUTION
(CGP)toluene = 0.52/49 = 0.0106
(CGP)phosgene = 0.52/0.0072 = 72.22

Potential: Corrosiveness as a Liquid or Solid

The corrosiveness as a liquid or solid potential for compound j is given by Equation (4.6).

$$(CLP)_j = \frac{LD_{50,\text{toluene}}}{LD_{50,j}}, \qquad (4.6)$$

if compound j is a potentially corrosive liquid
$(CLP)_j = 0$, if compound j is a non-corrosive liquid
where, $LD_{50(j)}$ is the lethal dose to kill 50% of population upon subcutaneous or dermal exposure to compound j
Reference compound: toluene; LD50(toluene, subcutaneous) = 2250 mg/kg
Human health hazard: skin burns

Example 4.5

Determine the corrosiveness potential as a liquid for the following substances.

Acetone, LD50(acetone, dermal) = 5000 mg/kg
Methanol, LD50(methanol, dermal) = 15800 mg/kg

SOLUTION
(CLP)acetone = 2250/5000 = 0.45
(CLP)methanol = 2250/15800 = 0.142

Potential: Dermal Absorption[30,31]

The dermal absorption potential for compound j is given by Equation (4.7).

$$(DAP)_j = \frac{(Fl^*)_j}{(Fl^*)_{toluene}} \qquad (4.7)$$

where $(Fl^*)_j$ is the critical flux for dermal absorption for compound j in mg/cm²/h
Reference compound: toluene

(Fl^*)toluene = 0.0563 mg/cm²/h (*reference* [30])
(Fl^*)toluene = 0.2829 mg/cm²/h (*reference* [31])

More reliable values given in reference [31].

Potential: Explosive Strength[32]

The explosive strength potential for compound j is given by Equation (4.8).

$$(XSP)_j = \frac{(LBT)_j}{(LBT)_{RDX}}, \qquad (4.8)$$

for potentially explosive compounds
$(XSP)_j = 0$, for non-explosive compounds
where $(LBT)_j$ is Trauzl lead block test for compound j (cm³ for a 10 g sample)

Dimensions of lead block:

Cylinder of length 200 mm, diameter 200 mm
Borehole in cylinder: depth 125 mm, 25 mm diameter
Volume of borehole: $\pi r^2 d$ = 3.14*(2.5/2 cm)²*12.5 cm = 61.36 cm³
Reference compound: RDX (cyclotrimethylenetrinitramine)
LBT(RDX) = 483 cm³ per 10 g sample
Human health hazard: maiming, death

Safety-Hazard Impact Metrics

Example 4.6

Determine the explosive strength potential for the following substances.

Nitroglycerin, LBT = 530
Trinitrotoluene, LBT = 300
Ammonium nitrate, LBT = 198

Solution
XSP(nitroglycerin) = 530/483 = 1.097
XSP(trinitrotoluene) = 300/483 = 0.621
XSP(ammonium nitrate) = 198/483 = 0.410

Potential: Explosive Vapor[6,7,33]

The explosive vapor potential for compound j is given by Equation (4.9).

$$(XVP)_j = \frac{(LEL)_{\text{diethyl-ether}}}{(LEL)_j}, \qquad (4.9)$$

if compound j is potentially explosive gas/vapor
$(XVP)_j = 0$, if compound j is non-explosive gas/vapor
where $(LEL)_j$ = lower explosive limit expressed as % by volume in air for compound j at 1 atm and 25°C (values quoted as upward flame propagation in standard 2-inch cylindrical tube)
Reference compound: diethyl ether; (LEL)diethyl ether = 1.7
Human health hazard: maiming, death

Example 4.7

Determine the explosive vapor potential for the following substances.

Carbon monoxide, (LEL)carbon monoxide = 12.5
Benzene, LEL(benzene) = 1.4
Hydrogen, LEL(hydrogen) = 4.1
(XVP)carbon monoxide = 1.7/12.5 = 0.136
(XVP)benzene = 1.7/1.4 = 1.214
(XVP)hydrogen = 1.7/4.1 = 0.415

Potential: Flammability[34,35]

The flammability potential for compound j is given by Equation (4.10).

$$(FP)_j = \frac{(FLP)_{\text{diethyl-ether}}}{(FLP)_j}, \qquad (4.10)$$

if compound j is potentially flammable
   $(FP)_j = 0$, if compound j is non-flammable
where $(FLP)_j$ = closed-cup flash point of compound j (degrees K); (FLP)diethyl ether = closed-cup flash point of diethyl ether

$F = 1.8\ C + 32$
$C = (F-32)/1.8$
$K = 273 + [(F-32)/1.8]$

If compound j has a lower flashpoint than Et_2O, then $(FLP)_j < (FLP)Et_2O$ and $(FP)_j > 1$.

If compound j has a higher flashpoint than Et_2O, then $(FLP)_j > (FLP)Et_2O$ and $(FP)_j < 1$.

Reference compound: diethyl ether
(FLP)diethyl ether = $-45°C = 228$ K
Human health hazard: skin burns

Example 4.8

Determine the flammability potential for the following substances.
Benzene, (FLP)benzene = $-11°C$
Ethyl alcohol, (FLP)ethyl alcohol = 55 F

Solution

(FLP)benzene = 262 K
(FP)benzene = 228/262 = 0.870
(FLP)ethyl alcohol = 285.8 K
(FP)ethyl alcohol = 228/285.8 = 0.798

Potential: Hydrogen Generation[9]

The hydrogen generation potential for compound j is given by Equation (4.11)

$$(HGP)_j = \frac{M_{H_2}/(MW)_j}{1/23} \quad (4.11)$$

$(HGP)_j = 0$, for compounds that do not produce H_2 upon reaction with water
where $M(H_2)$ = mass of hydrogen gas generated.
Reference reaction:

$$Na + H_2O \rightarrow NaOH + 1/2 H_2$$

If $(HGP)_j > 1$, then compound j generates more hydrogen gas than sodium when reacted with water.

If $(HGP)_j < 1$, then compound j generates less hydrogen gas than sodium when reacted with water.

Human health hazard: maiming, death.

Example 4.9

Determine the hydrogen generation potential for the following substances.

Compound
Lithium
Potassium
Sodium borohydride
Lithium aluminum borohydride
Diborane
Tributyltin hydride
Sodium cyanoborohydride
Borane etherate complex
Tin
Zinc

Solution

Compound	Formula	MW (g/mol)	Equivalents of H_2 generated	HGP
Lithium	Li	7	0.5	3.286
Potassium	K	39	0.5	0.590
Sodium borohydride	$NaBH_4$	78	4	2.359
Lithium aluminum borohydride	$LiAlH_4$	78	4	2.359
Diborane	B_2H_6	28	6	9.857
Tributyltin hydride	nBu_3SnH	290.69	1	0.158
Sodium cyanoborohydride	$NaBH_3CN$	63	3	2.190
Borane etherate complex	$BH_3 \, OEt_2$	88	3	1.568
Tin	Sn	118.69	2	0.775
Zinc	Zn	65.38	2	1.407

Reactions:

$$M+H_2O \rightarrow MOH+1/2H_2$$

for M = Li, Na, K

$$NaBH_4+4H_2O \rightarrow NaOH+B(OH)_3+4H_2$$

$$LiAlH_4+4H_2O \rightarrow LiOH+Al(OH)_3+4H_2$$

$$B_2H_6+6H_2O \rightarrow 2B(OH)_3+6H_2$$

$$nBu_3SnH+H_2O \rightarrow nBu_3SnOH+H_2$$

$$NaBH_3CN+3H_2O \rightarrow NaCN+B(OH)_3+3H_2$$

$$BH_3OEt_2 + 3H_2O \rightarrow B(OH)_3 + OEt_2 + 3H_2$$

$$M + 2HCl \rightarrow MCl_2 + H_2$$

for M = Zn, Sn

Potential: Impact Sensitivity[36,37]

The impact sensitivity potential for compound j is given by Equation (4.12).

$$(ISP)_j = \frac{(IS)_{RDX}}{(IS)_j} \qquad (4.12)$$

for potentially explosive compounds
$(ISP)_j = 0$, for non-explosive compounds
where $(IS)_j$ is impact sensitivity of compound j
Sample size = 20 mg
Drop height = 50 cm
Reference compound: RDX (cyclotrimethylenetrinitramine)

$(IS)_{RDX} = 0.75$ kp m $= 0.75*9.8067$ N m $= 7.36$ N m
Highly explosive compounds have a low impact sensitivity value.
Weakly explosive compounds have a high impact sensitivity value.
Human health hazard: maiming, death

Example 4.10

Determine the impact sensitivity potential for the following substances.

Nitroglycerin, IS = 0.2
Trinitrotoluene, IS = 14.71
Ammonium perchlorate, IS = 24.52

SOLUTION
Nitroglycerin, ISP = 7.36/0.2 = 36.8
Trinitrotoluene, ISP = 7.36/14.71 = 0.50
Ammonium perchlorate, ISP = 7.36/24.52 = 0.30

Potential: Maximum Allowable Concentration[38]

The maximum allowable concentration potential for compound j is given by Equation (4.13).

$$(MACP)_j = \frac{(MAC)_{toluene}}{(MAC)_j} \qquad (4.13)$$

where $(MAC)_j$ is the occupational exposure limit for compound j in mmol/m^3
Reference compound: toluene

$(MAC)_{toluene} = 100$ mg/m^3 = 1.086 mmol/m^3

Safety-Hazard Impact Metrics

Potential: Occupational Exposure Limit[39]

The occupational exposure limit potential for compound j is given by Equation (4.14).

$$(OELP)_j = \frac{(OEL)_{toluene}}{(OEL)_j} \qquad (4.14)$$

where $(OEL)_j$ is the occupational exposure limit for compound j in mmol/m³ (or ppm)
Reference compound: toluene

$(OEL)_{toluene} = 100$ ppm $= 375$ mg/m³ $= 4.08$ mmol/m³
Time weighted average (TWA) values are used for both definitions.

Unit conversions:

$$ppm = \frac{24.45 * 1000 * (g/m^3)}{(MW)} \text{(converting g/m}^3 \text{ to ppm)}$$

$$(g/m^3) = \frac{(ppm)*(MW)}{24.45*1000} \text{(converting ppm to g/m}^3\text{)}$$

Human health hazard: various

Example 4.11

Determine the occupational exposure limit potential for the following substances.

1. Dibutyl phthalate, OEL = 0.432 ppm
2. Copper dust, OEL = 1 mg/m³

SOLUTION

(a)

MW dibutyl phthalate = 278 g/mol
(OEL)dibutyl phthalate
 = (0.432*278)/(24.45*1000)
 = 0.00491 g/m³
 = 4.91 mg/m³
 = 4.91/278 mmol/m³
 = 0.0177 mmol/m³
(OELP)dibutyl phthalate = 4.08/0.0177 = 230.5 (based on mmol per m³ ratio)
(OELP)dibutyl phthalate = 100/0.432 = 231.5 (based on ppm ratio)

(b)

MW copper = 63.55 g/mol
(OEL)copper dust
 = 1 mg/m³
 = 1/63.55 mmol per m³
 = 0.0157 mmol per m³
(OELP)copper dust = 4.08/0.0157 = 259.2

Potential: Oxygen Balance[9]

If a compound is capable of undergoing a combustion or a calcination reaction, the oxygen balance potential for compound j is given by Equation (4.15a).

$$(OBP)_j = \frac{(OB)_j}{(OB)_{methane}} = \frac{16(n_{substrate} - n_{products})/(MW)_j}{-4} \qquad (4.15a)$$

where

$n_{substrate}$ is the number of oxygen atoms in substrate,
$n_{products}$ is the number of oxygen atoms in the products after complete combustion or calcination, and
$(MW)_j$ is the molecular weight of compound j

Reference compound: methane

This definition applies to compounds involved in complete combustion reactions where oxygen is consumed. Hence, $n_{substrate} < n_{products}$, and $(OB)_j < 0$. This indicates a negative oxygen balance, since a net consumption of oxygen occurs.

For complete combustion reactions (net minimum oxygen consumption):

- oxygen balance measures the degree to which *complete* oxidation is possible in a given compound
- oxygen balance corresponds to the *minimum* mass of extra oxygen that is needed from air, in addition to oxygen content in a given compound, in order for it to completely combust to carbon dioxide and water or to be completely calcinated
- in such reactions all elements in substrate are oxidized except for O and H.
  H (protic) $\rightarrow H_2O$
  C $\rightarrow CO_2$
  N $\rightarrow N_2$
  S $\rightarrow SO_2$
  M(0) $\rightarrow MO_x$ (highest oxidation of metal for complete combustion—corresponds to complete calcinations process)
  $M^{+n} \rightarrow M^{+m}$ where $m > n$ (oxidation of metal ions to higher oxidation numbers)
  X- $\rightarrow X_2$ (oxidation of halides to diatomic halogens)
- $P^{+n} \rightarrow P^{+m}$, where $m > n$ (oxidation of phosphorus to higher oxidation numbers)

The oxygen balance (OB) for compound j containing only C, H, and O is given by Equation (4.15b).

$$(OB)_j = \frac{16[z - (2x + 0.5y)]}{(MW)_j} = \frac{-16[n]}{(MW)_j} = \frac{16(n_{substrate} - n_{products})}{(MW)_j}, \qquad (4.15b)$$

Safety-Hazard Impact Metrics

where:
- x = number of carbon atoms in compound j
- y = number of hydrogen atoms in compound j
- z = number of oxygen atoms in compound j
- $(MW)_j$ = molecular weight of compound j
- $n_{substrate}$ = number of moles O in substrate
- $n_{products}$ = total number of moles O in all products from combustion reaction

$$n = \left[2x + (y/2) - z\right];\ n > 0$$

$$n_{substrate} < n_{products}$$

Example 4.12

Determine OB and OBP for the following substances. For each case write out the combustion or calcinations reaction as appropriate.
   (a) trinitrotoluene, (b) peroxyacetic acid, (c) aluminum

SOLUTION

(a)

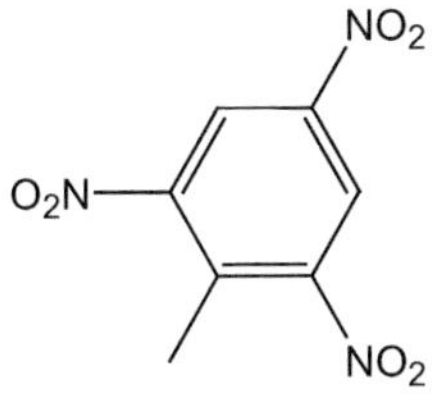

[structure of trinitrotoluene] + (14 + 2.5 - 6) [O] = 7 CO_2 + 2.5 H_2O + 1.5 N_2

$C_7H_5N_3O_6$
Mol. Wt.: 227

$n_{substrate} = 6$
$n_{products} = 14 + 2.5 = 16.5$
OB = 16*(6 − 16.5)/227 = −0.740
OBP = −4*(6 − 16.5)/227 = 0.185

(b)

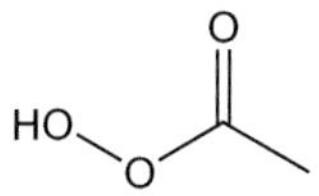

[structure of peroxyacetic acid] + (4 + 2 - 3) [O] = 2 CO_2 + 2 H_2O

$C_2H_4O_3$
Mol. Wt.: 76

$n_{substrate} = 3$
$n_{products} = 4 + 2 = 6$
OB = 16*(3−6)/76 = −0.632
OBP = −4*(3−6)/76 = 0.158

(c)

$2Al + 3[O] = Al_2O_3$
$n_{substrate} = 0$
$n_{products} = 3$
$OB = 16*(0-3)/(2*27) = -0.889$
$OBP = -4*(0-3)/(2*27) = 0.222$

If a compound is capable of undergoing a reduction reaction, the oxygen balance potential for compound j is given by Equation (4.16a).

$$(OBP)_j = \frac{(OB)_j}{(OB)_{hydrogen\text{-}peroxide}} = \frac{16(n_{substrate} - n_{products})/(MW)_j}{0.471} \quad (4.16a)$$

where $n_{substrate}$ is the number of oxygen atoms in substrate, $n_{products}$ is the number of oxygen atoms in the products after complete reduction, and $(MW)_j$ is the molecular weight of compound j
Reference compound: hydrogen peroxide

This definition applies to compounds that have the potential of being oxidizing agents would be involved in hypothetical complete reduction reactions where oxygen is liberated. Hence, $n_{substrate} > n_{products}$, and $(OB)_j > 0$. This indicates a positive oxygen balance, since a net liberation of oxygen occurs.

For complete reduction reactions (net maximum oxygen liberation):

- oxygen balance measures the degree to which *complete* reduction is possible in a given compound
- oxygen balance corresponds to the *maximum* mass of oxygen liberated in a reduction reaction with respect to substrate, in order for it to be completely reduced to its constituent elements
- in such reactions, all elements in substrate are reduced except for C, O, H, and X (halides).
  H (protic) → H_2O
  MO_x → M(0)
  X- → X_2

The oxygen balance (OB) for compound j is given by Equation (4.16b).

$$(OB)_j = \frac{16[n_{substrate} - n_{products}]}{(MW)_j} \quad (4.16b)$$

where:
$(MW)_j$ = molecular weight of compound j
$n_{substrate}$ = number of moles O in substrate
$n_{products}$ = total number of moles O in all products from combustion reaction

$$n_{substrate} > n_{products}$$

Safety-Hazard Impact Metrics

If a compound is not capable of undergoing either a combustion or a reduction reaction, the oxygen balance potential for compound j is given by Equation (4.17).

$$(OBP)_j = 0 \qquad (4.17)$$

where $n_{substrate} = n_{products}$

This expression applies to that are oxygen balanced, that is, those that cannot be oxidized (combusted) or reduced further.

Example 4.13

Determine OB and OBP for the following substances. For each case write out the reduction reaction as appropriate.

(a) ammonium nitrate, (b) sodium perchlorate, (c) chromic acid, (d) ozone, (e) potassium permanganate

Solution

(a)

$$NH_4NO_3 = 2H_2O + N_2 + 5[O]$$

$H_4N_2O_3$

Mol. Wt.: 80
$n_{substrate} = 3$
$n_{products} = 2$
$OB = 16*(3-2)/80 = 0.2$
$OBP = (16/0.471)*(3-2)/80 = 0.425$

(b)

$$NaClO_4 = Na + 0.5Cl_2 + 4[O]$$

$ClNaO_4$

Mol. Wt.: 122.45
$n_{substrate} = 4$
$n_{products} = 0$
$OB = 16*(4-0)/122.45 = 0.523$
$OBP = (16/0.471)*(4-0)/122.45 = 1.110$

(c)

$$CrO_3 = Cr + 3[O]$$

CrO_3

Mol. Wt.: 99.996
$n_{substrate} = 3$
$n_{products} = 0$
$OB = 16*(3-0)/99.996 = 0.480$
$OBP = (16/0.471)*(3-0)/99.996 = 1.019$

(d)

$O_3 = O_2 + [O]$

O_3

Mol. Wt.: 48
$n_{substrate} = 3$
$n_{products} = 2$
$OB = 16*(3-2)/48 = 0.333$
$OBP = (16/0.471)*(3-2)/48 = 0.708$

(e)

$KMnO_4 = K + Mn + 4[O]$

$KMnO_4$

Mol. Wt.: 157.94
$n_{substrate} = 4$
$n_{products} = 0$
$OB = 16*(4-0)/157.94 = 0.405$
$OBP = (16/0.471)*(4-0)/157.94 = 0.860$

Potential: Risk Phrase[9,40]

The risk phrase potential for compound j is given by Equation (4.18).

$$(RRP)_j = \frac{\log\left[\prod_{k=1}^{n} Q_{k,j}\right]}{\log\left[\prod_{k=1}^{n} Q_{k,toluene}\right]} = \frac{n + \log\left[\prod_{k=1}^{n}(Q - \text{Eissen})_{k,j}\right]}{2 + \log[2*1.5]} = \frac{n + \log\left[\prod_{k=1}^{n}(Q - \text{Eissen})_{k,j}\right]}{2.477} \quad (4.18)$$

where $Q(k,j) = Q\text{-Eissen}(k,j)*10$; k refers to kth R-phrase, j refers to compound j

(Note: Eissen Q-factors are multiplied by 10 to ensure whole numbers since fractions will diminish the overall product value; the overall risk potential is interpreted as a multiplicative quantity where all Q-factors for a given compound are multiplied together.)

n refers to the number of risk phrases associated with compound j
Reference compound is toluene with risk phrases: R11, R20

R11 = highly flammable ($Q_{Eissen} = 2$)
R20 = harmful by inhalation ($Q_{Eissen} = 1.5$)
$n = 2$
Q(toluene) = $(10^2)*(2*1.5) = 300$ and log Q(toluene) = 2.447

Risk phrases apply to a given compound based on the standard conditions for which it is commercially available. Table 4.2 summarizes various R-phrases used in material safety data sheets (MSDS).

TABLE 4.2
Summary of R-Phrases

Q (Eissen)	No.	R-phrase
4	R1	Explosive when dry.
4	R2	Risk of explosion by shock, friction, fire, or other sources of ignition.
5	R3	Extreme risk of explosion by shock, friction, fire, or other sources of ignition.
2	R4	Forms very sensitive explosive metallic compounds.
3	R5	Heating may cause an explosion.
4	R6	Explosive with or without contact with air.
3	R7	May cause fire.
4	R8	Contact with combustible material may cause fire.
3	R9	Explosive when mixed with combustible material.
1	R10	Flammable.
2	R11	Highly flammable.
3	R12	Extremely flammable.
3	R13	Extremely flammable liquid gas.
2	R14	Reacts violently with water.
3	R14/15	Reacts violently with water, liberating extremely flammable gases.
3	R15	Contact with water liberates extremely flammable gases.
3	R15/29	Contact with water liberates toxic, extremely flammable gas.
3	R16	Explosive when mixed with oxidizing substances.
4	R17	Spontaneously flammable in air.
3	R18	In use, may form flammable/explosive vapor-air mixture.
3	R19	May form explosive peroxides.
1.5	R20	Harmful by inhalation.
1.5	R20/21	Harmful by inhalation and in contact with skin.
1.5	R20/21/22	Harmful by inhalation, in contact with skin and if swallowed.
1.5	R20/22	Harmful by inhalation and if swallowed.
1.5	R21	Harmful in contact with skin.
1.5	R21/22	Harmful in contact with skin and if swallowed.
0.5	R22	Harmful if swallowed.
2.5	R23	Toxic by inhalation.
2.5	R23/24	Toxic by inhalation and in contact with skin.
2.5	R23/24/25	Toxic by inhalation, in contact with skin and if swallowed.
2.5	R23/25	Toxic by inhalation and if swallowed.
2.5	R24	Toxic in contact with skin.
2.5	R24/25	Toxic in contact with skin and if swallowed.
0.5	R25	Toxic if swallowed.
3.5	R26	Very toxic by inhalation.
3.5	R26/27	Very toxic by inhalation and in contact with skin.
3.5	R26/27/28	Very toxic by inhalation, in contact with skin and if swallowed.
3.5	R26/28	Very toxic by inhalation and if swallowed.
3.5	R27	Very toxic in contact with skin.

(Continued)

TABLE 4.2 (CONTINUED)
Summary of R-Phrases

Q (Eissen)	No.	R-phrase
3.5	R27/28	Very toxic in contact with skin and if swallowed.
0.5	R28	Very toxic if swallowed.
1.5	R29	Contact with water liberates toxic gas.
1.5	R30	Can become highly flammable in use.
1.5	R31	Contact with acids liberates toxic gas.
2.5	R32	Contact with acids liberates very toxic gas.
4	R33	Danger of cumulative effects.
2	R34	Causes burns.
3	R35	Causes severe burns.
1.5	R36	Irritating to eyes.
1.5	R36/37	Irritating to eyes and respiratory system.
1.5	R36/37/38	Irritating to eyes, respiratory system, and skin.
1.5	R36/38	Irritating to eyes and skin.
1.5	R37	Irritating to respiratory system.
1.5	R37/38	Irritating to respiratory system and skin.
1.5	R38	Irritating to skin.
3.5	R39	Danger of very serious irreversible effects.
3.5	R39/23	Toxic: danger of very serious irreversible effects through inhalation.
3.5	R39/23/24	Toxic: danger of very serious irreversible effects through inhalation and in contact with skin.
3.5	R39/23/24/25	Toxic: danger of very serious irreversible effects through inhalation, in contact with skin and if swallowed.
3.5	R39/23/25	Toxic: danger of very serious irreversible effects through inhalation and if swallowed.
3.5	R39/24	Toxic: danger of very serious irreversible effects in contact with skin.
3.5	R39/24/25	Toxic: danger of very serious irreversible effects in contact with skin and if swallowed.
0.5	R39/25	Toxic: danger of very serious irreversible effects if swallowed.
4	R39/26	Very toxic: danger of very serious irreversible effects through inhalation.
4	R39/26/27	Very toxic: danger of very serious irreversible effects through inhalation and in contact with skin.
4	R39/26/27/28	Very toxic: danger of very serious irreversible effects through inhalation, in contact with skin and if swallowed.
4	R39/26/28	Very toxic: danger of very serious irreversible effects through inhalation and if swallowed.
4	R39/27	Very toxic: danger of very serious irreversible effects in contact with skin.
4	R39/27/28	Very toxic: danger of very serious irreversible effects in contact with skin and if swallowed.
0.5	R39/28	Very toxic: danger of very serious irreversible effects if swallowed.
3	R40	Possible risks of irreversible effects.
3	R40/20	Harmful: possible risk of irreversible effects through inhalation.

(Continued)

TABLE 4.2 (CONTINUED)
Summary of R-Phrases

Q (Eissen)	No.	R-phrase
3	R40/20/21	Harmful: possible risk of irreversible effects through inhalation and in contact with skin.
3	R40/20/21/22	Harmful: possible risk of irreversible effects through inhalation, in contact with skin and if swallowed.
3	R40/20/22	Harmful: possible risk of irreversible effects through inhalation and if swallowed.
3	R40/21	Harmful: possible risk of irreversible effects in contact with skin.
3	R40/21/22	Harmful: possible risk of irreversible effects in contact with skin and if swallowed.
0.5	R40/22	Harmful: possible risk of irreversible effects if swallowed.
2	R41	Risk of serious damage to eyes.
4	R42	May cause sensitization by inhalation.
4	R42/43	May cause sensitization by inhalation and skin contact.
4	R43	May cause sensitization by skin contact.
4	R44	Risk of explosion if heated under confinement.
5	R45	May cause cancer.
5	R46	May cause heritable genetic damage.
5	R47	May cause deformities.
1.5	R48	Danger of serious damage to health by prolonged exposure.
1.5	R48/20	Harmful: danger of serious damage to health by prolonged exposure through inhalation.
1.5	R48/20/21	Harmful: danger of serious damage to health by prolonged exposure through inhalation and in contact with skin.
1.5	R48/20/21/22	Harmful: danger of serious damage to health by prolonged exposure through inhalation, in contact with skin and if swallowed.
1.5	R48/20/22	Harmful: danger of serious damage to health by prolonged exposure through inhalation and if swallowed.
1.5	R48/21	Harmful: danger of serious damage to health by prolonged exposure in contact with skin.
1.5	R48/21/22	Harmful: danger of serious damage to health by prolonged exposure in contact with skin and if swallowed.
0.1	R48/22	Harmful: danger of serious damage to health by prolonged exposure if swallowed.
2.5	R48/23	Toxic: danger of serious damage to health by prolonged exposure through inhalation.
2.5	R48/23/24	Toxic: danger of serious damage to health by prolonged exposure through inhalation and in contact with skin.
2.5	R48/23/24/25	Toxic: danger of serious damage to health by prolonged exposure through inhalation, in contact with skin and if swallowed.
2.5	R48/23/25	Toxic: danger of serious damage to health by prolonged exposure through inhalation and if swallowed.
2.5	R48/24	Toxic: danger of serious damage to health by prolonged exposure in contact with skin.

(Continued)

TABLE 4.2 (CONTINUED)
Summary of R-Phrases

Q (Eissen)	No.	R-phrase
2.5	R48/24/25	Toxic: danger of serious damage to health by prolonged exposure in contact with skin and if swallowed.
0.1	R48/25	Toxic: danger of serious damage to health by prolonged exposure if swallowed.
5	R49	May cause cancer by inhalation.
0.1	R50	Very toxic to aquatic organisms.
0.1	R50/53	Very toxic to aquatic organisms, may cause long-term adverse effects in the aquatic environment.
0.1	R51	Toxic to aquatic organisms.
0.1	R51/53	Toxic to aquatic organisms, may cause long-term adverse effects in the aquatic environment.
0.1	R52	Harmful to aquatic organisms.
0.1	R52/53	Harmful to aquatic organisms, may cause long-term adverse effects in the aquatic environment.
0.1	R53	May cause long-term adverse effects in the aquatic environment.
0.1	R54	Toxic to flora.
0.1	R55	Toxic to fauna.
0.1	R56	Toxic to soil organisms.
0.1	R57	Toxic to bees.
0.1	R58	May cause long-term adverse effects in the environment.
0.1	R59	Dangerous for the ozone layer.
5	R60	May impair fertility.
5	R61	May cause harm to the unborn child.
5	R62	Possible risk of impaired fertility.
5	R63	Possible risk of harm to the unborn child.
5	R64	May cause harm to breastfed babies.
0.5	R65	Harmful: may cause lung damage if swallowed.
0.1	R66	Repeated exposure may cause skin dryness or cracking.
0.1	R67	Vapors may cause drowsiness and dizziness.
0.1	R68	Possible risk of irreversible effects.
0.1	R68/20	Harmful: possible risk of irreversible effects through inhalation.
0.1	R68/21	Harmful: possible risk of irreversible effects in contact with skin.
0.1	R68/22	Harmful: possible risk of irreversible effects if swallowed.
0.1	R68/20/21	Harmful: possible risk of irreversible effects through inhalation and in contact with skin.
0.1	R68/20/22	Harmful: possible risk of irreversible effects through inhalation and if swallowed.
0.1	R68/21/22	Harmful: possible risk of irreversible effects in contact with skin and if swallowed.
0.1	R68/20/21/22	Harmful: possible risk of irreversible effects through inhalation, in contact with skin and if swallowed.
4	R69	Stored in pressurized gas cylinder. (new)
4	R70	Strong oxidizer. (new)

(Continued)

TABLE 4.2 (CONTINUED)
Summary of R-Phrases

Q (Eissen)	No.	R-phrase
4	R71	Asphyxiant. (new)
3	R72	Causes thermal burns due to liquefied cryogenic gas or dry ice. (new) (see R34 and R35)
0	R73	No perceived OHS hazard. (new)

[a] Italicized entries refer to risk phrases not counted in the safety-hazard impact potentials since they are considered in the environmental impact potentials (see Chapter 3).

Example 4.14

Determine the risk phrase potential for methanol.

Risk Phrase	Description	Q_{Eissen}
R11	Highly flammable.	2
R23/24/25	Toxic by inhalation, in contact with skin, and if swallowed.	2.5
R39	Danger of very serious irreversible effects.	3.5
R39/23/24/25	Toxic: danger of very serious irreversible effects through inhalation, in contact with skin and if swallowed.	3.5

SOLUTION

$n = 4$
(RPP)methanol = [****4 + log(2*2.5*3.5*3.5)]/2.477 = 5.787/2.477 = 2.336

Potential: Skin Dose[41]

The skin dose potential for compound j is given by Equation (4.19).

$$(SDP)_j = \frac{(K_p)_j (WS)_j}{(K_p)_{toluene} (WS)_{toluene}}, \quad (4.19)$$

where:
  $(K_p)_j$ is the transdermal permeation coefficient for compound j in cm/h
  $(WS)_j$ is the water solubility of compound j in mg/cm^3 or mg/mL

When $(SDP)_j > 1$, compound j penetrates skin to a greater degree than toluene.
When $(SDP)_j < 1$, compound j penetrates skin to a lesser degree than toluene.
Note: The skin dose in mg is based on an exposed surface area of 360 cm^2 for both hands exposed to the chemical for 8 h. Since the product of area and time appears in both the numerator and denominator of SDP, this quantity cancels out.

For compound j the skin dose is given by Equation (4.20).

$$(SD)_j = 8*360*(WS)_j*K_p \qquad (4.20)$$

The transdermal permeation coefficient, K_p, is calculated according to the Robinson model given by Equations (4.21a–d)[42].

$$K_p = \dfrac{1}{\dfrac{1}{K_{psc}+K_{pol}} + \dfrac{1}{K_{aq}}} \qquad (4.21a)$$

where:
K_{psc} = permeation coefficient in lipid fraction of *stratum corneum* (h/cm)
K_{pol} = permeation coefficient in protein fraction (h/cm)
K_{aq} = permeation coefficient in watery epidermal layer (h/cm)

$$\log K_{psc} = -1.326 + 0.6097 \log K_{ow} - 0.1786\sqrt{MW}$$

$$K_{pol} = 0.0001519/\sqrt{MW} \qquad (4.21b,c,d)$$

$$K_{aq} = 2.5/\sqrt{MW}$$

MW = molecular weight (g/mol); K_{ow} = octanol–water partition coefficient
Reference compound: toluene

(K_p)toluene = 0.0363 cm/h
(WS)toluene = 0.53 mg/mL
Compilations of K_{psc} can be found in references [42–44].

Example 4.15

Use the *template-skin-dose-potential.xls* spreadsheet to determine the skin dose potential and skin dose for the following substances: (a) aniline; (b) styrene.

SOLUTION

(a)

Substance	Aniline	
MW	93	g/mol
log Kow	0.98	
Water solubility	36	mg/cm³
log K(psc)	−2.45085	
K(psc)	0.003541	h/cm
K(pol)	1.58E−05	h/cm
K(aq)	0.259238	h/cm
Kp	0.003509	cm/h
SDP	6.5588	
SD	363.8	mg

(b)

Substance	Styrene	
MW	104	g/mol
log Kow	2.95	
Water solubility	0.3	mg/cm^3
log K(psc)	−1.34875	
K(psc)	0.044797	h/cm
K(pol)	1.49E−05	h/cm
K(aq)	0.245145	h/cm
Kp	0.037886	cm/h
SDP	0.5902	
SD	32.7	mg

The relative penetration of skin for the two compounds is aniline > styrene. The degree of penetration follows the degree of polarity and the degree of hydrophilicity of the compounds.

4.3 SAFETY-HAZARD INDEX

4.3.1 Terms, Definitions, and Examples

Safety-Hazard Impact[9]

This is a general term used in life cycle assessment literature that estimates the damage done by input materials emitted and waste materials produced in a chemical reaction via occupational exposure according to several criteria called potentials. Safety-hazard impact is quantified according to Equation (4.22).

$$\text{SHZI} = \frac{\sum_j f_j \Omega_j}{\sum_j \Omega_j} \qquad (4.22)$$

where j refers to the jth compound, Ω_j is the sum of all safety-hazard potentials and $f_j = $ (mass of compound j)/(total mass of compounds of type j involved in reaction). A value of 0 corresponds to complete greenness with respect to the selected safety-hazard impact potentials and a value of 1 corresponds to complete anti-greenness.

Safety-Hazard Index[9]

The safety-hazard index is a measure of the degree of greenness of a chemical reaction or synthesis with respect to a set of safety-hazard impact potentials. It can be applied to input materials or to waste materials. The mathematical definition is given by Equation (4.23).

$$\text{SHI} = 1 - \text{SHZI} = 1 - \frac{\sum_j f_j \Omega_j}{\sum_j \Omega_j} \quad (4.23)$$

where *SHI* is the environmental impact, *j* refers to the *j*th compound, Ω_j is the sum of all safety-hazard potentials, and f_j = (mass of compound *j*)/(total mass of compounds of type *j* involved in reaction). A value of 1 corresponds to complete greenness with respect to the selected safety-hazard impact potentials and a value of 0 corresponds to complete anti-greenness.

Example 4.16[45]

For a general chemical reaction given by

Input materials → Desired product + Waste materials

the safety-hazard indices (SHI) for input materials and waste materials is defined according to Equations (4.24) and (4.25) shown below.

$$\text{SHI(in)} = 1 - \text{SHZI(in)} = 1 - \frac{\sum_j f(\text{input})_j \Omega_j}{\sum_j \Omega_j} \quad (4.24)$$

and

$$\text{SHI(waste)} = 1 - \text{SHZI(waste)} = 1 - \frac{\sum_j f(\text{waste})_j \Omega_j}{\sum_j \Omega_j} \quad (4.25)$$

where SHZI(in) and SHZI(waste) are safety-hazard impacts of input and waste materials,

$$f(\text{input})_j = \frac{\text{mass of input compound } j}{\text{total mass of input compounds used in reaction}}$$

$$f(\text{waste})_j = \frac{\text{mass of waste compound } j}{\text{total mass of waste compounds produced in reaction}}$$

and Ω_j is the sum of all potentials pertaining to compound *j* given by

$$\Omega_j = (\text{FP})_j + (\text{CGP})_j + (\text{CLP})_j + (\text{OBP})_j + (\text{HGP})_j + (\text{XVP})_j + (\text{XSP})_j + (\text{ISP})_j$$
$$+ (\text{RPP})_j + (\text{OELP})_j + (\text{SDP})_j$$

Safety-Hazard Impact Metrics

CGP = corrosiveness as a gas potential
CLP = corrosiveness as a liquid potential
FP = flash point potential
HGP = hydrogen generation potential
ISP = impact sensitivity potential
OBP = oxygen balance potential
OELP = occupational exposure limit potential
RPP = risk phrase potential
SDP = skin dose potential
XVP = explosive vapor potential
XSP = explosive strength potential

An industrial synthesis for the production of aniline by catalytic hydrogenation of nitrobenzene is given in the scheme below.

$$\text{C}_6\text{H}_5\text{NO}_2 \xrightarrow[\text{CuCO}_3 \text{ (cat.)}]{\text{H}_2} \text{C}_6\text{H}_5\text{NH}_2$$

600 kg/h nitrobenzene
6000 cubic meters per hour of hydrogen gas at $T = 200°C$ and $p = 5$ psig
3500 kg aqueous copper carbonate solution (0.07 kg Cu per 100 kg aniline produced)
workup: 2500 kg 50 wt% sodium silicate aqueous solution
Yield of aniline = 98%

Using the *template-REACTION.xls* spreadsheet, the *calculator-safety-hazard-index.xls* spreadsheet, and the data given below determine SHI(in) and SHI(waste) for the reaction. Comment on which chemicals have the most impact.

Substance	LD50 (dermal) (mg/kg)	LC50 (g/m³, 4 h)	Flash pt. (°C)	Moles H2 generated	LEL (%)	OEL (mmol/m³)	SD (mg)	Q (R-phrases)
Nitrobenzene	800	2.8	88	0	1.8	0.041	3.88E+01	300
Hydrogen				0	4		2.20E−01	1200
Water				0				0
CuCO₃	NA	NA		0		0.001	3.90E−105	75
Na₄SiO₄	4640	NA		0		NA	NA	450

NA = not available

Solution

Balanced chemical equation:

$$\text{C}_6\text{H}_5\text{NO}_2 \xrightarrow[\text{CuCO}_3 \text{ (cat.)}]{3\,\text{H}_2,\; -2\,\text{H}_2\text{O}} \text{C}_6\text{H}_5\text{NH}_2$$

123   93

Reactants	MW(g/mol)	Stoich. Coeff. (SC)	Mass used (g)	Moles
Nitrobenzene	123	1	600000	4878
Dihydrogen	2	3	105174	52587

We assume hydrogen gas obeys the ideal gas law.

Number of moles of hydrogen $= (PV)/(RT)$
$p = 5$ psig $= 0.3402$ atm
$V = 6000$ cu m $= 6000000$ L
$T = 200 + 273 = 473$ K
Therefore, $n = (0.3402*6000000)/(0.08207*473) = 52587$ moles.

The limiting reagent is nitrobenzene (4878 moles).
Moles of aniline produced $= 0.98*4878 = 4780.5$
Mass of aniline produced $= 4780.5*93 = 444585$ g
Mass of water produced $= 2*4780.5*18 = 172098$ g
Mass of unreacted nitrobenzene $= (4878 - 4780.5)*123 = 11992.5$ g
Mass of unreacted hydrogen $= (52587 - 3*4780.5)*2 = 76491$ g

Check mass balance:

Input materials	Mass (g)	Output materials	Mass (g)
Nitrobenzene	600000	Aniline	444585
Hydrogen	105174	Water	172098
$CuCO_3$ (aq)	3500000	Nitrobenzene (unreacted)	11992.5
Na_4SiO_4 (aq)	2500000	Hydrogen (unreacted)	76491
SUM	6705174	$CuCO_3$ (aq)	3500000
		Na_4SiO_4 (aq)	2500000
		SUM	6705167

Safety-Hazard Impact Metrics

Input Materials

Substance	MW	Mass (g)	Proportion	CGP	CLP	FP	OBP	HGP	XVP	XSP	ISP	OELP	SDP	RPP	Total	SHZI contribution
Nitrobenzene	123	600000.0	0.0895	0.185714	2.8125	0.631579	0.406504	0.000000	0.944444	0.000000	0.000000	99.4146	0.583283	1.000000	105.9787	9.483303
Hydrogen	2	105174.0	0.0157	0	0	0.000000	2.000000	0.000000	0.425000	0.000000	0.000000	0.0000	0.003307	1.243048	3.671356	0.057587
Water	18	4749183.5	0.7083	0	0	0.000000	0.000000	0.000000	0.000000	0.000000	0.000000	0.0000	0.000000	0.000000	0	0
$CuCO_3$	123.55	816.5	0.0001	0	0	0.000000	0.000000	0.000000	0.000000	0.000000	0.000000	4076.0000	0.000000	0.756952	4076.757	0.496458
Na_4SiO_4	184	1250000.0	0.1864	0	0.484914	0.000000	0.000000	0.000000	0.000000	0.000000	0.000000	0.0000	0.000000	1.071087	1.556001	0.290075
TOTAL		6705174.0	1.0000											Total	4187.963	10.3274
														Overall SHZI		0.002466
														Safety-Hazard Index		0.9975

Waste Materials

Substance	MW	Mass (g)	Proportion	CGP	CLP	FP	OBP	HGP	XVP	XSP	ISP	OELP	SDP	RPP	Total	SHZI contribution
Water (by-product)	18	172098.0	0.0275	0	0	0.000000	0.000000	0.000000	0.000000	0.000000	0.000000	0.0000	0.000000	0.000000	0	0
Nitrobenzene (unreacted)	123	11992.5	0.0019	0.185714	2.8125	0.631579	0.406504	0.000000	0.944444	0.000000	0.000000	99.4146	0.583283	1.000000	105.9787	0.203008
Hydrogen (unreacted)	2	76491.0	0.0122	0	0	0.000000	2.000000	0.000000	0.425000	0.000000	0.000000	0.0000	0.003307	1.243048	3.671356	0.044856
Water	18	4749183.5	0.7586	0	0	0.000000	0.000000	0.000000	0.000000	0.000000	0.000000	0.0000	0.000000	0.000000	0	0
CuCO$_3$	123.55	816.5	0.0001	0	0	0.000000	0.000000	0.000000	0.000000	0.000000	0.000000	4076.0000	0.000000	0.756952	4076.757	0.531713
Na$_4$SiO$_4$	184	1250000.0	0.1997	0	0.484914	0.000000	0.000000	0.000000	0.000000	0.000000	0.000000	0.0000	0.000000	1.071087	1.556001	0.310674
TOTAL		6260581.5	1.0000											Total	4187.963	1.0903

Overall SHZI 0.00026

Safety-Hazard Index 0.9997

Safety-Hazard Impact Metrics

SHI(waste) slightly exceeds SHI(input) suggesting that the waste produced in the reaction is slightly less hazardous than the input materials used at the outset.

Nitrobenzene is the most hazardous input material. Copper carbonate and sodium silicate are the most hazardous waste materials. The occupational exposure limit potential is the dominant contributing potential at 91% and 66% for input and waste materials, respectively.

Uncertainty in SHI[46]

This is a measure of error in the SHI determinations when data are not available or are unreliable. The mathematical expression for uncertainty is given by Equation (4.26).

$$\%\text{uncertainty} = \left(\frac{x}{nC}\right)100 \qquad (4.26)$$

where x is the number of missing parameters for all the substances used in a given reaction, n is the number of parameters needed to estimate SHI ($n=11$) for each substance, and C is the total number of chemicals required for carrying out a given reaction including reagents, catalysts, additives, ligands, reaction solvents, workup materials, purification materials, and its associated by-products.

See *safety-hazard index*.

Example 4.17

In Example 10.24, determine the % uncertainties in SHI(input) and SHI(waste).

SOLUTION

For the input materials, we have $n=11$, $C=5$, and $x=6$. Hence the % uncertainty in SHI(input) is 100*(6/(11*5)) = 10.9%.
For the waste materials, we have $n=11$, $C=6$, and $x=6$. Hence the % uncertainty in SHI(waste) is 100*(6/(11*6)) = 9.1%.

4.4 EXCEL SPREADSHEET TOOLS: PART 3

calculator-safety-hazard index.xls
database-industrial-chemicals-safety-hazard parameters.xls
database-safety-hazard-metrics.xls
solubility-product-constants.xls
synthesis-plans-hazards-inputs-template.xls
synthesis-plans-hazards-waste-template.xls
template-HAZARDS.xls
template-skin-dose-potential.xls

For each spreadsheet we briefly describe the kind of inputs required and the results obtained.

The *calculator-safety-hazard-index.xls* spreadsheet determines the value of the SHI index for input materials (SHI(input)), or waste materials (SHI(waste)) for any chemical reaction. This calculator also determines the fractional contribution of each potential for each chemical substance so the most offending chemicals can be identified as well as the corresponding potentials that are responsible for those offences.

The *database-industrial chemicals-safety-hazard-parameters.xls* and *database-safety-hazard-metrics.xls* spreadsheets are a repository of relevant parameters for many thousands of chemicals that are available in a single resource. These database spreadsheets greatly increase the speed in carrying out determinations of the SHI indices.

The *solubility-product-constants.xls* spreadsheet lists solubility product constants, K_{sp}, for a variety of inorganic salts along with their water solubilities in g/100 mL calculated using Equation (4.4).

The *synthesis-plans-hazards-inputs-template.xls* and *synthesis-plans-hazards-waste-template.xls* spreadsheets determine SHI(input) and SHI(waste) respectively for multi-step synthesis plans using the following inputs: masses of all input materials, masses of by-products, and scaling factors for each reaction step; and safety-hazard potentials for each substance as determined from the *template-HAZARD.xls* spreadsheet.

The *template-HAZARDS.xls* spreadsheet determines for each reaction step the SHI(input), SHI(output), and uncertainty in these determinations based on the following input data: (a) balanced chemical equation; (b) reaction temperature; (c) reaction pressure; (d) molecular weights and masses of all input materials, by-products, and reaction product; (e) oxygen consumption or liberation equation for determination of the oxygen balance; (f) LD50(dermal); (g) LC50(inhalation); (h) flash point; (i) moles of hydrogen gas generated; (j) lower explosion limit; (k) lead block test parameter; (l) occupational exposure limit; (m) skin dose; and (n) risk phrase Q-factor.

The *template-skin-dose-potential.xls* spreadsheet is a calculator that determines skin dose potential (SDP) and skin dose (SD) in mg based on equations (4.19) and (4.20) using the following input data: molecular weight of the substance, log K_{ow}, and water solubility (mg/cm³).

4.5 PROBLEMS

PROBLEM 4.1
Prove that Equation (4.4) relating water solubility and solubility product constant for a general salt X_nY_m is true.

PROBLEM 4.2
Determine the water solubilities in g/100 mL of the following salts based on the given solubility product constants: (a) silver chromate (9.00E−12); (b) calcium fluoride (3.95E−11); (c) silver cyanide (2.20E−12).

In each case, write out the corresponding equilibrium equation.

PROBLEM 4.3
Determine the corrosiveness potential as a gas for acetone given that LC50(acetone) = 50.1 g/m³, 8 h.

PROBLEM 4.4
Determine the corrosiveness potential as a liquid for benzene given that LD50(benzene, dermal) = 5 mg/kg

PROBLEM 4.5
Determine the flammability potential for aniline given that the closed-cup flash point is 70°C.

PROBLEM 4.6
Determine OB and OBP for the following substances. For each case write out the combustion or calcinations reaction as appropriate.
(a) carbon, (b) benzyl chloride, (c) methyl amine

PROBLEM 4.7
Determine OB and OBP for the following substances. For each case write out the reduction reaction as appropriate.
(a) potassium persulfate, (b) sodium orthosilicate, (c) sodium tetraborate, (d) sulfuric acid, (e) nitric acid

PROBLEM 4.8
Use the *template-skin-dose-potential.xls* spreadsheet to determine the skin dose potential and skin dose for the following substances: (a) acetic acid; (b) methanol.

PROBLEM 4.9[47]
Themes: balancing chemical equations; experimental procedures; materials efficiency metrics; and environmental and safety-hazard comparison of solvents.

The following Suzuki reaction was carried out in a so-called "green solvent," 2-methyltetrahydrofuran, instead of the traditional solvent tetrahydrofuran.

Reagents/conditions above arrow:
$NiCl_2(PCy_3)_2$ (cat.)
K_3PO_4
2-methyltetrahydrofuran
97 %

Conditions:
1 mol % $NiCl_2(PCy_3)_2$
1 equiv. of arylbromide (0.9 mmol)
2.50 equiv. aryl boronic acid
4.50 equiv. K_3PO_4
2 mL water (as 1 M HCl solution)
reaction time: 12 h

Workup materials:
3 X 2 mL EtOAc
1 g $MgSO_4$
1 g silica plug
10 mL EtOAc

Part 1
Write out a balanced chemical equation for the reaction.

Part 2

From the information given under reaction conditions and the balanced chemical equation, determine the atom economy, kernel reaction mass efficiency, stoichiometric factor, turnover number, and turnover frequency.

Part 3

Use the *template-REACTION.xls* spreadsheet to determine the E-factor profile for the reaction and all material efficiency metrics. State any assumptions made in the calculations.

Part 4

Compare the environmental and safety-hazard impacts of 2-methyltetrahydrofuran with tetrahydrofuran, which is the traditional solvent for carrying out Suzuki couplings, according to the following criteria: LD50, LC50, SFP, log K_{ow}, Henry law constant, flashpoint, lower explosion limit, risk phrases, safety phrases, and occupational exposure limit. Use the *database-LCA-parameters.xls* and *database-safety-hazard-metrics.xls* spreadsheets to find appropriate values for these parameters. Which solvent has the better profile according to these criteria?

PROBLEM 4.10[48]

An industrial synthesis for the production of 2000 lbs of aniline by reducing nitrobenzene with iron filings is given in the scheme below. The authors stated that the by-product of the reaction is Fe_3O_4.

2780 lbs nitrobenzene
3200 lbs iron filings
2500 lbs 30 wt% HCl

Using the *template-REACTION.xls* spreadsheet, the *calculator-safety-hazard-index.xls* spreadsheet, and the data given below, determine SHI(in) and SHI(waste) for the reaction and their uncertainties. Comment on which chemicals have the most impact.

Substance	LD50 (dermal) (mg/kg)	LC50 (g/m³, 4 h)	Flash pt.(°C)	Moles H2 Generated	LEL (%)	OEL (mmol/m³)	SD (mg)	Q (R-phrases)
Nitrobenzene	800	2.8	88	0	1.8	0.041	3.88E+01	300
Iron	NA			0		0.09	5.85E−105	20
Water				0				0
HCl	18.8	1.16		0		0.192	3.43E+04	300
Fe_3O_4	NA	NA		0		0.022	2.87E−105	NA

NA = not available

Determine the % uncertainties in the SHI(input) and SHI(waste) estimates.

REFERENCES

1. Gupta, J. P.; Babu, B. S. *J. Haz. Mater.* 1999, *A67*, 1.
2. Edwards, D. W.; Lawrence, D. *Process Safety Environ. Protection* 1993, *71*, 252.
3. Registry of Toxic Effects of Chemical Substances (RTECS). (http://ccinfoweb.ccohs.ca/rtecs/search.html)
4. Material Safety Data Sheets (MSDS). (https://www.sigmaaldrich.com/safety-center.html)
5. Lange, N. A. *Handbook of Chemistry*, 10th ed., McGraw-Hill Book Co. Inc.: New York, 1961, pp. 32–49; 824.
6. Haynes, W. M. (ed.) *CRC Handbook of Chemistry & Physics*, 92nd ed., 2011–2012, CRC Press: Boca Raton, 2011, Chapter 16, pp. 16–31.
7. Proctor, L. D.; Warr, A. J. *Org. Proc. Res. Dev.* 2002, *6*, 884.
8. Lothrup, W. C.; Handrick, G. R. *Chem. Rev.* 1949, *44*, 419.
9. Andraos, J. *Org. Process Res. Dev.* 2013, *17*, 175.
10. Seidell, A. *Solubilities of Organic and Inorganic Compounds*, D. Van Nostrand: New York, 1941.
11. Yalkowsky, S. H.; He, Y.; Jain, P. *Handbook of Aqueous Solubility Data*, CRC Press: Boca Raton, 2010.
12. Stephen, H.; Stephen, T. *Solubilities of Inorganic and Organic Compounds*, Macmillan Book Co.: New York, 1963, Vol. 1, 2.
13. EPA/540/R95/128, Soil Screening Guidance: Technical Background Document, 1996, Part 5 Table 36. (www.epa.gov/superfund/health/conmedia/soil/pdfs/part_5.pdf)
14. Haynes, W. M. (ed.) *CRC Handbook of Chemistry and Physics*, 92nd ed., CRC Press: Boca Raton, 2011, pp. 149–152.
15. Lange, N. A. *Handbook of Chemistry*, 10th ed., McGraw-Hill Book Co., Inc.: New York, 1961, pp. 1091–1093.
16. Howard, P. H.; Meylan, W. M. (eds.) *Handbook of Physical Properties of Organic Chemicals*, Lewis Publishers: Boca Raton, 1997.
17. Mackay, D. *Env. Sci. Tech.* 1982, *16*, 224.
18. Suzuki, T. *J. Computer Aided Molec. Design* 1991, *5*, 149.
19. Lange, N. A. *Handbook of Chemistry*, 10th ed., McGraw-Hill Book Co., Inc.: New York, 1961, pp. 1088–1090.
20. www.quimica.aeok.org.ar/Archivos/Tablas/kps.pdf
21. Clever, H. L.; Johnston, F. J. *J. Phys. Chem. Ref. Data*, 1980, *9*, 751.
22. Marcus, Y. *J. Phys. Chem. Ref. Data*, 1980, *9*, 1307.
23. Clever, H. L.; Johnson, S. A.; Derrick, M. E. *J. Phys. Chem. Ref. Data*, 1985, *14*, 631.
24. Clever, H. L.; Johnson, S. A.; Derrick, M. E. *J. Phys. Chem. Ref. Data*, 1992, *21*, 941.
25. Mackay, D.; Shiu, W. Y.; Ma, K. C.; Lee, S. C. *Handbook of Physical-Chemical Properties and Environmental Fate for Organic Chemicals*, CRC Press: Boca Raton, 2006, Vol. 1–4.
26. Lewis, R. J. Jr. Sax's Dangerous Properties of Industrial Materials, Wiley: Hoboken, 2004, Vol. 1–3.
27. Allen, D. T.; Shonnard, D. R. *Green Engineering: Environmentally Conscious Design of Chemical Processes*, Prentice Hall: Upper Saddle River, 2002, p. 386.
28. Davis, G. A.; Kincaid, L.; Swanson, M.; Schultz, T.; Bartmess, J.; Griffith, B.; Jones, S. EPA-600-R-94-177 Chemical Hazard Evaluation for Management Strategies: a method for ranking and scoring chemicals by potential human health and environmental impacts, 1994.

29. EPA-540-R-97-036 Health Effects Assessment Summary Tables (HEAST), 1997. Report of the International Workshop on In Vitro Methods for Assessing Acute Systemic Toxicity, Appendix H, National Toxicity Program, 2006. (http://iccvam.niehs.nih.gov/docs/acutetox_docs/brdval031706/rbrd/appH.pdf)
30. Fiserova-Bergerova, V.; Pierce, T.; Droz, P. O. *Am. J. Ind. Med.* 1990, *17*, 617.
31. Kupczewska-Dobecka, M.; Jakubowski, M.; Czerczak, S. *Env. Toxicol. Pharmacol.* 2010, *30*, 95.
32. Köhler, J.; Meyer, R. *Explosives*, 4th ed., VCH: Weinheim, 1993, pp. 213–215.
33. Lange, N. A. *Handbook of Chemistry*, 10th ed., McGraw-Hill Book Co., Inc.: New York, 1961, pp. 32–49; 824.
34. Lange, N. A. *Handbook of Chemistry*, 10th ed., McGraw-Hill Book Co., Inc.: New York, 1961, pp. 32–49.
35. Haynes, W. M. (ed.) *CRC Handbook of Chemistry & Physics*, 92nd ed., 2011–2012, CRC Press, 2011, Chapter 15, pp. 13–22.
36. Köhler, J.; Meyer, R. *Explosives*, 4th ed., VCH: Weinheim, 1993, pp. 197–201.
37. Tuma, L. D. *Thermochimica Acta* 1994, *243*, 161.
38. Czerczak, S.; Kupczewska, M. *Appl. Occup. Env. Hyg.* 2002, *17*, 187.
39. NIOSH Pocket Guide to Chemical Hazards, National Institute for Occupational Safety and Health, September 2007, Publication No. 2005-149. (www.cdc.gov/niosh)
40. Eissen, M. *Bewertung der Umweltverträglichkeit organisch-chemischer Synthesen.* PhD thesis 2001, Universität Oldenburg.
41. Chen, C. P.; Ahlers, H. W.; Dotson, G. S.; Lin, Y. C.; Chang, W. C.; Maier, A.; Gadagbui, B. *Reg. Toxicol. Pharmacol.* 2011, *61*, 63.
42. Wilschut, A., ten Berge, W. F., Robinson, P. J., McKone, T. E. *Chemosphere* 1995, *30*, 1275.
43. Flynn, G. L. in Garrity, T. R.; Henry, C. J. (eds.) *Principles of Route-to-Route Extrapolation for Risk Assessment*, Elsevier: New York, 1990, pp. 93–127.
44. Patel, H.; ten Berge, W.; Cronin, M. T. D. *Chemosphere* 2002, *48*, 603.
45. Groggins, P. H. *Unit Operations in Organic Syntheses*, 5th ed., McGraw-Hill Book Co.: London, pp. 180–182.
46. Andraos, J.; Ballerini, E.; Vaccaro, L. *Green Chem.* 2015, *17*, 913.
47. Ramgren, S. D.; Hie, L.; Ye, Y.; Garg, N. K. *Org. Lett.* 2013, 15, 3950.
48. Faith, W. L.; Keyes, D. B.; Clark, R. L. *Industrial Chemicals*, 3rd ed., Wiley: Hoboken, 1966; p. 101.

5 Energy Metrics

In this chapter, we tackle the challenge of determining energy metrics for individual reactions and synthesis plans. Based on standard practices in reporting experimental procedures, the disclosure of essential data necessary for a proper determination of energy consumption for carrying out an individual chemical reaction is essentially relegated to reaction temperature and reaction pressure (the latter is particular for industrial reactions for first and second-generation feedstocks). A proper determination of energy consumption involves the following components: (a) a complete data set of thermodynamic parameters for all input materials involved in a chemical reaction: reactants, solvents, catalysts, and any other additives; (b) reaction temperature and reaction pressure; and (c) electricity consumption for all operational procedures requiring equipment such as heating and cooling apparatuses, mixers, rotatory evaporators, vacuum pumps, and so on. The latter contribution from operational procedures is completely absent from experimental procedures and so we focus our attention on estimating energy consumption from the first two contributors. The key thermodynamic parameter sought after is the change in enthalpy for each input material, $q_{\text{input}, i}$, which is made up of a temperature component and a pressure component. The temperature and pressure contributions arise from the respective temperature and pressure changes from ambient standard conditions of 298 K and 1 atm (state 1) to the reaction conditions of T_{rxn} and P_{rxn} (state 2). The temperature contribution requires knowledge of temperature and phase dependent heat capacity functions for each input material. The pressure contribution requires knowledge of an appropriate equation of state for fluids (gases and liquids) applicable to each input material. Figure 5.1 shows a flowchart that tracks the logic and strategy of arriving at the overall change of enthalpy for all input materials in a chemical reaction according to the relationships given in Equations (5.1) and (5.2).

$$q_{\text{input},i} = |q_{\text{temp},i}| + |q_{\text{press},i}| \tag{5.1}$$

$$\Delta H = \sum_{\text{input},i} q_{\text{input},i} = \sum_{\text{input},i} |q_{\text{temp},i}| + \sum_{\text{input},i} |q_{\text{press},i}| \tag{5.2}$$

Absolute values are applied to each component so that a maximum estimate of the overall change of enthalpy for input materials can be made. When $q_{\text{temp}} > 0$, this means that energy needs to be added to the system from the surroundings, that is, heating is required. When $q_{\text{temp}} < 0$ this means that energy needs to be removed from the system to the surroundings, that is, cooling is required. In both cases, clearly energy will be consumed though the movement of that energy is in opposite directions. The same logic is applied to q_{press}, where a positive value corresponds to a situation of expansion or decompression and a negative value corresponds to a situation of compression of a fluid.

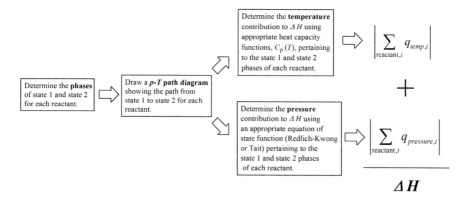

FIGURE 5.1 A flowchart showing how to determine the enthalpy change for each input material in a chemical reaction from state 1 (298 K, 1 atm) to state 2 (T_{rxn}, P_{rxn}).

We begin in Section 5.1.1 by giving definitions of key terms in the field of thermodynamics as a refresher with examples as necessary. These concepts are relevant to analyzing input energy requirements for running a chemical reaction. We remind the reader that the word "thermodynamics" means "powerful movement (Gk: *dynamikos*) of heat (Gk: *thermé, thermos*)" and that this branch of physics is based on two fundamental tenets: (a) it is based on the change of state variables (like enthalpy) from some initial state (state 1 or initial state corresponding to ambient conditions of 298 K and 1 atm) to some other final state (state 2 corresponding to the reaction conditions); and (b) the difference in state variables is independent of the path from state 1 to state 2. Also, by convention, the movement of energy from the system to the surroundings is negative valued and is termed "work done by the system"; and the movement of energy from the surroundings to the system is positive valued and is termed "work done on the system."

Next, in Section 5.1.2, we describe various computational methods used to estimate thermodynamic parameters for substances when they are not available in handbooks. We assume that such situations apply to cases when experimental data are not available. The most reputable sources of thermodynamic parameters are the DIPPR 801 database from the American Institute of Chemical Engineers and the *Yaws Handbooks of Chemical Properties* (see Section 5.1). The best source for group contribution data necessary for carrying out predictive computational methods is Section 2 of *Perry's Chemical Engineers' Handbook*.

In Section 5.2, we delve into the nuts and bolts of how to accomplish the task described by the flowchart in Figure 5.1. We note that this section is highly mathematical but is presented in a readable style that is to the point while maintaining full rigor in the treatment. Each subsection has accompanying worked examples illustrating the concepts. We note that the bulk of the mathematical gymnastics arises from handling cubic equations of state (EOS). This subject is extremely vast in the literature with more than 300 equations of state documented and most of the literature is just that, mathematical gymnastics, with very little practical value. For our purposes, we choose the Redlich–Kwong and Peng–Robinson EOS to do practical work since

Energy Metrics

they are accepted and are well used and well described in industrial chemistry and chemical engineering literature and textbooks. For historical importance, we also include the ideal gas and van der Waals equations of state. The prerequisite mathematical skills necessary for reading this section and working through the examples are elementary algebra, and derivative and integral calculus. Readers who are not confident in these topics can consult introductory textbooks to refresh their knowledge or may skip ahead to the next sections without loss of understanding. We highly recommend to the reader the excellent pedagogical textbook on thermodynamics by I. Tosun, which is written in clear English and is annotated with worked examples and problems with answers.[1] We used this textbook as a model for presenting the material in this chapter.

In Section 5.3, we describe various energy metrics that have been put forward in the green chemistry literature that are applicable to individual reactions and synthesis plans. This section is fully annotated with worked examples.

In Section 5.4, we describe Excel spreadsheets that automate much of the tedious computations described in the mathematical Section 5.2. This makes the entire task of estimating the change in enthalpy for input materials effortless once the input thermodynamic parameters and data are entered.

We conclude the chapter with a set of problems for the reader.

5.1 THERMODYNAMIC PRELIMINARIES

5.1.1 Terms, Definitions, and Examples

Acentric Factor[2–5]

The acentric factor of a compound, ω, is defined in terms of the reduced vapor pressure evaluated at a reduced temperature of 0.7 and is given by Equation (5.3).

$$\omega = -\log p_{\text{vap},r}\big|_{T_r=0.7} - 1 \qquad (5.3)$$

It is used as a measure of deviations from non-spherical molecular shape, hence the name, suggesting molecular interactions that are not between centers of molecules. The acentric factor increases slightly with increasing polarity for molecules of similar size and shape. Its value is close to zero for small, spherically shaped, non-polar molecules (argon, methane, etc.). It increases in value with larger deviations of molecular shape from spherical (longer chain lengths, less chain branching, etc.) and with increasing molecular polarity. This parameter was introduced by Kenneth Pitzer in 1955.

Adiabatic Process

A process that takes place with no exchange of heat between the system and its surroundings.

Critical Constants[4]

The critical constants refer to the values of temperature (T_c), pressure (P_c), and volume (V_c) of a substance at its critical point. These quantities are important in determining phase boundaries of a substance in a phase diagram, particularly a pressure–temperature diagram in which the critical point represents the point at which no phase boundaries exist. The critical temperature and critical pressure constants are required input parameters for most thermodynamic calculations and are derivable from particular equations of state under the conditions $(dp/dV)_T=0$ and $(d^2p/dV^2)_T=0$. The critical temperature of a compound is the temperature above which a liquid phase cannot be formed, no matter the pressure of the system. The critical pressure is the vapor pressure of the compound at the critical temperature. The critical volume is the volume occupied by a set amount of a compound (typically 1 mol) at its critical temperature and pressure.

Density[6]

Density is defined as the mass of a substance per unit volume. Typical units are kg/m³ or g/cm³.

DIPPR density correlation functions for liquids and solids are given by Equations (5.4) and (5.5), respectively.

Density (liquid): (eq. 105 in DIPPR database)

$$\rho = \frac{A}{B^{1+\left(1-\frac{T}{C}\right)^D}}; \text{Units: kmol m}^{-3} \tag{5.4}$$

where A, B, C, and D are constants specific to a given substance.

Density (solid): (eq. 100 in DIPPR database)

$$\rho = A + BT + CT^2 + DT^3 + ET^4; \text{Units: kmol m}^{-3} \tag{5.5}$$

where A, B, C, D, and E are constants specific to a given substance.

Yaws density correlation function for liquids is given by Equation (5.6).

$$\rho = \frac{A}{B^{(1-T_r)^n}}; \text{Units: g mL}^{-1} \tag{5.6}$$

where A and B are constants specific to a given substance.

Design Institute for Physical Property Data (DIPPR 801)[7]

The DIPPR 801 project is affiliated with the American Institute of Chemical Engineers (AIChE). The database collates reliable thermophysical property data on 2424 industrial chemicals including 34 constant and 15 temperature dependent properties. Among these the following are important in estimating input energy consumption in carrying out chemical reactions: acentric factor, normal boiling point, critical pressure, critical temperature, heat capacity of ideal gas, heat capacity of liquid, heat capacity of solid, heat of vaporization, liquid density, solid density, and vapor pressure of liquid.

Enthalpy[8]

Enthalpy represents the sum of the internal energy of a system, U, and the multiplicative product of pressure, p, and volume, V given by Equation (5.7).

$$H = U + pV; \text{ Units: J/mol} \tag{5.7}$$

Energy Metrics

In thermodynamics, enthalpy is considered an extensive property because it scales with the size of the physical system. Since enthalpy is a function of temperature and pressure, $H(T,p)$, the corresponding total differential is given by Equation (5.8).

$$dH = \left(\frac{\partial H}{\partial T}\right)_p dT + \left(\frac{\partial H}{\partial p}\right)_T dp \qquad (5.8)$$

Since $\left(\frac{\partial H}{\partial T}\right)_p = C_p$ (heat capacity at constant pressure definition) and $\left(\frac{\partial H}{\partial p}\right)_T = V + T\left(\frac{\partial S}{\partial p}\right)_T = V - T\left(\frac{\partial V}{\partial T}\right)_p$ (by Maxwell's relations), Equation (5.8) can be rewritten according to Equation (5.9), which is a key relationship used in determining the enthalpy change upon transformation of state 1 (p_1, T_1) to state 2 (p_2, T_2) (see Section 5.2.5).

$$dH = C_p dT + \left[V - T\left(\frac{\partial V}{\partial T}\right)_p\right]dp$$

$$\int_{H_1}^{H_2} dH = \int_{T_1}^{T_2} C_p(T) dT + \int_{p_1}^{p_2}\left[V - T\left(\frac{\partial V}{\partial T}\right)_p\right]dp \qquad (5.9)$$

$$\Delta H = q_{\text{input}} = \int_{T_1}^{T_2} C_p(T) dT + \int_{p_1}^{p_2}\left[V - T\left(\frac{\partial V}{\partial T}\right)_p\right]dp$$

$$q_{\text{input}} = q_{\text{temp}} + q_{\text{press}}$$

The first integral corresponds to the temperature change contribution and the second integral corresponds to the pressure change contribution to the enthalpy change.

Enthalpy of Formation (Heat of Formation)[9]

"The standard enthalpy (heat) of formation is the enthalpy change upon formation of 1 mol of the compound in its standard state from its constituent elements in their standard states. Two different standard enthalpies of formation are commonly defined based on the chosen standard state. The standard state enthalpy of formation ΔH_f^s uses the naturally occurring phase at 298.15 K and 1 bar as the standard state; the ideal gas standard enthalpy (heat) of formation ΔH_f^o uses the compound in the ideal gas state at 298.15 K and 1 bar as the standard state. In both cases, the standard state for the elements is their naturally occurring state of aggregation at 298.15 K and 1 atm." Note that 1 bar = 0.98692 atm and 1 atm = 1.0133 bar. If $\Delta H_f > 0$, then heat is absorbed in the process and if $\Delta H_f < 0$, then heat is evolved in the process. The heat of formation of the elements in their pure standard states is 0 by definition. Units: kJ/mol, J/mol.

Enthalpy of Fusion (Heat of Fusion)[10]

The enthalpy (heat) of fusion, ΔH_{fus}, is the difference between the molar enthalpies of the equilibrium liquid and solid at the melting temperature and 1.0 atm pressure. At the melting point, it represents the heat required to transition from the liquid phase to the solid phase, or vice versa, and is often called latent heat of fusion.

Enthalpy of Reaction (Heat of Reaction)

The enthalpy of reaction is the amount of heat that is liberated (applicable to exothermic reactions so that $\Delta H_{rxn} < 0$) or the amount of heat that is absorbed (applicable to endothermic reactions so that $\Delta H_{rxn} > 0$) when a chemical reaction takes place. Units: kJ/mol, J/mol.

The amount of heat liberated or absorbed depends on the bond breaking and bond making processes that take place as the reaction proceeds. The enthalpy of reaction is calculated from Hess's law using enthalpy of formation data for all reactants and all products based on a balanced chemical equation with appropriate stoichiometric coefficients. Since all of the enthalpy of formation quantities are determined under standard state conditions (298 K, 1 atm), then the resulting calculated enthalpy of reaction is also determined under the same standard state conditions.

Enthalpy of Vaporization (Heat of Vaporization)[11]

The enthalpy (heat) of vaporization, ΔH_{vap}, is the difference between the molar enthalpies of the saturated vapor and saturated liquid at a temperature between the triple point and critical point (at the corresponding vapor pressure). At the boiling point, it represents the heat required to transition from the liquid phase to the gas phase, or vice versa, and is often called latent heat of vaporization.

The DIPPR correlation function for enthalpy of vaporization (eq. 106 in database) is given by Equation (5.10).

$$\Delta H_{vap} = A(1-T_r)^{B+CT_r+DT_r^2+ET_r^3} ; \text{Units: J/kmol} \qquad (5.10)$$

where A, B, C, D, and E are constants specific to a given substance.

The Yaws correlation function for enthalpy of vaporization is given by Equation (5.11).

$$\Delta H_{vap} = A(1-T_r)^n ; \text{Units: kJ/mol} \qquad (5.11)$$

where A and n are constants specific to a given substance.

Equation Of State

Generally, an equation of state is any functional relationship between state variables. Specifically, an equation of state is a function of pressure, p, in terms of temperature and molar volume, given generally by $p(T, V)$. The functional forms of most commonly used equations of state are typically cubic functions in V. An equation of state is a function that describes the behavior of a fluid over a range of pressures,

Energy Metrics

temperatures, and volumes. In such a parameterization, the gas and liquid phases are treated as one continuous fluid.

Equation of State: Ideal Gas

The ideal gas equation of state is given by Equation (5.12).

$$p = \frac{RT}{V} \qquad (5.12)$$

where p=pressure (atm), R=universal gas constant=0.08207 L atm/(mol deg K), T=temperature (degrees K), and V=molar volume (L/mol).

Equation of State: Van Der Waals[12,13]

The van der Waals equation of state is given by Equation (5.13).

$$p = \frac{RT}{V-b} - \frac{a}{V^2} \qquad (5.13)$$

where a is the attractive constant and b is the repulsion constant given by Equations (5.14) and (5.15) involving the critical constants.

$$a = \frac{27}{64} \frac{R^2 T_c^2}{P_c} \qquad (5.14)$$

$$b = \frac{RT_c}{8P_c} \qquad (5.15)$$

Equation of State: Redlich–Kwong[14]

The Redlich–Kwong equation of state is given by Equation (5.16).

$$p = \frac{RT}{V-b} - \frac{a}{\sqrt{T}} \frac{1}{V(V+b)} \qquad (5.16)$$

where a is the attractive constant and b is the repulsion constant given by Equations (5.17) and (5.18) involving the critical constants.

$$a = \frac{R^2 T_c^{5/2}}{P_c} \left(\frac{1 + 2^{1/3}(2^{1/3}+1)}{9} \right) = 0.427480234 \frac{R^2 T_c^{5/2}}{P_c} \qquad (5.17)$$

$$b = \left(\frac{2^{1/3}-1}{3} \right) \left(\frac{RT_c}{P_c} \right) = 0.08664035 \frac{RT_c}{P_c} \qquad (5.18)$$

Equation of State: Peng–Robinson[15]

The Peng–Robinson equation of state is given by Equation (5.19).

$$p = \frac{RT}{V-b} - \frac{a\alpha}{V^2 + 2bV - b^2} \qquad (5.19)$$

where $a\alpha$ is the attractive constant and b is the repulsion constant given by Equations (5.20) to (5.24) involving the critical constants.

$$a\alpha = \frac{R^2 T_c^2}{p_c}\left[\frac{1+4\beta+10\beta^2}{3}\right] = 0.457235529 \frac{R^2 T_c^2}{p_c} \quad (5.20)$$

$$b = \beta\frac{RT_c}{p_c} = 0.077796074 \frac{RT_c}{p_c} \quad (5.21)$$

$$\beta = \frac{1}{2^5}\left\{3\left[13+(2^4)\sqrt{2}\right]^{1/3}+3\left[13-(2^4)\sqrt{2}\right]^{1/3}-1\right\} = 0.077796074 \quad (5.22)$$

$$\alpha = \left(1+\kappa\left(1-\sqrt{T_r}\right)\right)^2 \quad (5.23)$$

$$\kappa = 0.37464 + 1.54226\omega - 0.26992\omega^2 \quad (5.24)$$

where ω is the acentric factor.

The following steps are followed for the determination of constants in terms of critical constants for various equations of state:

1. Rewrite equation of state in expanded form.
2. Take the total differential at constant T.
3. Divide each term by dV.
4. Set $dp/dV = 0$ and simplify.
5. Again take the total differential at constant T.
6. Again set $dp/dV = 0$ and simplify.
7. From equations obtained in steps 4 and 6 isolate constant variables in terms of p, V, R, and T.
8. Substitute equations obtained in step 7 into original equation of state to obtain an expression for V in terms of R, T, and p.
9. At the critical point, determine expression for V_c in terms of R, T_c, and p_c.
10. Substitute equation from step 9 into equations obtained in step 7 to obtain expressions for the constant variables in terms of R, T_c, and p_c.

Example 5.1

Verify Equations (5.14) and (5.15) for the van der Waals equation of state.

SOLUTION

Given $p = \dfrac{RT}{V-b} - \dfrac{a}{V^2}$

rewrite as

$$p(V^3 - bV^2) = RTV^2 - aV + ab$$

Energy Metrics

Taking the total differential of at constant T yields

$$p(3V^2 - 2bV)dV + (V^3 - bV^2)dp = (2RTV - a)dV.$$

After dividing by dV and setting $dp/dV=0$, we have

$$bp2V + RT2V - 3pV^2 = a.$$

Taking the total differential again at constant T yields

$$2RTdV + 2b(pdV + Vdp) - 6VpdV - 3V^2 dp = 0.$$

After dividing by dV and setting $dp/dV=0$, we have

$$b = 3V - \frac{RT}{p}.$$

After substituting and simplifying we obtain

$$a = 3pV^2.$$

After substituting the previous two equations into the original van der Waals equation and simplifying we obtain

$$V = \frac{3}{8}\frac{RT}{p}.$$

Therefore, at the critical point

$$V_c = \frac{3}{8}\frac{RT_c}{p_c}, \quad a = 3p_c V_c^2 = \frac{27}{64}\frac{R^2 T_c^2}{p_c}, \quad \text{and} \quad b = 3V_c - \frac{RT_c}{p_c} = \frac{1}{8}\frac{RT_c}{p_c}.$$

PROBLEM 5.1
Verify Equations (5.17) and (5.18) for the Redlich–Kwong equation of state.

PROBLEM 5.2
Verify Equations (5.20) and (5.21) for the Peng–Robinson equation of state.

Heat Capacity at Constant Pressure, C_p[16–18]

Heat capacity is the energy required to change the temperature of a unit mass (specific heat) or mole (molar heat capacity) of the material by one degree. Units: J/(kg deg K); J/(mol deg K). The formal thermodynamic definition is given by the derivative shown in Equation (5.25).

$$C_p = \left(\frac{\partial H}{\partial T}\right)_p \tag{5.25}$$

Heat capacity is an extensive property since it is a physical property that scales with the size of a physical system.

DIPPR heat capacity correlation functions are given by Equations (5.26) to (5.29).
Heat capacity (gas): (eq. 107 in DIPPR database)

$$C_{p,g} = A + B\left[\frac{C/T}{\sinh(C/T)}\right]^2 + D\left[\frac{E/T}{\cosh(E/T)}\right]^2 ; \text{Units: J kmol}^{-1}\text{K}^{-1} \quad (5.26)$$

where A, B, C, D, and E are constants specific to a given substance.
Heat capacity (liquid): (eq. 100 in DIPPR database)

$$C_{p,l} = A + BT + CT^2 + DT^3 + ET^4 ; \text{Units: J kmol}^{-1}\text{K}^{-1} \quad (5.27)$$

where A, B, C, D, and E are constants specific to a given substance.
Heat capacity (solid): (eq. 100 or eq. 102 in DIPPR database)

$$C_{p,s} = A + BT + CT^2 + DT^3 + ET^4 ; \text{Units: J kmol}^{-1}\text{K}^{-1} \quad (5.28)$$

where A, B, C, D, and E are constants specific to a given substance.

$$C_{p,s} = \frac{AT^B}{1 + \frac{C}{T} + \frac{D}{T^2}} ; \text{Units: J kmol}^{-1}\text{K}^{-1} \quad (5.29)$$

where A, B, C, and D are constants specific to a given substance.

Yaws heat capacity correlation functions are given by Equations (5.30) to (5.32).
Heat capacity (gas):

$$C_{p,g} = A + BT + CT^2 + DT^3 + ET^4 ; \text{Units: J mol}^{-1}\text{K}^{-1} \quad (5.30)$$

where A, B, C, D, and E are constants specific to a given substance.
Heat capacity (liquid):

$$C_{p,l} = A + BT + CT^2 + DT^3 ; \text{Units: J mol}^{-1}\text{K}^{-1} \quad (5.31)$$

where A, B, C, and D are constants specific to a given substance.
Heat capacity (solid):

$$C_{p,s} = A + BT + CT^2 ; \text{Units: J mol}^{-1}\text{K}^{-1} \quad (5.32)$$

where A, B, and C are constants specific to a given substance.

Hess's Law[19,20]

The mathematical statement of Hess's law for a balanced chemical reaction is given by Equation (5.33).

Energy Metrics

$$\Delta H^o_{rxn} = \sum_{\text{products},i} \nu_i \left(\Delta H^o_f\right)_i - \sum_{\text{reactants},i} \nu_i \left(\Delta H^o_f\right)_i \tag{5.33}$$

where ν_i is the stoichiometric coefficient pertaining to chemical species i, and $\left(\Delta H^o_f\right)_i$ is the heat of formation of chemical species i.

If the reaction is carried out at a temperature different from 298 K, then the enthalpy of reaction calculated at 298 K must be corrected using temperature dependent heat capacity data pertaining to each reactant and each product appearing in the chemical reaction according to Equation (5.34).

$$\begin{aligned}
\Delta H^{T \ne 298}_{rxn} &= \Delta H^o_{rxn} + \int_{T=298}^{T \ne 298} C^o_{p,rxn}(T)\,dT \\
&= \Delta H^o_{rxn} + \int_{T=298}^{T \ne 298} \left[\sum_{\text{products},i} \nu_i C^o_{p,\text{products},i}(T) - \sum_{\text{reactants},i} \nu_i C^o_{p,\text{reactants},i}(T)\right] dT
\end{aligned} \tag{5.34}$$

Example 5.2

Calculate the enthalpy of reaction at 298 K for the combustion of methane according to the chemical equation shown below.

$$CH_4(g) + 2\,O_2(g) \to CO_2(g) + 2\,H_2O(g)$$

Solution

Chemical	ΔH^o_f (kJ/mol)
CH_4 (g)	-74.8
O_2 (g)	0
CO_2 (g)	-393.5
H_2O (g)	-241.8

Therefore, $\Delta H^o_{rxn} = \left[2(-241.8) + 1(-393.5)\right] - \left[1(-74.8) + 2(0)\right] = -802.3\,\text{kJ/mol}$.
Since the value is negative, the combustion reaction is exothermic.

Example 5.3

Calculate the enthalpy of reaction at 298 K for the combustion of methane according to the chemical equation shown below.

$$CH_4(g) + 2\,O_2(g) \to CO_2(g) + 2\,H_2O(\text{liq})$$

Solution

Chemical	ΔH_f^o (kJ/mol)
CH_4 (g)	−74.8
O_2 (g)	0
CO_2 (g)	−393.5
H_2O (liq)	−285.8

Therefore, $\Delta H_{rxn}^o = \left[2(-285.8)+1(-393.5)\right] - \left[1(-74.8)+2(0)\right] = -890.3\,\text{kJ/mol}$.
Since the value is negative, the combustion reaction is exothermic.

Example 5.4

If methane is combusted at 500°C in a furnace, determine the heat of reaction according to the chemical equation shown below.

$$CH_4(g) + 2O_2(g) \rightarrow CO_2(g) + 2H_2O(g)$$

The relevant temperature dependent heat capacity function is given below along with the coefficient data.

$$C_p^o(T) = A + BT + CT^2 + DT^3 + ET^4$$

Chemical	A	B	C	D	E
CH_4 (g)	34.942	−4.00E−02	1.19E−04	−1.53E−07	3.93E−11
O_2 (g)	29.526	−8.90E−03	3.81E−05	−3.26E−08	8.86E−12
CO_2 (g)	27.437	4.23E−02	−1.96E−05	4.00E−09	−2.99E−13
H_2O (g)	33.933	−8.42E−03	2.99E−05	−1.78E−08	3.69E−12

Solution

$T_1 = 298\,\text{K}$
$T_2 = 500\,\text{C} = 500 + 273 = 773\,\text{K}$

$$\int_{T_1}^{T_2} C_p^o(T)dT = \int_{T_1}^{T_2}\left(A + BT + CT^2 + DT^3 + ET^4\right)dT$$

$$= \left[AT + B\frac{T^2}{2} + C\frac{T^3}{3} + D\frac{T^4}{4} + E\frac{T^5}{5}\right]_{T_1}^{T_2}$$

$$= A(T_2 - T_1) + \frac{B}{2}(T_2^2 - T_1^2) + \frac{C}{3}(T_2^3 - T_1^3) + \frac{D}{4}(T_2^4 - T_1^4) + \frac{E}{5}(T_2^5 - T_1^5)$$

Heat capacity contributions:

Chemical	A	B	C	D	E	T1	T2	C_p Contribution (J/mol)
CH_4 (g)	34.942	−4.00E−02	1.19E−04	−1.53E−07	3.93E−11	2.98E+02	773	1.25E+04
O_2 (g)	29.526	−8.90E−03	3.81E−05	−3.26E−08	8.86E−12	2.98E+02	773	1.49E+04
CO_2 (g)	27.437	4.23E−02	−1.96E−05	4.00E−09	−2.99E−13	2.98E+02	773	2.13E+04
H_2O (g)	33.933	−8.42E−03	2.99E−05	−1.78E−08	3.69E−12	2.98E+02	773	1.70E+04

Therefore, overall heat capacity contribution is

$$2*(1.70E4)+1*(2.13E4)-\left((1)*1.25E4+(2)*1.49E4\right)=13000\,\text{J/mol}=12.9\,\text{kJ/mol}.$$

Therefore, heat of reaction is $-890.3+12.9=-877.4$ kJ/mol. Since the value is negative, the combustion reaction is exothermic.

Isobaric
A process that takes place at constant pressure.

Isochoric or Isometric
A process that takes place at constant volume.

Isothermal
A process that takes place at constant temperature.

Maxwell Equal Area Rule[21]
An important feature of original or reduced forms of cubic EOS is the so-called van der Waals loop occurring in isotherms below the critical temperature which, as shown by the sketch of a typical cubic p-V function in Figure 5.2, is characterized by a wiggle shape where the pressure decreases steeply to a sharp minimum value as the molar volume is increased, then increases to a broad maximum value upon further increase in molar volume, and finally gradually decreases slowly as a long trailing tail. Since this functional behavior has no physical sense as it is expected that the molar volume increase as the pressure steadily decreases, Maxwell in 1875 proposed an equal area rule to circumvent the problem of this wiggle. The basic idea is to draw a horizontal tie line, representing the saturation pressure, across the span of the wiggle in such a way that the area enclosed by the horizontal line and the lower part of the cubic curve (area A) is equal to the area enclosed by the horizontal line and the upper part of the cubic curve (area B). The physical interpretation is that for a given isotherm below the critical temperature, T_c, the curve preceding the tie line represents the liquid P-V line, the horizontal tie line represents the two-phase liquid-vapor region, and the curve following the tie line represents the vapor P-V line. The intersection points to the left and right of the saturation pressure tie line represent the

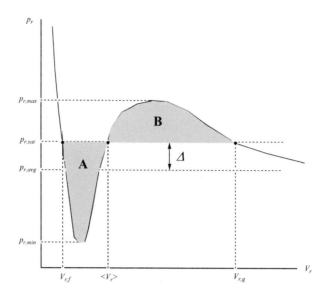

FIGURE 5.2 Generic plot of reduced pressure, p_r, versus reduced molar volume, V_r, for a general cubic equation of state function where $T_r < 1$ showing implementation of Maxwell equal area rule where *area A* equals *area B*. Reduced saturation pressure is the horizontal dashed line intersecting the curve in three points: saturated liquid molar volume, $V_{r,f}$, saturated vapor molar volume, $V_{r,g}$, and saturated virtual molar volume, $<V_r>$. Δ represents the difference between the reduced saturation pressure, $p_r(sat)$, and the reduced average pressure, $p_r(avg) = (p_r(max) + p_r(min))/2$.

molar volumes of the saturated liquid and saturated vapor, respectively. The central intersection point is a virtual point that has no physical meaning. These three points represent the three real roots of the cubic equation of state in a mathematical sense when the cubic is solved for molar volume at a given pressure. It should be noted that the central intersection point along the horizontal tie line does *not* represent a point of inflection along the cubic curve. Also, Maxwell's original 1875 sketch is unfortunately misleading since he drew a symmetric wiggle where the horizontal tie line appeared to pass through an inflection point. The correct shape of P-V curves based on cubic functions in fact shows a strongly asymmetric wiggle and the horizontal tie line always lies above both the point of inflection and the average between the minimum and maximum points (see Figure 5.2).

Example 5.5

Determine area A and area B for the reduced van der Waals equation of state.

$$p_r = \frac{8T_r}{3V_r - 1} - \frac{3}{V_r^2}$$

Solution

$$\text{Area A} = \int_{V_{r,f}}^{\langle V_r \rangle} \left\{ p_{r,\text{sat}} - \left[\frac{8T_r}{3V_r - 1} - \frac{3}{V_r^2} \right] \right\} dV_r$$

$$= p_{r,\text{sat}} \left[\langle V_r \rangle - V_{r,f} \right] - 8T_r \int_{V_{r,f}}^{\langle V_r \rangle} \frac{dV_r}{3V_r - 1} + 3 \int_{V_{r,f}}^{\langle V_r \rangle} \frac{dV_r}{V_r^2}$$

$$= p_{r,\text{sat}} \left[\langle V_r \rangle - V_{r,f} \right] - \frac{8T_r}{3} \ln(3V_r - 1) \Big|_{V_{r,f}}^{\langle V_r \rangle} - 3 \frac{1}{V_r} \Big|_{V_{r,f}}^{\langle V_r \rangle}$$

$$= p_{r,\text{sat}} \left[\langle V_r \rangle - V_{r,f} \right] - \frac{8T_r}{3} \ln \left[\frac{3\langle V_r \rangle - 1}{3V_{r,f} - 1} \right] - 3 \left[\frac{1}{\langle V_r \rangle} - \frac{1}{V_{r,f}} \right]$$

$$\text{Area B} = \int_{\langle V_r \rangle}^{V_{r,g}} \left\{ \left[\frac{8T_r}{3V_r - 1} - \frac{3}{V_r^2} \right] - p_{r,\text{sat}} \right\} dV_r$$

$$= 8T_r \int_{\langle V_r \rangle}^{V_{r,g}} \frac{dV_r}{3V_r - 1} - 3 \int_{\langle V_r \rangle}^{V_{r,g}} \frac{dV_r}{V_r^2} - p_{r,\text{sat}} \left[V_{r,g} - \langle V_r \rangle \right]$$

$$= \frac{8T_r}{3} \ln(3V_r - 1) \Big|_{\langle V_r \rangle}^{V_{r,g}} + 3 \frac{1}{V_r} \Big|_{\langle V_r \rangle}^{V_{r,g}} - p_{r,\text{sat}} \left[V_{r,g} - \langle V_r \rangle \right]$$

$$= \frac{8T_r}{3} \ln \left[\frac{3V_{r,g} - 1}{3\langle V_r \rangle - 1} \right] + 3 \left[\frac{1}{V_{r,g}} - \frac{1}{\langle V_r \rangle} \right] - p_{r,\text{sat}} \left[V_{r,g} - \langle V_r \rangle \right]$$

PROBLEM 5.3
Determine area A and area B for the reduced Redlich–Kwong equation of state.

$$p_r = \frac{3T_r}{V_r - 0.259921} - \frac{1}{\sqrt{T_r}} \frac{3.847322}{V_r(V_r + 0.259921)}$$

PROBLEM 5.4
Determine area A and area B for the reduced Peng–Robinson equation of state.

$$p_r = \frac{3T_r}{V_r(1-\beta) - 3\beta} - \frac{3(1 + 4\beta + 10\beta^2)}{V_r^2(1-\beta)^2 + 6V_r\beta(1-\beta) - 9\beta^2}$$

Figures 5.3 through 5.5 show example equal area determinations for the van der Waals, Redlich–Kwong, and Peng–Robinson equation of states.

The polynomial correlations given in Equations (5.35) to (5.37) for the reduced saturation pressure as a function of reduced temperature are applicable to the van der Waals, Redlich–Kwong, and Peng–Robinson equations of state, respectively.

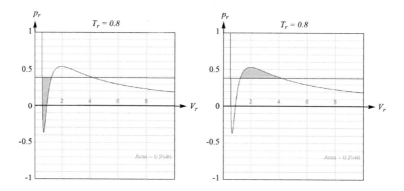

FIGURE 5.3 Maxwell equal area construction for the reduced van der Waals equation of state function with $T_r = 0.8$.

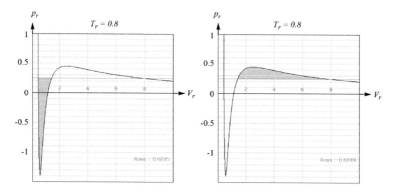

FIGURE 5.4 Maxwell equal area construction for the reduced Redlich–Kwong equation of state function with $T_r = 0.8$.

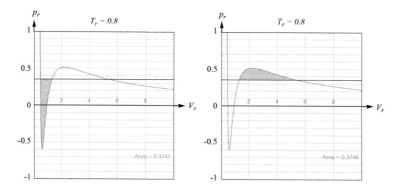

FIGURE 5.5 Maxwell equal area construction for the reduced Peng–Robinson equation of state function with $T_r = 0.8$.

$$p_{r,\text{sat}} = 2.6349T_r^3 - 2.1395T_r^2 + 0.5612T_r - 0.0471; r^2 = 1 \quad (5.35)$$

$$p_{r,\text{sat}} = 6.9385T_r^3 - 10.014T_r^2 + 4.8794T_r - 0.8012; r^2 = 1 \quad (5.36)$$

$$p_{r,\text{sat}} = 3.1792T_r^3 - 3.0580T_r^2 + 0.9871T_r - 0.1058; r^2 = 1 \quad (5.37)$$

As a check, note that at the critical point, when $T_r = 1$, $p_{r,\text{sat}} = 1$ in all cases.

Similarly, the polynomial correlations given in Equations (5.38) to (5.40) for the saturated liquid molar volume as a function of reduced temperature are applicable to the van der Waals, Redlich–Kwong, and Peng–Robinson equations of state, respectively.

$$V_{r,f} = 2.9170T_r^4 - 5.4003T_r^3 + 3.7752T_r^2 - 0.9587T_r + 0.4374; r^2 = 0.9991 \quad (5.38)$$

$$V_{r,f} = 14.8320T_r^4 - 38.3870T_r^3 + 37.5070T_r^2 - 16.0640T_r + 2.8354; r^2 = 0.9997 \quad (5.39)$$

$$V_{r,f} = 4.2945T_r^4 - 8.5115T_r^3 + 6.3390T_r^2 - 1.8709T_r + 0.4693; r^2 = 0.9990 \quad (5.40)$$

Similarly, the polynomial correlations given in Equations (5.41) to (5.43) for the saturated vapor molar volume as a function of reduced temperature are applicable to the van der Waals, Redlich–Kwong, and Peng–Robinson equations of state, respectively.

$$\ln V_{r,g} = 113.09T_r^4 - 318.53T_r^3 + 335.40T_r^2 - 163.87T_r + 34.598; r^2 = 0.9993 \quad (5.41)$$

$$\ln V_{r,g} = 42.211T_r^4 - 164.43T_r^3 + 238.55T_r^2 - 161.55T_r + 45.543; r^2 = 1 \quad (5.42)$$

$$\ln V_{r,g} = 79.629T_r^4 - 238.12T_r^3 + 267.74T_r^2 - 141.63T_r + 32.927; r^2 = 0.9998 \quad (5.43)$$

Molar Volume

Molar volume is the volume occupied by 1 mole of a substance under standard state conditions (25°C, 1 atm). Typical units are L/mol, cm³/mol, m³/mol, m³/kmol. The density of a substance is obtained when the molecular weight of a substance is divided by its molar volume. The *liquid* molar volume is calculated from the correlation equation for the liquid density of a substance at 298 K and 1 atm. If the substance is a solid at 298 K, the molar volume is calculated at its triple point temperature. If the substance is a gas at 298 K, the molar volume is calculated at its normal boiling point temperature.

Normal Boiling Point

The normal boiling temperature, T_b, is defined as the temperature at which the vapor pressure of a liquid equals 101.325 kPa (1.0 atm), or alternatively the temperature at

TABLE 5.1
Summary of Possible Pairwise Phase Transitions between the Solid, Liquid, and Gas Phases

	Solid	Liquid	Gas
Solid		solid -> liquid (melting)	solid -> gas (sublimation)
Liquid	liquid -> solid (freezing)		liquid -> gas (vaporization)
Gas	gas -> solid (deposition)	gas -> liquid (condensation)	

which a substance undergoes a phase transition from liquid to gas (vaporization) at atmospheric pressure ($p = 1$ atm).

Normal Melting Point

The normal melting point, T_m, is defined as the temperature at which a substance undergoes a phase transition from solid to liquid (melting) at atmospheric pressure ($p = 1$ atm). It is also known as the normal fusion (or freezing) point when the phase transition occurs in the opposite direction, namely from the liquid state to the solid state.

Phase Transition

A phase transition corresponds to a pathway that changes the phase of a substance either by applying a change of temperature or a change of pressure. In a typical pressure–temperature (p-T) phase diagram, a phase transition occurs if a vertical or a horizontal line crosses anyone of the three boundary lines representing the boundaries between the solid, liquid, and gas phase regions of a substance. Since there are three main phases, there are six possible pairwise phase transitions that are possible as summarized in Table 5.1. Figure 5.6 shows some phase transitions depicted on a typical p-T phase diagram.

Principle Of Corresponding States

The principle of corresponding states applies to conformal fluids. Two fluids are said to be conformal if their intermolecular interactions are equivalent when scaled in dimensionless form. For example, all conformal fluids obeying a given equation of state p-V function will have identical dimensionless properties because each fluid interacts through the identical scaled properties defined by the equation of state. Generalization of this scaling principle is commonly done using critical temperature T_c and critical pressure P_c as scaling factors. At the same reduced coordinates ($T_r = T/T_c$ and $P_r = P/P_c$), all conformal fluids will have the same dimensionless properties.

The following steps are followed to transform a given p-V equation of state into dimensionless form:

1. Rewrite original equation of state by making the following substitutions: $p = p_r p_c$, $T = T_r T_c$, and $V = V_r V_c$.
2. Substitute relationships for all constants and V_c in terms of R, T_c, and p_c.
3. Simplify until a function of $p_r(T_r, V_r)$ is obtained.

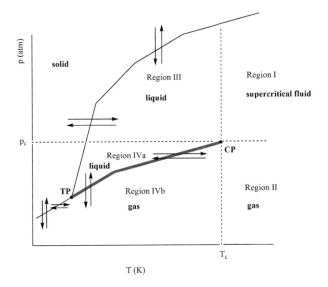

FIGURE 5.6 Diagrammatic representation of phase transitions on a typical p-T phase diagram. TP=triple point; CP=critical point; Red curve is vapor pressure curve, $p_{vap}(T)$.

Example 5.6

Transform the van der Waals EOS into a reduced form.

$$p = \frac{RT}{V-b} - \frac{a}{V^2}, \text{ where } a = \frac{27}{64}\frac{R^2(T_c)^2}{p_c} \text{ and } b = \frac{RT_c}{8p_c}$$

Solution

Rewrite the EOS as follows applying appropriate substitutions and simplification.

$$p_r p_c = \frac{RT_r T_c}{V_r V_c - b} - \frac{a}{V_r^2 V_c^2}$$

$$= \frac{RT_r T_c}{V_r \dfrac{3}{8}\dfrac{RT_c}{p_c} - \dfrac{RT_c}{8p_c}} - \frac{\dfrac{27}{64}\dfrac{R^2 T_c^2}{p_c}}{V_r^2 \left(\dfrac{3}{8}\dfrac{RT_c}{p_c}\right)^2}$$

$$= \frac{T_r}{V_r \dfrac{3}{8}\dfrac{1}{p_c} - \dfrac{1}{8p_c}} - \frac{\dfrac{27}{64}\dfrac{1}{p_c}}{V_r^2 \left(\dfrac{3}{8}\dfrac{1}{p_c}\right)^2}$$

$$= \frac{8p_c T_r}{3V_r - 1} - \frac{3p_c}{V_r^2}$$

Therefore, the reduced van der Waals EOS is $p_r = \dfrac{8T_r}{3V_r - 1} - \dfrac{3}{V_r^2}$. Note that comparison of this reduced form to the original EOS shows that the constants *a* and *b* disappear.

PROBLEM 5.5
Transform the Redlich–Kwong EOS into a reduced form.

PROBLEM 5.6
Transform the Peng–Robinson EOS into a reduced form.

Reduced Constants
Reduced constants are dimensionless quantities used to transform a given equation of state *p-V* function into a dimensionless form that does not depend on specific constants of a given fluid pertaining to that equation of state. Such a transformation is applicable to conformal fluids that obey the principle of corresponding states. Typical reduced constants are reduced temperature, T_r, defined as $T_r = T/T_c$, and reduced pressure, P_r, defined as $P_r = P/P_c$. A reduced boiling point is defined as $T_{br} = T_b/T_c$.

Saturation Temperature and Pressure
A liquid is said to be *saturated* if it contains as much thermal energy as it can without boiling. A vapor is said to be *saturated* if it contains as little thermal energy as it can without condensing.

The saturation temperature refers to the *boiling point*. The saturation temperature is the temperature for a corresponding saturation pressure at which a liquid boils into its vapor phase. The liquid is therefore saturated with thermal energy. Any addition of more thermal energy results in a phase transition to the vapor phase.

Saturation pressure is the pressure for a corresponding saturation temperature at which a liquid boils into its vapor phase. Saturation pressure and saturation temperature have a direct linear relationship: as saturation pressure is increased so is saturation temperature.

If the temperature in a system remains constant (an isothermal system), vapor at saturation pressure and temperature will begin to condense into its liquid phase as the system pressure is increased. Similarly, a liquid at saturation pressure and temperature will tend to flash into its vapor phase as system pressure is decreased.

Standard State Conditions
Standard state conditions pertain to a reference temperature and a reference pressure for a chemical system. The usual standard state conditions are 298 K and 1 atm, or 298 K and 1 bar corresponding to ambient conditions.

Tait Equation[22–24]
The Tait equation is an equation of state used to relate liquid density to pressure as given by Equation (5.44).

$$V = V_s \left[1 - C \ln\left(\frac{B+p}{B+p_s}\right)\right] \quad (5.44)$$

where V_s is the saturation molar volume (L/mole), p_s is the saturated vapor pressure (Pa), and the constants B and C are given by Equations (5.45) and (5.46).

$$\frac{B}{p_c} = -1 + a(1-T_r)^{1/3} + b(1-T_r)^{2/3} + d(1-T_r) + e(1-T_r)^{4/3} \quad (5.45)$$

$$C = 0.0861488 + 0.0344483\omega \quad (5.46)$$

where p_c is the critical pressure (Pa), T_r is the reduced temperature equal to T/T_c, T_c is the critical temperature (degrees K), ω is the acentric factor, $a = -9.070217$, $b = 62.45326$, $d = -135.1102$, $e = \exp\left[f + g\omega + h\omega^2\right]$, $f = 4.79594$, $g = 0.250047$, and $h = 1.14188$.

Thermal Expansion Coefficient (Isobaric)[25]

The thermal expansion coefficient of a substance is the ratio of the increase in volume for a given rise in temperature (usually from 0°C to 1°C) to the original volume. The formal definition applicable to liquids is given by Equation (5.47).

$$\phi = \frac{1}{V}\left(\frac{\partial V}{\partial T}\right)_p \quad (5.47)$$

The Yaws correlation functions for thermal expansion coefficient of a liquid are given by Equations (5.48) and (5.49).

$$\phi = a(1-T_r)^m; \text{ Units: } C^{-1} \quad (5.48)$$

$$\phi = \frac{1}{\dfrac{1}{a(1-T_r)^m} + 273}; \text{ Units: } K^{-1} \quad (5.49)$$

where a and m are constants specific to a given substance.

Triple Point

On a pressure–temperature (p-T) phase diagram, the triple point, T_t, is the temperature at which the gas, liquid, and solid states coincide at equilibrium. It represents the intersection point of the three boundary lines separating the three phases of a pure substance.

Vapor Pressure Of Liquids[26]

Vapor pressure is the pressure of a vapor (gas) of a substance that co-exists with its liquid phase at equilibrium at a specified temperature. The liquid vapor pressure curve (p versus T) of a substance is a monotonic function of temperature from its minimum value (the triple point pressure) at the triple point temperature T_t to its maximum value (the critical pressure) at T_c. On a p-T phase diagram the liquid vapor pressure curve represents the boundary line between the liquid and gas phases connecting the triple point to the critical point of a pure substance.

The DIPPR vapor pressure correlation function for liquids (eq. 101 in DIPPR database) is given by Equation (5.50).

$$p_s = \exp\left[A + \frac{B}{T} + C \ln T + DT^E\right]; \text{Units: Pa} \quad (5.50)$$

where A, B, C, D, and E are constants specific to a given substance.

The Yaws vapor pressure correlation function for liquids is given by Equation (5.51).

$$p_s = \left(\frac{101325}{760}\right)10 \wedge \left[A + \frac{B}{T} + C \log T + DT + ET^2\right]; \text{Units: Pa} \quad (5.51)$$

where A, B, C, D, and E are constants specific to a given substance.

Vapor Pressure of Solids

The vapor pressure of a solid is the pressure at which the solid and gas phases of a pure substance are at equilibrium. The solid vapor pressure (p versus T) of a substance is a monotonic function of temperature with a maximum at the triple point. On a p-T phase diagram the solid vapor pressure curve represents the boundary line between the solid and gas phases connecting temperatures below the triple point to the triple point of a pure substance.

The DIPPR vapor pressure correlation function for solids (eq. 101 in DIPPR database) is given by Equation (5.49).

5.1.2 Predicting Thermochemical Properties

When thermochemical property data are not available in the DIPPR database or in the *Yaws Handbook*, they may be estimated using various computational methods. These are described below showing in each case the relevant equation used, appropriate references where tables of group parameters may be found, and worked examples.

Acentric Factor (Edmister Formula)[27,28]

The Edmister formula for estimating the acentric factor is given by Equation (5.52).

$$\omega = \frac{3}{7}\left(\frac{T_{br}}{1-T_{br}}\right)\log P_c - 1; \text{Units: dimensionless} \quad (5.52)$$

where T_{br} = reduced normal boiling point, and P_c = critical pressure (atm).

Energy Metrics

Example 5.7

Estimate the acentric factor for acetaldehyde given the following data:
$T_b = 292$ K, $T_c = 455$, $P_c = 54.97$ atm

SOLUTION

$T_{br} = 292/455 = 0.6418$
$\omega = (3/7)*(0.6418/(1-0.6418))*\log(54.97)-1 = 0.336$
(Cf. DIPPR value of 0.291)

Example 5.8

Derive the Edmister formula from the Clausius–Clapeyron equation.

SOLUTION

Begin with Clausius–Clapeyron equation:

$$\log\left(\frac{p}{P_c}\right) = \frac{\Delta H_{vap}}{R}\left[\frac{1}{T_c} - \frac{1}{T}\right]$$

At the critical point, $p = p_c$ and $T = T_c$. Left and right-hand sides of Clausius–Clapeyron equation yield 0.
At the normal point, $p = 1$ and $T = T_b$.
Therefore,

$$\log\left(\frac{1}{P_c}\right) = \frac{\Delta H_{vap}}{R}\left[\frac{1}{T_c} - \frac{1}{T_b}\right]$$

Let $\theta = \dfrac{T_b}{T_c}$ be the reduced boiling point.

Then,

$$\log\left(\frac{1}{P_c}\right) = \frac{\Delta H_{vap}}{R}\left[\frac{1}{T_c} - \frac{1}{\theta T_c}\right]$$

$$-\log P_c = \frac{\Delta H_{vap}}{RT_c}\left[1 - \frac{1}{\theta}\right]$$

$$\frac{\Delta H_{vap}}{R} = T_c\left(\log P_c\right)\left[\frac{1}{\theta} - 1\right]^{-1}$$

Substituting back into Clausius–Clapeyron equation yields

$$\log\left(\frac{p}{P_c}\right) = T_c\left(\log P_c\right)\frac{\theta}{1-\theta}\left[\frac{1}{T_c} - \frac{1}{T}\right]$$

$$\log p_r = \left(\log P_c\right)\frac{\theta}{1-\theta}\left[1 - \frac{1}{T_r}\right]$$

where $T_r = T/T_c$ is the reduced temperature and $p_r = p/p_c$ is the reduced pressure.

When $T_r = 0.7$, $\log p_r (\text{at } T_r = 0.7) = (\log P_c) \dfrac{\theta}{1-\theta} \left[1 - \dfrac{10}{7} \right] = -\dfrac{3}{7} (\log P_c) \dfrac{\theta}{1-\theta}$.

The acentric factor is defined as

$$\omega = -1 - \log p_r (\text{at } T_r = 0.7)$$

Therefore,

$\omega = -1 + \dfrac{3}{7} (\log P_c) \dfrac{\theta}{1-\theta}$, which is the Edmister formula (see Equation (5.51)).

Note: The logarithm function is to the base 10.

Critical Constants (Joback Method)[29-31]

The Joback estimates of critical temperature and critical pressure are given by Equations (5.53) and (5.54).

$$T_c = T_b \left[0.584 + 0.965 \sum \Delta_T - \left(\sum \Delta_T \right)^2 \right]^{-1} ; \text{Units: degrees K} \quad (5.53)$$

where Δ_T represent temperature group contributions.

$$P_c = \left[0.113 + 0.0032 n_A - \sum \Delta_P \right]^{-2} ; \text{Units: bar} \quad (5.54)$$

where n_A is the number of atoms in the molecule and Δ_P represent pressure group contributions.

Example 5.9

Estimate the critical temperature and critical pressure for o-xylene.

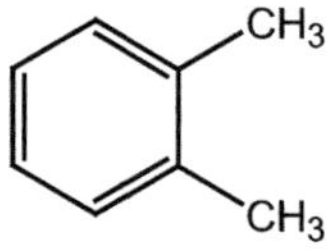

C_8H_{10}
Mol. Wt.: 106

SOLUTION

Group	n_i	Δ_T	Δ_P
=CH– (ring)	4	0.0082	0.0011
=C< (ring)	2	0.0143	0.0008
–CH$_3$	2	0.0141	–0.0012

$n_A = 18$; $T_b = 417.58$ K

$$\sum \Delta_T = 4(0.0082) + 2(0.0143) + 2(0.0141) = 0.0896$$

$$\sum \Delta_P = 4(0.0011) + 2(0.0008) + 2(-0.0012) = 0.0036$$

$T_c = 417.58*(0.584 + 0.965*0.0896 - (0.0896)^2)^{-1} = 630.37$ K (Cf. DIPPR value of 630.3 K)

$P_c = (0.113 + 0.0032*18 - 0.0036)^{-2} = 35.86$ bar (Cf. DIPPR value of 37.32 bar)

Enthalpy of Formation (Domalski–Hearing Method)[32,33]

The Domalski–Hearing estimate of heat of formation is given by Equation (5.55).

$$\Delta H_f^o = \sum_{i=1}^{N} n_i \left(\Delta H_f^o \right)_i; \text{ Units: kJ/mol} \quad (5.55)$$

where N = total number of groups in molecule; n = number of groups of type i, and $\left(\Delta H_f^o \right)_i$ is the enthalpy of formation group contribution of type i.

Example 5.10

Estimate the standard and ideal gas enthalpies of formation of o-toluidine.

C_7H_9N
Mol. Wt.: 107

SOLUTION

The melting point (256.8 K) and boiling point (473.49 K) given for o-toluidine in the DIPPR® 801 database bracket 298.15 K, and so the standard state phase at 298.15 K and 1 bar must be liquid.

Group	n_i	ΔH_f^o (gas)	ΔH_f^o (liq)
Cb-(H)(2Cb)	4	13.81	8.16
Cb-(C)(2Cb)	1	23.64	19.16
Cb-(N)(2Cb)	1	−1.30	1.50
C-(3H)(C)	1	−42.26	−47.61
N-(2H)(Cb)	1	19.25	−11.00
	Total	54.57	−5.31

Therefore, $\Delta H_f^o (\text{gas}) = 54.57$ kJ/mol and $\Delta H_f^s (\text{liq}) = -5.31$ kJ/mol. (Cf. DIPPR values of 53.20 and −4.72 kJ/mol, respectively.)

Enthalpy of Fusion at Melting Point (Chickos Method)[34,35]

The Chickos estimate of heat of fusion is given by Equation (5.56).

$$\Delta H_{\text{fus}} = T_m (a + b); \text{ Units: J/mol} \quad (5.56)$$

where T_m is the melting point (deg K),

$$a = \begin{cases} 0, & \text{if no non-aromatic rings exist} \\ 35.19 N_R + 4.289 (N_{CR} - 3 N_R), & \text{if non-aromatic rings exist} \end{cases}$$

$$b = \sum_{i=1}^{n_g} (N_g)_i \Delta s_i + \sum_{j=1}^{n_s} (N_s)_j \Delta s_j (C_s)_j + \sum_{k=1}^{n_f} (N_f)_k \Delta s_k (C_t)_k$$

N_R = number of non-aromatic rings
N_{CR} = number of -CH$_2$- groups in non-aromatic ring(s) required to form cyclic paraffin of same ring size(s)
$(N_g)_i$ = number of C-H groups of type i bonded to other carbon atoms
$(N_s)_j$ = number of C-H groups of type j bonded to at least one functional group or atom
$(N_f)_k$ = number of functional groups of type k
n_g = number of different non-ring or aromatic C-H groups bonded to other carbon atoms
n_s = number of different non-ring or aromatic C-H groups bonded to at least one functional group or atom
n_f = number of different functional groups or atoms
$(C_s)_j$ = value for C-H group j bonded to at least one functional group or atom
$(C_t)_k$ = value for non-C-H functional group k
Δs_i = contribution for C-H functional group i
Δs_k = contribution for non-C-H functional group k
Non-aromatic ring -CH$_2$ groups are accounted for in the a term and are *not* included in the b term.

Example 5.11

Estimate the heat of fusion for benzothiophene.

C$_8$H$_6$S
Mol. Wt.: 134

SOLUTION

$N_R = 1$, $N_{CR} = 5$, $a = 35.19 + (5-3)*(4.289) = 43.77$, $T_m = 304.5$ K

Group	Description	N	C	Δs	Group Total
=CH-	Aromatic (N_g type)	4	1	6.44	25.76
=C-	Ring (N_g type)	1	1	−11.72	−11.72
=C-	Ring (N_s type)	1	0.86	−11.72	−10.08
=CH-	Ring (N_g type)	1	1	−4.35	−4.35
=CH-	Ring (N_s type)	1	0.62	−4.35	−2.70
-S-	Ring	1	1	2.18	2.18
				Total	−0.91

$\Delta H_{fus} = (304.5)*(43.77 - 0.91) = 13050$ J/mol $= 13.05$ kJ/mol (Cf. DIPPR value of 11.83 kJ/mol)

Energy Metrics

Enthalpy of Vaporization at Boiling Point (Vetere Method)[36,37]

The Vetere estimate of heat of vaporization is given by Equation (5.57).

$$\Delta H_{vap} = RT_b \left[\frac{\tau_b^{0.38} \ln P_c - 0.513 + \dfrac{0.5066}{P_c(T_{br})^2}}{\tau_b + F(1-\tau_b^{0.38}) \ln T_{br}} \right]; \text{Units: kJ}/(\text{mol K}) \quad (5.57)$$

where $R = 8.314$ J/(mol K), T_b = boiling point (deg K), P_c = critical pressure (bar), T_{br} = reduced boiling point, $F=1$ for most compounds, $F=1.05$ for alcohols with more than two carbon atoms and for compounds that dimerize such as SO_2, NO, and NO_2, and

$$\tau_b = 1 - T_{br}.$$

Example 5.12

Estimate the enthalpy of vaporization for acetaldehyde given the following data:

$T_c = 466$ K, $T_b = 294$ K, $P_c = 55.5$ bar, $F = 1$

SOLUTION

$T_{br} = 294/466 = 0.631$
$\tau_b = 1 - 0.631 = 0.369$

$$\Delta H_{vap} = (8.314)(294) \left[\frac{0.369^{0.38} \ln 55.5 - 0.513 + \dfrac{0.5066}{55.5(0.631)^2}}{0.369 + 1(1-0.369^{0.38}) \ln 0.631} \right] = 26.36 \text{ kJ}/(\text{mol K})$$

(Cf. DIPPR value of 25.73 kJ/(mol K))

Heat Capacity of Gases (Benson Method)[38,39]

The Benson estimate of heat capacity of gases is given by Equation (5.58).

$$C_{p,g}^o = \sum_{i=1}^{N} n_i X_i; \text{Units: J}/(\text{mol K}) \quad (5.58)$$

where N = number of group types, n_i = number of occurrences of group i, X_i = individual group contribution

Example 5.13

Estimate the ideal gas heat capacity of isoprene.

C_5H_8
Mol. Wt.: 68

SOLUTION

Group	n_i	X_i	Group Total
=CH$_2$	2	26.62	53.24
=C<	1	19.3	19.3
-CH$_3$-(=C)	1	32.82	32.82
=CH-(C)	1	21.05	21.05
		TOTAL	126.41 J/(mol/K)

(Cf. DIPPR value of 130.4 J/(mol K))

Heat Capacity of Liquids (Ruzicka–Domalski Method)[40–43]

The Ruzicka–Domalski estimate of heat capacity of liquids is given by Equation (5.59).

$$C_{p,l} = R\left[A + B\left(\frac{T}{100}\right) + D\left(\frac{T}{100}\right)^2\right]; \text{ Units: J/(mol K)} \quad (5.59)$$

where $R = 8.314$ J/(mol K), T = temperature in degrees K,

$$A = \sum_{i=1}^{N} n_i a_i$$

$$B = \sum_{i=1}^{N} n_i b_i$$

$$D = \sum_{i=1}^{N} n_i d_i$$

n_i = number of occurrences of group i; a_i, b_i, and d_i = individual group contributions.

Example 5.14

Estimate the liquid heat capacity for 2-methyl-2-propanol at 340 K.

Energy Metrics

Solution

Group	n_i	α_i	b_i	d_i
C-(3C,O) (alcohol)	1	−44.690	31.769	−4.8791
O-(H)(C)	1	12.952	−10.145	2.6261
C-(3H)(C)	3	3.8452	−0.33997	0.19489
	Total	−20.202	20.604	−1.668

$C_p = 8.314*(-20.202 + 20.604*(340/100) - 1.668*(340/100)^2) = 254.16$ J/(mol K)
(Cf. DIPPR value of 252.40 J/(mol K))

Heat Capacity of Solids (Goodman Method)[44,45]

The Goodman estimate for the heat capacity of solids is given by Equation (5.60).

$$C_{p,s} = \left(\frac{A}{1000}\right) T^{0.79267}; \text{ Units: J/(mol K)} \quad (5.60)$$

where T = temperature in degrees K and

$$A = \exp\left[6.7796 + \sum_{i=1}^{N} n_i a_i + \sum_{i=1}^{N} n_i^2 b_i\right]$$

n_i = number of occurrences of group i, N = number of group types, a_i = individual group contribution, b_i = non-linear correction term for chain and aromatic carbon atoms

Example 5.15

Estimate the solid heat capacity for p-cresol at 307.93 K.

HO—⌬—CH₃ C_7H_8O
              Mol. Wt.: 108

Solution

Group	n	a_i	b_i
-CH₃	1	0.20184	0
Ar-CH=	4	0.082478	−0.00033
Ar>C=	2	0.012958	0
-OH	1	0.10341	0

$A = \exp(6.7796 + 0.20184 + 4*0.082478 + 2*0.012958 + 0.10341 + 4^2*(-0.00033)) = 1694.9$
$C_{p,g} = (1694.9/1000)*307.93^{0.79267} = 159.1$ J/(mol K)
(Cf. DIPPR value of 155.2 J/(mol K))

Normal Boiling Points (Nannoolal Method)[46,47]

The Nannoolal estimate of the normal boiling point is given by Equation (5.61).

$$T_b = \frac{\sum_{i=1}^{N} n_i X_i + \sum_{i=1}^{M} \frac{m_i}{n} Y_i}{n^{0.6583} + 1.6868} + 84.3395; \text{ Units: degrees K} \tag{5.61}$$

where n = number of non-hydrogen atoms, n_i = number of occurrences of group i, N = number of group types, X_i = group contribution, m_i = number of occurrences of intramolecular group–group interactions of type i, M = number of intramolecular group–group interactions, and Y_i = intramolecular group–group contribution.

Example 5.16

Estimate the normal boiling point of di-isopropanolamine.

$C_6H_{15}NO_2$
Mol. Wt.: 133

SOLUTION

$n = 9$
Number of pairwise intramolecular group–group interactions:

	-OH	-OH	-NH
-OH		-OH,-OH	-OH,-NH
-OH			-OH,-NH
-NH			

There are two -OH,-OH and one –OH,-NH interactions.

Group	n_i	X_i	Group total
-CH$_3$	2	177.3066	354.6132
>C(c)<(e)	4	266.8769	1067.5080
-OH sec	2	390.2446	780.4892
-NH-	1	223.0992	223.0992
		Total	2425.7096

Intramolecular Group–Group Interaction	m_i/n	Y_i	Group Total
-OH,-OH	1/9	291.7985	32.4221
-OH,-NH	2/9	286.9698	63.7711
		Total	96.1932

$T_b = (2425.7096 + 96.1932)/(9^{0.6583} + 1.6868) + 84.3395 = 509.3$ K
(Cf. DIPPR value of 521.9 K)

Normal Melting Point (Constantinou–Gani Method)[48,49]

The Constantinou–Gani estimate of the normal melting point is given by Equation (5.62).

$$T_m = 102.425 \ln \left(\sum_{i=1}^{N} N_i t_{m1,i} + \sum_{j=1}^{M} M_j t_{m2,j} \right); \text{ Units: degrees K} \quad (5.62)$$

where N_i = number of first-order groups of type i, M_j = number of second-order groups of type j, $t_{m1,i}$ = first-order group contributions, $t_{m2,j}$ = second-order group contributions, N = number of first-order groups, and M = number of second-order groups.

Example 5.17

Estimate the melting point for 2,6-dimethylpyridine.

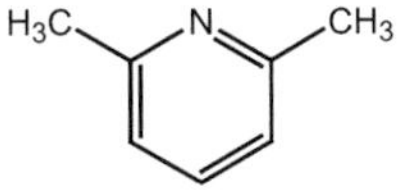

C_7H_9N
Mol. Wt.: 107

SOLUTION

Group	N	M	t_{m1}	t_{m2}
-CH$_3$	2		0.4640	
-C$_5$H$_3$(N)-	1		12.6275	
Six-membered ring		1		1.5656

$T_m = (102.425 \text{ K}) \ln [(2)(0.4640) + (1)12.6275 + (1)1.5656] = 278 \text{ K}$
(Cf. DIPPR value of 267 K)

Vapor Pressure Of Liquids (Riedel Method)[50–52]

The Riedel estimate of the vapor pressure of liquids is given by Equation (5.63).

$$\ln \left(\frac{P_{vap}}{P_c} \right) = A + \frac{B}{T_r} + C \ln T_r + D T_r^6; \text{ Units: bar} \quad (5.63)$$

where P_c = critical pressure (bar), T_r = reduced temperature, T_{br} = reduced normal boiling point,

$$A = 35D$$
$$B = -36D$$
$$C = \alpha_c - 42D$$
$$D = K(\alpha_c - 3.758)$$

$$\alpha_c = \frac{3.758 K \Psi + \ln\left(\dfrac{P_c}{1.01325}\right)}{K\Psi - \ln T_{br}}$$

$$\Psi = -35 + \frac{36}{T_{br}} + 42 \ln T_{br} - T_{br}^6$$

$$K = \begin{cases} -0.120 + 0.025h, & \text{acids} \\ 0.373 - 0.030h, & \text{alcohols} \\ 0.0838, & \text{all other compounds} \end{cases}$$

$$h = \left(\frac{T_{br}}{1 - T_{br}}\right) \ln\left(\frac{P_c}{1.01325}\right)$$

Example 5.18

Estimate the vapor pressure of chlorobenzene at 500 K given the following data:

$T_c = 632.35$ K, $T_b = 404.87$ K, $P_c = 45.1911$ bar.

Solution

$T_{br} = 404.87/632.35 = 0.640$
$K = 0.0838$

$$\Psi = -35 + \frac{36}{0.640} + 42\ln 0.640 - (0.640)^6 = 2.437$$

$$\alpha_c = \frac{(3.758)(0.0838)(2.437) + \ln\left(\dfrac{45.1911}{1.01325}\right)}{(0.0838)(2.437) - \ln 0.640} = 7.0179$$

$D = 0.0838\ (7.0179 - 3.758) = 0.2732$
$C = 7.0179 - 42\ (0.2732) = -4.4565$
$B = -36\ (0.2732) = -9.8352$
$A = 35\ (0.2732) = 9.5620$
$T_r = 500/632.35 = 0.7907$

$$\ln p_{vap} = \ln(45.1911) + 9.5620 - \frac{9.8352}{0.7907} - 4.4565 \ln(0.7907) + (0.2732)(0.7907)^6$$

$$= 2.0476$$

Therefore, $p_{vap} = \exp(2.0476) = 7.75$ bar
(Cf. DIPPR value of 7.67 bar)

Table 5.2 summarizes the percent uncertainties in each of the estimation methods discussed in this section.

TABLE 5.2
Summary of Computational Methods Used to Estimate Thermochemical Parameters along with Percent Uncertainties

Thermochemical Parameter	Method	% Uncertainty
T_m	Constantinou–Gani	25
T_b	Nannoolal	2
T_c	Joback	1
P_c	Joback	5
ΔH(fus)	Chickos	50
ΔH(vap)	Vetere	4
ΔH(formation)	Domalski–Hearing	3
p(vap)	Riedel	1, for $T > T_b$
		5 to 30, for $T > T_t$
C_p(gas)	Benson	4
C_p(liq)	Ruzicka–Domalski	4
C_p(solid)	Goodman	10
acentric factor, ω	Edmister	6

5.2 THERMODYNAMIC CALCULATIONS

5.2.1 Determination of Phases of Input Materials in a Given Chemical Reaction

Based on the regions shown in the p-T phase diagram of Figure 5.6, Table 5.3 summarizes the corresponding phases and applicable equations of state for the determination of the pressure contribution to the change in enthalpy for a given reaction (see Section 5.2.5).

5.2.2 Determination of Molar Volume of a Substance at a Given Pressure and Temperature for a Given Cubic Equation of State

The van der Waals, Redlich–Kwong, and Peng–Robinson equations of state are classified as cubic equations of state because if they are expanded algebraically they can be written as cubic functions with respect to the molar volume, V, variable.

Example 5.19

Show that the van der Waals equation is a cubic function in V.

Solution

Beginning with $p = \dfrac{RT}{V-b} - \dfrac{a}{V^2}$, we multiply both sides by $V^2(V-b)$ to obtain $pV^2(V-b) = RTV^2 - a(V-b)$. Expanding both sides and rearranging yields the cubic equation $pV^3 - (pb + RT)V^2 + aV - ab = 0$.

TABLE 5.3
Summary of Phase Regions and Equations of State for Determining Change in Enthalpy for a Given Reaction

Region	Phase	T_r Range	p_r Range	Equation of State Used for Pressure Contribution to ΔH
I	Supercritical fluid	$T_r > 1$	$p_r > 1$	Redlich–Kwong
II	Gas	$T_r > 1$	$p_r < 1$	Redlich–Kwong
III	Liquid	$T_r < 1$	$p_r > 1$	Tait
IVa	Liquid	$\dfrac{T_{TP}}{T_c} < T_r < 1$	$\dfrac{p_{vap}(T_{rxn})}{p_c} < p_r < 1$	Tait
IVb	Gas	$\dfrac{T_{TP}}{T_c} < T_r < 1$	$\dfrac{p_{vap}(T_{rxn})}{p_c} > p_r$	Redlich–Kwong
V	Critical point	$T_r = 1$	$p_r = 1$	Redlich–Kwong

PROBLEM 5.7
Show that the Redlich–Kwong equation is a cubic function in V.

PROBLEM 5.8
Show that the Peng–Robinson equation is a cubic function in V.

If we are given values of pressure and temperature for a fluid, then we can solve the cubic equation of state to obtain the corresponding molar volume. In order to perform this task we need to be reminded of how to solve cubic equations of the form $AV^3 + BV^2 + CV + D = 0$. If we make the substitution $V = y - \dfrac{B}{3A}$, we obtain the reduced or incomplete cubic equation $y^3 + Py + Q = 0$ where $P = -\dfrac{1}{3}\left(\dfrac{B}{A}\right)^2 + \dfrac{C}{A}$ and $Q = \dfrac{2}{27}\left(\dfrac{B}{A}\right)^3 - \dfrac{BC}{3A^2} + \dfrac{D}{A}$. The solution to the reduced cubic equation is given by the roots shown in Equations (5.64) to (5.66).

$$y_1 = u + v \tag{5.64}$$

$$y_2 = -\frac{1}{2}(u+v) + i\frac{\sqrt{3}}{2}(u-v) \tag{5.65}$$

$$y_3 = -\frac{1}{2}(u+v) - i\frac{\sqrt{3}}{2}(u-v) \tag{5.66}$$

where $u = \left(-\dfrac{Q}{2} + \sqrt{\phi}\right)^{1/3}$, $v = \left(-\dfrac{Q}{2} - \sqrt{\phi}\right)^{1/3}$, and $\phi = \left(\dfrac{P}{3}\right)^3 + \left(\dfrac{Q}{2}\right)^2$. The parameter ϕ is called the discriminant. When $\phi > 0$, there is one real root and two complex conjugate roots. When $\phi < 0$, there are three real roots. When $\phi = 0$, there is one real root and another real root of double multiplicity. From a practical point of view, the first condition having a positive discriminant means that one physically meaningful real value of the molar volume for the fluid in the gaseous state is obtained corresponding to $V = u + v - \dfrac{B}{3A}$. When the discriminant is negative, the three real roots are found using Equations (5.67) to (5.69).

$$y_1^* = 2\sqrt{-\dfrac{P}{3}} \cos\left(\dfrac{\alpha}{3}\right) \qquad (5.67)$$

$$y_2^* = -2\sqrt{-\dfrac{P}{3}} \cos\left(\dfrac{\alpha}{3} + \dfrac{\pi}{3}\right) \qquad (5.68)$$

$$y_3^* = -2\sqrt{-\dfrac{P}{3}} \cos\left(\dfrac{\alpha}{3} - \dfrac{\pi}{3}\right) \qquad (5.69)$$

where $\alpha = \arccos\left(-\dfrac{Q}{2\sqrt{-(P/3)^3}}\right)$.

These three real roots correspond to the molar volume of the liquid, $V_f = y_1^* - \dfrac{B}{3A}$, the molar volume of the gas, $V_g = y_3^* - \dfrac{B}{3A}$, and the virtual molar volume, $\langle V \rangle = y_2^* - \dfrac{B}{3A}$, having a value in between V_f and V_g.

Example 5.20

Determine the molar volume of ethylene having a pressure of 1 atm and a temperature of 260°C if it behaves as a van der Waals fluid. The critical constants for ethylene are $P_c = 49.66$ atm and $T_c = 282.36$ K.

Solution

The van der Waals constants are

$$a = \dfrac{27\, R^2 T_c^2}{64\, P_c} = \dfrac{27\, (0.08207)^2 (282.36)^2}{64} \cdot \dfrac{1}{49.66} = 4.5620\, L^2\, atm\, mol^{-2} \text{ and}$$

$$b = \dfrac{RT_c}{8 P_c} = \dfrac{(0.08207)(282.36)}{49.66} = 0.05833\, L/mol.$$

The values of the relevant coefficients are given below.

$A = P$	1
$B = -(Pb + RT)$	−43.8016
$C = a$	4.561969
$D = -ab$	−0.2661
P	−634.966
Q	−6158.63
Discriminant	453.1909

Since the discriminant is positive valued, there is only one real root corresponding to the molar volume of the gaseous state.

u	14.58194315	
v	14.51489032	
u + v	29.09683346	
y1 = u + v	29.09683346	root 1
B/3A	−14.60054662	
V1	43.69738	L/mol

Therefore, the molar volume of the gas is 43.70 L/mol.

Example 5.21

Determine the molar volume of ethylene having a pressure of 1 atm and a temperature of −100°C if it behaves as a van der Waals fluid. The critical constants for ethylene are $P_c = 49.66$ atm and $T_c = 282.36$ K.

Solution

The values of the relevant coefficients are given below.

$A = P$	1
$B = -(Pb + RT)$	−14.2564
$C = a$	4.561969
$D = -ab$	−0.2661
P	−63.1867
q	−193.221
Discriminant	−9.95877

Since the discriminant is negative, there are three real roots corresponding to the molar volumes of the gaseous, virtual, and liquid states.

Cos (alpha)	0.999467		
Alpha	0.032653		
y1	9.17818	root 1	
y2	−4.50257	root 2	
y3	−4.67561	root 3	
V1	13.93033	L/mol	molar volume of gas
V2	0.249575	L/mol	
V3	0.076539	L/mol	molar volume of liquid

Therefore, the molar volume of the gas is 13.93 L/mol and the molar volume of the liquid is 0.0765 L/mol.

PROBLEM 5.9

Part 1

Determine the molar volume of ethylene having a pressure of 1 atm and a temperature of 260°C if it behaves as a Redlich–Kwong fluid. Repeat the calculation for a temperature of −100°C. The critical constants for ethylene are $P_c = 49.66$ atm and $T_c = 282.36$ K.

Part 2

Determine the molar volume of ethylene having a pressure of 1 atm and a temperature of 260°C if it behaves as a Peng–Robinson fluid. Repeat the calculation for a temperature of −100°C. The critical constants for ethylene are $P_c = 49.66$ atm and $T_c = 282.36$ K.

Part 3

Summarize the values in a table and compare them with the results found in Example 5.21 using the van der Waals equation of state.

5.2.3 Derivation of $\int pdV$ Expressions for Various Cases in P-V Diagrams Related to Determining Work Done by Fluids During Reversible Isothermal Expansion or Contraction

Case I (Figure 5.7): $V_1, V_2 < V_f$ and $p_1, p_2 > p_{sat}$
The expression is given by Equation (5.70).

$$W = -n \int_{V_1}^{V_2} p(V) dV, \text{ where } n \text{ is the number of moles} \quad (5.70)$$

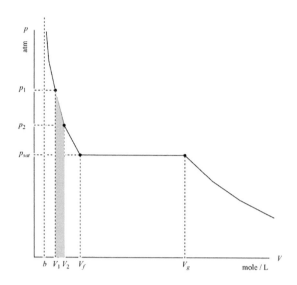

FIGURE 5.7 Pressure versus molar volume graph pertaining to Case I.

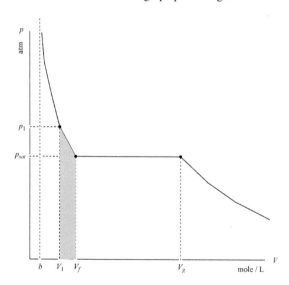

FIGURE 5.8 Pressure versus molar volume graph pertaining to Case II.

Case II (Figure 5.8): $V_1 < V_f$, $V_2 = V_f$, $p_1 > p_{sat}$, and $p_2 = p_{sat}$
The expression is given by Equation (5.71).

$$W = -n \int_{V_1}^{V_f} p(V) dV, \tag{5.71}$$

where n is the number of moles

Energy Metrics

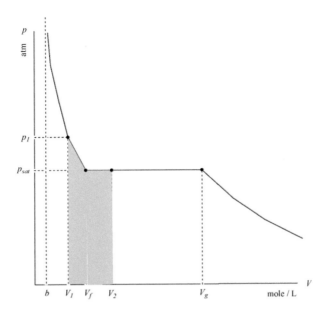

FIGURE 5.9 Pressure versus molar volume graph pertaining to Case III.

Case III (Figure 5.9): $V_1 < V_f$, $V_g > V_2 > V_f$, $p_1 > p_{sat}$, and $p_2 = p_{sat}$
The expression is given by Equation (5.72).

$$W = -n\left[\int_{V_1}^{V_f} p(V)dV + (V_2 - V_f)p_{sat}\right], \quad (5.72)$$

where n is the number of moles

Case IV (Figure 5.10): $V_g > V_1 > V_f$, $V_g > V_2 > V_f$, and $p_1 = p_2 = p_{sat}$
The expression is given by Equation (5.73).

$$W = -n(V_2 - V_1)p_{sat}, \quad (5.73)$$

where n is the number of moles

Case V (Figure 5.11): $V_g > V_1 > V_f$, $V_2 > V_g$, $p_1 = p_{sat}$, and $p_2 < p_{sat}$
The expression is given by Equation (5.74).

$$W = -n\left[\int_{V_g}^{V_2} p(V)dV + (V_g - V_1)p_{sat}\right], \quad (5.74)$$

where n is the number of moles

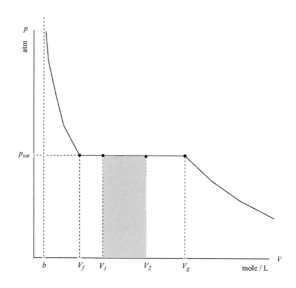

FIGURE 5.10 Pressure versus molar volume graph pertaining to Case IV.

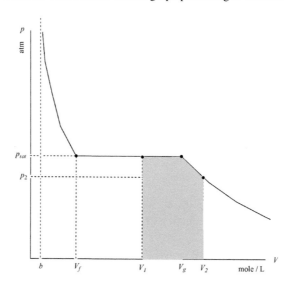

FIGURE 5.11 Pressure versus molar volume graph pertaining to Case V.

Case VI (Figure 5.12): $V_1 = V_g$, $V_2 > V_g$, $p_1 = p_{sat}$, and $p_2 < p_{sat}$
The expression is given by Equation (5.75).

$$W = -n \int_{V_g}^{V_2} p(V) dV, \quad (5.75)$$

where n is the number of moles

Energy Metrics

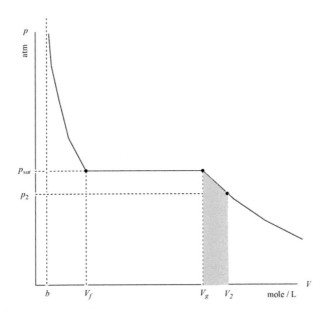

FIGURE 5.12 Pressure versus molar volume graph pertaining to Case VI.

Case VII (Figure 5.13): $V_1, V_2 > V_g$ and $p_1, p_2 < p_{sat}$
The expression is given by Equation (5.76).

$$W = -n \int_{V_1}^{V_2} p(V) dV, \qquad (5.76)$$

where n is the number of moles

Case VIII (Figure 5.14):
The expression is given by Equation (5.77).

$$W = -n \left[\int_{V_1}^{V_f} p(V) dV + \int_{V_g}^{V_2} p(V) dV + (V_g - V_f) p_{sat} \right], \qquad (5.77)$$

where n is the number of moles

Table 5.4 summarizes the eight cases for expressions for work done by a fluid during reversible expansion. The convention is that work done by a system has $W < 0$ and work done on a system has $W > 0$.

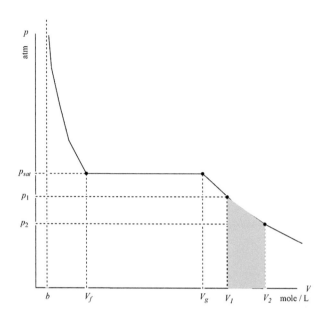

FIGURE 5.13 Pressure versus molar volume graph pertaining to Case VII.

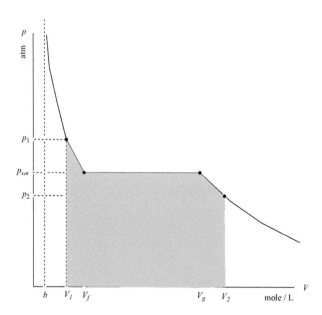

FIGURE 5.14 Pressure versus molar volume graph pertaining to Case VIII.

Energy Metrics

TABLE 5.4
Summary of Case Results for Expressions for Work Done by a Fluid during Reversible Isothermal Expansion

Case	Range	Work Expression
I	$V_1, V_2 < V_f$ and $p_1, p_2 > p_{sat}$	$W = -n \int_{V_1}^{V_2} p(V)dV$
II	$V_1 < V_f$, $V_2 = V_f$, $p_1 > p_{sat}$, and $p_2 = p_{sat}$	$W = -n \int_{V_1}^{V_f} p(V)dV$
III	$V_1 < V_f$, $V_g > V_2 > V_f$, $p_1 > p_{sat}$, and $p_2 = p_{sat}$	$W = -n \left[\int_{V_1}^{V_f} p(V)dV + (V_2 - V_f)p_{sat} \right]$
IV	$V_g > V_1 > V_f$, $V_g > V_2 > V_f$, and $p_1 = p_2 = p_{sat}$	$W = -n(V_2 - V_1)p_{sat}$
V	$V_g > V_1 > V_f$, $V_2 > V_g$, $p_1 = p_{sat}$, and $p_2 < p_{sat}$	$W = -n \left[\int_{V_g}^{V_2} p(V)dV + (V_g - V_1)p_{sat} \right]$
VI	$V_1 = V_g$, $V_2 > V_g$, $p_1 = p_{sat}$, and $p_2 < p_{sat}$	$W = -n \int_{V_g}^{V_2} p(V)dV$
VII	$V_1, V_2 > V_g$ and $p_1, p_2 < p_{sat}$	$W = -n \int_{V_1}^{V_2} p(V)dV$
VIII	$V_1 < V_f$, $V_2 > V_g$, $p_1 > p_{sat}$, and $p_2 < p_{sat}$	$W = -n \left[\int_{V_1}^{V_f} p(V)dV + \int_{V_g}^{V_2} p(V)dV + (V_g - V_f)p_{sat} \right]$

5.2.4 Derivation of Expressions for $\int p(V)dV$ for Various Equations of State

Equations (5.78) to (5.80) are the respective expressions for the van der Waals, Redlich–Kwong, and Peng–Robinson equations of state.

5.2.4.1 van der Waals EOS

$$p = \frac{RT}{V-b} - \frac{a}{V^2}$$

$$\int p(V)dV = RT \int \frac{dV}{V-b} - a \int \frac{dV}{V^2} = RT \ln(V-b) + a\frac{1}{V} \quad (5.78)$$

5.2.4.2 Redlich–Kwong EOS

$$p = \frac{RT}{V-b} - \frac{a}{\sqrt{T}} \frac{1}{V(V+b)}$$

$$\int p(V)dV = RT \int \frac{dV}{V-b} - \frac{a}{\sqrt{T}} \int \frac{dV}{V(V+b)} = RT\ln(V-b) - \frac{a}{\sqrt{T}}\left[\frac{1}{b}\int \frac{dV}{V} - \frac{1}{b}\int \frac{dV}{V+b}\right]$$

$$= RT\ln(V-b) - \frac{a}{b\sqrt{T}}\left[\ln V - \ln(V+b)\right]$$

$$= RT\ln(V-b) - \frac{a}{b\sqrt{T}}\ln\left(\frac{V}{V+b}\right)$$

(5.79)

5.2.4.3 Peng–Robinson EOS

$$p = \frac{RT}{V-b} - \frac{a\alpha}{V^2 + 2bV - b^2}$$

$$\int p(V)dV = RT \int \frac{dV}{V-b} - a\alpha \int \frac{dV}{V^2 + 2bV - b^2} = RT\ln(V-b) - a\alpha \int \frac{dV}{V^2 + 2bV - b^2}$$

Evaluation of second integral:

Let $V^2 + 2bV - b^2 = (V+b)^2 - 2b^2$.

Let $u = V + b$. Then, $du = dV$
Therefore,

$$\int \frac{dV}{V^2 + 2bV - b^2} = \int \frac{du}{u^2 - 2b^2} = \int \frac{du}{(u+b\sqrt{2})(u-b\sqrt{2})}$$

$$= -\frac{1}{2b\sqrt{2}} \int \frac{du}{u+b\sqrt{2}} + \frac{1}{2b\sqrt{2}} \int \frac{du}{u-b\sqrt{2}}$$

$$= \frac{1}{2b\sqrt{2}}\left[-\ln(u+b\sqrt{2}) + \ln(u-b\sqrt{2})\right]$$

$$= \frac{1}{2b\sqrt{2}} \ln\left(\frac{u-b\sqrt{2}}{u+b\sqrt{2}}\right)$$

$$= \frac{1}{2b\sqrt{2}} \ln\left(\frac{V+b-b\sqrt{2}}{V+b+b\sqrt{2}}\right)$$

$$= \frac{1}{2b\sqrt{2}} \ln\left(\frac{V+b(1-\sqrt{2})}{V+b(1+\sqrt{2})}\right)$$

Therefore,

$$\int p(V)dV = RT\ln(V-b) - \frac{a\alpha}{2b\sqrt{2}}\ln\left[\frac{V+b(1-\sqrt{2})}{V+b(1+\sqrt{2})}\right]. \tag{5.80}$$

5.2.5 Determination of Expressions for Enthalpy Change due to Temperature and Pressure Contributions

In each case shown below, a p-T path diagram is drawn first to guide the calculations. Figures 5.15 to 5.18 are the corresponding path diagrams for the four common state transition cases encountered.

5.2.5.1 Case I (Figure 5.15): State 1 (gas) to State 2 (gas)

The general relationship for the change of enthalpy along the path shown is given by Equation (5.81).

$$\Delta H = \int_{p_1}^{0}\left[V - T\left(\frac{\partial V}{\partial T}\right)_p\right]_{T_1} dp + \int_{T_1}^{T_2}\left[C_{p,g}(T)\right]_{p=0} dT + \int_{0}^{p_2}\left[V - T\left(\frac{\partial V}{\partial T}\right)_p\right]_{T_2} dp \tag{5.81}$$

p
p2
state 2 (p2, T2)
p1
state 1 (p1, T1)
0
T
T1
T2
p1 = 1 atm; T1 = 298 K
At p = 0 is the ideal gas regime.

FIGURE 5.15 Pressure–temperature path diagram pertaining to Case I.

p
p2
state 2 (p2, T2)
state 1 (p1, T1)
p1
T
T1
Tb
T2
p1 = 1 atm; T1 = 298 K

FIGURE 5.16 Pressure–temperature path diagram pertaining to Case II.

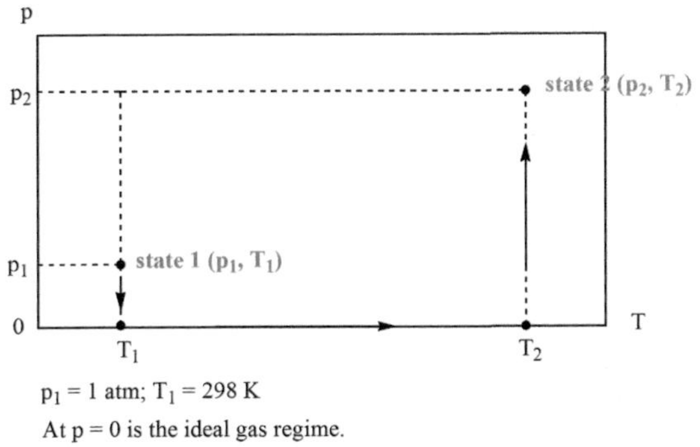

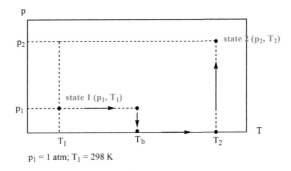

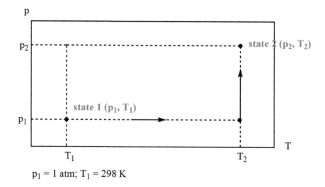

FIGURE 5.17 Pressure–temperature path diagram pertaining to Case III.

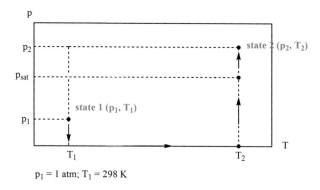

FIGURE 5.18 Pressure–temperature path diagram pertaining to Case IV.

Heat capacity temperature dependent functions at constant pressure for gases in the DIPPR database are given for the ideal gas case.

Since the first and third integrals are integrated over pressure, it is desirable that the molar volume and $(\partial V / \partial T)_p$ functions are written in terms of pressure and temperature. However, the equations of state functions are written as $p(V,T)$ rather than $V(p,T)$. Hence, it is required to transform the integrals as follows.

We may write the total differential of pressure for a function $p(V,T)$ as Equation (5.82).

$$dp = \left(\frac{\partial p}{\partial V}\right)_T dV + \left(\frac{\partial p}{\partial T}\right)_V dT \tag{5.82}$$

Since both integrals follow isotherms (constant T), then Equation (5.82) reduces to Equation (5.83)

$$dp = \left(\frac{\partial p}{\partial V}\right)_T dV \tag{5.83}$$

Energy Metrics

Therefore, we have Equation (5.84).

$$\int_{p_1}^{0}\left[V-T\left(\frac{\partial V}{\partial T}\right)_p\right]_{T_1} dp = \int_{V_1}^{\infty}\left[V\left(\frac{\partial p}{\partial V}\right)_T - T\left(\frac{\partial V}{\partial T}\right)_p\left(\frac{\partial p}{\partial V}\right)_T\right]_{T_1} dV \quad (5.84)$$

Since by the chain rule relation given in Equation (5.85)

$$\left(\frac{\partial V}{\partial T}\right)_p\left(\frac{\partial T}{\partial p}\right)_V\left(\frac{\partial p}{\partial V}\right)_T = -1 \quad (5.85)$$

we have Equation (5.86).

$$\left(\frac{\partial V}{\partial T}\right)_p\left(\frac{\partial p}{\partial V}\right)_T = -\frac{1}{\left(\frac{\partial T}{\partial p}\right)_V} = -\left(\frac{\partial p}{\partial T}\right)_V \quad (5.86)$$

Substitution of Equation (5.87) into Equation (5.84) yields Equation (5.87) for the first integral.

$$\int_{p_1}^{0}\left[V-T\left(\frac{\partial V}{\partial T}\right)_p\right]_{T_1} dp = \int_{V_1}^{\infty}\left[V\left(\frac{\partial p}{\partial V}\right)_T + T\left(\frac{\partial p}{\partial T}\right)_V\right]_{T_1} dV \quad (5.87)$$

From a specified equation of state function $p(T,V)$, at a given isotherm T_1, the molar volume V_1 is evaluated corresponding to a pressure p_1.

Similarly, the second integral is transformed to Equation (5.88).

$$\int_{0}^{p_2}\left[V-T\left(\frac{\partial V}{\partial T}\right)_p\right]_{T_2} dp = \int_{\infty}^{V_2}\left[V\left(\frac{\partial p}{\partial V}\right)_T + T\left(\frac{\partial p}{\partial T}\right)_V\right]_{T_2} dV \quad (5.88)$$

Substitution of Equations (5.86) and (5.87) into Equation (5.80) yields Equation (5.89) for the change of enthalpy.

$$\Delta H = \int_{V_1}^{\infty}\left[V\left(\frac{\partial p}{\partial V}\right)_T + T\left(\frac{\partial p}{\partial T}\right)_V\right]_{T_1} dV + \int_{T_1}^{T_2}\left[C_{p,g}(T)\right]_{p=0} dT$$

$$+ \int_{\infty}^{V_2}\left[V\left(\frac{\partial p}{\partial V}\right)_T + T\left(\frac{\partial p}{\partial T}\right)_V\right]_{T_2} dV \quad (5.89)$$

From the Redlich–Kwong equation of state for gases given by Equation (5.16) we are able to work out the following.

$$\left(\frac{\partial p}{\partial T}\right)_V = \frac{R}{V-b} + \frac{a}{2V(V+b)}T^{-3/2}$$

$$\left(\frac{\partial p}{\partial V}\right)_T = -RT(V-b)^{-2} + \frac{a}{\sqrt{T}}(V^2+bV)^{-2}(2V+b)$$

$$\int\left[V\left(\frac{\partial p}{\partial V}\right)_T + T\left(\frac{\partial p}{\partial T}\right)_V\right]dV$$

$$= \frac{a}{\sqrt{T}}\int\left[\frac{2V+b}{V(V+b)^2}\right]dV - RT\int\left[\frac{V}{(V-b)^2}\right]dV + RT\int\left[\frac{1}{V-b}\right]dV$$

$$+ \frac{a}{2\sqrt{T}}\int\left[\frac{1}{V(V+b)}\right]dV$$

$$= \frac{2a}{\sqrt{T}}\int\left[\frac{1}{(V+b)^2}\right]dV + \frac{ab}{\sqrt{T}}\int\left[\frac{1}{V(V+b)^2}\right]dV - RT\int\left[\frac{V}{(V-b)^2}\right]dV$$

$$+ RT\int\left[\frac{1}{V-b}\right]dV + \frac{a}{2\sqrt{T}}\int\left[\frac{1}{V(V+b)}\right]dV$$

$$= \frac{2a}{\sqrt{T}}\left(-\frac{1}{V+b}\right) + \frac{ab}{\sqrt{T}}\left(\frac{1}{b(V+b)} - \frac{1}{b^2}\ln\left(\frac{V+b}{V}\right)\right)$$

$$-RT\left(\ln(V-b) - \frac{b}{V-b}\right) + RT\ln(V-b) - \frac{a}{2b\sqrt{T}}\ln\left(\frac{V+b}{V}\right)$$

$$= -\frac{a}{\sqrt{T}}\left(\frac{1}{V+b}\right) + RT\left(\frac{b}{V-b}\right) - \frac{3a}{2b\sqrt{T}}\ln\left(\frac{V+b}{V}\right)$$

$$\Delta H = \int_{T_1}^{T_2} C_{p,g}(T)dT + \left[\frac{RT_1 b}{V-b} - \frac{a}{\sqrt{T_1}}\left(\frac{1}{V+b}\right) - \frac{3a}{2b\sqrt{T_1}}\ln\left(\frac{V+b}{V}\right)\right]_{V_1}^{\infty}$$

$$+ \left[\frac{RT_2 b}{V-b} - \frac{a}{\sqrt{T_2}}\left(\frac{1}{V+b}\right) - \frac{3a}{2b\sqrt{T_2}}\ln\left(\frac{V+b}{V}\right)\right]_{\infty}^{V_2}$$

Hence, the enthalpy change is given by Equation (5.90).

$$\Delta H = \int_{T_1}^{T_2} C_{p,g}(T)dT - \left[\frac{RT_1 b}{V_1-b} - \frac{a}{\sqrt{T_1}}\left(\frac{1}{V_1+b}\right) - \frac{3a}{2b\sqrt{T_1}}\ln\left(\frac{V_1+b}{V_1}\right)\right]$$

$$+ \left[\frac{RT_2 b}{V_2-b} - \frac{a}{\sqrt{T_2}}\left(\frac{1}{V_2+b}\right) - \frac{3a}{2b\sqrt{T_2}}\ln\left(\frac{V_2+b}{V_2}\right)\right] \quad (5.90)$$

Energy Metrics

Therefore, the temperature and pressure contributions to the change in enthalpy are given by Equations (5.91) and (5.92).

$$q_{\text{temp}} = \int_{T_1}^{T_2} C_{p,g}(T) dT \qquad (5.91)$$

$$\begin{aligned} q_{\text{press}} &= \left[\frac{RT_2 b}{V_2 - b} - \frac{a}{\sqrt{T_2}} \left(\frac{1}{V_2 + b} \right) - \frac{3a}{2b\sqrt{T_2}} \ln\left(\frac{V_2 + b}{V_2} \right) \right] \\ &- \left[\frac{RT_1 b}{V_1 - b} - \frac{a}{\sqrt{T_1}} \left(\frac{1}{V_1 + b} \right) - \frac{3a}{2b\sqrt{T_1}} \ln\left(\frac{V_1 + b}{V_1} \right) \right] \end{aligned} \qquad (5.92)$$

5.2.5.2 Case II (Figure 5.16): State 1 (liquid) to State 2 (gas)

$$\Delta H = \int_{T_1}^{T_b} [C_{p,\text{liq}}(T)]_{p=1} dT + \int_{p_1}^{0} \left[V - T\left(\frac{\partial V}{\partial T}\right)_p \right]_{T_b} dp$$

$$+ \int_{T_b}^{T_2} [C_{p,g}(T)]_{p=0} dT + \Delta H_{\text{vap}} + \int_{0}^{p_2} \left[V - T\left(\frac{\partial V}{\partial T}\right)_p \right]_{T_2} dp$$

$$= \int_{T_1}^{T_b} [C_{p,\text{liq}}(T)]_{p=1} dT + \int_{T_b}^{T_2} [C_{p,g}(T)]_{p=0} dT + \Delta H_{\text{vap}}$$

$$+ \int_{V_1}^{\infty} \left[V\left(\frac{\partial p}{\partial V}\right)_T + T\left(\frac{\partial p}{\partial T}\right)_V \right]_{T_b} dV + \int_{\infty}^{V_2} \left[V\left(\frac{\partial p}{\partial V}\right)_T + T\left(\frac{\partial p}{\partial T}\right)_V \right]_{T_2} dV$$

Applying the Redlich–Kwong equation of state for gases we obtain Equation (5.93).

$$\Delta H = \int_{T_1}^{T_b} [C_{p,\text{liq}}(T)]_{p=1} dT + \int_{T_b}^{T_2} [C_{p,g}(T)]_{p=0} dT + \Delta H_{\text{vap}}$$

$$- \left[\frac{RT_b b}{V_1 - b} - \frac{a}{\sqrt{T_b}} \left(\frac{1}{V_1 + b} \right) - \frac{3a}{2b\sqrt{T_b}} \ln\left(\frac{V_1 + b}{V_1} \right) \right] \qquad (5.93)$$

$$+ \left[\frac{RT_2 b}{V_2 - b} - \frac{a}{\sqrt{T_2}} \left(\frac{1}{V_2 + b} \right) - \frac{3a}{2b\sqrt{T_2}} \ln\left(\frac{V_2 + b}{V_2} \right) \right]$$

Therefore, the temperature and pressure contributions to the change in enthalpy are given by Equations (5.94) and (5.95).

$$q_{\text{temp}} = \int_{T_1}^{T_b} \left[C_{p,\text{liq}}(T) \right]_{p=1} dT + \int_{T_b}^{T_2} \left[C_{p,\text{g}}(T) \right]_{p=0} dT + \Delta H_{\text{vap}} \qquad (5.94)$$

$$q_{\text{press}} = \left[\frac{RT_2 b}{V_2 - b} - \frac{a}{\sqrt{T_2}} \left(\frac{1}{V_2 + b} \right) - \frac{3a}{2b\sqrt{T_2}} \ln \left(\frac{V_2 + b}{V_2} \right) \right]$$
$$- \left[\frac{RT_b b}{V_1 - b} - \frac{a}{\sqrt{T_b}} \left(\frac{1}{V_1 + b} \right) - \frac{3a}{2b\sqrt{T_b}} \ln \left(\frac{V_1 + b}{V_1} \right) \right] \qquad (5.95)$$

5.2.5.3 Case III (Figure 5.17): State 1 (liquid) to State 2 (liquid)

$$\Delta H = \int_{T_1}^{T_2} \left[C_{p,\text{liq}}(T) \right]_{p=1} dT + \int_{p_1}^{p_2} \left[V - T \left(\frac{\partial V}{\partial T} \right)_p \right]_{T_2} dp$$

$$= \int_{T_1}^{T_2} \left[C_{p,\text{liq}}(T) \right]_{p=1} dT + \int_{p_1}^{p_2} \left[V - TV\phi \right]_{T_2} dp$$

$$= \int_{T_1}^{T_2} \left[C_{p,\text{liq}}(T) \right]_{p=1} dT + (1 - \phi T_2) \int_{p_1}^{p_2} \left[V \right]_{T_2} dp$$

where $\phi = \frac{1}{V} \left(\frac{\partial V}{\partial T} \right)_p$ is the thermal expansion coefficient for liquids.

The Tait equation of state for liquids (see Equation (5.43)) is used for the second integral leading to Equation (5.96).

$$\Delta H = \int_{T_1}^{T_2} \left[C_{p,\text{liq}}(T) \right]_{p=101325}$$
$$+ (1 - \phi T_2) V_s \left[\begin{array}{l} (p_2 - 101325)(1+C) - BC \ln \left(\dfrac{B + p_2}{B + 101325} \right) \\ -C \left[p_2 \ln \left(\dfrac{B + p_2}{B + p_s} \right) - 101325 \ln \left(\dfrac{B + 101325}{B + p_s} \right) \right] \end{array} \right] \qquad (5.96)$$

Therefore, the temperature and pressure contributions to the change in enthalpy are given by Equations (5.97) and (5.98).

$$q_{\text{temp}} = \int_{T_1}^{T_2} \left[C_{p,\text{liq}}(T) \right]_{p=101325} \qquad (5.97)$$

$$q_{press} = (1 - \phi T_2) V_s \begin{bmatrix} (p_2 - 101325)(1+C) - BC \ln\left(\dfrac{B+p_2}{B+101325}\right) \\ -C\left[p_2 \ln\left(\dfrac{B+p_2}{B+p_s}\right) - 101325 \ln\left(\dfrac{B+101325}{B+p_s}\right) \right] \end{bmatrix} \quad (5.98)$$

The thermal expansion coefficient is also temperature dependent (see equations 5.99a and 5.99b):

$$\phi = \lambda(1-T_r)^m \quad \text{(units: per degree C)} \quad (5.99a)$$

$$\phi = \dfrac{1}{\dfrac{1}{\lambda(1-T_r)^m} + 273} \quad \text{(units: per degree K)} \quad (5.99b)$$

where $T_r = \dfrac{T_2}{T_c}$, λ and m are constants specific to the liquid.

The molar volume at temperature T_2 is determined from Equation (5.100).

$$V_s = \dfrac{MW}{1000\left[A_{11} B_{21}^{-(1-T_r)^n} \right]} \quad (5.100)$$

where A_{11}, B_{21}, and n are constants specific to the liquid and MW is the molecular weight of the liquid.

The vapor pressure at temperature T_2 is given by Equation (5.101).

$$p_s = \left(\dfrac{101325}{760}\right) 10^{A_{12} + \frac{B_{22}}{T} + C_{32} \log_{10} T + D_{42} T + E_{52} T^2} \quad \text{(units: Pa)} \quad (5.101)$$

where A_{12}, B_{22}, C_{32}, D_{42}, and E_{52} are constants specific to the liquid.

5.2.5.4 Case IV (Figure 5.18): State 1 (gas) to State 2 (liquid)

$$\Delta H = \int_{p_1}^{0} \left[V - T\left(\dfrac{\partial V}{\partial T}\right)_p \right]_{T_1} dp + \int_{T_1}^{T_2} \left[C_{p,g}(T) \right]_{p=0} dT$$

$$+ \int_{0}^{p_{sat}} \left[V - T\left(\dfrac{\partial V}{\partial T}\right)_p \right]_{T_2} dp + \int_{p_{sat}}^{p_2} \left[V - T\left(\dfrac{\partial V}{\partial T}\right)_p \right]_{T_2} dp$$

$$\Delta H = \int_{V_1}^{\infty} \left[V\left(\dfrac{\partial p}{\partial V}\right)_T - T\left(\dfrac{\partial p}{\partial T}\right)_V \right]_{T_1} dV + \int_{T_1}^{T_2} \left[C_{p,g}(T) \right]_{p=0} dT$$

$$+ \int_{\infty}^{V_{sat}} \left[V\left(\dfrac{\partial p}{\partial V}\right)_T - T\left(\dfrac{\partial p}{\partial T}\right)_V \right]_{T_2} dV + (1 - \phi T_2) \int_{p_{sat}}^{p_2} [V]_{T_2} dp$$

Applying the Redlich–Kwong equation of state for gases we obtain Equation (5.102).

$$\Delta H = \int_{T_1}^{T_2} \left[C_{p,g}(T)\right]_{p=0} dT - \left[\frac{RT_1 b}{V_1-b} - \frac{a}{\sqrt{T_1}}\left(\frac{1}{V_1+b}\right) - \frac{3a}{2b\sqrt{T_1}} \ln\left(\frac{V_1+b}{V_1}\right)\right]$$

$$+\left[\frac{RT_2 b}{V_{sat}-b} - \frac{a}{\sqrt{T_2}}\left(\frac{1}{V_{sat}+b}\right) - \frac{3a}{2b\sqrt{T_2}} \ln\left(\frac{V_{sat}+b}{V_{sat}}\right)\right] + (1-\phi T_2)\int_{p_{sat}}^{p_2} [V]_{T_2} dp$$

(5.102)

where $\phi = \dfrac{1}{V}\left(\dfrac{\partial V}{\partial T}\right)_p$ is the thermal expansion coefficient for liquids.

The Tait equation of state for liquids is used for the second integral as was done for Case III.

Therefore, we have Equation (5.103).

$$\Delta H = \int_{T_1}^{T_2} \left[C_{p,g}(T)\right]_{p=0} dT - \left[\frac{RT_1 b}{V_1-b} - \frac{a}{\sqrt{T_1}}\left(\frac{1}{V_1+b}\right) - \frac{3a}{2b\sqrt{T_1}} \ln\left(\frac{V_1+b}{V_1}\right)\right]$$

$$+\left[\frac{RT_2 b}{V_{sat}-b} - \frac{a}{\sqrt{T_2}}\left(\frac{1}{V_{sat}+b}\right) - \frac{3a}{2b\sqrt{T_2}} \ln\left(\frac{V_{sat}+b}{V_{sat}}\right)\right] + (1-\phi T_2)V_s \cdot$$

$$\left\{(p_2 - p_{sat})(1+C) - BC\ln\left(\frac{B+p_2}{B+p_{sat}}\right) - C\left[p_2 \ln\left(\frac{B+p_2}{B+p_s}\right) - p_{sat}\ln\left(\frac{B+p_{sat}}{B+p_s}\right)\right]\right\}$$

(5.103)

Therefore, the temperature and pressure contributions to the change in enthalpy are given by Equations (5.104) and (5.105).

$$q_{temp} = \int_{T_1}^{T_2} \left[C_{p,g}(T)\right]_{p=0} dT \tag{5.104}$$

$$q_{press} = \left[\frac{RT_2 b}{V_{sat}-b} - \frac{a}{\sqrt{T_2}}\left(\frac{1}{V_{sat}+b}\right) - \frac{3a}{2b\sqrt{T_2}} \ln\left(\frac{V_{sat}+b}{V_{sat}}\right)\right]$$

$$-\left[\frac{RT_1 b}{V_1-b} - \frac{a}{\sqrt{T_1}}\left(\frac{1}{V_1+b}\right) - \frac{3a}{2b\sqrt{T_1}} \ln\left(\frac{V_1+b}{V_1}\right)\right]$$

$$+(1-\phi T_2)V_s \cdot$$

$$\left\{(p_2 - p_{sat})(1+C) - BC\ln\left(\frac{B+p_2}{B+p_{sat}}\right) - C\left[p_2 \ln\left(\frac{B+p_2}{B+p_s}\right) - p_{sat}\ln\left(\frac{B+p_{sat}}{B+p_s}\right)\right]\right\}$$

(5.105)

Energy Metrics

The saturation pressure and molar volume are evaluated as follows.

At $T_r = \dfrac{T_2}{T_c}$, the reduced saturation pressure is obtained from Equation (5.106).

$$p_{r,\text{sat}} = 6.9385 T_r^3 - 10.014 T_r^2 + 4.8794 T_r - 0.8012 \tag{5.106}$$

Therefore, $p_{\text{sat}} = p_c p_{\text{sat},r}$.

The molar saturation volume is obtained by solving the Redlich–Kwong equation of state written in cubic form as shown in Equation (5.107).

$$V^3 - \left(\dfrac{RT_2}{p_{\text{sat}}}\right) V^2 - \left(b^2 + \dfrac{bRT_2}{p_{\text{sat}}} - \dfrac{a}{p_{\text{sat}}\sqrt{T_2}}\right) V - \dfrac{ab}{p_{\text{sat}}\sqrt{T_2}} = 0 \tag{5.107}$$

Note that in all of the preceding cases, the enthalpy changes along the temperature paths follow the ideal gas regime as shown by the arrows at the bottom of the p-T path diagrams. Hence, the tabulated heat capacity temperature correlation functions in the DIPPR or Yaws databases that are used are all determined under ideal gas conditions. Also, the magnitude of the q_{press} contribution to the enthalpy change is negligible for pressure changes less than 100 atm. The major contributor in all cases is q_{temp}.

When there are no pressure contributions to consider, the following cases given by Equations (5.108) to (5.117) illustrate the possible calculation scenarios for determining the temperature change contribution to enthalpy.

Case I

Heating a liquid at 298 K from 298 K to T_{rxn} (T_{rxn} is above boiling point, T_b) where the liquid undergoes a phase transition from liquid to gas:

$$q = \int_{298}^{T_b} C_{p,\text{liq}}(T) dT + \int_{T_b}^{T_{\text{rxn}}} C_{p,\text{gas}}(T) dT + \Delta H_{\text{vap}} \tag{5.108}$$

$$C_{p,\text{liq}}(T) = A + BT + CT^2 + DT^3 + ET^4 \tag{5.109}$$

$$C_{p,\text{gas}}(T) = A + B\left[\dfrac{C/T}{\sinh(C/T)}\right]^2 + D\left[\dfrac{E/T}{\cosh(E/T)}\right]^2 \text{ (DIPPR)} \tag{5.110}$$

$$C_{p,\text{gas}}(T) = A + BT + CT^2 + DT^3 + ET^4 \text{ (Yaws)} \tag{5.111}$$

where ΔH_{vap} is the heat of vaporization, the functions $C_p(T)$ represent the temperature dependent heat capacity functions at constant pressure for liquids and gases, and the parameters A, B, C, D, and E are constants specific to a given substance.

Case II
Heating a liquid at 298 K from 298 K to T_{rxn} (T_{rxn} is below boiling point, T_b) where the liquid does not undergo a phase transition:

$$q = \int_{298}^{T_{rxn}} C_{p,\text{liq}}(T)\,dT \qquad (5.112)$$

Case III
Heating a gas at 298 K from 298 K to T_{rxn}:

$$q = \int_{298}^{T_{rxn}} C_{p,\text{gas}}(T)\,dT \qquad (5.113)$$

Case IV
Heating a solid at 298 K from 298 K to T_{rxn} (T_{rxn} is above both the boiling point, T_b, and the melting point, T_m) where the solid undergoes phase transitions from solid to liquid and from liquid to gas:

$$q = \int_{298}^{T_m} C_{p,\text{sol}}(T)\,dT + \int_{T_m}^{T_b} C_{p,\text{liq}}(T)\,dT + \int_{T_b}^{T_{rxn}} C_{p,\text{gas}}(T)\,dT + \Delta H_{\text{fus}} + \Delta H_{\text{vap}} \qquad (5.114)$$

$$C_{p,\text{sol}}(T) = A + BT + CT^2 + DT^3 + ET^4 \qquad (5.115)$$

where ΔH_{fus} is the heat of fusion and $C_{p,\text{sol}}(T)$ is the temperature dependent heat capacity function at constant pressure for solids.

Case V
Heating a solid at 298 K from 298 K to T_{rxn} (T_{rxn} is above melting point, T_m) where the solid undergoes a phase transition from solid to liquid:

$$q = \int_{298}^{T_m} C_{p,\text{sol}}(T)\,dT + \int_{T_m}^{T_{rxn}} C_{p,\text{liq}}(T)\,dT + \Delta H_{\text{fus}} \qquad (5.116)$$

Case VI
Heating a solid at 298 K from 298 K to T_{rxn} (T_{rxn} is below melting point, T_m) where the solid does not undergo a phase transition:

$$q = \int_{298}^{T_{rxn}} C_{p,\text{sol}}(T)\,dT \qquad (5.117)$$

Energy Metrics

5.2.6 Worked Examples

Example 5.22[53]

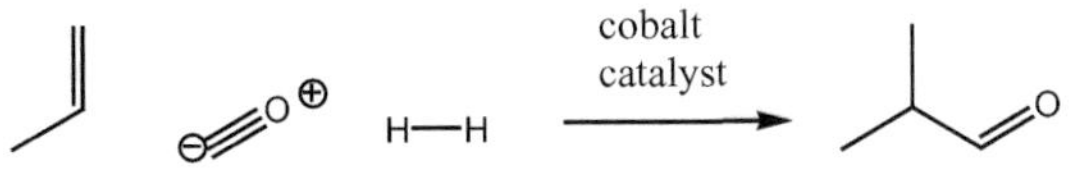

T1 = 298 K, p1 = 1 atm
T2 = 443 K, p2 = 500 atm
Input masses: 3163 kg propene, 5124.4 kg carbon monoxide, 179.6 kg hydrogen

Part 1
Determine the phases of the initial and final states of each input substance.

Solution

			State 1
Input Chemical	T1 (deg K)	p1 (atm)	Phase
Propene	298	1	gas
Carbon monoxide	298	1	gas
Hydrogen	298	1	gas

							State 2
Input Chemical	T2 (deg K)	p2 (atm)	Tc (deg K)	pc (atm)	Tr	pr	Phase
Propene	443	500	365	45.5	1.214	10.989	gas
Carbon monoxide	443	500	133	34.5	3.331	14.493	gas
Hydrogen	443	500	33.5	12.8	13.224	39.063	gas

Part 2
Determine the temperature contribution to the change of enthalpy for the reaction.

Solution

Heat capacity coefficients:

Input chemical	Phase changes	A	B	C	D	E
Propene	gas => gas	31.298	7.24E-02	1.95E-04	-2.16E-07	6.30E-11
Carbon monoxide	gas => gas	29.556	-6.58E-04	2.01E-05	-1.22E-08	2.26E-12
Hydrogen	gas => gas	25.399	2.02E-02	-3.85E-05	3.19E-08	-8.76E-12

Input chemical	H at 298 K	H at T2	q (kJ/mol)
Propene	1.39E+01	2.48E+01	1.09E+01
Carbon monoxide	8.93E+00	1.35E+01	4.57E+00
Hydrogen	8.18E+00	1.24E+01	4.21E+00

Input chemical	Raw mass (g)	Moles	q (kJ/mol)	q (kJ)
Propene	3163000	75310	10.9	820874
Carbon monoxide	5124400	183014	4.57	836375
Hydrogen	179600	89800	4.21	378058

Therefore, 820.9 + 836.4 + 378.1 = 2035.4 MJ is the temperature contribution.

Part 3
Determine the pressure contribution to the change of enthalpy for the reaction.

SOLUTION

Input Chemical	State 1 Phase	Tc (deg K)	pc (atm)	Tr	pr	State 2 Phase
Propylene	gas	364.76	45.52	1.214	10.984	gas: use RK EOS
Carbon monoxide	gas	132.92	34.53	3.333	14.480	gas: use RK EOS
Hydrogen	gas	33.18	12.8	13.351	39.063	gas: use RK EOS

Case I: State 1 (gas) to State 2 (gas): assume Redlich–Kwong equation of state

$$q_{press} = \left[\frac{RT_2 b}{V_2 - b} - \frac{a}{\sqrt{T_2}}\left(\frac{1}{V_2 + b}\right) - \frac{3a}{2b\sqrt{T_2}}\ln\left(\frac{V_2 + b}{V_2}\right)\right]$$
$$- \left[\frac{RT_1 b}{V_1 - b} - \frac{a}{\sqrt{T_1}}\left(\frac{1}{V_1 + b}\right) - \frac{3a}{2b\sqrt{T_1}}\ln\left(\frac{V_1 + b}{V_1}\right)\right]$$

Input Chemical	a (L^2 atm deg. K^0.5/ mol ^2)	b (L/mol)	V1 (L/mol) at p1, T1	V2 (L/mol) at p2, T2	Term 1	Term 2	q (kJ/mol)
Propylene	161.07411	0.05708	24.1291	0.0764	−6.29541	−0.09192	−6.203
Carbon monoxide	17.02609	0.02743	24.444	0.0836	−0.21111	−0.00743	−0.204
Hydrogen	1.46119	0.01862	24.472	0.08703	0.826238	0.001011	0.825

Energy Metrics

Input Chemical	Raw Mass (g)	Moles	q* (kJ/mol)	q* (kJ)
Propylene	3163000	75310	−6.203	−467145
Carbon monoxide	5124400	183014	−0.204	−37335
Hydrogen	179600	89800	0.825	74085

Therefore, −467.1 − 37.3 + 74.1 = −430.3 MJ is the pressure contribution.

Example 5.23[53]

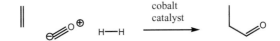

T1 = 298 K, p1 = 1 atm
T2 = 393 K, p2 = 780 atm

Input masses: 1999.8 kg ethylene, 9598.8 kg carbon monoxide, 1464.1 kg hydrogen, 3792.4 kg diethyl ether solvent

Part 1
Determine the phases of the initial and final states of each input substance.

Solution

			State 1
Input Chemical	T1 (deg K)	p1 (atm)	Phase
Ethylene	298	1	gas
Carbon monoxide	298	1	gas
Hydrogen	298	1	gas
Diethyl ether	298	1	liquid

							State 2
Input Chemical	T2 (deg K)	p2 (atm)	Tc (deg K)	pc (atm)	Tr	pr	Phase
Ethylene	393	780	282.36	49.66	1.392	15.707	gas
Carbon monoxide	393	780	132.92	34.53	2.957	22.589	gas
Hydrogen	393	780	33.25	12.8	11.820	60.938	gas
Diethyl ether	393	780	466.7	35.9	0.842	21.727	liquid

Part 2
Determine the temperature contribution to the change of enthalpy for the reaction.

SOLUTION
Heat capacity coefficients:

Input Chemical	Phase Changes	A	B	C	D	E
Ethylene	gas => gas	32.083	−1.48E−02	2.48E−04	−2.38E−07	6.83E−11
Carbon monoxide	gas => gas	29.556	−6.58E−04	2.01E−05	−1.22E−08	2.26E−12
Hydrogen	gas => gas	25.399	2.02E−02	−3.85E−05	3.19E−08	−8.76E−12
Diethyl ether	liquid => liquid	75.939	7.73E−01	−2.79E−03	4.44E−06	

Input Chemical	H at 298 K	H at T2	q (kJ/mol)
Ethylene	1.07E+01	1.52E+01	4.54E+00
Carbon monoxide	8.93E+00	1.19E+01	2.97E+00
Hydrogen	8.18E+00	1.09E+01	2.75E+00
Diethyl ether	4.11E+01	5.95E+01	1.84E+01

Input Chemical	Raw mass (g)	Moles	q (kJ/mol)	q (kJ)
Ethylene	1998900	71389	4.54	324107
Carbon monoxide	9598800	342814	2.97	1018158
Hydrogen	1464100	732050	2.75	2013138
Diethyl ether	3792400	51249	18.4	942975

Therefore, $324.1 + 1018.2 + 2013.1 + 943.0 = 4298.4$ MJ is the temperature contribution.

Part 3
Determine the pressure contribution to the change of enthalpy for the reaction.

SOLUTION

Input Chemical	State 1 Phase	Tc (deg K)	pc (atm)	Tr	pr	State 2 Phase
Ethylene	gas	282.36	49.66	1.392	15.707	gas: use RK EOS
Carbon monoxide	gas	132.92	34.53	2.957	22.589	gas: use RK EOS
Hydrogen	gas	33.18	12.8	11.844	60.938	gas: use RK EOS
Diethyl ether	liquid	466.7	35.9	0.842	21.727	liquid; use Tait EOS

Energy Metrics

Case I: State 1 (gas) to State 2 (gas): assume Redlich–Kwong equation of state

$$q_{\text{press}} = \left[\frac{RT_2 b}{V_2 - b} - \frac{a}{\sqrt{T_2}}\left(\frac{1}{V_2 + b}\right) - \frac{3a}{2b\sqrt{T_2}}\ln\left(\frac{V_2 + b}{V_2}\right)\right]$$

$$- \left[\frac{RT_1 b}{V_1 - b} - \frac{a}{\sqrt{T_1}}\left(\frac{1}{V_1 + b}\right) - \frac{3a}{2b\sqrt{T_1}}\ln\left(\frac{V_1 + b}{V_1}\right)\right]$$

Input Chemical	a (L^2 atm deg K^0.5/mol^2)	b (L/mol)	V1 (L/mol) at p1, T1	V2 (L/mol) at p2, T2	term 1	term 2	q (kJ/mol)
Ethylene	77.6795	0.04046	24.3129	0.0641	−5.40631	−0.0427	−5.36
Carbon monoxide	16.9855	0.02739	24.4441	0.0619	−0.12034	−0.00741	−0.11
Hydrogen	1.434	0.01848	24.472	0.05901	1.233446	0.001013	1.23

Case III: State 1 (liquid) to State 2 (liquid): assume Tait equation of state

$$q_{\text{press}} = (1 - \phi T_2) V_s \left[\begin{array}{l}(p_2 - 101325)(1 + C) - BC\ln\left(\dfrac{B + p_2}{B + 101325}\right) \\ -C\left[p_2 \ln\left(\dfrac{B + p_2}{B + p_s}\right) - 101325\ln\left(\dfrac{B + 101325}{B + p_s}\right)\right]\end{array}\right]$$

Input Chemical	diethyl ether	
MW	74	g/mol
ω	0.281	
Tc	466.7	K
pc	3.64E+06	Pa
T2	393	K
p2	7.90E+07	Pa
Tr	0.842082708	
1 − Tr	0.157917292	
Thermal expansion coefficient calculation		
Lambda	8.10E−04	
m	−0.7064	
phi	0.00164382	K^−1
Molar volume calculation at T2, 1 atm		
A11	0.2727	
B21	0.2761	

n	0.2936	
Density	0.576483371	g/mL
Vs	0.128364501	L/mol

Liquid vapor pressure calculation at T2

A12	41.752	
B22	−2741	
C32	−12.27	
D42	−3.19E−10	
E52	5.98E−06	
log ps	3.867884484	
ps	7377.079841	mm Hg
ps	9.706684001	atm
ps	983529.7564	Pa

Enthalpy change pressure contribution calculation

C	0.095828772	
B	11402716.98	
$(1 - phi*T2)$ Vs	0.045438297	
$(p2 - 101325)(1 + C)$	8.65E+07	
$B\ C\ \ln[(B+p2)/(B+101325)]$	2253106.052	
$C\ (p2\ \ln(B+p2) - 101325\ \ln(B+101325))$	1.39E+08	
$C\ (101325 - p2)\ \ln(B+ps)$	−1.24E+08	
q	3.14E+06	Pa L/mol
q	3.14E+00	kJ/mol

Approximate calculation (molar volume does not change with pressure)

$(1 - phi*T2)$ Vs	0.045438297	L/mol
p2 − 101325	7.89E+07	Pa
q	3.59E+06	Pa L/mol
q	3.59E+00	kJ/mol

Input Chemical	Raw Mass (g)	Moles	q* (kJ/mol)	q* (kJ)
Ethylene	1998900	71389	−5.36	−382647
Carbon monoxide	9598800	342814	−0.11	−37710
Hydrogen	1464100	732050	1.23	900422
Diethyl ether	3792400	51249	3.14	160921

Therefore, −382.6 − 37.7 + 900.4 + 160.9 = 641.0 MJ is the pressure contribution.

Example 5.24[54]

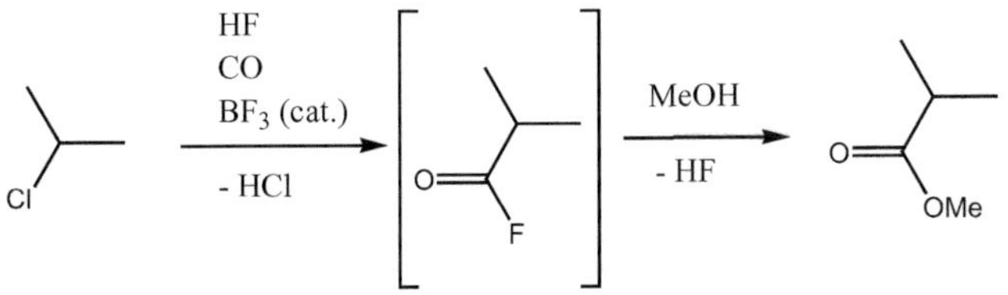

HF
CO
BF₃ (cat.)
Cl — ⟶ (−HCl) [O=/F] ⟶ MeOH (−HF) ⟶ O=/OMe

T1 = 298 K, p1 = 1 atm
T2 = 288 K, p2 = 400 atm

Input masses: 3256.5 isopropyl chloride, 2442.4 kg HF, 60228.6 kg CO, 12895.7 kg methanol, and 8141.2 kg boron trifluoride.

Part 1
Determine the phases of the initial and final states of each input substance.

			State 1
Input chemical	T1 (deg K)	p1 (atm)	Phase
Isopropyl chloride	298	1	liquid
HF	298	1	gas
CO	298	1	gas
Methanol	298	1	liquid
BF₃	298	1	gas

							State 2
Input Chemical	T2 (deg K)	p2 (atm)	Tc (deg K)	pc (atm)	Tr	pr	Phase
Isopropyl chloride	288	400	489	45	0.589	8.889	liquid
HF	288	400	461.15	64	0.625	6.250	liquid
CO	288	400	132.92	34.53	2.167	11.584	gas
Methanol	288	400	512.58	79.9	0.562	5.006	liquid
BF₃	288	400	261	49.2	1.103	8.130	gas

Part 2
Determine the temperature contribution to the change of enthalpy for the reaction.

SOLUTION

Heat capacity coefficients:

Input Chemical	Phase Changes	ΔH(vap) at Tb	A	B	C	D	E
Isopropyl chloride	liquid => liquid	26.69	64.547	5.23E−01	−1.82E−03	2.82E−06	
HF	gas => Tb condensation	7.52	29.085	9.61E−04	−4.47E−06	6.78E−09	−2.20E−12
	Tb => liquid		24.415	2.04E−01	−9.43E−04	1.90E−06	
CO	gas => gas	5.91	29.556	−6.58E−04	2.01E−05	−1.22E−08	2.26E−12
Methanol	liquid => liquid	35.14	40.152	3.10E−01	−1.03E−03	1.46E−06	
BF_3	gas => gas	17.24	22.487	1.18E−01	−8.71E−05	2.23E−08	1.22E−13

Input Chemical	H at T1	H at T2	q (kJ/mol)
Isopropyl chloride	3.20E+01	3.07E+01	−1.33E+00
HF	8.68E+00	8.53E+00	−1.55E−01
			−7.52E+00
	1.15E+01	1.12E+01	−2.36E−01
		sum	−7.91E+00
CO	8.93E+00	8.62E+00	−3.08E−01
Methanol	1.96E+01	1.88E+01	−7.95E−01
BF_3	1.12E+01	1.07E+01	−5.02E−01

Input Chemical	Raw Mass (g)	Moles	q (kJ/mol)	q (kJ)
Isopropyl chloride	3256500	41511	−1.33	−55209
HF	2442400	122120	−7.91	−965969
CO	60228600	2151021	−0.308	−662515
Methanol	12895700	402991	−0.795	−320378
BF_3	8141200	119724	−0.502	−60101

Therefore, −55.2 − 966.0 − 662.5 − 320.4 − 60.1 = −2064.2 MJ is the temperature contribution.

Part 3

Determine the pressure contribution to the change of enthalpy for the reaction.

Energy Metrics

SOLUTION

Input Chemical	State 1 Phase	Tc (deg K)	pc (atm)	Tr	pr	State 2 Phase
Isopropyl chloride	liquid	489	45	0.589	8.889	liquid; use Tait EOS
Carbon monoxide	gas	132.92	34.53	2.167	11.584	gas: use RK EOS
Methanol	liquid	512.58	79.9	0.562	5.006	liquid; use Tait EOS
HF	gas	461.15	64	0.625	6.250	liquid; use Tait EOS
BF_3	gas	261	49.2	1.103	8.130	gas: use RK EOS

Case I: State 1 (gas) to State 2 (gas): assume Redlich–Kwong equation of state

$$q_{press} = \left[\frac{RT_2 b}{V_2 - b} - \frac{a}{\sqrt{T_2}}\left(\frac{1}{V_2 + b}\right) - \frac{3a}{2b\sqrt{T_2}}\ln\left(\frac{V_2 + b}{V_2}\right)\right]$$
$$- \left[\frac{RT_1 b}{V_1 - b} - \frac{a}{\sqrt{T_1}}\left(\frac{1}{V_1 + b}\right) - \frac{3a}{2b\sqrt{T_1}}\ln\left(\frac{V_1 + b}{V_1}\right)\right]$$

Input Chemical	a (L^2 atm deg K^0.5/ mol^2)	b (L/mol)	V1 (L/mol) at p1, T1	V2 (L/mol) at p2, T2	Term 1	Term 2	q (kJ/mol)
Carbon monoxide	16.9857	0.02739	24.4441	0.0625	−1.2783	−0.00741	−1.271
BF_3	64.4082	0.03775	24.3418	0.052	−6.28012	−0.03494	−6.245

Case III: State 1 (liquid) to State 2 (liquid): assume Tait equation of state

$$q_{press} = (1 - \phi T_2) V_s \left[\begin{array}{l}(p_2 - 101325)(1 + C) - BC\ln\left(\dfrac{B + p_2}{B + 101325}\right) \\ -C\left[p_2 \ln\left(\dfrac{B + p_2}{B + p_s}\right) - 101325\ln\left(\dfrac{B + 101325}{B + p_s}\right)\right]\end{array}\right]$$

Input Chemical	Isopropyl Chloride	
MW	78.45	g/mol
ω	0.215	
Tc	489	K
pc	4.56E+06	Pa
T2	288	K
p2	4.05E+07	Pa
Tr	0.588957055	
1 − Tr	0.411042945	
Thermal expansion coefficient calculation		
Lambda	7.54E−04	
m	−0.7143	
phi	0.001025002	K^−1
Molar volume calculation at T2, 1 atm		
A11	0.31887	
B21	0.27503	
n	0.28571	
Density	0.867913511	g/mL
Vs	0.090389191	L/mol
Liquid vapor pressure calculation at T2		
A12	115.1259	
B22	−4.63E+03	
C32	−4.22E+01	
D42	2.53E−02	
E52	−9.89E−13	
log ps	2.553045112	
ps	357.3099515	mm Hg
ps	0.470144673	atm
ps	47637.409	Pa
Enthalpy change pressure contribution calculation		
C	0.093555185	
B	56492846.72	
(1 − phi*T2) Vs	0.063706254	
(p2 − 101325) (1 + C)	4.42E+07	
B C ln[(B + p2)/(B + 101325)]	2848936.028	
C (p2 ln(B+p2) − 101325 ln(B+101325))	6.96E+07	
C (101325 - p2) ln(B + ps)	−6.75E+07	
q	2.50E+06	Pa L/mol
q	2.50E+00	kJ/mol
Approximate calculation (molar volume does not change with pressure)		
(1 − phi*T2) Vs	0.063706254	L/mol
p2 − 101325	4.04E+07	Pa
q	2.58E+06	Pa L/mol
q	2.58E+00	kJ/mol

Energy Metrics

Input Chemical	Methanol	
MW	32	g/mol
ω	0.577	
Tc	512.58	K
pc	8.10E+06	Pa
T2	288	K
p2	4.05E+07	Pa
Tr	0.561863514	
1 − Tr	0.438136486	
Thermal expansion coefficient calculation		
lambda	5.92E−04	
M	−0.7669	
phi	0.000854868	K^-1
Molar volume calculation at T2, 1 atm		
A11	0.27197	
B21	0.27192	
n	0.2331	
Density	0.796365882	g/mL
Vs	0.040182535	L/mol
Liquid vapor pressure calculation at T2		
A12	45.6171	
B22	−3.24E+03	
C32	−1.40E+01	
D42	6.64E−03	
E52	−1.05E−13	
log ps	1.860110428	
ps	72.46201858	mm Hg
ps	0.095344761	atm
ps	9660.807938	Pa
Enthalpy change pressure contribution calculation		
C	0.106025469	
B	299388479.2	
(1 − phi*T2) Vs	0.030289519	
(p2 − 101325) (1 + C)	4.47E+07	
B C ln[(B+p2)/(B+101325)]	4019446.757	
C (p2 ln(B+p2) − 101325 ln(B+101325))	8.42E+07	
C (101325 − p2) ln(B + ps)	−8.37E+07	
q	1.22E+06	Pa L/mol
q	1.22E+00	kJ/mol
Approximate calculation (molar volume does not change with pressure)		
(1 − phi*T2) Vs	0.030289519	L/mol
p2 − 101325	4.04E+07	Pa
q	1.22E+06	Pa L/mol
q	1.22E+00	kJ/mol

Input Chemical	HF	
MW	20	g/mol
ω	0.345	
Tc	461.15	K
pc	6.48E+06	Pa
T2	288	K
p2	4.05E+07	Pa
Tr	0.6245256	
1 − Tr	0.3754744	
Thermal expansion coefficient calculation		
Lambda	1.40E−03	
m	−0.6267	
phi	0.0015186	K^−1
Molar volume calculation at T2, 1 atm		
A11	0.29041	
B21	0.1766	
n	0.3733	
Density	0.9669346	g/mL
Vs	0.0206839	L/mol
Liquid vapor pressure calculation at T2		
A12	23.7347	
B22	−1.80E+03	
C32	−6.18E+00	
D42	−5.00E−10	
E52	6.15E−06	
log ps	2.8060026	
ps	639.73864	mm Hg
ps	0.8417614	atm
ps	85291.471	Pa
Enthalpy change pressure contribution calculation		
C	0.098033464	
B	98357221.15	
(1 − phi*T2) Vs	0.011652732	
(p2 − 101325) (1 + C)	4.44E+07	
B C ln[(B + p2)/(B + 101325)]	3317207.713	
C (p2 ln(B+p2) − 101325 ln(B+101325))	7.43E+07	
C (101325 − p2) ln(B + ps)	−7.29E+07	
q	4.62E+05	Pa L/mol
q	4.62E−01	kJ/mol
Approximate calculation (molar volume does not change with pressure)		
(1 − phi*T2) Vs	0.011652732	L/mol
p2 − 101325	4.04E+07	Pa
q	4.71E+05	Pa L/mol
q	4.71E−01	kJ/mol

Energy Metrics

Input Chemical	Raw Mass (g)	Moles	q* (kJ/mol)	q* (kJ)
Isopropyl chloride	3256500	41511	2.5	103776
HF	2442400	122120	0.463	56542
CO	60228600	2151021	−1.271	−2733948
Methanol	12895700	402991	1.22	491649
BF_3	8141200	119724	−6.245	−747673

Therefore, $103.8 + 56.5 - 2733.9 + 491.6 - 747.7 = -2829.7$ MJ is the pressure contribution.

5.3 TYPES OF ENERGY METRICS

5.3.1 TERMS, DEFINITIONS, AND EXAMPLES

Cumulative Energy Demand[55–60]

The cumulative energy demand (CED) of a product represents the direct and indirect energy use in units of MJ throughout the life cycle, including the energy consumed during the extraction, manufacturing and disposal of the raw and auxiliary materials. The total CED is composed of the fossil cumulative energy demand (i.e., from hard coal, lignite, peat, natural gas, and crude oil) and the CED of nuclear, biomass, water, wind, and solar energy in the life cycle. The CED therefore represents the total energy usage over the entire life cycle of a product covering its manufacture from raw materials and its end of life disposal. CED is also known as "primary energy consumption." Units: MJ-eq. (equivalents).

Energy Efficiency[61]

For a complete chemical process, energy efficiency is defined as the ratio of theoretical energy demand based on thermodynamic calculations to actual energy spent based on experimental determinations. For an individual reaction, energy efficiency is defined as the ratio of energy spent on that reaction to total energy spent on a synthesis plan of which that reaction was a particular step.

Energy Index (EI)[62]

The energy index is given by Equation (5.118).

$$EI = \begin{cases} 1, & Q_{plan} \leq 92.62 \text{ kJ mol}^{-1} \\ \dfrac{q_{water}}{Q_{plan} - q_{water}} & Q_{plan} > 92.62 \text{ kJ mol}^{-1} \end{cases} \quad (5.118)$$

where q_{water} is the energy input required to heat and evaporate 1 mol of liquid water from a starting state of 25°C and 1 atm (46.31 kJ/mol) and Q_{plan} is the energy input consumption in kJ for the synthesis of 1 mole of target product including contributions from temperature and pressure changes beyond ambient conditions of 25°C and 1 atm.

Energy Input (Absolute)

The absolute energy input consumption for an individual reaction or an entire chemical process is determined based on two main considerations: thermodynamic and operational. The thermodynamic component depends on reaction temperature, reaction pressure, the heat capacities of all input materials used, and the masses of all input materials used. The operational component depends on electricity consumption needed to run all apparatuses and equipment arising from heating, cooling, or pressurization procedures. This includes energy inputs for operating any equipment during the operation of running the reaction, and during the workup, and purification stages. A key parameter in determining the operational component of energy input consumption is the time required for running each of the apparatuses involved. In industry this is often called process energy. The usual units are kJ or MJ. For a synthesis plan involving N steps that is appropriately mass scaled, the total energy input is the sum of the energy inputs for each step.

Energy Input per Mass of Target Product

The absolute energy input consumption divided by the mass of target product produced in a chemical reaction. The units are kJ/kg or MJ/kg. For a synthesis plan involving N steps that is appropriately mass scaled, the sum of all energy inputs for all steps is divided by the mass of the final target product.

Energy Intensity[63]

Energy intensity is the ratio of total process energy needed to produce a given product to the mass of that product. This quantity can be applied to an individual reaction or an entire synthesis plan. Units are kJ/kg or MJ/kg.

Primary Energy Usage[64]

Primary energy usage is the ratio of mass of fuel consumed to mass of product. This definition is based on a relative mass ratio though strictly speaking energy units are expressed in MJ. A better definition is the ratio of input MJ required from all energy sources to manufacture 1 kg of product.

Solvent Recovery Energy

Solvent recovery energy is the ratio of total solvent recovery energy requirements to mass of product made.

Waste Treatment Energy

Waste treatment energy is the ratio of energy used to treat waste from a process to mass of product made.

Example 5.25[65]

Themes: balancing chemical equations, reaction yield, atom economy, E-factor, process mass intensity, stoichiometric factor, conversion, selectivity, energy intensity.

Energy Metrics

n-Butanol is obtained industrially from hydroformylation of propylene, carbon monoxide, and hydrogen according to the following reaction.

$$\diagup\!\!\!\!\diagdown + CO + 2H_2 \longrightarrow \diagup\!\!\!\!\diagdown\!\!\!\!\diagup\!\!\!\!\diagdown OH$$

Isobutanol is also formed in the reaction as a side product.

The process uses 609 kg of propylene, 399 kg/h of CO, and 58 kg/h of H_2 to produce 1000 kg/h of 99% pure n-butanol and 50 kg/h of 85% isobutanol. The process also requires 749 MJ/h of electricity and 2270 MJ/h of steam.

Part 1

If n-butanol is the target product, determine the % yield, % AE, E-factor, PMI, stoichiometric factor, % conversion of propylene, % selectivity to n-butanol, and energy intensity.

Solution

$$\diagup\!\!\!\!\diagdown + CO + 2H_2 \longrightarrow \diagup\!\!\!\!\diagdown\!\!\!\!\diagup\!\!\!\!\diagdown OH$$

$$42 \qquad 28 \qquad 2(2) \qquad\qquad 74$$

Reactants: $42 + 28 + 2*2 = 74$
Products: 74

Reactant	MW (g/mol)	Mass (g)	Exp. moles	Stoich. Moles
Propylene	42	609000	14500	14250
CO	28	399000	14250	14250
Hydrogen	2	58000	29000	28500

The limiting reagent is carbon monoxide.

Since 1 mole of carbon monoxide produces 1 mole of n-butanol, then 14250 moles of carbon monoxide are expected to produce 14250 moles of n-butanol.

Yield of n-butanol is $1000*0.99 = 990$ kg $= 990000$ g or $990000/74 = 13378.38$ moles.

% yield of n-butanol $= 100*(13378.38/14250) = 93.9\%$
MW n-butanol product $= 74$
MW of reactants $= 74$
% AE $= 100*(74/74) = 100\%$
Mass of n-butanol product $= 990$ kg
Mass of input materials $= 609 + 399 + 58 = 1066$ kg
Mass of waste $= 1066 - 990 = 76$ kg
E-factor $= 76/990 = 0.077$
PMI $= 1066/990 = 1.077$

Reactant	MW (g/mol)	Mass (g)	Exp. Moles	Stoich. Moles	Excess Moles
Propylene	42	609000	14500	14250	250
CO	28	399000	14250	14250	0
Hydrogen	2	58000	29000	28500	500

SF = 1 + (excess mass of reagents/stoichiometric mass of reagents)
SF = 1 + ((250*42 + 500*2)/(14500*42 + 14250*28 + 29000*2)) = 1.011
Note that 1/(AE*Y/SF) = 1/(1*0.939/1.011) = 1.077 = kernel PMI.
14500 moles of propylene are converted to 13378.38 moles of *n*-butanol and 0.85*50000/74 = 574.32 moles of isobutanol.
Therefore, % conversion of propylene is 100*(13378.38 + 574.32)/14500 = 96.2%
% selectivity to *n*-butanol = 100*(fractional yield of *n*-butanol)/(fractional conversion of propylene to products)
Therefore, % selectivity to *n*-butanol = 100*(0.939/0.962) = 97.6%.
Energy intensity to produce *n*-butanol = (749 + 2270)/990 = 3.05 MJ/h/kg

Part 2

Repeat the same analysis if isobutanol were the target product.

SOLUTION

The balanced chemical equation is given below.

$$\text{CH}_2=\text{CHCH}_3 + CO + 2H_2 \longrightarrow (CH_3)_2CHCH_2OH$$

42   28   2(2)   74

Yield of isobutanol is 0.85*50 = 42.5 kg or 42.5*1000/74 = 574.32 moles
% yield of isobutanol = 100*(574.32/14250) = 4.03%
% AE = 100%
Mass of isobutanol product = 42.5 kg
Mass of input materials = 609 + 399 + 58 = 1066 kg
Mass of waste = 1066 − 42.5 = 1023.5 kg
E-factor = 1023.5/42.5 = 24.08
PMI = 1066/42.5 = 25.08
SF = 1.011
Note that 1/(AE*Y/SF) = 1/(1*0.0403/1.011) = 25.09 = kernel PMI
% conversion of propylene = 96.2%
% selectivity to isobutanol = 100*(fractional yield of isobutanol)/(fractional conversion of propylene to products)
Therefore, % selectivity to isobutanol = 100*(0.0403/0.962) = 4.19%.
Energy intensity to produce isobutanol = (749 + 2270)/42.5 = 71.04 MJ/h/kg

SUMMARY

Metric	n-Butanol	Isobutanol
% Y	93.9	4.03
% AE	100	100
SF	1.011	1.011
E-factor	0.077	24.08
PMI	1.077	25.08
% Selectivity	97.6	4.19
Energy intensity (MJ/h/kg)	3.05	71.04

Example 5.26[66]

Themes: balancing chemical equations, ring construction strategies, mechanism, input energy.

Condition	Rxn Time	% Yield
Stirring at rt	5 h	37
Ultrasound 230 W	7 min	84
Microwaves 640 W	20 min	76

rt = room temperature

Part 1
Write out a mechanism for this transformation and determine the ring construction strategy.

Part 2
Write out a balanced chemical equation for the reaction and determine its atom economy.

Part 3
Using the experimental data given, determine the kernel reaction mass efficiency and input energy required to carry out the reaction under the three conditions.

SOLUTION
Part 1
Mechanism

Ring construction strategy: [3 + 2]

Part 2

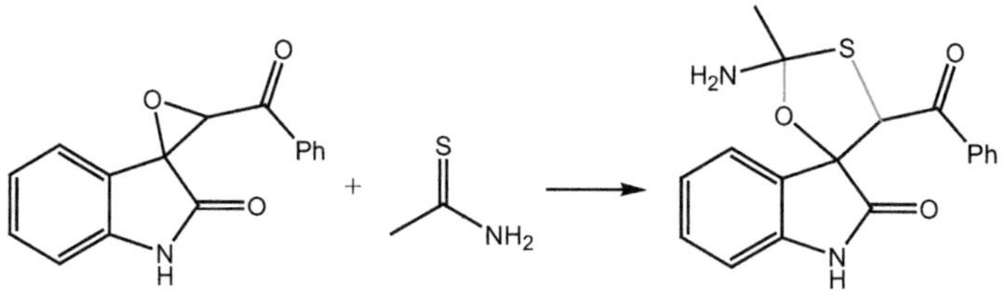

265 + 75 → 340

Reactants: 265 + 75 = 340
Products: 340
AE = 1, or 100%

Part 3

Condition	Rxn Time	% Yield	AE %	Kernel RME %	Energy Input (kJ)
Stirring at rt	5 h	37	1	37	0
Ultrasound (230 W)	7 min	84	1	84	96.6
Microwaves (640 W)	20 min	76	1	76	768

Energy = Power*Time
J = W*s

Example 5.27[67]

Themes: balancing chemical equations, material efficiency metrics, energy metrics.

The O-thiocarbamates can be transformed to S-thiocarbamates via the Newman–Kwart rearrangement. Various microwave irradiation experiments under batch conditions were conducted on the same test reaction using different apparatuses. Based on the data presented below, determine the following parameters: % yield, PMI, process time, space-time-yield, and energy consumption (kJ/kg product). State any assumptions made in the calculations.

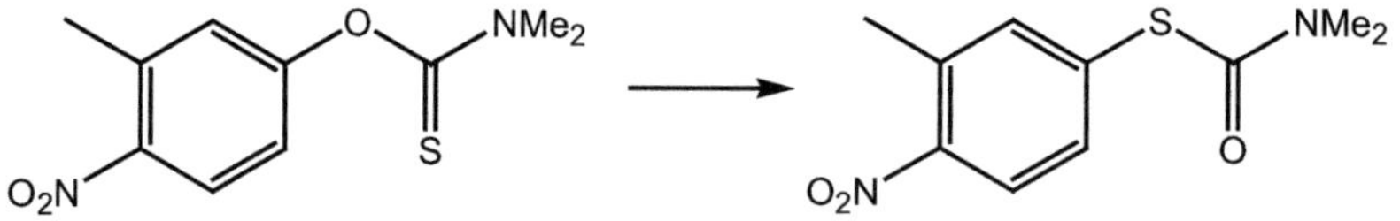

EXPERIMENT #1

Apparatus: Anton Paar Synthos 3000
Operation: Multi-mode
8 × 12.5 g 3-Methyl-4-nitrophenyl-O-thiocarbamate
8 × 50 mL dimethylacetamide
92.8 g 3-Methyl-4-nitrophenyl-S-thiocarbamate

Reaction temperature = 220°C
MW Power = 1400 W
Residence time = 13 min
Holding time = 20 min
Cooling time = 40 min

EXPERIMENT #2

Apparatus: CEM MARS
Operation: Single-mode
82.5 g 3-Methyl-4-nitrophenyl-O-thiocarbamate
850 mL N-methylpyrrolidone
2200 mL water in work-up
71.1 g 3-Methyl-4-nitrophenyl-S-thiocarbamate
Reaction temperature = 200 C
MW Power = 1200 W
Residence time = 9 min
Holding time = 40 min
Cooling time = 30 min

EXPERIMENT #3

Apparatus: CEM Voyager
Operation: Multi-mode
5 × 10 g 3-Methyl-4-nitrophenyl-O-thiocarbamate
5 × 40 mL dimethylacetamide
600 mL water in work-up
41.2 g 3-Methyl-4-nitrophenyl-S-thiocarbamate
Reaction temperature = 210°C
MW Power = 300 W
Residence time = 3.5 min
Holding time = 20 min
Cooling time = 5 min

EXPERIMENT #4

Apparatus: Milestone MicroSYNTH
Operation: Single-mode
50 g 3-Methyl-4-nitrophenyl-O-thiocarbamate
200 mL dimethylacetamide
1100 mL water in work-up
46.4 g 3-Methyl-4-nitrophenyl-S-thiocarbamate
Reaction temperature = 210°C
MW Power = 800 W
Residence time = 4 min
Holding time = 20 min
Cooling time = unspecified

Solution

Assumptions:
Density of dimethylacetamide = 0.937 g/mL
Density of N-methylpyrrolidone = 1.028 g/mL

The volume of 3-methyl-4-nitrophenyl-O-thiocarbamate is not counted since it is dissolved in a solvent and is expected to contribute a negligible increase to the volume of the solvent used.

Process time = residence time + holding time + cooling time
STY = mass of product (kg)/(residence time (s) * reaction volume (m^3))

MW Apparatus	% Yield	PMI	Rxn Vol (L)	Res. Time (s)	Process Time (h)	STY (kg/m^3 s)	Energy (kJ/kg)
Anton Paar Synthos 3000	92.8	5.1	0.4	780	1.22	0.3	66,078
CEM MARS	83.5	44.4	0.85	540	1.32	0.15	80,000
CEM Voyager	82.4	20.3	0.2	210	0.48	0.98	12,451
Milestone MicroSYNTH	92.8	28.8	0.2	240	>0.4	0.97	24,828

Example 5.28[68,69]

Themes: balancing chemical equations, material efficiency metrics, space-time yield, energy metrics.

1-Methylbenzimidazole can be made by condensing o-phenylenediamine and acetic acid under various thermal conditions. Experimental conditions for conventional heating and various microwave heating conditions are given below.

Temp. (deg C)	Pressure (atm)	Heating Mode	Power (W)	Rxn Time	% Yield
100	1	water bath		2 h	68
200	1	MW single-mode	1000	5 min	96
200	1	MW multi-mode	1000	5 min	98
200	130	MW continuous flow	1000	0.5 min	94

Conventional Heating Conditions (water bath):

Reaction vessel (glass) = 500 mL
54 g o-phenylenediamine
45 g acetic acid

Work-up: 10 wt% sodium hydroxide, ice cold water, 50 mL water

Purification: 750 mL boiling water, 2 g decolorizing charcoal, 50 mL cold water

Yield: 68%

Microwave Single-Mode Conditions:
Reaction vessel (pyrex) = 1 × 10 mL
216 mg o-phenylenediamine
2 mL of 1 M acetic acid
Work-up: saturated potassium carbonate solution, 3 × EtOAc, sodium sulfate drying
Yield: 253 mg 2-methylbenzimidazole

Microwave Multi-Mode Conditions:
Reaction vessel (PTFE and PEEK) = 16 × 100 mL
16 × 24.3 g o-phenylenediamine
16 × 45 mL of 5 M acetic acid
Work-up: saturated potassium carbonate solution, 3 × EtOAc, sodium sulfate drying
Yield: 465.7 g 2-methylbenzimidazole

Microwave Continuous Flow Conditions (single tube):
Reaction vessel (stainless steel coil) = 4 mL
43.2 g o-phenylenediamine
400 mL of 1 M acetic acid flow rate = 8 mL/min
Work-up: saturated potassium carbonate solution, 3 X EtOAc, sodium sulfate drying
Yield: 50.7 g 2-methylbenzimidazole
PTFE = polytetrafluoroethylene
PEEK = polyether ether ketone

Part 1
Using the *template-REACTION.xls* spreadsheet and the details of the experimental procedures determine the materials efficiency metrics for each thermal condition. State any assumptions made in the calculations.

SOLUTION
Conventional Heating Conditions (water bath)
Assumptions:

100 mL 10 wt% NaOH solution, 150 mL water
44.9 g 1-methylbenzimidazole product collected corresponding to 68% yield

Microwave Single-Mode Conditions
Assumptions:

5 mL saturated potassium carbonate, 5 mL ethyl acetate, 0.5 g sodium acetate drying agent
Microwave Multi-Mode Conditions
Assumptions: scale-up factor relative to microwave single-mode conditions = 1800

This scaling factor is obtained by dividing the mole scale of multi-mode operation by the mole scale of the single-mode operation, that is, 3.6/0.002 = 1800.

9000 mL saturated potassium carbonate, 9000 mL ethyl acetate, 900 g sodium acetate drying agent

Microwave Continuous Flow Conditions (single tube)
Assumptions: scale-up factor relative to microwave single-mode conditions = 200
This scaling factor is obtained by dividing the mole scale of continuous flow operation by the mole scale of the single-mode operation, that is, 0.4/0.002 = 200.
1000 mL saturated potassium carbonate, 1000 mL ethyl acetate, 100 g sodium acetate drying agent

Method	% AE	% Yield	PMI
Conventional heating	78.6	68	25.88
Microwave single-mode	78.6	96	49.99
Microwave multi-mode	78.6	98	42.71
Microwave continuous flow	78.6	96	49.89

Part 2

Determine the space-time yield (STY) for each experimental condition according to the relationship shown in Equation (5.119).

$$\text{space-time yield} = \frac{\text{mass of product}}{(\text{total volume of reaction materials})(\text{reaction time})}, \quad (5.119)$$

kg per cubic meter per second

Solution

	Conventional Heating	MW Single-Mode	MW Multi-Mode	MW Continuous Flow (Single Tube)
mass product (kg)	0.0449	0.000253	0.4657	0.0507
reaction volume (m^3)	8.56E−5	2.17E−6	1.03E−3	4.34E−4
reaction time (sec)	7200	300	300	30
space-time yield (kg/m^3 sec)	0.073	0.389	1.511	3.893

These data show that the STY increases steadily from the conventional heating mode to the continuous flow mode.
Relative ranking: 1: 5.3: 20.7: 53.3

Part 3

Given the following thermodynamic data use the *energy-input-analysis-synthesis-plans.xls* spreadsheet to determine the input energy required for carrying out the reaction under each of the thermal conditions.

Energy Metrics

Input Material	Tm (°C)	Tb (°C)	Δ H(fus), kJ/mol	Δ H(vap), kJ/mol
o-phenylenediamine	103	256	22.6	55.86
acetic acid	16	117	11.715	23.33
water	0	100	6.002	39.5

Heat capacity coefficients to determine C_p (J/(mol deg K)):

Input Material	Phase	A	B	C	D	E
o-Phenylenediamine	Solid[a]	135.62	0.5717	1.1893E−03	1.0768E−05	
	Liquid	−56.502	1.6935	−3.4068E−03	2.6393E−06	
Acetic acid	Liquid	−18.944	1.0971	−2.8921E−03	2.9275E−06	
	Gas	34.85	3.7626E−02	2.8311E−04	−3.0767E−07	9.2646E−11
Water	Liquid	92.053	−3.995E−02	−2.1103E−04	5.3469E−07	
	Gas	33.933	−8.4186E−03	2.9906E−05	−1.7825E−08	3.6934E−12

[a] Coefficients correspond to temperature in degrees C.

Functional form of temperature dependence of Cp:

$$C_p = A + BT + CT^2 + DT^3 + ET^4$$

Solution

Thermal Method	Energy Input, kJ/kg Product
Conventional	319
MW single-mode	24600
MW multi-mode	4690
MW continuous flow	24600

Part 4
Based on the operation of each type of microwave irradiation experiment, determine the respective energy inputs in kJ per kg of product.

Solution
Single-mode:

Temperature	473	K
Pressure	1	atm
Flow rate	0	L/h
MW power	1000	W
Residence time	300	s
Reaction volume	0.00217	L
Process time	0.08	h
Energy consumption	300	kJ
Mass of product	0.000253	kg
Energy consumption	1185771	kJ/kg

Multi-mode:

Temperature	473	K
Pressure	1	atm
Flow rate	0	L/h
MW power	1000	W
Residence time	300	s
Reaction volume	1.027	L
Process time	0.08	h
Energy consumption	300	kJ
Mass of product	0.4657	kg
Energy consumption	644	kJ/kg

Continuous flow:

Temperature	473	K
Pressure	130	atm
Flow rate	0.48	L/h
MW power	1000	W
Residence time	30	s
Reaction volume	0.434	L
Process time	0.83	h
Energy consumption	3000	kJ
Mass of product	0.0507	kg
Energy consumption	64222	kJ/kg

Note that for non-flow (batch) conditions process time = residence time. For flow conditions the following relations apply.

process time = total volume of reaction materials * flow rate
residence time = reactor volume * flow rate.

Part 5

Based on Parts 3 and 4, determine the respective overall energy input estimates for each microwave irradiation condition.

SOLUTION

Overall energy input = energy input contribution from thermodynamics + energy input contribution from instrument operation

Single-mode:

Overall energy input = 24600 + 1185771 = 1210371 kJ/kg product

Multi-mode:

Overall energy input = 4690 + 644 = 5,334 kJ/kg product

Continuous flow:

Overall energy input = 24600 + 64222 = 88822 kJ/kg product

Energy Metrics

Example 5.29[70]

Themes: balancing chemical equations, energy metrics.
The performance of the Diels–Alder reaction between anthracene and maleic anhydride under conventional thermal and microwave irradiation are compared.

Thermal Condition	% Yield	T(rxn) (C)	Rxn time (min)
Conventional	90	138	10
Microwave irradiation	92	187	3

Assuming that the reaction is run at the same scale and with no excess reagents under both thermal conditions, determine the respective energy inputs using the thermodynamic data given below. State any other assumptions made in the calculations.

Substance	Formula	MW (g/mol)	Tm (deg C)	Tb (deg C)	ΔH(fus) kJ/mol	ΔH(vap) kJ/mol
Maleic anhydride	$C_4H_2O_3$	98	51	200	13.55	47.22
Anthracene	$C_{14}H_{10}$	178	216	340	29.37	54.92
p-Xylene	C_8H_{10}	240	13.25	138	17.12	35.67

Heat capacity coefficients:

Substance	Phase	A	B	C	D	E	Tmin (K)	Tmax (K)
Maleic anhydride	Gas	−72.015	1.0423	−1.87E−03	1.65E−06	−5.56E−10	298	1000
	Liquid	−12.662	1.0564	−2.32E−03	2.05E−06		327	649
	Solid	32.5	2.10E−01	2.73E−04			89	320
Anthracene	Gas	−68.18	1.0788	−8.13E−04	2.90E−07	−3.82E−11	200	1500
	Liquid	−269.578	3.1196	−5.34E−03	3.36E−06		490	786
	Solid	11.1	5.82E−01	2.79E−05			30	489
p-Xylene	Gas	−17.36	5.65E−01	−2.63E−04	1.12E−08	1.65E−11	200	1500
	Liquid	−11.035	1.5158	−3.90E−03	3.92E−06		287	555
	Solid	0.872	8.08E−01	−9.54E−04			153	286

Energy Metrics

Functional form of temperature dependence of C_p:

$$C_p = A + BT + CT^2 + DT^3 + ET^4, \text{ J/mol deg K}$$

SOLUTION

Assume 1 mole of maleic anhydride and 1 mole of anthracene are reacted in 1 L of p-xylene.
Density of p-xylene = 0.862 g/mL

Phase transitions under conventional heating:

anthracene => solid
maleic anhydride => solid to liquid
p-xylene => liquid

Phase transitions under microwave heating:

anthracene => solid
maleic anhydride => solid to liquid

p-xylene => liquid to gas

The *energy-input-analysis-synthesis-plans.xls* spreadsheet yields the following results.
Conventional heating: 622 kJ per kg product
Microwave heating: 1519 kJ per kg product

Example 5.30

Themes: synthesis tree diagram, energy consumption.
Reaction conditions for a 4-step linear sequence are shown in the synthesis tree diagram and table below. Assume that reaction solvents are designated as waste, that the reaction scale of target product is 1 mol, and that the reaction solvents are heated from a temperature of 25°C.

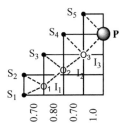

Step	Exp. Scale (mol)	AE	Solvent	Volume Used (100 mL)	Solvent Density (g/mL)	Molar Heat Capacity (J/mol deg C)	Rxn Temp. (deg C)
1	0.1	0.8	methanol	100	0.792	223.2	63
2	0.1	1	toluene	100	0.866	162.1	80
3	0.1	0.5	water	100	1	75.3	50
4	0.1	0.7	none	none	–	–	25

Part 1

Determine the total input energy required for heating solvents for this reaction sequence. Express the energy units in kWh, where 1 kWh = 3600 kJ, this is the energy required to heat 10328 g of water from 16.7°C to 100°C.

Use the relation $q = n\, C\, \Delta T$ to calculate the heat transfer required for solvents for each reaction, where n is the number of moles of solvent, C is the molar heat capacity of the solvent, and ΔT is the change in temperature. Assume that heat contributions from the reactions themselves are negligible.

Part 2

Determine the fraction of total input energy in Part 1 that is directed toward making the target product.

Determine the fraction of the total input energy in Part 1 that is directed toward making waste.

SOLUTION

Part 1

Step	Scale Based on Tree (mol)	Scaling Factor	Adjusted Volume of Solvent (mL)
1	1/(0.7*0.8*0.7)	1/(0.7*0.8*0.7*1)	100/(0.7*0.8*0.7*1) = 255.102
2	1/(0.7*0.8)	1/(0.7*0.8*1)	100/(0.7*0.8*1) = 178.571
3	1/0.7	1/(0.7*1)	100/(0.7*1) = 142.857
4	1	1/1	0

Step	Adjusted Volume of Solvent (mL)	Adjusted Mass of Solvent (g)	MW Solvent (g/mol)	Adjusted Moles of Solvent	ΔT (deg C)
1	255.102	202.041	32	6.314	38
2	178.571	154.643	92	1.681	55
3	142.857	142.857	18	7.937	25
4	0	0	0	0	0

Step	q (kJ)
1	6.314*38*223.2/1000 = 53.553
2	1.681*55*162.1/1000 = 14.987
3	7.937*25*75.3/1000 = 14.941
4	0

Total input energy = 53.553 + 14.987 + 14.941 = 83.481 kJ = 0.0232 kWh

Energy Metrics

Part 2

For each step, the amount of energy that is directed toward making the product in that step = RME(kernel) * q

Step 1: $S_1 + S_2 \rightarrow I_1$
Step 2: $S_3 + I_1 \rightarrow I_2$
Step 3: $S_4 + I_2 \rightarrow I_3$
Step 4: $S_5 + I_3 \rightarrow P$

$$\text{RME}_{step1} = \frac{MW_{I_1}\left(\frac{1}{1*0.7*0.8}\right)}{(MW_{S_1} + MW_{S_2})\left(\frac{1}{1*0.7*0.8*0.7}\right)} = (AE)_1 0.7 = 0.8*0.7 = 0.56$$

$$\text{RME}_{step2} = \frac{MW_{I_2}\left(\frac{1}{1*0.7}\right)}{(MW_{S_3} + MW_{I_1})\left(\frac{1}{1*0.7*0.8}\right)} = (AE)_2 0.8 = 1*0.8 = 0.8$$

$$\text{RME}_{step3} = \frac{MW_{I_3}\left(\frac{1}{1}\right)}{(MW_{S_4} + MW_{I_2})\left(\frac{1}{1*0.7}\right)} = (AE)_3 0.7 = 0.5*0.7 = 0.35$$

$$\text{RME}_{step4} = \frac{MW_P(1)}{(MW_{S_5} + MW_{I_3})\left(\frac{1}{1}\right)} = (AE)_4 1 = 0.7*1 = 0.7$$

Total input energy directed to making product P:

$$\sum_j (\text{RME})_{kernel,j} \, q_j = 0.56*53.553 + 0.8*14.987 + 0.35*14.941 + 0.7*0$$

$$= 47.197 \, kJ$$

Fraction of total input energy directed to making product P = 47.197/83.481 = 0.565, or 56.5%.

Fraction of total input energy directed to making waste = 1 − 0.565 = 0.435, or 43.5%.

5.4 EXCEL SPREADSHEET TOOLS: PART 4

change-in-enthalpy-pressure-contribution-template.xls
change-in-enthalpy-temperature-contribution-template.xls
database-industrial-chemicals-thermodynamic-parameters.xls
database-thermodynamics-parameters-special.xls
energy-input-analysis-synthesis-plans.xls
EOS-Peng-Robinson-template.xls
EOS-Redlich-Kwong-template.xls
EOS-van-der-Waals-template.xls
heat-of-reaction-template.xls

For each spreadsheet we briefly describe the kind of inputs required and the results obtained.

The *change-in-enthalpy-pressure-contribution-template.xls* spreadsheet is divided into two main parts: (a) the determination of the phases of state 1 and state 2 for each input material used in a chemical reaction based on the standard state conditions (298 K, 1 atm) and the reaction conditions (T_{rxn}, P_{rxn}); and (b) the determination of the pressure contribution to the enthalpy change, ΔH, for input materials. The determination of the phases of each input material is made using the following code in the Excel program: IF(AND(J27>1,K27>1), "gas: use RK EOS",IF(AND(J27>1,K27<1), "gas: use RK EOS", IF(AND(J27<1,K27>1), "liquid; use Tait EOS", "Case IVa or IVb–check vapor pressure curve"))), where J27 refers to reduced temperature, T_r, and K27 refers to reduced temperature, P_r. These criteria pertain to the inequalities given in Table 5.3. The determination of the enthalpic pressure contribution is subdivided into the following cases: (a) gas-to-gas transition using the Redlich–Kwong EOS (see Equation (5.91) with $T_1 = 298$ and $T_2 = T_{rxn}$); (b) liquid-to-gas transition using the Redlich–Kwong EOS (Equation (5.91) with $T_1 = T_b$ and $T_2 = T_{rxn}$); (c) gas-to-liquid transition using the Redlich–Kwong EOS (Equation (5.91) with $T_1 = 298$, $T_2 = T_{rxn}$, and $V_2 = V_{sat}$); and (d) liquid-to-liquid transition using the Tait EOS (see Equations (5.43) to (5.45)).

The *change-in-enthalpy-temperature-contribution-template.xls* spreadsheet is divided into two main parts: (a) the determination of the phases of state 1 and state 2 as in the *change-in-enthalpy-pressure-contribution-template.xls* spreadsheet; and (b) the determination of the temperature contribution to the enthalpy change, ΔH, for input materials. The latter part consists of the following data entries for each input material: normal melting point, T_m; normal boiling point, T_b; enthalpy of fusion at T_m; enthalpy of vaporization at T_b; and the heat capacity coefficients for either the gas, liquid, or solid phases as appropriate so that integrated forms of fourth-order polynomial expressions as found in the *Yaws Handbook* are used (see Equation (5.30) and Equations (5.107) to (5.116)).

The *database-industrial-chemicals-thermodynamic-parameters.xls* spreadsheet contains the following four sheets pertaining to 314 industrial commodity chemicals: (a) property list including molecular formula, CAS number, T_m, T_b, phase at standard temperature and pressure, enthalpy of fusion, enthalpy of vaporization, critical temperature (T_c), critical pressure (P_c), acentric factor, Redlich–Kwong coefficients *a* and *b*, thermal expansion coefficients (see Equations (5.47) and (5.98)), molar volume coefficients (see Equation (5.99)), and liquid vapor pressure coefficients (see Equation (5.50)); (b) heat capacity coefficients for the gas phase; (c) heat capacity coefficients for the liquid phase; and (d) heat capacity coefficients for the solid phase.

The *database-thermodynamics-parameters-special.xls* spreadsheet is composed of six sheets containing data for various industrial chemicals covering critical constants, heat of formation, thermal expansion coefficients, acentric factor, liquid density molar volume coefficients, and liquid vapor pressure coefficients. The following correlations are determined for the heats of formation and heat capacities at constant pressure given in Equations (5.119) to (5.122) obtained from data found in the *CRC Handbook of Physics and Chemistry*, 87th ed., Chapter 5, pp. 4–42.

$$\Delta H^o_{f,\text{gas}} = 0.981 \Delta H^o_{f,\text{liquid}} + 40.129; \; r^2 = 0.9970; \; N = 747 \quad (5.120)$$

$$\Delta H^o_{f,\text{gas}} = 0.9871 \Delta H^o_{f,\text{solid}} + 96.316; \; r^2 = 0.9945; \; N = 187 \quad (5.121)$$

$$C^o_{p,\text{gas}} = 0.7423 C^o_{p,\text{liquid}} - 18.416; \; r^2 = 0.9372; \; N = 57 \quad (5.122)$$

$$C^o_{p,\text{gas}} = 0.8044 C^o_{p,\text{solid}} + 1.6673; \; r^2 = 0.9549; \; N = 97 \quad (5.123)$$

The *energy-input-analysis-synthesis-plans.xls* spreadsheet determines the total enthalpy change for input materials for a chemical reaction based on the scaled masses of those materials and the q_{temp} and q_{press} contributions determined in the *change-in-enthalpy-pressure-contribution-template.xls* and *change-in-enthalpy-temperature-contribution-template.xls* spreadsheets. When consecutive reactions are entered as in a synthesis plan, the total enthalpy change for all input materials for the entire plan is determined using a basis scale of 1 ton of final product.

The *EOS-Peng-Robinson-template.xls*, *EOS-Redlich-Kwong-template.xls*, *EOS-van-der-Waals-template.xls* spreadsheets determine the molar volumes of liquid and vapor of a fluid at a given temperature and pressure according to the cubic equation forms of the respective equations of state following the protocol of solving cubic equations given in Section 5.2.2.

The *heat-of-reaction-template.xls* spreadsheet determines the heat of reaction, ΔH_{rxn}, for a chemical reaction according to Hess's law (see Equations (5.33) and (5.34)).

5.5 PROBLEMS

PROBLEM 5.10
Estimate the normal boiling point of dicyclohexyl urea by the Nannoolal method.

$C_{13}H_{24}N_2O$
Mol. Wt.: 224

PROBLEM 5.11
Based on the boiling estimate of Problem 5.10, estimate the critical constants of dicyclohexyl urea by the Joback method.

PROBLEM 5.12[71]
Another method to estimate the vapor pressure of liquids in addition to the Riedel method is the Lee-Kesler method given by the equation below.[71]

$$\ln\left(\frac{P_{\text{vap}}}{P_c}\right) = f_0 + \omega f_1; \; \text{Units: kPa}$$

where

$$f_0 = 5.92714 - \frac{6.09648}{T_r} - 1.28862 \ln T_r + 0.169347 T_r^6$$

$$f_1 = 15.2518 - \frac{15.6875}{T_r} - 13.4721 \ln T_r + 0.43577 T_r^6$$

Estimate the vapor pressure of chlorobenzene at 500 K given the following data:

$$T_c = 632.35 \, K, \, P_c = 45.1911 \, bar, \, \omega = 0.2499$$

Compare the result with the found in Example 5.18 by the Riedel method.

PROBLEM 5.13
A mass of 500 g of ammonia is contained in a 30,000 cubic centimeter vessel immersed in a constant bath at 65°C. Determine the pressure inside the vessel assuming ammonia behaves as an ideal gas. Repeat the calculation assuming that ammonia obeys the Redlich–Kwong equation of state.

PROBLEM 5.14
What pressure in atmospheres is generated when 1 pound-mole of methane is stored in a 2 cubic foot volume at 122°F? Assume methane obeys the Peng–Robinson equation of state. Use the following as the only input data for methane gas: $T_c = 190.58$ K, $P_c = 45.44$ atm, $T_b = 111.66$ K.

PROBLEM 5.15
Given that the vapor pressure of methyl chloride at 60°C is 13.76 bar, use the *EOS-Redlich-Kwong-template.xls* spreadsheet to calculate the molar volumes of the saturated vapor and saturated liquid. For methyl chloride: $T_c = 416.3$ K, $P_c = 66.8$ bar.

PROBLEM 5.16
Calculate the heat of reaction of methanol synthesis at 800°C using the data below and the *heat-of-reaction-template.xls* spreadsheet.

$$CO(g) + 2H_2(g) \rightarrow CH_3OH(g)$$

Chemical	ΔHf (kJ/mol) at 298 K
Methanol	−200.6
Carbon monoxide	−110.525
Hydrogen	0

Heat capacity coefficients:

Chemical	A	B	C	D	E
Methanol	40.046	−3.83E−02	2.45E−04	−2.17E−07	5.99E−11
Carbon monoxide	29.556	−6.58E−04	2.01E−05	−1.22E−08	2.26E−12
Hydrogen	25.399	2.02E−02	−3.85E−05	3.19E−08	−8.76E−12

PROBLEM 5.17

Part 1

200 g of water is heated from 300 to 350 K in a closed vessel at 1 atm. Assuming no heat losses how much energy is required?

Part 2

200 g of water is heated from 300 to 1000 K in a closed vessel. Assuming no heat losses how much energy is required?

Part 3

1000 kg of steam is cooled from 800 K to 300 K at 1 atm. Assuming no heat losses how much energy needs to be removed?

Part 4

50 moles of oxygen gas are cooled from 150°C to 25°C at 1 atm. How much energy needs to be removed?

Part 5

10 kg methanol are heated from 25°C to 130°C at 1 atm. Assuming no heat losses how much energy is required?

Part 6

1 kg of water is cooled from 25°C to −10°C at 1 atm. Assuming no heat losses how much energy needs to be removed?

PROBLEM 5.18

For the following state changes for water determine the corresponding enthalpy change. In each case draw a *p-T* path diagram showing the connection between the two states to guide the calculation. Use the *change-in-enthalpy-pressure-contribution-template.xls* and *change-in-enthalpy-temperature-contribution-template.xls* spreadsheets to facilitate the calculations.

1. State 1: 1 atm and 25°C; State 2: 1000 atm and 50°C.
2. State 1: 1 atm and 25°C; State 2: 1000 atm and 500°C.

3. State 1: 1 atm and 25°C; State 2: 10 atm and 50°C.
4. State 1: 1 atm and 25°C; State 2: 10 atm and 500°C.

PROBLEM 5.19[72,73]

For the Berthelot equation of state shown below, determine the values of the constants a and b in terms of the critical constants.

$$p = \frac{RT}{V-b} - \frac{a}{TV^2}$$

Derive the reduced form of this equation of state.

PROBLEM 5.20[74]

For the Clausius equation of state shown below, determine the values of the constants a, b, and c in terms of the critical constants.

$$p = \frac{RT}{V-b} - \frac{a}{T(V+c)^2}$$

Derive the reduced form of this equation of state.

PROBLEM 5.21[75]

Alfred Wohl proposed a number of cubic equations of state shown below. Derive reduced forms for these expressions given the values of the constants in terms of the critical constants.

$$p = \frac{RT}{V} - \frac{a}{V^2 + b^2} \qquad a = \frac{8}{3} P_c V_c^2, b = \frac{V_c}{3}, V_c = \frac{RT_c}{3P_c}$$

$$p = \frac{RT}{V} - \frac{a}{V^2}\frac{V-b}{V} \qquad a = 3P_c V_c^2, b = \frac{V_c}{3}, V_c = \frac{RT_c}{3P_c}$$

$$p = \left(\frac{RT}{V-b}\right)\left(\frac{V}{V-b}\right) - \frac{a}{V^2} \qquad a = \frac{23}{4} P_c V_c^2, b = \frac{V_c}{2}, V_c = \frac{RT_c}{3P_c}$$

$$p = \frac{RT}{V-b} - \frac{a}{(V+b)^2} \qquad a = \frac{108}{25} P_c V_c^2, b = \frac{V_c}{5}, V_c = \frac{5}{16}\frac{RT_c}{P_c}$$

$$p = \frac{RT}{V} - \frac{a}{(V+b)^2} \qquad a = \frac{27}{4} P_c V_c^2, b = \frac{V_c}{2}, V_c = \frac{RT_c}{4P_c}$$

PROBLEM 5.22[76]

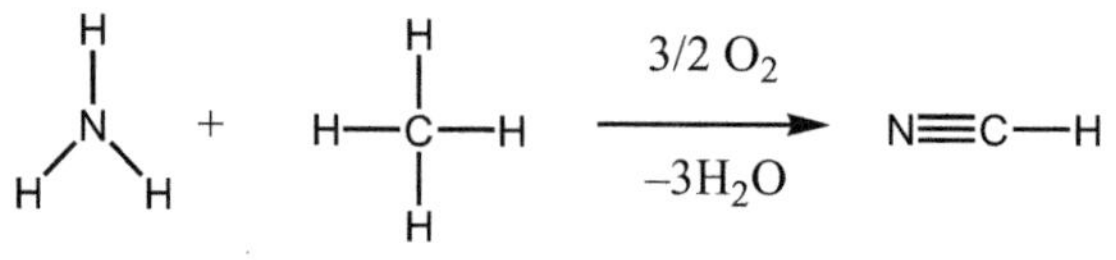

$$\text{NH}_3 + \text{CH}_4 \xrightarrow[-3\text{H}_2\text{O}]{3/2\ \text{O}_2} \text{N}\equiv\text{C}-\text{H}$$

T1 = 298 K, p1 = 1 atm
T2 = 1363 K, p2 = 1.7 atm

Input materials: 746.2 kg ammonia, 252.7 kg methane, 3282.0 kg oxygen, 3.1 kg

Part 1
Determine the phases of the initial and final states for each input material.

Part 2
Determine the temperature contribution to the change of enthalpy for the reaction.

Part 3
Determine the pressure contribution to the change of enthalpy for the reaction.

PROBLEM 5.23[77]

T1 = 298 K, p1 = 1 atm
T2 = 321 K, p2 = 34 atm

Input materials: 1786.8 kg *n*-propanol, 4583.5 kg carbon monoxide, 6082.7 kg sulfuric acid, and 301.0 kg water

Part 1
Determine the phases of the initial and final states for each input material.

Part 2
Determine the temperature contribution to the change of enthalpy for the reaction.

Part 3
Determine the pressure contribution to the change of enthalpy for the reaction.

REFERENCES

1. Tosun, I. *The Thermodynamics of Phase and Reaction Equilibria*, Elsevier: Amsterdam, 2013.
2. Pitzer, K.S.; Lippmann, D.Z.; Curl, R.F. Jr.; Huggins, C.M.; Petersen, D.E. *J. Am. Chem. Soc.* 1955, 77, 3433.
3. Pitzer, K.S. *ACS Symp. Ser.* 1977, 60, 1.
4. Yaws, C.L. *Chemical Properties Handbook*, McGraw-Hill: New York, 1999, pp. 1–29.

5. Poling, B.E.; Thomson, G.H.; Friend, D.G.; Rowley, R.L.; Wilding, W.V. *Perry's Chemical Engineers' Handbook*, 8th ed., McGraw-Hill: New York, 2008, Section 2, p. 473.
6. Yaws, C.L. *Chemical Properties Handbook*, McGraw-Hill: New York, 1999, pp. 185–211.
7. www.aiche.org/dippr/projects/801
8. Smith, J.M.; van Hess, H.C. *Introduction to Chemical Engineering Thermodynamics*, 4th ed., McGraw-Hill Book Company: New York, 1987, pp. 166–173.
9. Poling, B.E.; Thomson, G.H.; Friend, D.G.; Rowley, R.L.; Wilding, W.V. *Perry's Chemical Engineers' Handbook*, 8th ed., McGraw-Hill: New York, 2008, Section 2, p. 478.
10. Yaws, C.L. *Chemical Properties Handbook*, McGraw-Hill: New York, 1999, pp. 135–158.
11. Yaws, C.L. *Chemical Properties Handbook*, McGraw-Hill: New York, 1999, pp. 109–134.
12. van der Waals, J.D. *On the Continuity of the Gaseous and Liquid States*. PhD thesis, Universiteit Leiden, 1873.
13. Rowlinson, J.S. *Studies in Statistical Physics*, Vol. 14, North-Holland: Amsterdam, 1988.
14. Redlich, O.; Kwong, J.N.S. *Chem. Rev.* 1949, *44*, 233.
15. Peng, D.Y.; Robinson, D.B. *Ind. Eng. Chem. Fund.* 1976, *15*, 59.
16. For gases: Yaws, C.L. *Chemical Properties Handbook*, McGraw-Hill: New York, 1999, pp. 30–55.
17. For liquids: Yaws, C.L. *Chemical Properties Handbook*, McGraw-Hill: New York, 1999, pp. 56–82.
18. For solids: Yaws, C.L. *Chemical Properties Handbook*, McGraw-Hill: New York, 1999, pp. 83–108.
19. Hess, G.H. *Bull. Sci. Acad. Imper. Sci. (St. Petersburg)* 1840, *8*, 257.
20. Hess, G.H. *Ann. Physik Chem.* 1840, *50*, 385.
21. Maxwell, J.C. *Nature* 1874, *10*, 477.
22. Tait, P.G. *Phys. Chem.* 1888, *2*, 1.
23. Li, Y.H. *J. Geophys. Res.* 1967, *72*, 2665.
24. Dymond, J.H.; Malhotra, R. *Int. J. Thermophysics* 1988, *9*, 941.
25. Yaws, C.L. *Chemical Properties Handbook*, McGraw-Hill: New York, 1999, pp. 616–642.
26. Yaws, C.L. *Chemical Properties Handbook*, McGraw-Hill: New York, 1999, pp. 159–184.
27. Edmister, W.C. *Petroleum Refinery* 1958, *37*, 173.
28. Prasad, D.H.L. *Chem. Eng. Res. Design* 1994, *72*, 123.
29. Joback, K.G. *A Unified Approach to Physical Property Estimation Using Multivariate Statistical Techniques*. MSc thesis, Massachusetts Institute of Technology, 1984.
30. Joback K.G.; Reid R.C. *Chem. Eng. Commun.* 1987, *57*, 233.
31. Poling, B.E.; Thomson, G.H.; Friend, D.G.; Rowley, R.L.; Wilding, W.V. *Perry's Chemical Engineers' Handbook*, 8th ed., McGraw-Hill: New York, 2008, Section 2, pp. 470–471.
32. Domalski, E.S.; Hearing, E.D. *J. Phys. Chem. Ref. Data* 1993, *22*, 805.
33. Poling, B.E.; Thomson, G.H.; Friend, D.G.; Rowley, R.L.; Wilding, W.V. *Perry's Chemical Engineers' Handbook*, 8th ed., McGraw-Hill: New York, 2008, Section 2, pp. 478–484.
34. Chickos, J.S.; Braton, C.M.; Hesse, D.G.; Liebman, J.F. *J. Org. Chem.* 1991, *56*, 927.
35. Poling, B.E.; Thomson, G.H.; Friend, D.G.; Rowley, R.L.; Wilding, W.V. *Perry's Chemical Engineers' Handbook*, 8th ed., McGraw-Hill: New York, 2008, Section 2, pp. 487–488.

36. Vetere, A. *Fluid Phase Equilib.* 1995, *106*, 1.
37. Poling, B.E.; Thomson, G.H.; Friend, D.G.; Rowley, R.L.; Wilding, W.V. *Perry's Chemical Engineers' Handbook*, 8th ed., McGraw-Hill: New York, 2008, Section 2, p. 487.
38. Benson, S.W.; Cruikshank, F.R.; Golden, D.M.; Haugen, G.R.; O'Neal, H.E.; Rodgers, A.S.; Shaw, R.; Walsh, R. *Chem. Rev.* 1969, *69*, 279.
39. Poling, B.E.; Thomson, G.H.; Friend, D.G.; Rowley, R.L.; Wilding, W.V. *Perry's Chemical Engineers' Handbook*, 8th ed., McGraw-Hill: New York, 2008, Section 2, pp. 490–495.
40. Ruzicka, V.; Domalski, E.S. *J. Phys. Chem. Ref. Data* 1993, *22*, 597.
41. Ruzicka, V.; Domalski, E.S. *J. Phys. Chem. Ref. Data* 1993, *22*, 619.
42. Domalski, E.S.; Hearing, E.D. *J. Phys. Chem. Ref. Data* 1996, *25*, 1.
43. Poling, B.E.; Thomson, G.H.; Friend, D.G.; Rowley, R.L.; Wilding, W.V. *Perry's Chemical Engineers' Handbook*, 8th ed., McGraw-Hill: New York, 2008, Section 2, pp. 495–497.
44. Goodman, B.T.; Wilding, V.W.; Oscarson, J.L.; Rowley, R.L. *J. Chem. Eng. Data* 2004, *49*, 24.
45. Poling, B.E.; Thomson, G.H.; Friend, D.G.; Rowley, R.L.; Wilding, W.V. *Perry's Chemical Engineers' Handbook*, 8th ed., McGraw-Hill: New York, 2008, Section 2, pp. 495, 497.
46. Nannoolal, Y.; Rarey, J.; Ramjugernath, D.; Cordes, W. *Fluid Phase Equilib.* 2004, *226*, 45.
47. Poling, B.E.; Thomson, G.H.; Friend, D.G.; Rowley, R.L.; Wilding, W.V. *Perry's Chemical Engineers' Handbook*, 8th ed., McGraw-Hill: New York, 2008, Section 2, pp. 473–476.
48. Constantinou, L.; Gani, R. *AIChE J.* 1994, *40*, 1697.
49. Poling, B.E.; Thomson, G.H.; Friend, D.G.; Rowley, R.L.; Wilding, W.V. *Perry's Chemical Engineers' Handbook*, 8th ed., McGraw-Hill: New York, 2008, Section 2, p. 471.
50. Riedel, L. *Chem. Ing. Technik* 1954, *26*, 679.
51. Danner, R.P.; Daubert, T.E. *Manual for Predicting Chemical Process Design Data: Data Prediction Manual*, American Institute of Chemical Engineers: New York, 1983.
52. Poling, B.E.; Thomson, G.H.; Friend, D.G.; Rowley, R.L.; Wilding, W.V. *Perry's Chemical Engineers' Handbook*, 8th ed., McGraw-Hill: New York, 2008, Section 2, p. 477.
53. Gresham, W.F.; Brooks, R.E. US 2437600 (duPont, 1948).
54. Gresham, W.F.; Webb, I.D. US 2570793 (duPont, 1951).
55. Klöppfer, W. *Int. J. Life Cycle Assess.* 1997, *2*, 61.
56. Anonymous (Ed.): VDI-Richtlinien "Cumulative Energy Demand - Terms, Definitions, Methods of Calculation." Verein Deutscher Ingenieure-guideline 4600, Düsseldorf, 1997.
57. Frischknecht, R.; Heijungs, R.; Hofstetter, P. *Int. J. Life Cycle Assess.* 1998, *3*, 266.
58. Chapman, P. F.; Leach, G.; Slesser, M. *Energy Policy* 1974, *2*, 231.
59. Huijbregts, M.J.; Hellweg, S.; Frischknecht, R.; Hendriks, H.M.; Hungerbühler, K.; Hendriks, A.J. *Environ. Sci. Technol.* 2010, *44*, 2189.
60. Frischknecht, R.; Wyss, F.; Knöpfel, S.B.; Lützkendorf, T.; Balouktsi, M. *Int. J. Life Cycle Assess.* 2015, *20*, 957.
61. Lapkin, A.; Joyce; L.; Crittenden, B. *Environ. Sci. Technol.* 2004, *38*, 5815.
62. Andraos, J. *ACS Sust. Chem. Eng.* 2016, *4*, 312.
63. Curzons, A.D.; Constable, D.J.C.; Mortimer, D.N.; Cunningham, V.L. *Green Chem.* 2001, *3*, 1.
64. McElroy, C.R.; Constantinou, A.; Jones, L.C.; Summerton, L.; Clark, J.H. *Green Chem.* 2015, *17*, 3111.

65. Jimenez-Gonzalez, C.; Constable, D.J.C. *Green Chemistry and Engineering: A Practical Design Approach*, Wiley: Hoboken, 2011, p. 98.
66. Dandia, A.; Singh, R.; Bhaskaran, S. *Ultrason. Sonochem.* 2010, *17*, 399.
67. Moseley, J.D.; Lenden, P.; Lockwood, M.; Ruda, K.; Sherlock, J.P.; Thomson, A.D.; Gilday, J.P. *Org. Process Res. Dev.* 2008, *12*, 30.
68. Damm, D.; Glasnov, T.N.; Kappe, C.O. *Org. Process Res. Dev.* 2010, *14*, 215.
69. Wagner, E.C.; Millett, W.H. *Org. Synth. Coll.* 1943, *2*, 65.
70. Giguere, R.J.; Bray, T.L.; Duncan, S.M. *Tetrahedron Lett.* 1986, *27*, 4945.
71. Lee, B.I.; Kesler, M.G. *AIChE J.* 1975, *21*, 510.
72. Berthelot, D. *J. Phys.* 1899, *8*, 263.
73. Berthelot, D. *Arch. Néerl.* 1900, *5(Ser. 2)*, 417.
74. Clausius, R. *Ann. Physik* 1880, *169*, 337.
75. Wohl, A. *Z. Physik. Chem.* 1914, *87*, 1.
76. Lowenheim, F.A.; Moran, M.K. *Faith, Keyes, & Clark's Industrial Chemicals*, 4th ed., Wiley: Hoboken, 1975, p. 482.
77. De Benedictis, A.; Furman, K.E. US 2913489 (Shell, 1959).

6 Algorithms

In this chapter, we compare seven available algorithms in the literature on example industrial syntheses of aniline shown in Figure 6.1 with the following experimental details.

Procedure #1[1,2]

  2780 lbs nitrobenzene
  3200 lbs iron borings
  2500 lbs 30 wt% hydrochloric acid
  Product: 2000 lbs aniline

Procedure #2[1,2]

  2700 lbs nitrobenzene
  60,221 g hydrogen gas
  680.4 g copper(II)carbonate
  Product: 2000 lbs aniline

Procedure #3[2,3]

  2500 lbs chlorobenzene
  7450 lbs 28 wt% ammonium hydroxide solution
  350 lbs copper(I)oxide
  Product: 2000 lbs aniline

For each algorithm, a brief description is given along with the results obtained for the aniline examples as an illustration.

6.1 ENVIRONMENTAL ASSESSMENT TOOL FOR ORGANIC SYNTHESIS[4,5]

6.1.1 Description of Program

EATOS (*Environmental Assessment Tool for Organic Synthesis*) is a program first made available in 2001 and is the pioneering work on automated material efficiency green metrics calculations for individual reactions and synthesis plans. It runs on various Windows operating systems and is freely available online through registration; however, it requires a separate JavaScript program (Java Run Time Environment (version 1.4)) to run it. This software has not been updated since its launch. Features of the program include:

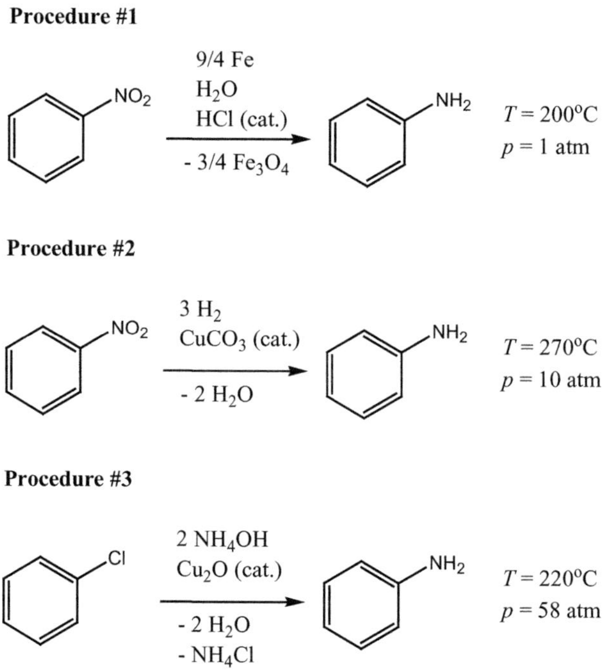

Procedure #1

PhNO$_2$ $\xrightarrow[\text{- 3/4 Fe}_3\text{O}_4]{\text{9/4 Fe, H}_2\text{O, HCl (cat.)}}$ PhNH$_2$ $T = 200°C$, $p = 1$ atm

Procedure #2

PhNO$_2$ $\xrightarrow[\text{- 2 H}_2\text{O}]{\text{3 H}_2, \text{CuCO}_3 \text{ (cat.)}}$ PhNH$_2$ $T = 270°C$, $p = 10$ atm

Procedure #3

PhCl $\xrightarrow[\substack{\text{- 2 H}_2\text{O} \\ \text{- NH}_4\text{Cl}}]{\text{2 NH}_4\text{OH, Cu}_2\text{O (cat.)}}$ PhNH$_2$ $T = 220°C$, $p = 58$ atm

FIGURE 6.1 Three industrial procedures to synthesize aniline.

1. calculation of global process mass intensities (PMIs) and global E-factors for individual reactions and entire synthesis plans including environmental impact corrections;
2. breakdown of the above metrics into their constituent contributions according to substrates, catalysts, solvents, auxiliary materials, coupled products, and by-products;
3. program performance is more reliable if masses of isolated intermediate products in each step are entered rather than reaction yields, which have to be worked out independently by the user;
4. use of a chain method of importing data sheets containing masses of intermediate products and input materials in the forward sense beginning from the first step and working toward the last step in a plan; and
5. visual output of the program is a histogram of four bars pertaining to global E-factor and global PMI outputs with and without environmental impact factor corrections.

The program uses the following terminologies:

1. "atom selectivity" means "atom economy";
2. "mass index (S^{-1})" means "process mass intensity";
3. "mass efficiency" means "global reaction mass efficiency";

4. "by-products" refers to products arising from competing reactions other than the entered balanced chemical equation, (what we call *"side products"* in this book); and
5. "coupled products" refers to products arising as a consequence of producing the desired product according to the balanced chemical equation (what we call *"by-products"* in this book).

EATOS uses Q-factors to amplify waste material E-factor contributions according to Sheldon's original concept of environmental quotient, where the E-factor contribution of a single waste product j, E_j, is adjusted by multiplying it by a factor Q that reflects its environmental impact or risk. Equation (6.1) shows the expression for the overall adjusted E-factor for a reaction, EI_out, covering all waste materials; namely, excess substrates, coupled products, by-products, auxiliaries, and solvents.

$$\text{EI\_out} = \sum_j (E_j)(Q_{j,\text{avg}}) = \sum_j \left[(E_j) \frac{Q_{j,\text{hum.tox.}} + Q_{j,\text{accum.}} + \ldots + Q_{j,n}}{n} \right] \quad (6.1)$$

where each waste material Q-factor is an arithmetic average of individual Q-factors for human toxicity, bioaccumulation, and other contributions. A Q value equal to 1 implies no added impact whereas a value larger than 1 suggests that a given waste material has potential to cause environmental harm, thereby amplifying its effective mass of waste. This measure was taken to distinguish the kinds of waste produced even though they may have the same mass. For example, a reaction producing 1 g each of mercury and sodium chloride would have its coupled products distinguishable beyond mass. EATOS extends the same idea to input materials as well; hence, individual input material PMI contributions are adjusted by a similar multiplication process. Equation (6.2) shows an analogous expression for overall adjusted PMI for a reaction, EI_in, covering all input materials; namely, substrates, auxiliaries, and solvents.

$$\text{EI\_in} = \sum_j (PMI)_j (Q_{j,\text{avg}}) \quad (6.2)$$

The Q-factor values are arbitrarily based on risk phrases and ranges of other factors such as human toxicity parameters (LD50(oral), LD50(dermal), LC50(inhalation)), occupational exposure limits (OEL), ozone depletion potential (ODP), global warming potential (GWP), and acidification potential (AP). Since the program was originally written in German, OEL values are entered under the heading MAK for "Maximale Arbeitsplatzkonzentration." All of these data are entered in the "weighting" folder in the seven-sheet window for substrates-catalysts-solvents- auxiliary materials-product-coupled products-by-products. Table 4.2 in Chapter 4 gives Q values associated with various risk phrases found in MSDS sheets. Risk phrases are entered as, for example, 36/37/38-40 if they appear in MSDS sheets as R36/37/38 and R40. Other Q-factor scales based on other potentials are given in the EATOS manual. It should be noted that only one of either LD50(oral), LD50(dermal), or

LC50(inhalation) can be entered under the acute toxicity column. This is restrictive for compounds that pose multiple acute toxicities, so one is left to choose the parameter associated with the greatest risk.

Following the inverse Hodge–Sterner scale of toxicity, it is advisable to select the parameter with the least value since that corresponds to maximum toxicity. The outputted histogram compares raw PMIs and E-factors based solely on mass versus amplified environmental impact values EI_in and EI_out. The same computational breakdown of step PMI and step E-factor contributions is possible in addition to individual material contributions. EATOS also considers costs of input materials which in this discussion we do not consider. Greener procedures are characterized as having low values for EI_in and EI_out, and most input and waste materials with Q-factors equal to 1.

6.1.2 Instructions for Using Program

Analysis of a single reaction

STAGE 1

>To initiate program double click EATOS executable JAR file.
>Click NEW to begin a new project.
>A pop-up NEW Eatos.project file appears.
>Click on this option and select RENAME.
>Enter the project name: single-reaction.project
>Click on single-reaction.project and select OPEN.
>A pop-up window appears that prompts you to enter a name for the synthesis files. Select NEW and enter the name "single reaction."

STAGE 2

>After entering the name, another window appears showing three sheets to enter data for substrates, product, and coupled products.
>Under the substrates sheet a default of two entries is given. If the reaction involves more than two reactants, click MORE as necessary to introduce more entries.
>For each substrate entry, fill in the chemical name of the reactant (Name), stoichiometric coefficient (Coef.), and molecular formula (Formula). Formulas are entered with capital letters for the elements followed by numbers. Lower case letters result in a syntax error. The molecular weight will be automatically calculated for each substrate. Fractional stoichiometric coefficients are entered as ½ instead of 0.5. When done click OK.
>Go to the product sheet and repeat the data entries for the target product. Note that only one entry is allowed in the product sheet. When done click OK.
>Go to the coupled products sheet and repeat the data entries for the reaction by-products.

Note: This program uses the term "coupled product" to mean "by-product," which refers to products formed as a consequence of producing the target product according to the balanced chemical equation. If all entries in the three sheets are correctly done, you will be able to proceed to the next stage; otherwise, you will get an error message indicating that the entered chemical equation is not balanced. Make adjustments as necessary if an error occurs.

STAGE 3

When complete, a window appears showing the name of the synthesis (e.g., single reaction). Select it in the window and click OPEN.

A new window appears showing seven sheets: substrates; catalysts; solvents; auxiliary materials; product; coupled products; and by-products. The mass amounts of each material used in the reaction are entered accordingly in the appropriate category.

Note: "product" refers to the target product; "coupled products" refers to other products appearing in balanced chemical equation; "by-products" refer to side products that arise from competing reactions whose balanced chemical equations are different from the one that results in the desired product.

For the catalyst, solvents, and auxiliaries sheets click NEW to open its own sheet. You will be prompted to enter the name of the catalyst, solvent, or auxiliary. The substrate, product, and coupled product sheets are already set since they are automatically linked to the entered balanced chemical equation in STAGE 1.

In the product sheet there are two options: you can enter either the mass of product or the percent yield. Note the percent yield is calculated with respect to the limiting reagent. For the product sheet, enter the mass collected since this is the experimental variable. In the coupled products sheet you can either enter the masses or percent yield or leave them blank. Normally the masses of coupled products are not determined from an experiment since the target product is the one that is isolated. The percent yield of a coupled product is the same as the percent yield of the target product provided the stoichiometric coefficients are the same for both products. If not, then the percent yield of coupled product is given by Equation (6.3).

$$\%\text{yield} = \left(\frac{\text{moles}_{\text{product}}}{\text{moles}_{\text{limiting reagent}}}\right)\left(\frac{SC_{\text{limiting reagent}}}{SC_{\text{product}}}\right)100 \qquad (6.3)$$

where SC = stoichiometric coefficient

At the top of the sheet menu select LITERATURE to enter the literature details of the source of the reaction in the appropriate fields. When done click OK.

Once all entries are complete, to save your work select FILE → SAVE AS → filename.project.

STAGE 4

To see the result of the metrics calculation, select EDIT → CALCULATE.

A window pops up showing a histogram with four bars.

Select VIEW → Mass balance region → Reaction.

The **first** bar refers to mass index, S^{-1} (or PMI), which pertains to the mass of input materials used in the reaction divided by the mass of target product. The bar is subdivided according to the substrate, auxiliaries, and solvent categories.

The **second** bar refers to E-factor, which pertains to the mass of waste produced divided by the mass of target product. The bar is subdivided according to the coupled products, by-products, substrates, auxiliaries, and solvent categories. Recall that the coupled products refer to those other products appearing in the balanced chemical equation beside the target product. The by-products refers to unreacted starting materials.

The **third** bar refers to EI_in, environment impact of the input materials. If no weighting factors are entered in STAGE 3 that amplify the masses of input materials, the third bar will be identical to the first bar.

The **fourth** bar refers to EI_out, environment impact of the waste materials. If no weighting factors are entered in STAGE 3 that amplify the masses of waste materials, the third bar will be identical to the second bar.

The histogram diagram can be exported as a bitmap file (filename.bmp). Select DIAGRAM → Export graphic. This image can be entered into Word via INSERT → PICTURE → From File.

To get a numerical report of the results for any of the sections of the histogram bars you need to bring the cursor over the appropriate section and click on it. A pop-up window appears with text which can be selected and copied into Word. To get a full report of the results you need to do this for each color coded section of each histogram. For the first bar you will need to do this three times, for the second bar five times, for the third bar three times, and for the fourth bar, five times.

To get a report of the atom economy of the reaction, from the histogram window select Miscellaneous → Display equations of syntheses. This text can also be selected and copied into Word.

Note: The term "atom selectivity" or AS is used to represent atom economy.

Analysis of linear synthesis (three steps) (see Problem 6.2)
Analysis for step 1
STAGE 1

To initiate program double click EATOS executable JAR file.
Click NEW to begin a new project.
A pop-up NEW Eatos.project file appears.
Click on this option and select RENAME.
Enter the project name: linear.project
Click on linear.project and select OPEN.
A pop-up window appears that prompts you to enter a name for the synthesis files. Select NEW and enter the name "step-1."
Follow the steps as in the analysis for a single reaction for STAGES 2, 3, and 4.
Click FILE → SAVE PROJECT to save your work.

Analysis for step 2
STAGE 2

After saving the histogram as a bitmap file (step-1.bmp) and extracting the numerical results, close the histogram window.

Algorithms

In the "step-1" window go to the top menu and select FILE → Change synthesis.

The original linear.project window appears again. Click NEW and name a second file as "step-2."

In the 3-sheet window for substrates, product, and coupled products for step 2 select IMPORT for substrate 1 → linear.project → step-1 → Import. The entry for thiete 1,1-dioxide (product of step 1) automatically appears in the "name" and "formula" fields. Enter the appropriate stoichiometric coefficient in the "coef" field.

Continue as before to complete all entries for all reaction materials for step 2.

In the linear.project window select "step-2" and OPEN.

Continue as before to complete all masses of input materials for step 2.

Click FILE → SAVE PROJECT to save your work.

Analysis for step 3

STAGE 3

After saving the histogram as a bitmap file (step-1.bmp) and extracting the numerical results, close the histogram window.

In the "step-2" window, go to the top menu and select FILE → Change synthesis.

The original linear.project window appears again. Click NEW and name a second file as "step-3."

In the 3-sheet window for substrates, product, and coupled products for step 3 select IMPORT for substrate 1 → linear.project → step-2 → Import. The entry for 3-chlorothietane 1,1-dioxide (product of step 2) automatically appears in the "name" and "formula" fields. Enter the appropriate stoichiometric coefficient in the "coef" field.

Continue as before to complete all entries for all reaction materials for step 3.

In the linear.project window select "step-3" and OPEN.

Continue as before to complete all masses of input materials for step 3.

Click FILE → SAVE PROJECT to save your work.

Analysis of convergent synthesis (two parallel steps, then a convergent step) (see Problem 6.3)

Analysis for step 1A

STAGE 1

To initiate program double click EATOS executable JAR file.

Click NEW to begin a new project.

A pop-up NEW Eatos.project file appears.

Click on this option and select RENAME.

Enter the project name: convergent.project

Click on convergent.project and select OPEN.

A pop-up window appears that prompts you to enter a name for the synthesis files. Select NEW and enter the name "step-1A."

Follow the steps as in Part 1 for STAGES 2, 3, and 4.

Click FILE → SAVE PROJECT to save your work.

Analysis for step 1B
STAGE 2
> After saving the histogram as a bitmap file (step-1A.bmp) and extracting the numerical results, close the histogram window.
> In the "step-1A" window go to the top menu and select FILE → Change synthesis.
> The original convergent.project window appears again. Click NEW and name a second file as "step-1B."
> Follow the steps as in Part 1 for STAGES 2, 3, and 4.
> Click FILE → SAVE PROJECT to save your work.

Analysis for step 2
STAGE 3
> After saving the histogram as a bitmap file (step-1B.bmp) and extracting the numerical results, close the histogram window.
> In the "step-1B" window go to the top menu and select FILE → Change synthesis.
> The original linear.project window appears again. Click NEW and name a second file as "step-2."
> In the 3-sheet window for substrates, product, and coupled products for step 2 select IMPORT for substrate 1 → convergent.project → step-1B → Import. The entry for benzyl cyanide (product of step 1B) automatically appears in the "name" and "formula" fields. Enter the appropriate stoichiometric coefficient in the "coef" field.
> In the 3-sheet window for substrates, product, and coupled products for step 2 select IMPORT for substrate 2 → convergent.project → step-1A → Import. The entry for diethyl oxalate (product of step 1A) automatically appears in the "name" and "formula" fields. Enter the appropriate stoichiometric coefficient in the "coef" field.

Note: The first imported substrate in the convergent step must be the limiting reagent for that step in order for the calculation to proceed.

> Continue as before to complete all entries for all reaction materials for step 2.
> In the convergent.project window select "step-2" and OPEN.
> Continue as before to complete all masses of input materials for step 2.
> Click FILE → SAVE PROJECT to save your work.

Note: Only the longest linear chain is displayed in the histogram depicting the overall performance of the plan. The steps corresponding to convergent branches are not shown.

Summary
> 1. EATOS calculations are correct and agree entirely with results from *REACTION-template.xls* and *SYNTHESIS-template.xls* spreadsheets. The program works out overall PMI and E-factor for individual reactions and entire synthesis plans, and it breaks down these overall parameters into

their constituent contributions according to substrates, catalysts, solvents, auxiliary materials, coupled products, and by-products. PMI and E-factor are scale invariant.
2. Scaling factors linking reaction steps in a synthesis plan are not reported; they are hidden in the background. The mole scale of the product of the last step in a plan as entered in the experimental data sheet for "product" is the basis scale for the entire plan; all other mole scales of input materials in preceding steps are adjusted accordingly. However, an arbitrarily chosen target basis scale for the final product cannot be chosen and the masses of all input materials in the entire plan cannot be worked out accordingly. This is a problem for practical reasons when wanting to purchase correct quantities of ingredients for the production of a chosen mass of final product.
3. Number of moles of reagents or products is not calculated in data sheets. This is problematic in designating which reagent is the limiting reagent, especially when dealing with convergent steps that begin with substrates from different branches. One needs to do separate calculations outside the program to determine number of moles to ensure proper assignment of limiting reagents for entry in Substrate 1 field in substrate-product-coupled products sheet.
4. Entries in the substrate-product-coupled products sheet cannot be modified later. There is only one opportunity to enter the data. However, changes can be made to masses of all materials in the 7-sheet window for substrates-catalysts-solvents-auxiliary materials-product-coupled products-by-products.
5. Auxiliary materials for work-up and purification are lumped together so it is tedious to work out separate E-workup and E-purification contributions. Hence, E-auxiliaries (EATOS) = E-workup + E-purification.
6. For linear and convergent sequences, EATOS works in the forward direction beginning from step 1 onward.
7. If a contribution to the overall E-factor is small, its slice in the histogram is too small to be picked up by the cursor and hence its numerical report is not accessible (see points 19 and 20).
8. A full report is not possible with one keystroke or menu selection. It is awkward to cut and paste several entries by passing cursor of histograms (see point 20).
9. Reaction yields are not calculated when masses of products are entered.
10. Overall yields for synthesis plans are not calculated.
11. Different nomenclature is used for by-products and side products.
12. E-kernel = E-by-products (EATOS) + E-coupled products (EATOS)
13. E-aux = E-auxiliaries (EATOS) + E-solvents (EATOS) + E-catalyst (EATOS)
14. E-solvents = E-solvents (EATOS)
15. E-excess = E-substrates (EATOS)
16. For linear and convergent plans, the IMPORT function is used to enter substrates for successive steps in a chain fashion. The imported substrate for the Nth step is the same as the product from the (N − 1)th step.
17. For convergent plans, two substrates must be imported—one from the main branch and the other from the convergent branch; however, the first

substrate entered as "substrate 1" MUST always be the one that is the limiting reagent in the convergent step, otherwise the calculation is aborted and cannot proceed.
18. For convergent plans, when the histogram according to synthesis steps is displayed it shows only steps in the main branch where each imported substrate in the chain is the limiting reagent. All steps in convergent branches are not shown in the histogram (see point 21).
19. Histogram scale cannot be expanded or shrunk (no zoom in or out function) in order to circumvent the cursor problem when obtaining numerical results for small (less than 1%) contributions to S^{-1} or E-factor bars, for example, from catalysts and excess reagents. This is problematic when the contribution by auxiliary materials overwhelms other contributions.
20. The best way to get a numerical report of S^{-1} or E-factor contributions for a given reaction is to follow the sequence: Edit → Calculate → Diagram → Order → according to synthesis steps. Click step number on histogram to get a full report. This circumvents the cursor problem mentioned in point 19.
21. EATOS does not keep proper track of step counts in convergent plans. The program artificially compresses all steps in a branch leading up to a convergence point as a single step.

Application of environmental impacts

In the 7-sheet window, click on "Weighting." There appears a window with 13 folders and the following associated parameters:

Ozone depletion	ODP
Nutrification	NP
Acidification	AP
Degradability	half-life [hours]
Greenhouse effect	GWP
Ecotoxicology	LC50(fish, 96 h) [mg/L]
Ozone creation	POCP
Air pollution	LRV
Accumulation	log P_{ow} or BCF
Claiming of resources	price, density [g/mL], quantity (mass or volume)
Risk	R-phrases
Human toxicity	MAK [mg/m³], LD50(oral) [mg/kg] or LC50(inh, 4 h) [mg/L] or LD50(dermal) [mg/kg]
Chronic toxicity	carcinogen, mutagen, teratogen (no units)

The key parameters are shown in bold text.
ODP, AP, and GWP values based on BI-index analysis can be entered.
Log P_{ow} is the same as log K_{ow}.
Enter R-phrases as follows. If phrases are R36/37/38 and R40, then enter values as 36/37/38-40.
MAK corresponds to OEL.

Only one of either LD50(oral), LD50(dermal), or LC50(inh) can be entered under acute toxicity column. This is restrictive for compounds that pose multiple acute toxicities. One is left to choose the parameter with the greatest risk; that is, pick the lowest value.

For input materials, the assigned Q-factor to each material is determined from the risk phrases and toxicity data. For output materials, the assigned Q-factor for each material is determined as the average of Q-factors for human toxicity and accumulation. The Q-factors are multiplied by the E-factor and PMI contributions of materials.

Example 6.1

Figure 6.2 and Table 6.1 summarize the EATOS results for the three procedures to prepare aniline.

6.2 ECOSCALE[6]

6.2.1 Description of Algorithm

EcoScale is a semi-quantitative tool suitable for introductory undergraduate education on green chemistry that is used to assess environmental and hazard impacts of chemicals used in a chemical reaction. It uses an arbitrary penalty point system out of an ideal value of 100 covering the following categories: reaction yield, cost of reaction components (based on producing 10 mmol of final product), safety of reaction components, technical setup (type of equipment used), reaction temperature and reaction time, and workup and purification components. Greener procedures have high EcoScale values. The algorithm applies only to single reactions, not synthesis plans, and only to reaction input materials. It has limited coverage of actual toxicity and hazard parameters and is heavily weighted toward simplified WHMIS and NFPA-704 labeling systems (see Appendix) and qualitative information found in MSDS sheets. There is no visual display, and the EcoScale does not account for relative masses of input or waste materials in the assignment of penalty points. For example, 1 g of mercury used as reagent is assigned the same penalty points as if 100 g were used. The algorithm also does not consider waste reaction by-products. EcoScale is not recommended for work beyond teaching purposes at the introductory level.

6.2.2 EcoScale-Template.xls

The *EcoScale-template.xls* spreadsheet requires the following input data: a balanced chemical equation; molecular weights and stoichiometric coefficients for reactants, product, and by-products; masses of all reactants, target product, and all auxiliary materials (reaction solvent, catalyst, workup materials, and purification materials); safety designation symbols (N=dangerous for environment; T=toxic; F=highly flammable; E=explosive; F+=extremely flammable; T+=extremely toxic) for all materials used in the reaction; cost per gram of all materials used in the reaction; reaction temperature; reaction pressure; reaction time; and "yes/no" designations for assessing penalty points for technical setup, reaction temperature and time, and workup and purification. The spreadsheet has a built-in feature to check if the entered equation is balanced. The total

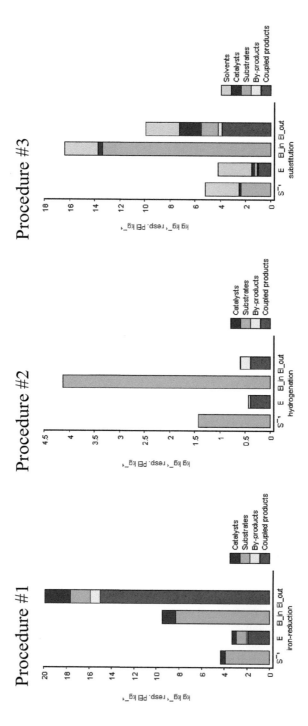

FIGURE 6.2 EATOS histograms for the three procedures to produce aniline: procedure #1 (iron reduction of nitrobenzene); procedure #2 (hydrogenation of nitrobenzene); and procedure #3 (substitution of chlorobenzene).

Algorithms

**TABLE 6.1
Summary of Results of EATOS Algorithm for Aniline Industrial Plans**

Plan	PMI	E-factor	EI_in	EI_out
Iron reduction	4.24	3.24	9.37	19.83
Hydrogenation	1.42	0.42	4.12	0.59
Substitution	5.15	4.15	16.37	9.88

The hydrogenation procedure has the lowest PMI, E-factor, EI_in, and EI_out values.
The best performing plan is hydrogenation of nitrobenzene.
E-factor rank: hydrogenation > iron reduction > substitution
EI_in rank: hydrogenation > iron reduction > substitution
EI_out rank: hydrogenation > substitution > iron reduction

**TABLE 6.2
Summary of EcoScale Results for Aniline Industrial Plans**

Plan	EcoScale score
Iron reduction	72.6
Hydrogenation	69.0
Substitution	68.4

Hydrogenation and substitution reactions have similar scores.
The best performing plan is iron reduction.
Score rank: iron reduction > hydrogenation > substitution

penalty point score and EcoScale score are automatically calculated with respect to a basis scale of product of 10 mmol. There are no visual displays for the results.

Example 6.2

Table 6.2 summarizes the EcoScale results for the three industrial aniline plans.

6.3 EDWARDS–LAWRENCE ALGORITHM[7]

6.3.1 DESCRIPTION OF ALGORITHM

In the original Edwards–Lawrence algorithm, an inherent safety index (ISI) is determined. It is calculated based on an arbitrary penalty point scoring system with no mass weighting of scores and a limited number of parameters. The ISI is composed of a sum of a chemical score and a process score. The chemical score is based on the following categories: inventory, flash point, boiling point, flammability, upper and lower exposure limits, explosiveness, and toxicity; and the process score is based on reaction temperature, reaction pressure, and % yield. Here we have modified this analysis to account for the masses of input materials used or waste materials produced so that the resulting chemical score component of the ISI is mass weighted. The relevant relationship for the chemical score, CS, for a given input or waste material j is given below in Equation (6.4).

$$(CS)_{j,i} = \frac{\text{moles}_j}{\text{moles}_{\text{target product}}} \left[(FS)_{j,i} + (ES)_{j,i} + (CRS)_{j,i} + (TS)_{j,i} \right] \quad (6.4)$$

where the j counter refers to the jth input or waste material (unreacted reagent, by-product, reaction solvent, catalyst, workup material, or purification material), the i counter refers to the ith step in the synthesis plan, FS=flammability score, ES=explosiveness score, CRS=corrosiveness score, and TS=toxicity score.

The process score, PS, is given by Equation (6.5).

$$(PS)_i = T_i + P_i + Y_i \quad (6.5)$$

where the i counter refers to the ith step in the synthesis plan, T=temperature score, P=pressure score, and Y=yield score.

The adjusted ISI scores for input materials and waste materials are given by Equations (6.6) and (6.7), respectively.

$$(ISI)_{adj,\text{input}} = \sum_{i,j} (CS)_{j\,\text{input},i} + \sum_i (PS)_i \quad (6.6)$$

$$(ISI)_{adj,\text{waste}} = \sum_{i,j} (CS)_{j\,\text{waste},i} + \sum_i (PS)_i \quad (6.7)$$

Table 6.3 and Table 6.4 summarize the scoring schemes for each category in the determination of the chemical and process scores, respectively.

6.3.2 EDWARDS-LAWRENCE-TEMPLATE.XLS

The *Edwards-Lawrence-template.xls* spreadsheet is composed of two sheets: an input materials sheet and a waste materials sheet from which the respective ISI indices are determined. The input data for the input materials sheet include: (a) the molecular weights of the target product in the synthesis plan and the input materials in each reaction step; (b) the basis mass of target product; (c) synthesis tree diagram (see Chapter 2); (d) scaled masses of all input materials for each reaction that will match the basis mass of the target product; (e) flash point, boiling point, % UEL (upper explosion limit), % LEL (lower explosion limit), OEL (occupational exposure limit), and SD (skin dose) for each input material in each reaction step;

TABLE 6.3
Summary of Scoring Scheme for Determining the Chemical Score in the Edwards–Lawrence Algorithm

Flammability Scores		Explosiveness Scores	
Flammability[a,b]	**Score**	**S = UEL % − LEL %**[c]	**Score**
Non-combustible	0	$0 \leq S < 10$	1
FP > 140 F	1	$10 \leq S < 20$	2
100 F < FP < 140 F	2	$20 \leq S < 30$	3
FP < 100 F and BP > 100 F	3	$30 \leq S < 40$	4
FP < 100 F and BP < 100 F	4	$40 \leq S < 50$	5
		$50 \leq S < 60$	6
		$60 \leq S < 70$	7
		$70 \leq S < 80$	8
		$80 \leq S < 90$	9
		$90 \leq S < 100$	10

Toxicity scores		Corrosivity	
OEL or TLV (ppm)[d]	**Score**	**Scores SD (mg)**[e]	**Score**
OEL < 0.001	8	SD < 0.001	0
$0.001 \leq OEL < 0.01$	7	$0.001 \leq SD < 0.01$	1
$0.01 \leq OEL < 0.1$	6	$0.01 \leq SD < 0.1$	2
$0.1 \leq OEL < 1$	5	$0.1 \leq SD < 1$	3
$1 \leq OEL < 10$	4	$1 \leq SD < 10$	4
$10 \leq OEL < 100$	3	$10 \leq SD < 100$	5
$100 \leq OEL < 1000$	2	$100 \leq SD < 1000$	6
$1000 \leq OEL < 10000$	1	$1000 \leq SD < 10000$	7
$10000 \leq OEL$	0	$10000 \leq SD < 100000$	8
		$100000 \leq SD$	9

[a] FP = flash point in degrees Fahrenheit
[b] BP = boiling point in degrees Fahrenheit
[c] UEL = upper explosion limit; LEL = lower explosion limit
[d] OEL = occupational exposure limit; TLV = threshold limit value
[e] SD = skin dose

and (f) reaction temperature, reaction pressure, and reaction yield for each reaction step. The same input data for the waste materials sheet are identical to those for the input materials sheet except that the masses of certain waste materials are different. The differences refer to the masses of unreacted reagents and by-products; whereas, those referring to auxiliary materials are unchanged. As more steps are involved in a synthesis plan, new reaction cell blocks can be entered with the proviso that formula entries embedded in cells for determining the chemical and process scores and ISI indices are adjusted accordingly. The ISI values are determined according to Equations (6.5) and (6.6).

TABLE 6.4
Summary of Scoring Scheme for Determining the Chemical Score in the Edwards–Lawrence Algorithm

Temperature Scores		Pressure Scores	
T (deg C)	Score	P (psi)[a]	Score
T < −25	10	0 < P < 90	1
−25°C ≤ T < −10°C	3	91 < P < 140	2
−110°C ≤ T < 10°C	1	141 < P < 250	3
10°C ≤ T < 30°C	0	251 < P < 420	4
30°C ≤ T < 100°C	1	421 < P < 700	5
100°C ≤ T < 200°C	2	701 < P < 1400	6
200°C ≤ T < 300°C	3	1401 < P < 3400	7
300°C ≤ T < 400°C	4	3401 < P < 4800	8
400°C ≤ T < 500°C	5	4801 < P < 6000	9
500°C ≤ T < 600°C	6	6001 < P < 8000	10
600°C ≤ T < 700°C	7		
700°C ≤ T < 800°C	8		
800°C ≤ T < 900°C	9		
900°C ≤ T	10		

Yield scores	
% Yield	Score
100	0
90 < Y < 99	1
80 < Y < 89	2
70 < Y < 79	3
60 < Y < 69	4
50 < Y < 59	5
40 < Y < 49	6
30 < Y < 39	7
20 < Y < 29	8
10 < Y < 19	9
0 < Y < 10	10

[a] PSI = pounds per square inch.

Example 6.3

Table 6.5 summarizes the Edwards–Lawrence algorithm results for the three procedures to prepare aniline.

TABLE 6.5
Summary of Adjusted Inherent Safety Index (ISI) Parameters for Three Industrial Procedures to Prepare Aniline With Respect to Input Materials, Waste Materials, and Combined Input and Waste Materials

Input materials		
Plan	Adjusted Overall ISI	Rank
Iron reduction	37.65	1
Hydrogenation	62.51	3
Substitution	45.26	2

Waste Materials		
Plan	Adjusted Overall ISI	Rank
Iron reduction	18.65	2
Hydrogenation	7.51	1
Substitution	26.26	3

Input + Waste Materials		
Plan	Adjusted Overall ISI	Rank
Iron reduction	56.3	1
Hydrogenation	70.02	2
Substitution	71.52	3

Based on an analysis of combined materials, the iron reduction procedure has the best performance.
Score rank: iron reduction > hydrogenation ~ substitution

6.4 GREENSTAR[8]

6.4.1 DESCRIPTION OF ALGORITHM

GreenStar is a more advanced point-based system to ascertain the degree of greenness of a chemical process as compared to the EcoScale because it attempts to include all 12 principles of green chemistry in a semiquantitative manner. Hence, it has been advertised as a "holistic approach" to assess the degree of greenness of a chemical reaction, however it excludes green principles 4 and 11, which refer to designing benign products and real-time monitoring of reactions to prevent pollution, respectively. Like the EcoScale, GreenStar is exclusively applied to individual reactions and not to synthesis plans. The main differences are that GreenStar is based on a positive merit point system as opposed to a negative demerit point system and that it takes into account reaction waste products; namely, reaction by-products. Nevertheless, the point system arbitrarily assigns a minimum value of 1 for non-green performance and a maximum value of 3 for benign performance to each green principle selected. The criteria scores for health, environmental impact, flammability, reactivity, degradability, and renewability characteristics, assigned to each chemical are summed in order to determine the green principle scores. The set of green principle scores in turn allows determination of the GreenStar area index (GSAI) parameter based on relative areas of radial dodecagons pertaining to the scores accumulated based on the ten green principles considered.

The GSAI is given by Equation (6.8).

$$\begin{aligned} \text{GSAI} &= 100 * \left(\frac{\text{Area}_{\text{actual}}}{\text{Area}_{\text{ideal}}} \right) \\ &= 100 * \left(\frac{\text{Area Radial Dodecagon}_{\text{actual}} - \text{Area Radial Dodecagon}_{\text{non-ideal}}}{\text{Area Radial Dodecagon}_{\text{ideal}} - \text{Area Radial Dodecagon}_{\text{non-ideal}}} \right) \end{aligned} \qquad (6.8)$$

where the dodecagons are based on green principles 1, 2, 3, 5, 6, 7, 8, 9, 10, and 12; "actual" refers to the dodecagon having 10 axes with values 1, 2, or 3; "non-ideal" refers to the dodecagon having 10 axes each with a value of 1; and "ideal" refers to the dodecagon having 10 axes each with a value of 3. Table 6.6 summarizes the scoring summary for GreenStar.

6.4.2 GREENSTAR-TEMPLATE.XLS

The mass input data for the *GreenStar-template.xls* spreadsheet is identical to the *template-REACTION.xls* spreadsheet for analyzing the material efficiency of a single reaction. The penalty point scores according to Table 6.6 are entered in three major categories for every input material used in a reaction: (a) Table 1 in spreadsheet covers green principles 1, 3, 5, and 9; (b) Table 2 in spreadsheet covers green principle 12; and (c) Table 3 in spreadsheet covers green principles 7 and 10. The GSAI score is determined according to Equation (6.8) and compared with ideal and non-ideal score scenarios.

Example 6.4

Figure 6.3 and Table 6.7 summarize the results of the GreenStar algorithm applied to the three industrial syntheses of aniline.

6.5 MULTIVARIATE METHOD[9–11]

6.5.1 DESCRIPTION OF ALGORITHM

The multivariate metric exercise developed at Queen's University is a truncated life cycle assessment (LCA) method based on the concept of defining a risk potential, P_j, for substance j as the ratio of a standard environmental impact parameter value, X_j, based on some property of the substance, to its value for an arbitrarily chosen reference compound, X_{ref}, according to Equation (6.9).

$$P_j = \frac{X_j}{X_{\text{ref}}}. \qquad (6.9)$$

TABLE 6.6
Revised Scoring Summary for Green Principles Used in GreenStar

GC Principle	Criteria	Score
P1: prevention	Waste is innocuous (S=3)	3
	Waste involves a moderate risk to human health and environment (S=2, for at least one substance)	2
	Waste involves a high risk to human health and environment (S=1, for at least one substance)	1
P2: atom economy	Reactions without excess of reagents (<10%) and without formation of by-products	3
	Reactions without excess of reagents (<10%) and with formation of by-products	2
	Reactions with excess of reagents (>10%) and without formation of by-products	2
	Reactions with excess of reagents (>10%) and with formation of by-products	1
P3: less hazardous chemical synthesis	All substances involved are innocuous (S=3)	3
	Substances involved have a moderate risk to human health and environment (S=2, for at least one substance)	2
	At least one substance involved has a high risk to human health and environment (S=1)	1
P4: designing of safer products	Not applicable	NA[a]
P5: safer solvents and auxiliary materials	Solvents and auxiliary substances are not used, but if used are innocuous (S=3)	3
	Solvents or/and auxiliary substances are used but have a moderate risk to human health and environment (S=2, for at least one substance)	2
	At least one solvent or auxiliary substance has a high risk to human health and environment (S=1)	1
P6: increasing energy efficiency	Room temperature and pressure	3
	Room pressure and temperature between 0 and 100°C when cooling or heating is needed	2
	Pressure different from room pressure and/or temperature >100°C or less than 0°C	1
P7: use renewable feedstocks	All substances involved are renewable (S=3)	3
	At least one substance involved is renewable, water is not considered (S=2)	2
	None of substances involved are renewable, water is not considered S=1)	1

(Continued)

TABLE 6.6 (CONTINUED)
Revised Scoring Summary for Green Principles Used in GreenStar

GC Principle	Criteria	Score
P8: reduce derivatives	Derivatizations are not used	3
	Only one derivatization or similar operation is used	2
	More than one derivatization or similar operations are used	1
P9: catalysts	Catalysts are not used and if used are innocuous (S=3)	3
	Catalysts are used but have a moderate risk to human health and environment (S=2)	2
	Catalysts are used and have a high risk to human health and environment (S=1)	1
P10: design degradation	All substances are degradable and break down to innocuous products (S=3)	3
	All substances not degradable may be treated to render them degradable to innocuous products (S=2)	2
	At least one substance is not degradable nor may be treated to render it degradable to innocuous products (S=1)	1
P11: real-time monitoring for pollution prevention	Not applicable	NA
P12: safer chemistry for accident prevention	Substances used have a low risk to cause chemical accidents (S=3)	3
	Substances used have a moderate risk to cause chemical accidents (S=2, for at least one substance)	2
	Substances used have a high risk to cause chemical accidents (S=1, for at least one substance	1

[a] NA = not applicable.

The exercise used the following seven potentials: acidification (AP); ozone depletion (ODP); smog formation (SFP); global warming (GWP); human toxicity by ingestion (INGTP); human toxicity by inhalation (INHTP); and abiotic resource depletion (ARDP). The global warming potential also incorporated a contribution from energy consumption in the form of CO_2 equivalents from heating, distillation, and refluxing operations. Energy consumptions from cooling and pressurization procedures were neglected. In addition, the degrees of bioaccumulation and persistence were estimated using octanol–water partition coefficients and the Boethling index given by Equation (6.10).

$$\text{Boethling index} = 3.199 - 0.00221(MW) + \sum_j f_j a_j \quad (6.10)$$

where MW is the molecular weight of a substance, a_j is an empirically determined parameter based on a particular functional group j in the molecule, and f_j is the number of times that functional group appears in the structure. The following Boethling index values correspond to the designated persistence times as shown: (a) 5 = hours; (b) 4 = days; (c) 3 = weeks; (d) 2 = months; and (e) 1 = long-lived.

Algorithms

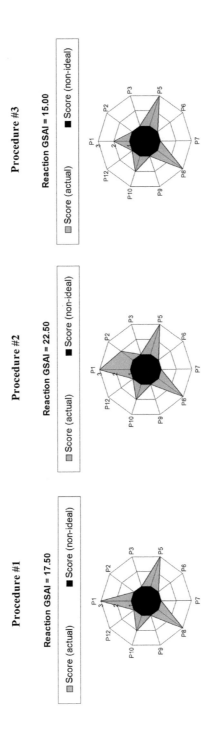

FIGURE 6.3 Radial diagrams based on GreenStar algorithm for the three procedures to produce aniline.

TABLE 6.7
Summary of GSAI Scores for Three Industrial Plans to Synthesize Aniline

Plan	GSAI
Iron reduction	17.50
Hydrogenation	22.50
Substitution	15.00

The hydrogenation procedure has the highest score.
The best performing plan is hydrogenation.
GSAI rank: hydrogenation > iron reduction > substitution

For a given reaction, the above seven potentials are determined for each substance that contributes to overall waste, namely, by-products, and all auxiliary materials (reaction solvents, catalysts, workup materials, and purification materials). Waste contributions from unreacted reagents are not considered in the analysis.

The risk indices pertaining to each property are then summed as shown in Equation (6.11).

$$\left[\sum_j P_j m_j\right]_{AP} + \left[\sum_j P_j m_j\right]_{ODP} + \left[\sum_j P_j m_j\right]_{SFP} + \left[\sum_j P_j m_j\right]_{GWP}$$

$$+ \left[\sum_j P_j m_j\right]_{INGTP} + \left[\sum_j P_j m_j\right]_{INHTP} + \left[\sum_j P_j m_j\right]_{ADP} \quad (6.11)$$

$$= \sum_j \left[I_{j,A} + I_{j,OD} + I_{j,SF} + I_{j,GW} + I_{j,INGT} + I_{j,INHT} + I_{j,AD} \right]$$

$$= I_A + I_{OD} + I_{SF} + I_{GW} + I_{INGT} + I_{INHT} + I_{AD}$$

When comparing results for different reactions leading to the same target compound at a common basis scale (usually 1 kg), tables are constructed showing these summed risk indices in a head-to-head fashion, and a red–yellow–green color-coding scheme is used to make decisions on which reaction is relatively greener. For each risk index category, red is assigned to the highest value, green is assigned to the lowest value, and yellow is assigned to intermediate values. Relatively greener plans are associated with a higher frequency of green-colored risk indices. Table 6.8 summarizes the key information pertaining to the environmental impact potentials used in the multivariate method.

TABLE 6.8
Multivariate Method Environmental Impact Potentials

Name	Abbreviation	Multiply by Mass	Reference Compound or State	Reference Value	Definition, P_j
Acidification potential	AP	Yes	SO_2	2/64	$\dfrac{\alpha_j/MW_j}{2/64}$
Ozone depletion potential	ODP	Yes	CCl_3F	1	$\dfrac{ODP_j}{ODP_{CFC11}}$
Smog formation potential	SFP	Yes	Standard mixture of reactive organic gases	3.1	$\dfrac{MIR_j}{3.1}$
Global warming potential	GWP	Yes	CO_2	1/44	$\dfrac{NC_j/MW_j}{1/44}$
Human toxicity by ingestion	INGTP	Yes	toluene	LD50(toluene) = 636 mg/kg (ppm) C(w,tol) 4.24E–10 mol/m³	$\dfrac{C_{w,j}/LD_{50,j}}{C_{w,tol}/LD_{50,tol}}$ LD50 units: Rat oral, mg/kg w = water
Human toxicity by inhalation	INHTP	Yes	toluene	LC50(toluene) = 49 g/m³, 4 h C(a,tol) 1.07E–10 mol/m³	$\dfrac{C_{a,j}/LC_{50,j}}{C_{a,tol}/LC_{50,tol}}$ LC50 units: Rat, g/m³, 4 h a = air
Persistence	PER	No	No reference	No value	Boethling index
Bioaccumulation	ACCU	No	No reference	No value	$\log K_{ow}$
Abiotic resource depletion potential	ADP	Input elements only	Sb	1	$\dfrac{ADP_j}{ADP_{Sb}}$

6.5.2 MULTIVARIATE-TEMPLATE.XLS

The *multivariate-template.xls* spreadsheet is composed of five sheets: (a) main; (b) MCM (multi-compartment model) calculator; (c) Boethling index calculator; (d) ADP values for each element; and (e) materials sheet. The main sheet is divided into four parts: (a) material efficiency metrics calculation; (b) potentials calculation according to entries in column 6 in Table 6.8; (c) risk indices calculation according to Equation (6.11); and (d) carbon dioxide equivalents calculation for input energy consumption which requires heat capacity inputs for each material used and CO_2 mass equivalents (kg) = overall input energy (kJ) * 0.042/1000.

Example 6.5

Table 6.9 summarizes the results of the multivariate method for the three industrial procedures to synthesize aniline.
The best performing plan is hydrogenation.

6.6 UTGCI METHOD[12]

6.6.1 DESCRIPTION OF ALGORITHM

An algorithm (*U Toronto Green Chemistry Initiative algorithm*) developed by the University of Toronto Green Chemistry Initiative (GCI) for accessing the degree of greenness of a chemical reaction according to material efficiency, environmental impact, and safety-hazard impact that is a hybridized method that combined the easy to understand attributes of the EcoScale and GreenStar methods with the ideas of mass weighted parameters and uncertainties advanced in the BI and SHI method.

For a given chemical reaction, the following parameters were considered for assessing greenness: reaction temperature, reaction pressure, LD50(oral), LD50(dermal), LC50(inhalation), OEL, log K_{ow}, GWP, acidification potential, LEL, flammability, corrosivity, explosiveness, reaction with water, oxidizing potential, and pyrophoricity. A red–yellow–green–gray color-coding scheme was implemented where each color was associated with a particular range of values for each parameter. Red indicated a notable hazard, yellow an intermediate hazard, green a mild hazard, and gray an uncertain hazard. Instead of simply counting the number of each kind of color for each waste substance, mass weighted color scores were determined according to Equations (6.12a) to (6.12d).

$$\text{red score} = \frac{\sum_j \phi_j r_j}{\sum_j r_j}, \quad \text{yellow score} = \frac{\sum_j \phi_j y_j}{\sum_j y_j},$$

$$\text{green score} = \frac{\sum_j \phi_j g_j}{\sum_j g_j}, \quad \text{gray score} = \frac{\sum_j \phi_j u_j}{\sum_j u_j}.$$

(6.12a,b,c,d)

TABLE 6.9
Summary of Multivariate Method Results for the Three Methods to Synthesize Aniline

Plan	I(A) (kg)	I(OD) (kg)	I(SF) (kg)	I(GW) (kg)	I(INHT) (kg)	I(INGT) (kg)	PER	log Kow	I(AD) (kg)	Sum[a] (kg)
Iron reduction	329.22	0	0.0013	0.28	21.06	56.48	months	1.85	0	407.04
Hydrogenation	0	0	0	0.11	0.11	5.03	months	1.85	0	5.25
Substitution	0	0	0	0.13	0.04	260.75	months	2.84	0	260.92

[a] Sum = I(A) + I(OD) + I(SF) + I(GW) + I(INHT) + I(INGT) + I(AD)

where ϕ_j represents the fractional mass contribution of waste substance j to the total waste and r_j, y_j, and g_j represent the number of red, yellow, and green cells accrued for each substance j, respectively.

The scores can be applied to input materials only and to waste materials only.

Pie charts showing the percent contributions of each color score are constructed to compare respective gains and losses on going from input materials to output materials, excluding the desired target product, for a given reaction. Essentially what is tracked is whether or not a reaction produces materials that are relatively greener than the input materials used at the outset. Equations (6.13a) and (6.13b) focus on the percent changes in the raw red and green mass weighted scores.

$$\%\Delta_{red} = 100\left[\left(\frac{\text{red score}}{\text{total color score}}\right)_{waste} - \left(\frac{\text{red score}}{\text{total color score}}\right)_{input}\right] \quad (6.13a)$$

$$\%\Delta_{green} = 100\left[\left(\frac{\text{green score}}{\text{total color score}}\right)_{waste} - \left(\frac{\text{green score}}{\text{total color score}}\right)_{input}\right] \quad (6.13b)$$

Worst and best case scenarios are also determined by adding any uncertainty contributions to the red and green scores, respectively, as shown in Equations (6.14a) and (6.14b).

$$\%\Delta_{red,worst} = 100\left[\left(\frac{\text{red score} + \text{gray score}}{\text{total color score}}\right)_{waste} - \left(\frac{\text{red score} + \text{gray score}}{\text{total color score}}\right)_{input}\right] \quad (6.14a)$$

$$\%\Delta_{green,best} = 100\left[\left(\frac{\text{green score} + \text{gray score}}{\text{total color score}}\right)_{waste} - \left(\frac{\text{green score} + \text{gray score}}{\text{total color score}}\right)_{input}\right] \quad (6.14b)$$

Equation (6.14b) represents the maximum change in the green score whereas Equation (6.13b) represents the corresponding minimum change. Similarly, Equation (6.14a) represents the maximum change in the red score whereas Equation (6.13a) represents the corresponding minimum change. A bar graph showing the worst and best case percent gains and losses is constructed to better visualize and interpret the results of Equations (6.14a) and (6.14b). Greener reactions have higher positive proportional changes in their green scores and higher negative proportional changes in their red scores when comparing output materials (waste produced) and input materials. The method can be extended to handle synthesis plans where appropriate scaling factors are used to scale masses of input or waste materials for each reaction according to the same procedure used to determine overall BI and overall SHI. The concept of tracking changes in mass weighted color scores proved more reliable in decision-making than interpreting the associated environmental-safety-hazard (ESH) scores as defined in Equations (6.15a) and (6.15b) for input and waste materials, respectively.

$$ESH_{input} = \left[(\text{green score})_{input} + (\text{yellow score})_{input} - (\text{red score})_{input}\right] \pm (\text{gray score})_{input} \quad (6.15a)$$

$$ESH_{waste} = \left[(\text{green score})_{waste} + (\text{yellow score})_{waste} - (\text{red score})_{waste}\right] \pm (\text{gray score})_{waste} \quad (6.15b)$$

Table 6.10 summarizes the green-yellow-red-gray code for the UTGCI method for each parameter considered.

6.6.2 UTGCI-Template.xls

Example 6.6

Figure 6.4 and Tables 6.11 and 6.12 summarize the results of the UTGCI method applied to the three industrial synthesis plans to aniline.

6.7 ANDRAOS ALGORITHM[13,14]

6.7.1 Description of Algorithm

The methods described in Chapters 3 and 4 are used to determine the benign index (BI) for waste, and safety-hazard indices (SHI) for waste and input materials. The vector magnitude ratio given by Equation (6.16) is used to rank the plans according to three material efficiency metrics (atom economy, yield, and reaction mass efficiency), and three LCA metrics (BI(waste), SHI(waste), and SHI(inputs)).

$$VMR = \frac{1}{\sqrt{6}}\left[\sqrt{(AE)^2 + (RY)^2 + (RME)^2 + (BI_w)^2 + (SHI_w)^2 + (SHI_{in})^2}\right] \quad (6.16)$$

Example 6.7

Figure 6.5 and Table 6.13 summarize the results of the Andraos algorithm applied to the three industrial syntheses of aniline.

TABLE 6.10
Code for the UTGCI Method

Parameter	Green	Yellow	Red	Gray
Rxn T (deg C)	$-10 < T < 100$	$100 < T < 300$ or $-25 < T < -10$	$T \leq -25$ or $T \geq 300$	not known
Rxn p (atm)	1	$1 < p < 9$	$p \geq 9$ or $p < 1$	not known
LD50(oral) (ppm)	LD50(oral) ≥ 500	$50 \leq$ LD50(oral) < 500	LD50(oral) < 50	not known
LD50(dermal) (ppm)	LD50(dermal) ≥ 500	$50 \leq$ LD50(dermal) < 500	LD50(dermal) < 50	not known
LC50(inh) (ppm)	LC50(inh) ≥ 500	$50 \leq$ LC50(inh) < 500	LC50(inh) < 50	not known
OEL (ppm)	OEL ≥ 500	$50 \leq$ OEL < 500	OEL < 50	not known
log K_{ow}	log $K_{ow} \leq 2$	$2 < \log K_{ow} < 4$	log $K_{ow} \geq 4$	not known
GWP	GWP ≤ 1	$1 <$ GWP < 4	GWP ≥ 4	not known
AP, for p$K_a < 7$	AP $= 0$	$0 <$ AP < 1	AP ≥ 1	not known
LEL (%)	LEL ≥ 7	$4 <$ LEL < 7	LEL ≤ 4	not known
Flammability	non-combustible or FP > 194 C	$122°C <$ FP $< 194°C$ OR FP $< 122°C$ and BP $> 122°C$	flammable OR FP $< 122°C$ and BP $< 122°C$	not known
Corrosivity	non-corrosive	R36, R37, R38; H311, H312, H313, H315	R34, R35; H290, H314	not known
Explosiveness	non-explosive		R1, R2, R3, R4, R5, R6, R9, R19	not known
Violent reaction with water	no reaction		R14, R14/15, R15, R15/29; H260	not known
Strong oxidizer	no		yes; H270, H272	not known
Pyrophoric	no		R17, R18	not known

(Continued)

TABLE 6.10 (CONTINUED)
Code for the UTGCI Method

Abbreviations

AP	Acidification potential
BP	Boiling point
FP	Flash point
GWP	Global warming potential
LC	Lethal concentration
LD	Lethal dose
LEL	Lower explosion limit
n/a	Not applicable (counted as green)
OEL	Occupational exposure limit

GWP = 44*(no. C atoms)/(MW)
AP = 32*(no. acidic Hs)/(MW)

Key Risk Phrases

Explosiveness

R1	Explosive when dry.
R2	Risk of explosion by shock, friction, fire, or other sources of ignition.
R3	Extreme risk of explosion by shock, friction, fire, or other sources of ignition.
R4	Forms very sensitive explosive metallic compounds.
R5	Heating may cause an explosion.
R6	Explosive with or without contact with air.
R9	Explosive when mixed with combustible material.
R19	May form explosive peroxides.

Reacts violently with water

R14	Reacts violently with water.
R14/15	Reacts violently with water, liberating extremely flammable gases.

(*Continued*)

TABLE 6.10 (CONTINUED)
Code for the UTGCI Method

R15	Contact with water liberates extremely flammable gases.
R15/29	Contact with water liberates toxic, extremely flammable gas.
H260	In contact with water releases flammable gases which may ignite spontaneously.
Corrosiveness	
R34	Causes burns.
R35	Causes severe burns.
R36	Irritating to eyes.
R36/37	Irritating to eyes and respiratory system.
R36/37/38	Irritating to eyes, respiratory system and skin.
R36/38	Irritating to eyes and skin.
R37	Irritating to respiratory system.
R37/38	Irritating to respiratory system and skin.
R38	Irritating to skin.
H290	May be corrosive to metals.
H311	Toxic in contact with skin.
H312	Harmful in contact with skin.
H313	May be harmful in contact with skin.
H314	Causes severe skin burns and eye damage.
H315	Causes skin irritation.
Pyrophoric	
R17	Spontaneously flammable in air.
R18	In use, may form flammable/explosive vapor–air mixture.
H250	Catches fire spontaneously if exposed to air.
Oxidizer	
H270	May cause or intensify fire; oxidizer.
H272	May intensify fire; oxidizer.

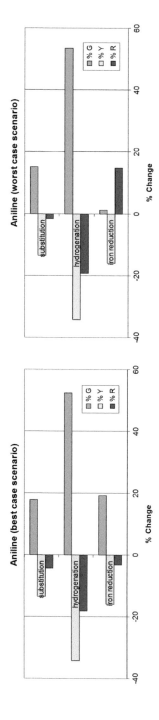

FIGURE 6.4 Percent changes in green, yellow, and red scores for the three industrial procedures to make aniline based on best- and worst-case scenarios.

TABLE 6.11
Summary of ESH(input) and ESH(output) Scores for Three Procedures to Make Aniline

Plan	ESH Input	ESH Output	Difference
Iron reduction	0.254 ± 0.377	0.093 ± 0.434	−0.161
Hydrogenation	0.656 ± 0.027	0.308 ± 0.003	−0.348
Substitution	0.285 ± 0.125	0.158 ± 0.069	−0.127

Hydrogenation has highest ESH scores for input and output.
ESH input rank: hydrogenation > substitution > iron reduction
ESH output rank: hydrogenation > substitution > iron reduction
Difference rank: hydrogenation > iron reduction > substitution

TABLE 6.12
Summary of Percent Color Scores for Three Procedures to Make Aniline

Waste − Inputs	% R Score	% Y Score	% G Score	% U Score
Iron reduction	−3.2	−16	19.2	18
Hydrogenation	−18.2	−34.2	53.5	−1.1
Substitution	−4.3	−13.5	15.1	2.6

% G score rank: hydrogenation > iron reduction > substitution
Best plan is hydrogenation.

Best scenario: % U transferred to % G

	Difference = waste − input		
	% R	% Y	% G
Iron reduction	−3.23	−15.967	19.198
Hydrogenation	−18.198	−34.206	52.406
Substitution	−4.284	−13.479	17.763

Worst scenario: % U transferred to % R

	Difference = Waste − Input		
	% R	% Y	% G
Iron reduction	14.808	−15.967	1.16
Hydrogenation	−19.252	−34.206	53.46
Substitution	−1.655	−13.479	15.134

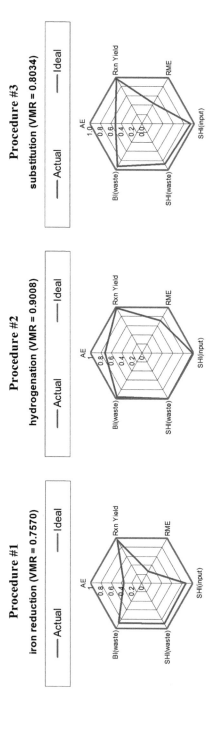

FIGURE 6.5 Radial diagrams based on Andraos algorithm for the three procedures to produce aniline.

TABLE 6.13
Summary of Results of Andraos Algorithm for Aniline Industrial Plans

Plan	BI Waste	SHI Input	SHI Waste	VMR
Iron reduction	0.889 ± 0%	0.864 ± 5%	0.899 ± 5%	0.7540
Hydrogenation	0.967 ± 3%	0.980 ± 7%	0.997 ± 7%	0.9008
Substitution	0.925 ± 2%	0.927 ± ± 7%	0.879 ± 7%	0.8034

The hydrogenation procedure has the highest BI waste, SHI input, and SHI waste scores.
The best performing plan is hydrogenation.
BI waste rank: hydrogenation > substitution > iron reduction
SHI input rank: hydrogenation > substitution > iron reduction
SHI waste rank: hydrogenation > substitution > iron reduction
VMR rank: hydrogenation > substitution > iron reduction

6.8 ROWAN SOLVENT GREENNESS INDEX[15]

6.8.1 DESCRIPTION OF ALGORITHM

The Rowan solvent greenness index (RSGI) is given by Equation (6.17)

$$RSGI = \sum_i m_i (OSI_{10})_i \tag{6.17}$$

where m_i is the mass of solvent i and the overall solvent index (OSI) normalized to be a number between 0 and 10 is given by Equation (6.18).

$$(OSI_{10})_i = 10 \left(\frac{OSI_i - OSI_{min}}{OSI_{max} - OSI_{min}} \right) \tag{6.18}$$

where OSI_i is the overall solvent index for solvent i, OSI_{min} is the minimum value of the overall solvent index over the entire set of solvents in the database, and OSI_{max} is the maximum value of the overall solvent index over the entire set of solvents in the database. For a given set of solvents, a value of 0 for OSI_{10} implies a comparatively benign solvent with minimal impact whereas a value of 10 implies a comparatively non-benign solvent with maximal impact.

For a given solvent i, the overall solvent index is given by Equation (6.19)

$$OSI_i = 2(M_{OEL,i} + M_{LD50,i} + M_{LC50,i}) + M_{GWP,i} + M_{SFP,i} + M_{ODP,i} + M_{ABP,i} \\ + M_{BCP,i} + M_{PER,i} + M_{soil,i} + M_{half\text{-}life,i} + M_{aqua,i} + M_{Q\text{-}phrase,i} + M_{SD,i} \tag{6.19}$$

where the metric parameters (M) cover OEL (ppm), LD50 (ingestion toxicity, mg/kg), LC50 (inhalation toxicity, g m^{-3} for 4 h), GWP (unitless), SFP (unitless), ODP (unitless), acidity-basicity potential (ABP, unitless), bioconcentration potential (BCP, unitless), persistence potential (PER, unitless), soil sorption coefficient (soil, K_{oc}), half-life of solvent in environment (half-life, h), aquatic toxicity to fish (aqua, mg/L for 96 h), Q-phrase potential (Q-phrase, unitless), and SD (mg). See Chapters 3 and 4 for descriptions of each of these parameters.

For a given parameter x for solvent i, the metric parameter is given by either of Equations (6.20a) to (6.20c).

$$M_{x,i} = 1 - \frac{\log x_i - \log x_{min}}{\log x_{max} - \log x_{min}} \quad (6.20a)$$

$$M_{x,i} = \frac{\log x_i - \log x_{min}}{\log x_{max} - \log x_{min}} \quad (6.20b)$$

$$M_{x,i} = 0 \quad (6.20c)$$

Equation (6.20a) is used if the parameter x_i increases with increasing greenness, Equation (6.20b) is used if the parameter x_i decreases with increasing greenness, and Equation (6.20c) is used if the parameter x_i is unknown or equal to zero.

For the acidity-basicity potential, absolute values are used for the x_i parameters to avoid computational errors using the logarithm function in cases when negative numbers arise.

The parameters applicable to Equation (6.20a) are: OEL, LD50, LC50, and aquatic toxicity where high values are consistent with benign conditions.

The parameters applicable to Equation (6.20b) are: GWP, SFP, ODP, ABP, BCP, PER, soil, half-life, aqua, Q-phrase, and SD, where low values are consistent with benign conditions.

It should be noted that expanding the list of solvents in a given database may result in changes to the maximum and minimum values for OSI and OSI_{10} and therefore to the rest of the OSI values for the intervening solvents depending on whether or not the additional solvents have higher or lower values than the existing maximum and minimum values. If the OSI values of the added solvents fall in between the existing limits then no changes to the magnitudes of the OSI values result in the rest of the solvents listed in the database.

As the name suggests the Rowan solvent greenness analysis applies to all solvents used in a chemical process including reaction solvents, workup solvents, and purification solvents. Generally, aqueous acidic and basic solvents are excluded since they are considered to be benign. The main focus is on the impacts of organic solvents.

Example 6.8

Two pharmaceutical filtration processes have the following solvent consumption:

Process 1 utilizes 120 kg acetone, 100 kg acetonitrile, and 90 kg of methyl ethyl ketone.
Process 2 utilizes 80 kg n-butanol, 100 kg n-heptane, and 80 kg n-pentane.

Based on the parameter values shown in the table below determine the Rowan solvent greenness indices for each process and decide which process is comparatively greener.

Algorithms

Parameter	Acetone	Acetonitrile	Methyl ethyl ketone	Butanol-n	Heptane-n	Pentane-n	Min. value	Max. value
OEL (ppm)	500	20	150	152	300	600	0.01	1000
GWP	2.276	2.146	2.444	2.378	3.08	3.056	0.286	3.385
SFP	0.13	unknown	unknown	unknown	0.41	0.65	0.02	2.75
LD50 (mg/kg)	3000	2460	2737	790	unknown	unknown	100	90000
LC50 (g/m^3, 4 h)	99.6	25.3	47	24	103	364	0.2	454
ODP	0	0	0	0	0	0	0	1.2
ABP (absolute value)	0	0	0	0.648	0	0	4.699	20.67
BCP	0.00553	0.0464	0.01398	0.03926	29.295	3.525	0.000287	29.295
PER	0.0857	0.0552	0.0974	0.0361	1.67	4.13	0.00949	26
K(oc)	1.981	4.5	3.827	2.443	274.7	80.77	1	1181
half-life (h)	321	327	298	239	108	90.2	87.2	636
aquatic toxicity (mg/L, 96 h)	5540	1640	400	1840	4	unknown	0.03	34000
Q-phrase	300	500	300	15000	40	40	5	75000000
SD (mg)	3325	4773	1411	849	2	12	0.7	8710

OSI(min) = 1.458
OSI(max) = 9.471

Solution

Metric	Acetone	Acetonitrile	Methyl ethyl ketone	Butanol-n	Heptane-n	Pentane-n
M(OEL)	0.060	0.340	0.165	0.164	0.105	0.044
M(GWP)	0.839	0.816	0.868	0.857	0.962	0.959
M(SFP)	0.38	0	0	0	0.613	0.707
M(LD50)	0.5	0.529	0.513	0.696	0	0
M(LC50)	0.196	0.374	0.293	0.380	0.192	0.029
M(ODP)	0	0	0	0	0	0
M(ABP)	0	0	0	0.648	0	0
M(BCP)	0.256	0.241	0.337	0.426	1	0.816
M(PER)	0.278	0.222	0.294	0.169	0.653	0.768
M(soil)	0.097	0.213	0.190	0.126	0.794	0.621
M(half-life)	0.656	0.665	0.618	0.507	0.108	0.017
M(aqua)	0.130	0.217	0.313	0.209	0.649	0
M(Q-phrase)	0.248	0.279	0.319	0.485	0.126	0.126
M(SD)	0.898	0.936	0.248	0.753	0.111	0.301
OSI	5.295	6.075	5.624	5.365	5.609	4.461
OSI(10)	4.789	5.762	5.200	4.877	5.181	3.747

RSGI (process 1) = 120*4.789 + 100*5.762 + 90*5.200 = 1618.88
RSGI (process 2) = 80*4.887 + 100*5.181 + 80*3.747 = 1208.02
Based on this analysis, process 2 is comparatively greener.

6.8.2 Rowan-Solvent-Greenness-Index.xls

The *Rowan-solvent-greenness-index.xls* spreadsheet includes M values corresponding to the 14 environmental and safety-hazard impacts for each of the 75 solvents commonly used in the chemical industry that are calculated using the method described in Section 6.8.1. The resulting OSI_{10} values for these solvents arranged in alphabetical order are summarized in Table 6.14.

Table 6.15 summarizes the same set of solvents in descending order of OSI_{10} values with the following color code: red ($OSI_{10} \geq 6.9$), yellow ($4 < OSI_{10} < 6.9$), and green ($OSI_{10} \leq 4$). The interpretation is that green colored solvents are considered benign and red colored solvents have the most environmental and safety-hazard impacts, which are therefore encouraged to be avoided as far as possible.

Example 6.9

The original medicinal chemistry route and an optimized process route to sildenafil citrate required the following amounts of solvents to manufacture 1 kg of product.

Medicinal Chemistry Route		
Solvent	L/kg API	d (g/mL)
Diethyl ether	20	0.708
Methanol	70	0.791
Acetone	130	0.792
Dichloromethane	970	1.336
Ethanol	80	0.789
Ethyl acetate	10	0.902
Methyl ethyl ketone	10	0.805
Optimized Process Route		
Solvent	L/kg API	d (g/mL)
t-Butanol	2	0.775
Toluene	0.3	0.865
Ethyl acetate	1.5	0.902
Methyl ethyl ketone	2.3	0.805

API = active pharmaceutical ingredient

Use the overall solvent indices $(OSI)_{10}$ values from the Rowan solvent greenness analysis to determine the Rowan solvent greenness indices for both routes to decide which route is comparatively greener based on solvent usage.

TABLE 6.14
Summary of OSI_{10} Values for 75 Industrially Important Solvents

Solvent	OSI(10)
Acetic acid	6.573
Acetic anhydride	7.012
Acetone	4.789
Acetonitrile	5.762
Amyl acetate	4.520
Amyl-t alcohol	4.406
Anisole	4.797
Benzene	10.000
Butanol-n	4.877
Butanol-sec	4.480
Butanol-tert	6.002
Carbon disulfide	5.813
Carbon tetrachloride	7.349
Chlorobenzene	8.844
Chloroform	7.041
Cyclohexane	6.572
Cyclopentanone	4.417
Cyclopentyl methyl ether	6.156
Dichlorobenzene-1,2	8.367
Dichloroethane-1,2	8.021
Dichloromethane	5.926
Diethyl ether	5.835
Diglyme	3.321
Dimethyl carbonate	3.016
Dimethyl sulfoxide	2.351
Dimethylacetamide	7.273
Dimethylformamide	6.104
Dimethylisosorbide	1.895
Dioxane-1,4	6.015
Ethanol	2.718
Ethyl acetate	5.553
Ethylene glycol	4.159
Ethylene glycol dimethyl ether	2.295
Ethylene glycol monomethyl ether	4.531
Formaldehyde	7.505
Glycol diacetate	3.697
Heptane-n	5.181
Heptanol-1	5.637
Hexamethylphosphoric triamide	7.097

(*Continued*)

TABLE 6.14 (CONTINUED)
Summary of OSI$_{10}$ Values for 75 Industrially Important Solvents

Solvent	OSI(10)
Hexane-n	6.620
Isoamyl acetate	4.498
Isoamyl alcohol	4.775
Isobutyl acetate	4.172
Isooctane	3.340
Isopropanol	4.682
Isopropyl acetate	5.539
Isopropyl ether	3.560
Methanol	3.393
Methyl acetate	4.539
Methyl ethyl ketone	5.200
Methyl formate	4.685
Methyl t-butyl ether	5.273
Methylcyclohexane	6.658
Nitrobenzene	9.162
Nitromethane	5.582
N-methylpyrrolidinone	4.826
Octanol-1	4.554
Pentane-n	3.747
Petroleum ether	5.433
Propanol-1	4.863
Propylene carbonate	2.675
Pyridine	8.346
Sulfolane	3.694
Tetrahydrofuran	5.980
Thionyl chloride	4.348
Toluene	8.107
Trichloroethylene	5.601
Triethylamine	8.154
Triethylene glycol monomethyl ether	2.903
Trifluoroacetic acid	4.500
Trifluorotoluene	5.665
Water	0
Xylene-m	7.843
Xylene-o	7.905
Xylene-p	7.992

TABLE 6.15
Summary of OSI_{10} Values for Various Industrial Solvents Listed in Descending Order

OSI(10)	Solvent
10.000	Benzene
9.162	Nitrobenzene
8.844	Chlorobenzene
8.367	Dichlorobenzene-1,2
8.346	Pyridine
8.154	Triethylamine
8.107	Toluene
8.021	Dichloroethane-1,2
7.992	Xylene-p
7.905	Xylene-o
7.843	Xylene-m
7.505	Formaldehyde
7.349	Carbon tetrachloride
7.273	Dimethylacetamide
7.097	Hexamethylphosphoric triamide
7.041	Chloroform
7.012	Acetic anhydride
6.658	Methylcyclohexane
6.620	Hexane-n
6.573	Acetic acid
6.572	Cyclohexane
6.156	Cyclopentyl methyl ether
6.104	Dimethylformamide
6.015	Dioxane-1,4
6.002	Butanol-tert
5.980	Tetrahydrofuran
5.926	Dichloromethane
5.835	Diethyl ether
5.813	Carbon disulfide
5.762	Acetonitrile
5.665	Trifluorotoluene
5.637	Heptanol-1
5.601	Trichloroethylene
5.582	Nitromethane
5.553	Ethyl acetate
5.539	Isopropyl acetate

(Continued)

TABLE 6.15 (CONTINUED)
Summary of OSI_{10} Values for Various Industrial Solvents Listed in Descending Order

OSI(10)	Solvent
5.433	Petroleum ether
5.273	Methyl t-butyl ether
5.200	Methyl ethyl ketone
5.181	Heptane-n
4.877	Butanol-n
4.863	Propanol-1
4.826	N-methylpyrrolidinone
4.797	Anisole
4.789	Acetone
4.775	Isoamyl alcohol
4.685	Methyl formate
4.682	Isopropanol
4.554	Octanol-1
4.539	Methyl acetate
4.531	Ethylene glycol monomethyl ether
4.520	Amyl acetate
4.500	Trifluoroacetic acid
4.498	Isoamyl acetate
4.480	Butanol-sec
4.417	Cyclopentanone
4.406	Amyl-t alcohol
4.348	Thionyl chloride
4.172	Isobutyl acetate
4.159	Ethylene glycol
3.747	Pentane-n
3.697	Glycol diacetate
3.694	Sulfolane
3.560	Isopropyl ether
3.393	Methanol
3.340	Isooctane
3.321	Diglyme
3.016	Dimethyl carbonate
2.903	Triethylene glycol monomethyl ether
2.718	Ethanol
2.675	Propylene carbonate
2.351	Dimethyl sulfoxide
2.295	Ethylene glycol dimethyl ether
1.895	Dimethylisosorbide
0	Water

Solution

Medicinal Chemistry Route

Solvent	L/kg API	d (g/mL)	Mass (kg)	OSI(10)	
Diethyl ether	20	0.708	14.16	5.835	82.62
Methanol	70	0.791	55.37	3.393	187.87
Acetone	130	0.792	102.96	4.789	493.08
Dichloromethane	970	1.336	1295.92	5.926	7679.62
Ethanol	80	0.789	63.12	2.718	171.56
Ethyl acetate	10	0.902	9.02	5.553	50.09
Methyl ethyl ketone	10	0.805	8.05	5.200	41.86
					8706.70 RGSI

Optimized Process Route

Solvent	L/kg API	d (g/mL)	mass (kg)	OSI(10)	
t-Butanol	2	0.775	1.55	6.002	9.30
Toluene	0.3	0.865	0.26	8.107	2.10
Ethyl acetate	1.5	0.902	1.35	5.553	7.51
Methyl ethyl ketone	2.3	0.805	1.85	5.200	9.63
					28.55 RGSI

The medicinal chemistry route has an RGSI value of 8707 and the optimized process route has an RGSI value of 29 suggesting that the latter route is significantly greener with respect to solvent consumption than the former route.

6.9 EPILOGUE

After having examined how the seven algorithms operate, we can summarize their findings for the example aniline plans to see how well they agree on which synthesis plan has relatively more "green" attributes. Table 6.16 summarizes the conclusions obtained for each algorithm along with advantages and disadvantages on their use. We see that the two simplest algorithms (EcoScale and Edwards–Lawrence) that use the least number of LCA parameters declare that the iron reduction method is optimal; whereas, the remaining five algorithms, which use larger data sets of LCA parameters, predict that the hydrogenation method is optimal. Based on the list of advantages and disadvantages given in Table 6.16, we recommend that the Edwards–Lawrence algorithm be used as a fast first-pass LCA assessment and then an advanced method such as the Andraos or UTGCI algorithms to obtain more reliable results. We also recommend that EcoScale and GreenStar be used only for teaching purposes in introductory green chemistry training courses, but not for professional work. We suggest that any algorithm that uses arbitrarily chosen penalty or merit point systems is far less reliable than one that uses actual values of physical and toxicological parameters in the LCA analysis. Finally, the reliability of conclusions drawn from any algorithm in the green decision-making process improves significantly when the data set size of physical, toxicological, and safety-hazard parameters that are used in the LCA analysis increases.

TABLE 6.16
Summary of Algorithm Results for the Three Industrial Procedures to Synthesize Aniline

Algorithm	"Greenest" Aniline Procedure	Advantages	Disadvantages
EATOS	Hydrogenation	Uses actual physical and toxicological parameters Applied to input and waste materials	Extraction of numerical data from visual graphs is tedious Requires specialized software to run
EcoScale	Iron reduction	Fast and easy to use Applied to input materials only	Uses arbitrary penalty point scoring Uses limited number of LCA parameters Does not account for masses of materials used Applied to single reactions only Has no visual aids
Edwards–Lawrence	Iron reduction	Fast and easy to use	Uses arbitrary penalty point scoring Uses limited number of LCA parameters Has no visual aids
GreenStar	Hydrogenation	Attempts to quantify all 12 principles of green chemistry using a merit point system	Uses arbitrary merit point scoring Does not account for masses of materials used Applied to single reactions only
Multivariate method	Hydrogenation	Easy to use Uses actual physical and toxicological parameters Applied to input and waste materials	Uses the sums of risk indices for ranking
UTGCI method	Hydrogenation	Easy to use Applied to input and waste materials	Uses a color code that is based on arbitrary penalty point scoring
Andraos	Hydrogenation	Uses actual physical and toxicological parameters Applied to input and waste materials	Requires separate reaction and synthesis spreadsheets apart from those used for material efficiency metrics

6.10 PROBLEMS

For each problem shown below, evaluate the reaction scheme according to the seven algorithms.

PROBLEM 6.1[16]

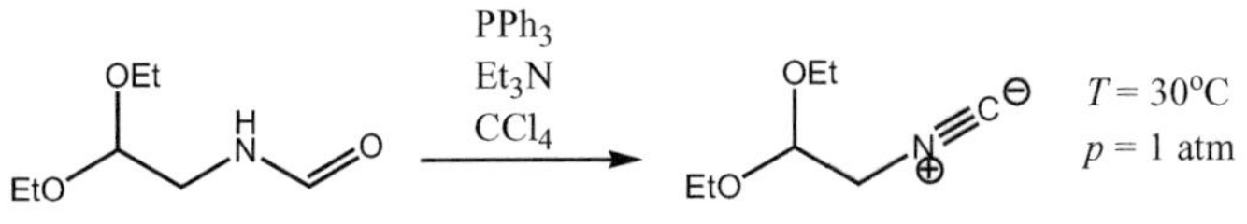

EtO-CH(OEt)-CH$_2$-NH-CHO →(PPh$_3$, Et$_3$N, CCl$_4$)→ EtO-CH(OEt)-CH$_2$-N≡C $T = 30°C$, $p = 1$ atm

Basis mass for final product = 13.5 g

Reagents:

24.2 g N-2,2-diethoxyethyl formamide
24.6 g tetrachloromethane
44.59 g triphenylphosphine
17.2 g triethylamine

Solvents:

150 mL dichloromethane

Workup:

150 mL diethyl ether
160 mL pentanes

Product:

13.5 g 2,2-diethoxy-1-isocyanoethane

PROBLEM 6.2[17]

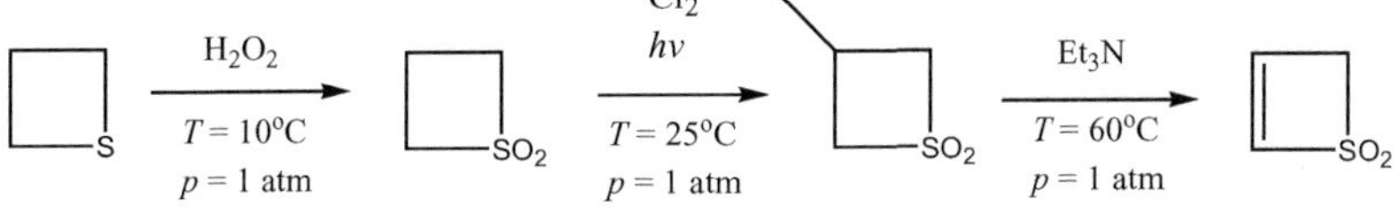

thietane →(H$_2$O$_2$, $T = 10°C$, $p = 1$ atm)→ -SO$_2$ →(Cl$_2$, hv, $T = 25°C$, $p = 1$ atm)→ Cl-/-SO$_2$ →(Et$_3$N, $T = 60°C$, $p = 1$ atm)→ -SO$_2$

Basis mass for final product = 100 g

Reaction #1:

47.5 g thietane
189 mL 30 wt% hydrogen peroxide
1.1 g tungstic acid
280 mL water

500 mL chloroform extraction
10 g magnesium sulfate drying agent
Product: 60.3 g thietane 1,1-dioxide

Reaction #2

14 g thietane 1,1-dioxide
10 g chlorine gas
300 mL carbon tetrachloride
Product: 5.4 g 3-chlorothietane 1,1-dioxide

Reaction #3

8 g 3-chlorothietane 1,1-dioxide
28.7 g triethylamine
300 mL toluene
100 mL toluene extraction
100 mL diethyl ether recrystallization solvent
Product: 4.5 g thiete 1,1-dioxide

PROBLEM 6.3

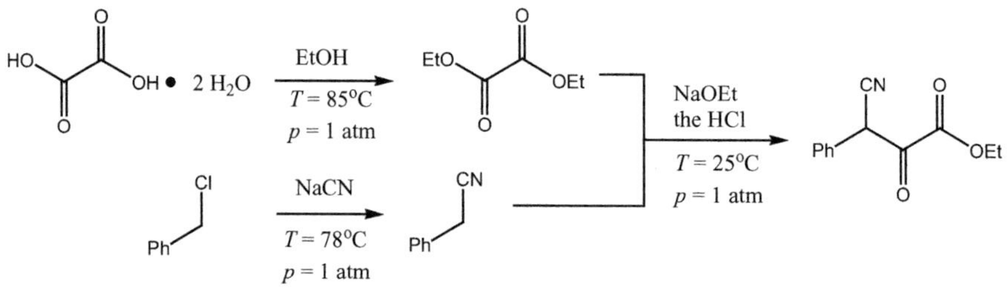

Basis mass for final target product = 1000 g.

Reaction #1[18]

1000 g oxalic acid dihydrate
1660 g 95 wt% ethanol
1330 g carbon tetrachloride
Product: 920 g diethyl oxalate

Reaction #2[19]

1000 g benzyl chloride
500 g sodium cyanide
450 mL water
1000 g 95 wt% ethanol
Product: 740 g benzyl cyanide

Reaction #3[20]

234 g benzyl cyanide
312 g diethyl oxalate
46 g sodium metal
197 g 37 wt% hydrochloric acid
650 mL ethanol
300 mL water wash
200 mL 60 wt% ethanol for recrystallization
Product: 300 g ethyl phenylcyanopyruvate

REFERENCES

1. Faith, W.L.; Keyes, D.B.; Clark, R.L. *Industrial Chemicals,* 3rd ed., Wiley: Hoboken, 1966, pp. 101, 541.
2. Shreve, R.N. *Chemical Process Industries,* 3rd ed., McGraw-Hill Book Co.: New York, 1967, p. 812.
3. Faith, W.L.; Keyes, D.B.; Clark, R.L. *Industrial Chemicals,* 3rd ed., Wiley: Hoboken, 1966, p. 101.
4. Eissen, M. (2001) *Bewertung der Umweltverträglichkeit organisch-chemischer Synthesen.* PhD thesis, Universität Oldenburg.
5. Environmental Assessment Tool for Organic Synthesis, www.metzger.chemie.uni-ol denburg.de/eatos/english.htm
6. Van Aken, K.; Strekowski, L.; Patiny, L. *Beilstein J. Org. Chem.* 2006, *2*, 3.
7. Edwards, D.W.; Lawrence, D. *Process Safety Environ. Protection* 1993, *71*, 252.
8. Ribeiro, M.G.T.C.; Costa, D.A.; Machado, A.A.S.C. *Green Chem. Lett. Rev.* 2010, *3*, 149.
9. Guinée, J.B., Ed., *Handbook on Life Cycle Assessment,* Kluwer Academic Publishers: Dordrecht, 2002.
10. Mercer, S.M.; Andraos, J.; Jessop, P.G. *J. Chem. Educ.* 2012, *89*, 215.
11. Boethling, R.S.; Howard, P.H.; Meylan, W.; Stiteler, W.; Beauman, J.; Tirado, N. *Environ. Sci. Technol.* 1994, *28*, 459.
12. Andraos, J.; Mastronardi, M.L.; Hoch, L.B.; Hent, A. *ACS Sust. Chem. Eng.* 2016, *4*, 1934.
13. Andraos, J. *Org. Process Res. Dev.* 2012, *16*, 1482.
14. Andraos, J. *Org. Process Res. Dev.* 2013, *17*, 175.
15. Slater, C.S.; Savelski, M. *J. Environ. Sci. Health* 2007, 42A, 1595.
16. Amato, F.; Marcaccini, S. *Org. Synth.* 2005, 82, 18; *Coll. Vol.* 2009, 11, 778.
17. Sedergran, T.C.; Dittmer, D.C. *Org. Synth. Coll. Vol.* 1990, 7, 491.
18. Clarke, H.T.; Davis, A.W. *Org. Synth. Coll.* 1941, 1, 261; 1925, 5, 59.
19. Adams, R.; Thal, A.F. *Org. Synth. Coll. Vol.* 1941, 1, 107; 1922, 2, 9.
20. Adams, R.; Calvery, H.O. *Org. Synth. Coll.* 1943, 2, 287; 1931, 11, 40.

7 Examples from the Chemical Industry

7.1 TERMS, DEFINITIONS, AND EXAMPLES

Process Time

The length of time elapsed to carry out a chemical reaction from the point of adding all materials to the reaction vessel to isolating the purified target product.

In batch operations, process time = residence time (reaction time) + workup time + purification time. Process time does not depend on reaction scale.

In continuous flow operations using a single tube, process time = total reaction volume/flow rate. The reaction volume is composed of the volume of reactants and the volume of reaction solvents. Process time depends on reaction scale.

In continuous flow operations using multiple tubes in parallel, process time = (total reaction volume/flow rate)*(1/number of parallel tubes). This operation is called numbering up or scaling out.

Process Solvent Mass Intensity[1]

Process solvent mass intensity is the total mass of solvent used (excluding water) per unit mass of product.

Process Water Mass Intensity

Process water mass intensity is the mass difference between freshwater usage and recycled water usage per unit mass of product made.

Process Water Use

Process water use is the total mass of water used in a process per unit mass of product made.

Residence Time

The length of time a set of reactant spends in a reaction vessel or chamber.

In batch operations, residence time = reaction time and does not change with reaction scale.

In continuous flow operations, residence time = reactor volume/flow rate and does not change with reaction scale. The reactor volume is determined from the geometry of the reaction vessel, usually cylindrical.

Solvent Intensity

Solvent intensity is the ratio of total mass of solvents used (including water) to mass of product made.

See *process solvent mass intensity*.

Throughput

A metric often used by process chemists to measure synthesis production efficiency is throughput given by Equation (7.1).

$$\text{throughput} = \frac{\text{mass of product}}{\text{process time}} \tag{7.1}$$

The units are kg/h.

7.2 PROBLEMS

In this chapter, we present 40 culminating problems covering all of the ladder concepts learned so far from the previous chapters in this book. These are applied to determining the material efficiency, environmental impact, safety-hazard impact, and energy efficiency of synthesis plans to various pharmaceuticals and industrially important commodity chemicals. The problems cover the following kinds of techniques: preparation of solutions, batch processes, microchannel technology, biofeedstocks, continuous flow microwave, scale-up versus scale-out, and synthesis plan analysis.

PROBLEM 7.1[2]
Themes: mixing solutions.

Part 1

A chemist needs to make 100 kg of a 10 wt% sodium hypochlorite solution from a 5 wt% and a 30 wt% stock solution. How can this be done?

Part 2

A chemist needs to make M kg of a A wt% sodium hypochlorite solution from a B wt% and a C wt% stock solution. How can this be done?

PROBLEM 7.2[3]
Themes: balancing chemical equations, material efficiency metrics.

Tetra(*n*-butyl)ammonium bromide is a quaternary ammonium salt used as a phase transfer catalyst. The process for its manufacture is summarized in the graphic below (Scheme 7.1).

The following output materials are fugitive losses in the process:

24.3 kg cyclohexane
4.44 kg n-butylbromide

Part 1

Write out a balanced chemical equation for the process.

Part 2

Determine the atom economy for the reaction.

Examples from the Chemical Industry

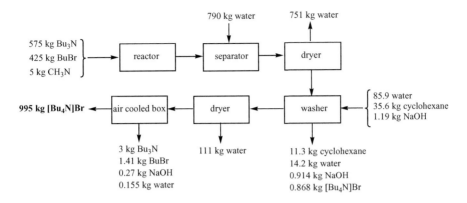

SCHEME 7.1 Process flowsheet for manufacturing Tetra(*n*-butyl)ammonium bromide phase transfer catalyst.

Part 3
Do a mass balance analysis.

Part 4
Determine the reaction yield and process mass intensity.

Part 5
Based on the following energy requirements (MJ/1000 kg product) determine the energy intensity for the process.

Electricity	31.7
Heating	2487
Energy input requirement	2519
Cooling	−751

PROBLEM 7.3[4]
Themes: balancing chemical equations, side reactions, mass balancing, process mass intensity.

A process to manufacture 40.2 wt% solutions of hypochlorous acid from chlorine and 40 wt% sodium hydroxide has a reaction yield of 79.14%.

$$Cl_2 + NaOH \rightarrow HOCl + NaCl \qquad (7.2)$$

Part 1
How much 40 wt% sodium hydroxide solution is needed to produce 1000 kg of 40.2 wt% hypochlorous acid solution?

Part 2

There are two side reactions that occur that produce sodium chlorate and dichlorine monoxide as additional side products.

$$Cl_2 + NaOH \rightarrow NaClO_3 + NaCl \quad (7.3)$$

$$Cl_2 + NaOH \rightarrow Cl_2O + NaCl \quad (7.4)$$

If 79.23% of the sodium hydroxide is used in the main reaction (1), 20.04% in side reaction (2), and the rest in side reaction (3), how much sodium chlorate and sodium chloride are produced in the process?

Part 3

How much chlorine gas is needed for all three reactions?

Part 4

What is the PMI for the production of hypochlorous acid?

PROBLEM 7.4[5]

Themes: synthesis tree diagrams, scheduling, Gantt diagrams.
   Given the following schedule for a synthesis plan (Scheme 7.2).

Part 1

Suppose there is change of protocol in step 4 of B1 that reduces the workup time by 4 h. How does this impact the schedule of the final product P?

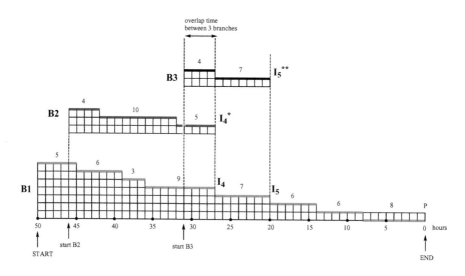

SCHEME 7.2 Gantt scheduling diagram.

Part 2

Suppose there is an unforeseen delay of 3 h in the procedure of step 4 in B1 that occurs 1 h before the synthesis of B3 is scheduled to begin. How does this impact the schedule of the final product P?

Part 3

Suppose there is an unforeseen delay of 3 h in the procedure of step 4 in B1 that occurs 1 h after the synthesis of B3 begins. How does this impact the schedule of the final product P?

PROBLEM 7.5[6]

Themes: balancing chemical equations, material efficiency metrics, synthesis tree diagram.

A two-step experimental procedure for the manufacture of alizarin dye is given below.

Step 1

To 125 g of oleum containing 18% SO_3 is added, with stirring, 100 g of anthraquinone of highest purity. The mixture is heated gradually to 135°C, while being stirred slowly, and held at this temperature for 3 h. The mixture is then allowed to cool to 50°C, and 80 g of 66% oleum is added during the course of 5 minutes, after which the whole is heated for 4 h more at 110°C. The sulfonation mixture is then cooled and poured into 1.5 L water, the aqueous suspension is heated to 80°C, and the unchanged anthraquinone is filtered off and washed with hot water. The hot filtrate is treated with 150 g of salt and cooled with stirring. After at least 10 h, the sodium anthraquinone-2-sulfonate, which separates as plates, is filtered off, washed with a small amount of 10% salt solution, pressed out, and dried. The yield is about 73 g, and about 37 g of anthraquinone is recovered.

Sodium anthraquinone-2-sulfonate, which is called "silver salt" in the industry because of its appearance, can be recrystallized from water to produce a material which is almost chemically pure. The air-dried material contains one molecule of water of crystallization.

Step 2

A mixture containing the equivalent of 100 g of 100% "silver salt," 260 g of 100% sodium hydroxide, 28 g of sodium chlorate, and enough water to make a total volume of 670 cc. is heated at 185°C. with continuous stirring in an autoclave. The pressure increases to 5 or 6 atmospheres. After 48 h, the apparatus is cooled, and a test is made to determine whether the fusion is completed. To this end, 2 cc. of the melt is removed, the alizarin is precipitated with the necessary amount of concentrated hydrochloric acid, and the filtrate is extracted twice with ether. The solution, thus freed from alizarin, is diluted to 15 cc. and examined for fluorescence caused by unchanged silver salt or monohydroxyanthraquinonesulfonic acid. Only a very weak, if any, fluorescence should be visible. If necessary, the mixture is heated for an additional 24 h at 190°C. The melt is then diluted with 2 L water, the mixture is heated to boiling, and the alizarin is precipitated with 50% sulfuric acid. The dye is filtered off at 50°C and washed until the washings are free from salts. The alizarin is not dried since then it does not dye well. The yield is determined by drying a test sample.

Usually, the dye is made up to a 20% preparation. About 70 g of pure alizarin is obtained from 100 g of pure silver salt.

Part 1

Write out balanced chemical equations for each reaction step and determine the corresponding atom economies.

Part 2

Write out the redox couple for the second step.

Part 3

Draw a synthesis tree diagram for the plan.

Part 4

Using the details of the experimental procedure and the *template-REACTION.xls* spreadsheet, determine the material efficiencies of each step and the overall plan. State any assumptions made in the calculations.

Part 5

Suggest a mechanism for step 2.

Part 6[7]

A modern procedure to make alizarin is given below.

A mixture of anhydrous $AlCl_3$ (5 mmol, 0.667 g) and pre-baked NaCl (2.5 mmol, 0.145 g) was heated (110°C) in an oil bath till molten. A homogenous mixture of phthalic anhydride (1 mmol) and catechol (1 mmol) were reacted with $AlCl_3$/NaCl melt. The temperature was slowly increased and maintained at 165°C. for 4 h. The reaction mixture was cooled to 0°C, 10 mL of 10% HCl added, stirred for 15 min at 0°C and refluxed at 100°C for 30 min. The reaction mixture was cooled to room temperature and extracted with ethyl acetate. The resulting product was purified by C18 MPLC column using acetonitrile:water (1:1) as the mobile phase. Yield: 55%.

Repeat the same analysis as above to determine the material efficiency of this synthesis.

PROBLEM 7.6[8,9]

Themes: balancing chemical equations, side reactions, batch versus microchannel synthesis.

Phenyl boronic acid can be made from phenyl magnesium bromide and trimethyl borate either in a batch reactor or a microtubular reactor (Scheme 7.3). The performances of each method are summarized below. In the microtubular reactor the production of diphenylboronic acid side product is effectively suppressed.

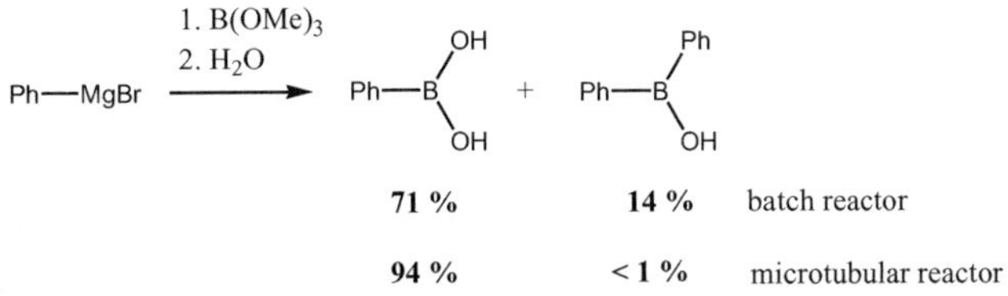

Examples from the Chemical Industry

Ph—MgBr → 1. B(OMe)$_3$ 2. H$_2$O → Ph—B(OH)$_2$ + Ph—B(Ph)(OH)

| | 71 % | 14 % | batch reactor |
| | 94 % | <1 % | microtubular reactor |

SCHEME 7.3 Synthesis of phenylboronic acid and diphenylboronic acid.

Part 1
Write out balanced chemical equations for the production of both products and determine their respective atom economies.

Part 2
Show a mechanism that accounts for the formation of both products.

Part 3
Explain why the yield of phenylboronic acid is higher in the microtubular reactor versus the batch reactor.

Part 4
Determine the kernel PMI for production of phenylboronic acid and diphenylborinic acid under batch and microtubular reactor conditions.

PROBLEM 7.7
Themes: balancing chemical equations, biofeedstocks, fermentation, material efficiency metrics.

Itaconic acid is a useful building block for making polymers that is obtainable by fermentation processes from biofeedstocks such as molasses and starch. Biopolymers made from itaconic acid are possible substitutes for petrochemical based polymers made from acrylic acid or methacrylic acid. It is used at as a co-monomer in the manufacture of resins and also in the manufacture of synthetic fibers, in coatings, adhesives, thickeners, and binders.

Part 1
An incomplete biochemical pathway is given below showing the transformation from D-glucose to itaconic acid via the glycolysis and tricarboxylic acid (TCA) cycles (Scheme 7.4). For each step show balanced equations and highlight the target bonds made.

E_1 = pyruvate carboxylase
E_2 = citrate synthase
E_3 = aconitase
E_4 = cis-aconitic decarboxylase (CAD)

NAD^+ = nicotinamide adenine dinucleotide
ADP = adenosine diphosphate
ATP = adenosine triphosphate

Determine the overall atom economy of the biochemical transformation.

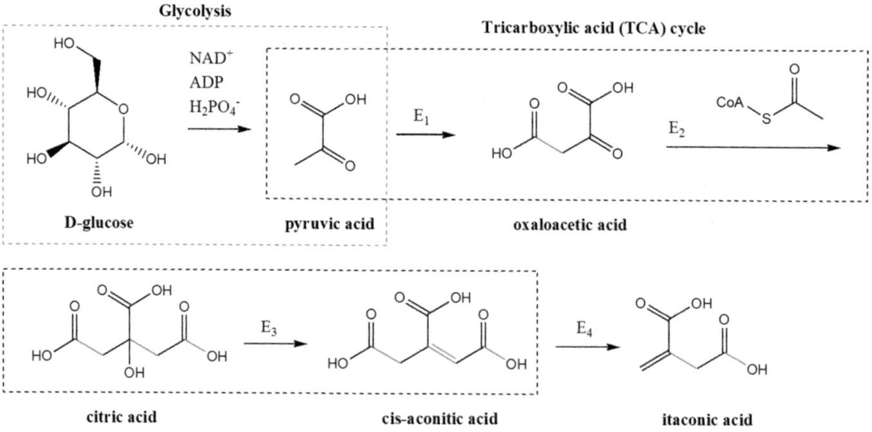

SCHEME 7.4 Biosynthetic pathways leading to itaconic acid from D-glucose.

Part 2[10]

Itaconic acid can be industrially produced by fermenting sugar containing solutions with *Aspergillus terreus* according to the procedure given below (Scheme 7.5).

Input materials:
2 L diluted molasses solution containing 196 g sucrose
30 g sodium nitrate
1 g sodium alginate
2.5 mL aqueous solution of *Aspergillus terreus*

Output after 7 days of fermentation:
48 g itaconic acid

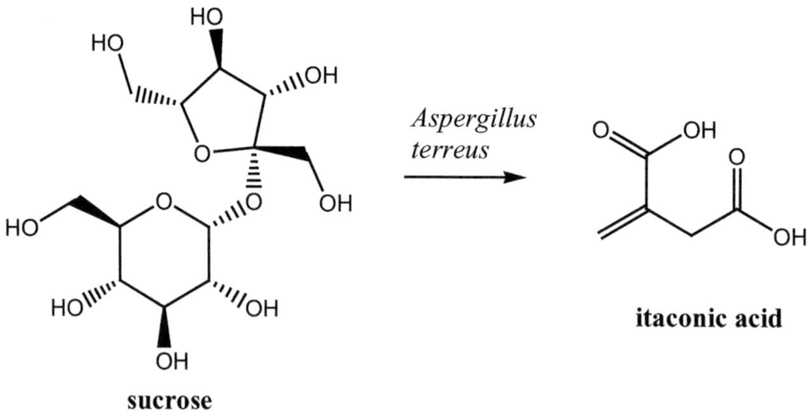

SCHEME 7.5 Biosynthetic pathways leading to itaconic acid from sucrose.

… From these data, determine the process mass intensity and % mass yield of product with respect to starting sucrose. State any assumptions made in the calculations.

Part 3

Itaconic acid has also been made by chemical synthesis according to the three routes shown in the Scheme 7.6. For each route, balance all chemical equations showing by-products and highlight the target bonds made in the itaconic acid structure. Based on the experimental information given for each reaction step determine the overall process mass intensity (PMI), overall yield, and overall atom economy (AE) for each route. State any assumptions made in the calculations. Which chemical route is most material efficient? How does compare with the fermentation route?

Chemical Synthesis #1[11]

10 g propargyl chloride in 50 mL acetone
   3 mL nickel tetracarbonyl in 200 mL acetone and 20 mL water
   50 L carbon monoxide per hour at p=1 atm and T=40°C
   Reaction time: 3 h
   Ether extraction
   Yield of itaconic acid=1 g

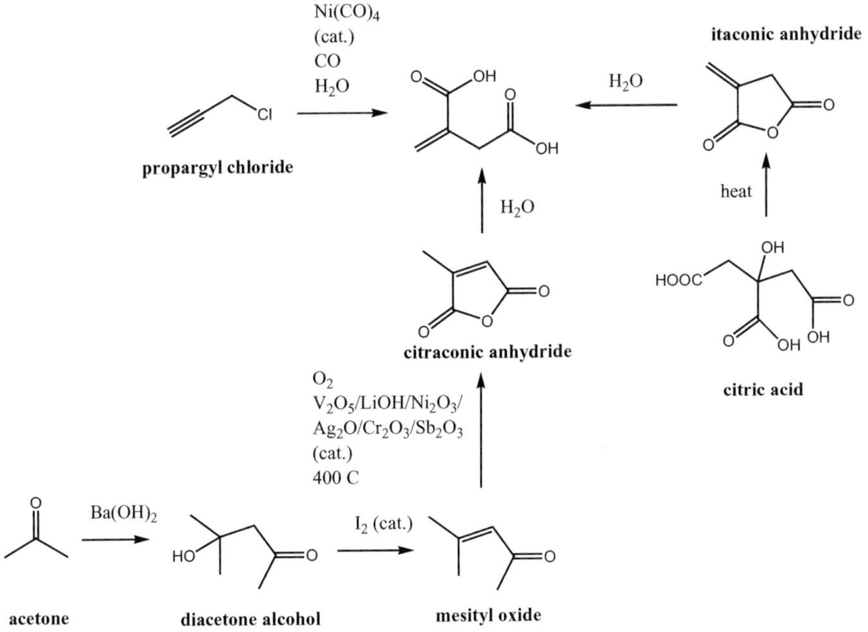

SCHEME 7.6 Chemical Syntheses of Itaconic Acid.

Chemical Synthesis #2
Step 1:[12]
1190 g acetone
   barium hydroxide catalyst
   Yield of diacetone alcohol = 850 g

Step 2:[13]
1100 g diacetone alcohol
   0.1 g iodine catalyst
   Yield of mesityl oxide = 650 g

Step 3:[14]
27 g vanadium pentoxide
   0.1 g lithium hydroxide
   0.1 g silver oxide
   2.7 g nickel(III)oxide
   0.1 g chromic oxide
   3.3 g antimony trioxide
   air to mesityl oxide volume ratio = 99.2 to 0.8
   contact time = 0.08 seconds
   T = 400°C
   Yield of citraconic anhydride = 36%

Step 4:[15]
1 kg of 25 wt% citraconic anhydride aqueous solution
   Yield of itaconic acid = 45%

Chemical Synthesis #3
Step 1:[16]
200 g citric acid monohydrate
   Yield of itaconic anhydride = 40 g

Step 2:[16]
40 g itaconic anhydride
   100 mL water
   Yield of itaconic acid = 11 g

PROBLEM 7.8[17]
Themes: balancing chemical equations, catalysis, material efficiency metrics.
    When perfluorinated alkanes are used in biphasic solvent systems with other non-miscible hydrocarbon solvents in catalytic reactions, it facilitates the separation of the catalysts from the reaction product due to preferential solubility of the catalyst in the fluorous phase. In such separations, the lower denser phase corresponds to the perfluorinated solvent. Reactions begin as a biphasic system but then become a

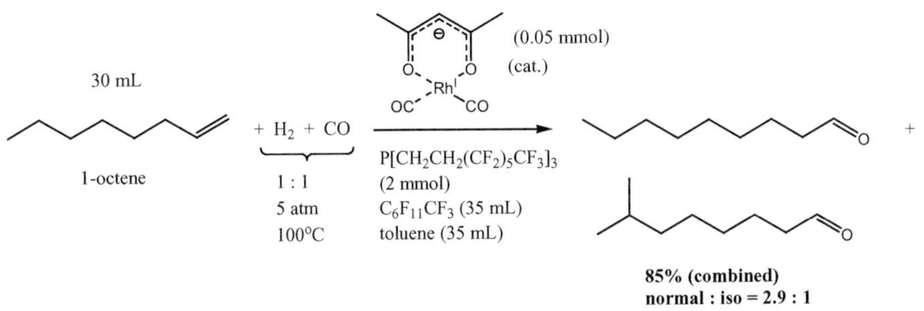

Examples from the Chemical Industry

SCHEME 7.7 Synthesis of n-nonanal and iso-nonanal via the hydroformylation reaction in perfluorinated alkane solvents.

single phase at the elevated reaction temperature. When cooled back down to ambient temperature, the system returns to a biphasic solvent system where the reaction product is mainly in the upper hydrocarbon layer and the catalyst is in the lower fluorous layer. This property discovered by Hungarian scientists at Exxon Research Laboratories in New Jersey has the potential of reducing the solvent demand used in extractions in workup procedures thereby reducing the E-workup contribution to the overall E-factor for a reaction. Fluorinated solvents are also exploited in extracting heavy metals often used in organometallic catalysts from hydrocarbon or aqueous solvents.

An important industrial reaction that demonstrates the utility of perfluorinated alkanes is the hydroformylation of olefins to aldehydes as shown by the experimental details given below in Scheme 7.7.

Autoclave volume = 300 mL

Part 1
Using the *template-REACTION.xls* spreadsheet and the details of the experiment determine the material efficiency metrics for production of n-nonanal and for iso-nonanal. State any assumptions made in the calculations.

Part 2
Repeat the analysis of Part 1 under the condition that the perfluoro(methylcyclohexane) ($C_6F_{11}CF_3$) is reclaimed. What is the percent reduction in PMI in each case?

Part 3
Show by means of a mechanism how iso-nonanal could be formed in this reaction.

PROBLEM 7.9[18]
Themes: balancing chemical equations, material efficiency metrics, energy metrics.

Continuous flow microwave synthesis under single-mode operation was demonstrated on an industrial scale for the following four reactions. Based on the experimental data given, determine the material efficiency metrics, process time, space-time yields, and energy consumption for each reaction. State any assumptions made in the calculations.

Example 1

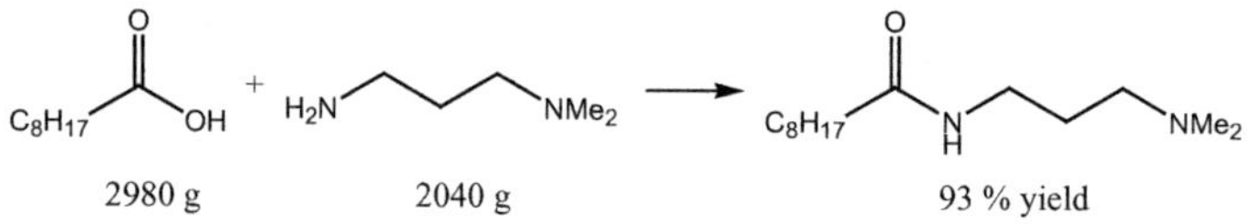

C₈H₁₇-C(O)-OH + H₂N-CH₂CH₂CH₂-NMe₂ → C₈H₁₇-C(O)-NH-CH₂CH₂CH₂-NMe₂

2980 g + 2040 g → 93 % yield

T = 280°C
p = 35 atm
flow rate = 5.6 L/h
power = 3100 W
residence time = 30 s

Example 2

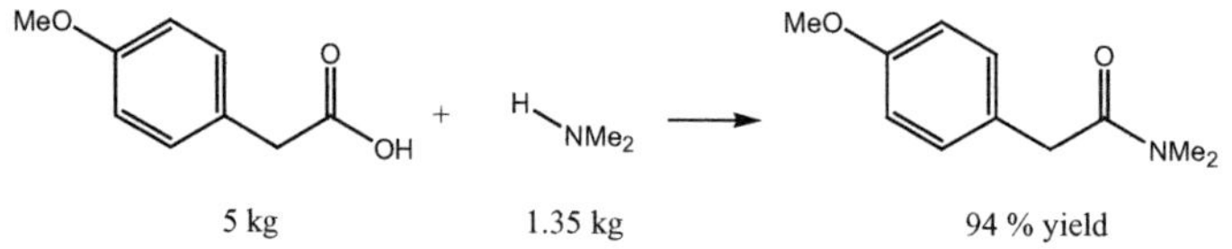

MeO-C₆H₄-CH₂-C(O)-OH + H-NMe₂ → MeO-C₆H₄-CH₂-C(O)-NMe₂

5 kg + 1.35 kg → 94 % yield

T = 245°C
p = 25 atm
flow rate = 3.5 L/h
power = 1950 W
residence time = 48 s

Example 3

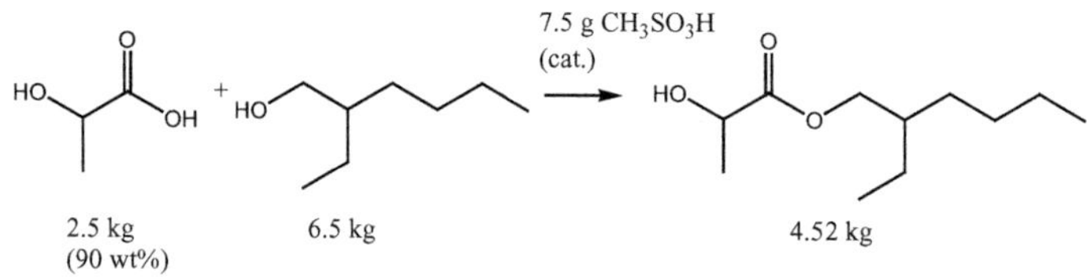

HO-CH(CH₃)-C(O)-OH + HO-CH₂-CH(C₂H₅)-C₄H₉ → HO-CH(CH₃)-C(O)-O-CH₂-CH(C₂H₅)-C₄H₉

(7.5 g CH₃SO₃H cat.)

2.5 kg (90 wt%) + 6.5 kg → 4.52 kg

T = 271°C
p = 25 atm
flow rate = 6 L/h
power = 3200 W
residence time = 28 s

Example 4

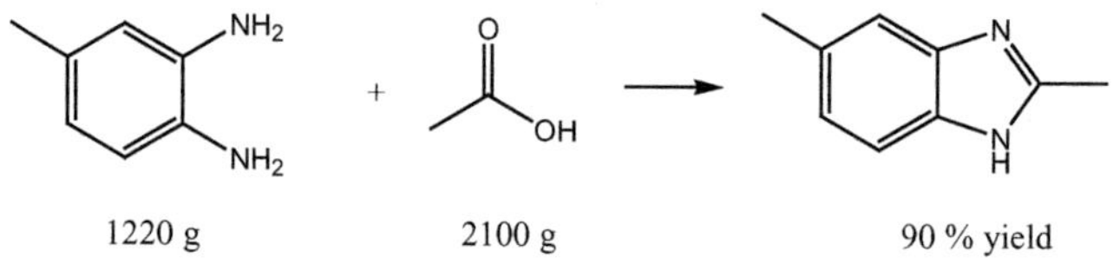

[4-methyl-1,2-phenylenediamine] + CH₃-C(O)-OH → [2,5-dimethylbenzimidazole]

1220 g + 2100 g → 90 % yield

Work-up: ice water, 1 M NaOH for neutralization

T = 267°C
p = 21 atm
flow rate = 5 L/h
power = 3000 W
residence time = 34 s

PROBLEM 7.10[19]

Themes: balancing chemical equations, continuous flow versus batch synthesis, material efficiency metrics, energy consumption.

The Bohlmann–Rahtz synthesis of pyridines has been carried out under three microwave irradiation conditions: batch synthesis in a sealed tube, continuous flow in a heating coil reactor, and continuous flow in a glass tube reactor (Scheme 7.8). Both continuous flow methods were run in single-mode operation.

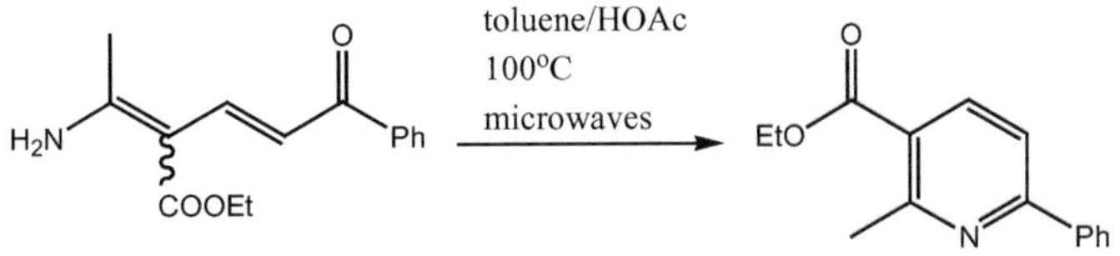

Examples from the Chemical Industry

SCHEME 7.8 Example Bohlmann–Rahtz synthesis of substituted pyridine.

The details of the three experimental procedures are given below.

Batch Synthesis in a Sealed Tube

A solution of 2-(1-amino-ethylidene)-5-oxo-5-phenyl-pent-3-enoic acid ethyl ester (80 mg, 0.3 mmol) in PhMe-AcOH (5:1) (3 mL) was irradiated for 2 min at 100°C (150 W) in a sealed pressure-rated glass tube. The reaction mixture was cooled by a flow of compressed air then partitioned between saturated aqueous $NaHCO_3$ and EtOAc, and the aqueous layer was further extracted with EtOAc. The combined organic extracts were washed with brine, dried ($MgSO_4$), and evaporated *in vacuo* to give the title compound (75 mg, >98%) as a yellow oil.

Continuous Flow Synthesis in a Heating Coil Reactor

A solution of 2-(1-amino-ethylidene)-5-oxo-5-phenyl-pent-3-enoic acid ethyl ester (1.3 g, 5.1 mmol) in PhMe-AcOH (5:1) (50 mL) was irradiated at 100°C (300 W) in a Teflon heating coil in a MW CF reactor at a flow rate of 1 mL per min and quenched in a solution of saturated aqueous $NaHCO_3$. The mixture was extracted three times with EtOAc and the organic extracts were combined, dried ($MgSO_4$), and evaporated *in vacuo* to give the title compound (1.2 g, >98%) as a yellow solid.

Continuous Flow Synthesis in a Glass Tube Reactor

A solution of 2-(1-amino-ethylidene)-5-oxo-5-phenyl-pent-3-enoic acid ethyl ester (80 mg, 0.3 mmol) in PhMe-AcOH (5:1) (3 mL) was irradiated at 100°C (300 W) in a pressure-rated glass tube (10 mL) filled with sand (about 12 g) in a MW CF reactor at a flow rate of 1.5 mL per min, while simultaneously cooling the tube in a flow of compressed air. The mixture was quenched immediately in a solution of saturated aqueous $NaHCO_3$ and extracted with EtOAc. The organic extract was dried ($MgSO_4$) and evaporated *in vacuo* to give the title compound (75 mg, >98%) as a yellow solid.

Part 1

Show a balanced chemical equation for the transformation, determine the atom economy, and highlight the target bonds made in the product structure.

Part 2

Using the *template-REACTION.xls* spreadsheet and the details of the three experimental procedures determine the material efficiency metrics for each method. State any assumptions made in the calculations. Which method has the best material efficiency profile?

Part 3

Estimate the energy consumption for each method. Which method requires the least energy input?

PROBLEM 7.11[20]

Themes: balancing chemical equations, reaction comparison, flow chemistry, material efficiency metrics.

Experimental procedures for the diazotization of a sulfoxide under batch versus flow conditions are given below in Schemes 7.9 and 7.10. The flow method is advertised with the following advantages: increased reaction yield, decreased reaction time, limiting exposure of product to basic conditions which can cause decomposition, increased safety profile, and scale-up potential.

Batch Experimental Procedure:

Triethylamine (0.29 mL, 1.95 mmol, 1 eq.) was added to a stirring solution of the sulfoxide **1** (0.39 g, 2.07 mmol) in acetonitrile (20 mL). Tosyl azide (0.40 g, 1.95 mmol) was then added dropwise at 0°C under nitrogen, and the solution was stirred overnight while slowly returning to room temperature to give a red colored solution. The mixture was concentrated under reduced pressure to give the crude product as an orange oil. Purification by column chromatography provided the pure α-diazosulfoxide **2**. Yield: 34%.

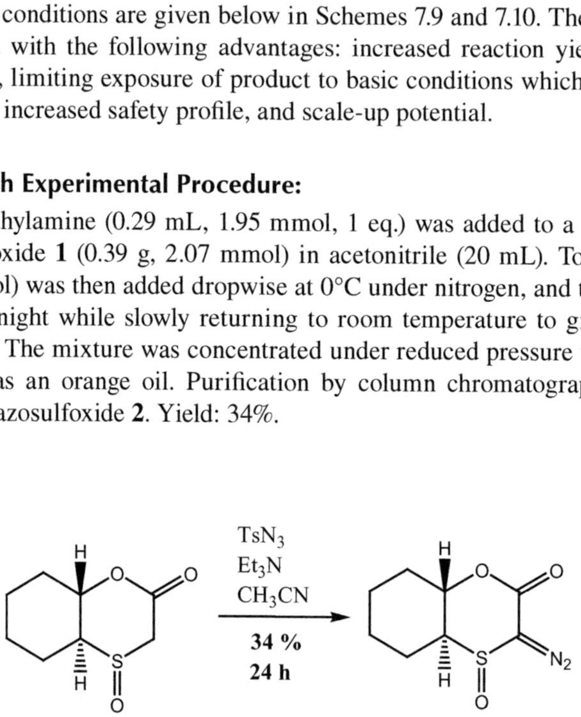

1 **2** **batch procedure**

SCHEME 7.9 Batch procedure for the synthesis of α-diazosulfoxide 2.

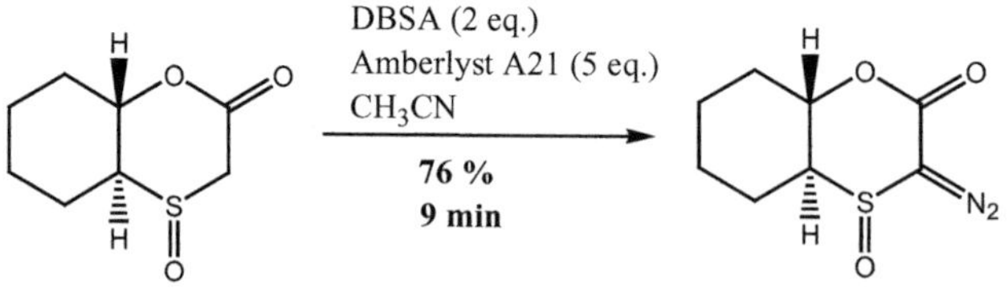

1 **2** **flow procedure**

SCHEME 7.10 Flow procedure for the synthesis of α-diazosulfoxide 2.

Flow Experimental Procedure:

General:

A packed bed reactor consisting of a fritted low pressure 10 mm ID×100 mm long Omnifit® glass column was packed with Amberlyst A21 (5 eq.) dispersed among acid washed sand (approx. 4.5 g) and mounted vertically. Acetonitrile was pumped through the column at a flow rate of 5 mL/min for 10 min to prepare the system by means of a peristaltic pump. The sulfoxide (1 eq.) was added to 5 mL of acetonitrile in a 10-mL volumetric flask. Dodecylbenzenesulfonyl azide (DBSA, 2 eq.) was added to the flask, and the solution was made up to the graduation mark with acetonitrile. The solution was pumped through the reactor with a residence time of 9 min at room temperature. The volume of the reactor was established by weighing the packed bed reactor while dry and again following saturation with acetonitrile. The system was fitted with an 8-bar back pressure regulator. The crude solution of product was concentrated under reduced pressure without heating, and conversion was established by ^{1}H NMR spectroscopy. Purification by column chromatography (ethyl acetate–hexane 50:50–100% ethyl acetate) gave the pure α-diazosulfoxides in good to excellent yields.

Specific details

A 1:0.7 diastereomeric mixture of the sulfoxides **1** (0.100 g, 0.53 mmol, 1 eq.) and dodecylbenzenesulfonyl azide (DBSA) (0.373 g, 1.062 mmol, 2 eq.) in acetonitrile (10 mL) was pumped through a 10-mm ID packed bed reactor containing Amberlyst A21 0.552 g, 2.65 mmol, 5 eq) and acid washed sand (approx. 4.6 g) with a residence time of 9 min. The crude product was a thick yellow oil which showed conversion to be 86% by ^{1}H NMR spectroscopy. Purification by column chromatography (ethyl acetate–hexane 50:50) gave the pure α-diazosulfoxides as a mixture of diastereomers (1:0.7) as a yellow crystalline solid (0.085 g, 76%).

Part 1

Use the *template-REACTION.xls* spreadsheet and the details of each experimental procedure to determine the respective material efficiency metrics. State any assumptions made in the calculations. Which method is more material efficient?

Part 2

The diazotization was repeated under batch conditions using DBSA as diazo group transfer agent for a direct comparison with the flow procedure that used this azide. This is a fairer comparison since identically the same reaction is used in both methods, which means the effect of AE cancels out.

A 1:1 diastereomeric mixture of the sulfoxides 1a (0.400 g, .002 mol, 1 eq) was dissolved in acetonitrile (40 mL). DBSA (1.49 g, .004 mol, 2 eq) was added to the solution. Amberlyst A21 (2.20 g, 5 eq) was added in one portion. After a reaction time of 16 h, 100% conversion was achieved with an isolated yield of 61% after column chromatography.

Repeat the metrics analysis to determine the PMI for this procedure and compare the results with the flow results.

PROBLEM 7.12[21]

Themes: balancing chemical equations, biofeedstock, material efficiency metrics.

Levoglucosan, obtained by pyrolysis of biomass, is a starting material that can be converted to itaconic acid via aerobic fermentation in the presence of *Aspergellus terreus*.

A 100 mL aqueous broth containing 8 g levoglucosan, 0.3 g ammonium nitrate, 0.2 g magnesium sulfate heptahydrate, and 0.2 g corn steep liquor is fermented at 35°C for 5 days. The fermentation yield determined by titration is 0.85 g itaconic acid per 100 mL of broth per day.

Part 1

Write out a balanced chemical equation for the aerobic fermentation process and determine its atom economy.

Part 2

Using the *template-REACTION.xls* spreadsheet and the experimental details given to determine the material efficiency metrics for the production of itaconic acid. State any assumptions made in the calculations.

PROBLEM 7.13[22]

Themes: process comparison.

A chemical versus a fermentation process are compared for the manufacture of a pharmaceutical. The table below summarizes the relevant metrics. Which process has the advantage?

What additional information is needed to make a firmer decision?

Mass:

Metric	Chemical Process	Fermentation Process
Purity (%)	98	98
Yield (%)	19	39
Cycle time (months/kg API)	18	3
PMI (excluding water)	350	468
Solvent intensity (kg/kg API)	332	452
% solvent	95	97
Water intensity (kg/kg API)	118	259
Solvent recovery (kg/kg API)	114	405
E-factor (kg waste/kg API)	467	726

Energy:

Metric	Chemical Process	Fermentation Process
Process energy (MJ/kg API)	693	1829
Waste treatment energy (MJ/kg API)	1382	2121
Life cycle energy (MJ/kg API)	7281	15211

Occupational Hazards:

Chemical Process

Material	OEL (ppm)
CH_2Cl_2	50
iPr_2NH_2	5
HCl (g)	2
Hexane	50
MTBE	50
THF	50
NaOH	2 $\mu g/m^3$
H_2SO_4	0.2 $\mu g/m^3$

Fermentation Process

Material	OEL (ppm)
HOAc	10
Glycerol	10 $\mu g/m^3$
Peracetic acid	1
NaOH	2 $\mu g/m^3$
H_2SO_4	0.2 $\mu g/m^3$

PROBLEM 7.14[23]
Themes: cost analysis.

Two methods to prepare an industrial commodity chemical are compared with respect to input materials, output materials, and emissions according to data summarized below. The basis unit for comparison is 100 pounds of product.

Input Material	Method A Mass (lb)	Method B Mass (lb)
Sand	31.9	31.9
Coal	105.4	105.7
Limestone	3.9	3.9
Oil	84.8	85.2
Ore rock	405	405
Natural gas	15.1	15.3
Water	1630.7	1630.7
Iron scrap	0.0	2.2

	Method A	Method B
Output Material	Mass (lb)	Mass (lb)
Product	100	100
Solid waste	228	232
Water waste	2	2
Airborne waste	391	394

Pollutant	Cost ($/ton)
Non-hazardous solid waste	35
Hazardous waste	250
CO_2	86
NO_x	2500
VOCs	1750
SO_x	135
Particulate matter	530

Which method is cheaper?

PROBLEM 7.15[24]
Themes: balancing chemical equations, material efficiency metrics, scale-up versus scale-out.

It is desired to make 1000 kg of epichlorohydrin by the procedure shown below. One option is to repeat the reaction several times at the scale of the procedure. This is called a scale-out method. A second option is to carry out the reaction once but scaled up appropriately for a target mass of product of 1000 kg. This is called a scale-up method. However, for this second option the reaction yield drops to 60%.

Experimental Procedure
In a 5 L flask equipped with a mechanical stirrer are added 3 L anhydrous diethyl ether and 1290 g glycerol α,γ-dichlorohydrin. The flask is surrounded by a cold water bath and 440 g of sodium hydroxide is added in small portions so that the temperature of the reaction is kept at 25–30°C. After addition is complete (about 20 minutes) the cold water bath is replaced with a warm water bath at 40°C and the mixture is gently boiled for 4 h. After cooling the ether is decanted from the solid residue which is rinsed with 2×250 mL diethyl ether. The combined ethereal solutions are fractionally distilled. The portion boiling at 115–117°C is pure epichlorohydrin (705 g, 76% yield).

Part 1
Compare and contrast the material efficiency metrics for both options assuming that the necessary equipment and quantities of all materials are available.

Part 2
Provide some advantages and disadvantages when the scale-out option may be the desired choice.

Provide some advantages and disadvantages when the scale-up option may be the desired choice.

PROBLEM 7.16[25]
Themes: balancing chemical equations, synthesis analysis, material efficiency metrics.

A recently claimed green synthesis of vanillin was reported that involved three steps starting from p-cresol (Scheme 7.11). The first step is an ortho oxybromination, the second is a catalytic aerobic oxidation of a methyl group to an aldehyde, and the third is a nucleophilic aromatic substitution reaction.

Experimental Procedure

Step 1
A three-necked flask was charged with 4-cresol (1.08 g, 10 mmol), CH_2Cl_2 (10 mL), and H_2O_2 (30%, 0.58 mL, d=1.11 g/mL, 5.5 mmol), and then the solution was cooled to 0°C. At this temperature, to the solution was slowly added a solution of bromine (0.26 mL, d=3.12 g/mL, 5.1 mmol) and CH_2Cl_2 (5 mL) over 4 h through a syringe pump. Afterwards, the mixture was stirred for another 4 h. Aqueous $NaHSO_3$ (10 mL, 2%) was added to the mixture at 0°C, and the reaction solution was allowed to stir for 1 h at room temperature. Furthermore, the solution was partitioned into two layers, and the aqueous phase was extracted with CH_2Cl_2 (5 mL×3). Finally, the combined organic layers were dried over anhydrous Na_2SO_4, and concentrated to give a crude oil, which was purified via column chromatography on silica gel (eluent: petroleum ether–ethyl acetate 20:1) to provide the desired product 2-bromo-4-methyl-phenol. Yellow oil, 1.80 g (96%).

Step 2
A three-necked flask was charged with ethylene glycol (10 mL), 2-bromo-4-methyl-phenol (0.94 g, 5.0 mmol), $Co(OAc)_2 \cdot 4 H_2O$ (12.5 mg, 0.05 mmol), and solid NaOH (0.80 g, 20 mmol), and then the solution was heated to 80°C. The molecular oxygen was continuously supplied to the reaction through a top tube inlet for 9 h. Hydrochloric acid (10 mL, 10%) and methyl tert-butyl ether (MTBE, 15 mL) were

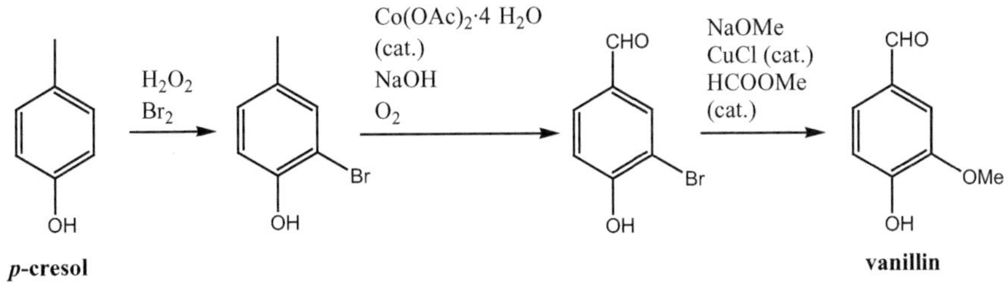

SCHEME 7.11 Three-step synthesis of vanillin from p-cresol.

successively added to the reaction mixture at room temperature. The MTBE phase was separated and the aqueous phase was further extracted with MTBE (15 mL × 2). The combined organic layers were dried over anhydrous Na_2SO_4 and concentrated in vacuo to give a residue, which was purified via column chromatography on silica gel (eluent: petroleum ether–ethyl acetate 20:1) to provide the desired product 3-bromo-4-hydroxy-benzaldehyde. Pale yellow solid, 0.90 g (90%).

Step 3

The Teflon-lined autoclave (25 mL) was charged with 3-bromo-4-hydroxy-benzaldehyde (0.80 g, 4.0 mmol), MeOH (5 mL), MeONa (0.65 g, 12 mmol), CuCl (15.8 mg, 0.16 mmol), and HCOOMe (0.10 mL, d = 0.97 g/mL, 1.6 mmol). The autoclave was heated to 115°C and stirred for 2 h. After the completion of the reaction, the reactor was cooled to room temperature. The reaction mixture was stirred for 0.5 h in an open system and then concentrated to recover pure MeOH. To the residue was added MTBE (5 mL) and diluted hydrochloric acid (1.0 M, 8 mL) to adjust the pH to 2.0–3.0. Furthermore, the solution was partitioned into two layers, and the aqueous phase was extracted with MTBE (5 mL × 3). The combined organic layers were dried over anhydrous Na_2SO_4, and concentrated in vacuo to give a solid, which was purified via column chromatography on silica gel (eluent: petroleum ether–ethyl acetate 15:1) to provide the desired vanillin.

White solid, 0.59 g (98%), m.p. 82–83°C.

Part 1

Based on the information given determine the material efficiency metrics for each step and for the entire plan. State any assumptions made in the calculations. Critically comment on the "green claims" made by the authors in their paper.

Part 2[26,27]

Below is the one-step aerobic biodegradation of ferulic acid to vanillin developed by Givaudan-Roure, a leading flavour and fragrance company (Scheme 7.12). Ferulic acid is sourced from biomass such as sugar beet waste.

The process description given in the patent is as follows:

> A preculture of *Streptomyces setonii* was grown in a shake flask at pH 7.2, 37°C, 190 rpm, for 24 h. The shake flask medium contained 5 g/L glucose, 4 g/L Na_2HPO_4, 1 g/L KH_2PO_4, 10 g/L yeast extract, and 0.2 g/L $MgSO_4$. A bioreactor was filled with 10 L of a medium containing 32 g/L glucose,

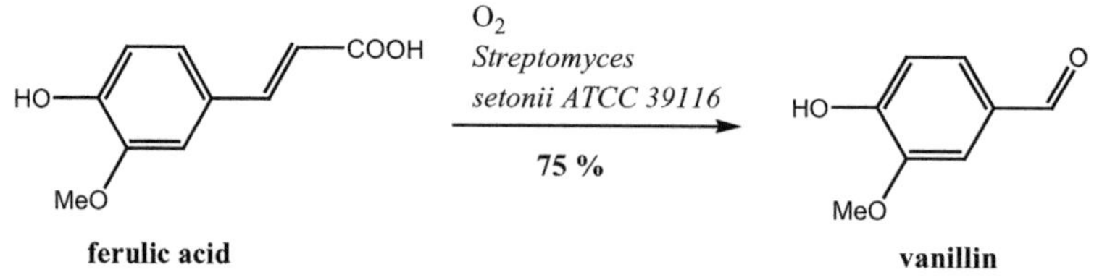

SCHEME 7.12 Biosynthetic route to vanillin from ferulic acid.

8 g/L yeast extract, 0.8 g/L $MgSO_4$ and 0.2 g/L antifoam agent (Dow Corning AF 1520). After thermal sterilization the reactor was inoculated with the previously grown shake flask preculture. The amount of inoculum used was 3%. The process conditions were 37°C, pH 7.2, airflow rate 1.0 vvm, 800 rpm. After 24 h of growth phase a remaining glucose concentration of 4.6 g/L was measured. Subsequently, the pH was shifted to 8.5 using NaOH (30%) and 24.5 h after inoculation 2.25 L of a 10% w/w solution of ferulic acid in 0.5 M NaOH was added to the fermentation broth. At the time of the feed, the glucose concentration was down to 4.0 g/L. Three to four hours after the addition of the precursor the beginning of the biotransformation of ferulic acid to vanillin was observed. Seventeen hours after the precursor feed, concentrations of 13.9 g/L vanillin and 0.4 g/L guaiacol were measured in the fermentation broth by GC. At that time, the ferulic acid was completely converted. A yield of vanillin of 75 mol % was calculated.

Based on the information given, estimate the material efficiency metrics for the one-step bioreaction. State any assumptions made in the calculations. How do the results compare with the three-step *p*-cresol plan to vanillin?

Part 3
Suggest a mechanism for the aerobic oxidation of 2-bromo-4-methyl-phenol to 3-bromo-4-hydroxy-benzaldehyde.

Part 4
Suggest a mechanism for the oxybromination in step 1.

PROBLEM 7.17[28,29]
Themes: synthesis plan analysis, material efficiency metrics.

Part 1
In a recent announcement of the 2017 Green Chemistry Challenge Awards, Merck process chemists received the Greener Synthetic Pathways Award for their improved synthesis of the antiviral drug letermovir. This drug is currently in phase III clinical trials for the treatment of human cytomegalovirus infections in organ transplant recipients. The claimed highlights of their synthesis that fulfill green chemistry principles are: (a) a 73% reduction in the overall process mass intensity; (b) a stable and fully recyclable chiral bistriflamide organocatalyst used in an asymmetric aza-Michael cyclization step; (c) a high atom economy; (d) elimination of hazardous dioxane and chlorobenzene solvents; (e) a 60% increase in the overall yield; (f) a 93% reduction in raw material costs; and (g) a waste reduction of 15000 tonnes.

The synthesis scheme is shown below (Scheme 7.13) which involves three sequences of telescoped steps where intermediates are not isolated after a given chemical transformation. These are shown in square brackets in the scheme. In such cases a solution of a given intermediate using the solvent of the previous step is

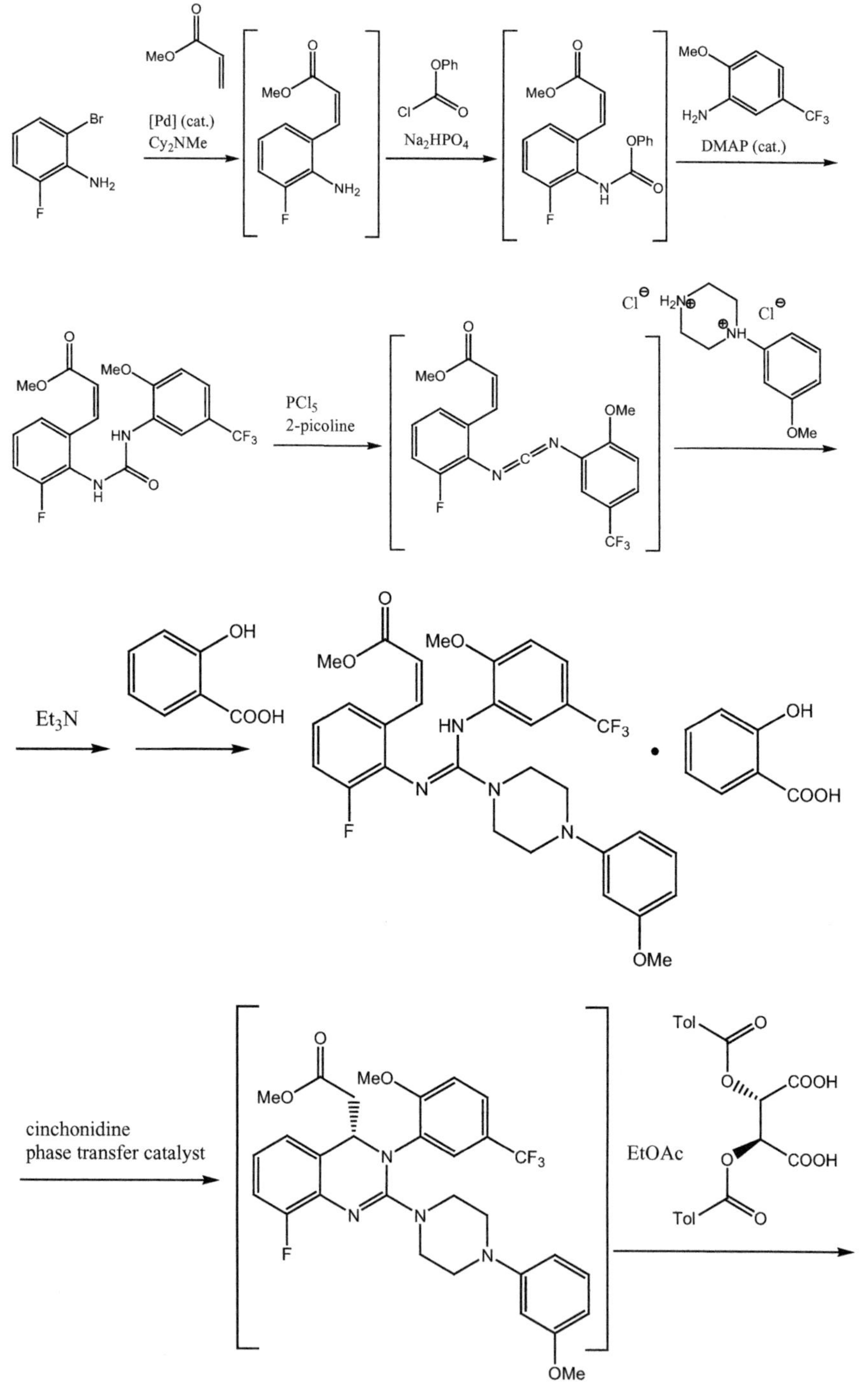

SCHEME 7.13 Merck process synthesis for letermovir. (continued)

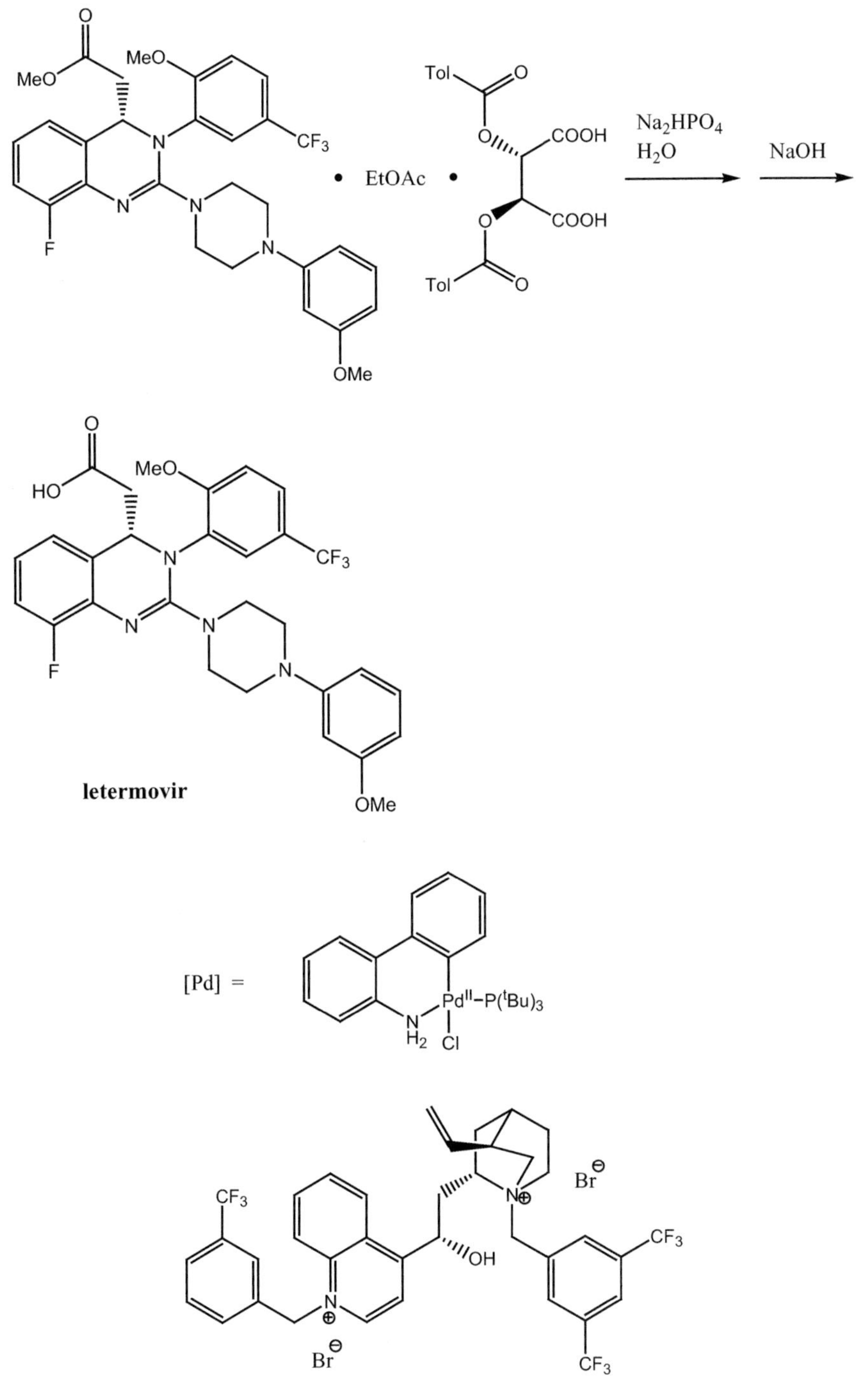

Examples from the Chemical Industry

letermovir

[Pd] =

cinchonidine derived organocatalyst

SCHEME 7.13 (CONTINUED) Merck process synthesis for letermovir.

carried forward into the next step without isolation. Isolations are reserved for those intermediates that are crystalline and have high purity.

Using the *template-REACTION.xls* spreadsheet and the details of the experimental procedure given in the *OPRD* paper, determine the material efficiency metrics for the synthesis of letermovir. State any assumptions made in the calculations. Draw a target bond map for the product structure showing the ring construction strategy used in the synthesis plan.

Part 2[30]

The discovery synthesis of letermovir was accomplished by AiCuris GmbH in Germany and is shown in Scheme 7.14. The plan involves six linear steps and the key cyclization step is a combined Heck coupling and aza-Michael cycloaddition that produces a racemic product which is resolved by classical resolution.

Using the *template-REACTION.xls* spreadsheet and the details of the experimental procedure given in the patent, determine the material efficiency metrics for the synthesis of letermovir. State any assumptions made in the calculations. Draw a target bond map for the product structure showing the ring construction strategy used in the synthesis plan.

Compare the results with the Merck plan. Are the Merck claims verifiable from the metrics analysis?

PROBLEM 7.18[22]

Themes: balancing chemical equations, patent literature, material efficiency metrics.

Experimental details from a patent for the production of hydrogen peroxide directly from hydrogen gas and oxygen gas are given below.

Reactor volume = 208 mL
Reactor temperature = 42°C to 52°C

Liquid medium contains:

Methanol
0.4 wt% sulfuric acid
1.0 wt% phosphoric acid
6 ppm sodium bromide
unspecified amount of Pd catalyst
feed rate = 72.8 g per hour

Gaseous medium contains:

Oxygen gas, feed rate = 48.6 L (STP) per hour
Hydrogen gas, feed rate = 5.5 L (STP) per hour
% conversion of hydrogen gas = 76%
% selectivity for hydrogen peroxide = 82%

After 100 h of operation, titration with potassium permanganate showed that a 7 wt% methanolic solution of hydrogen peroxide was obtained.

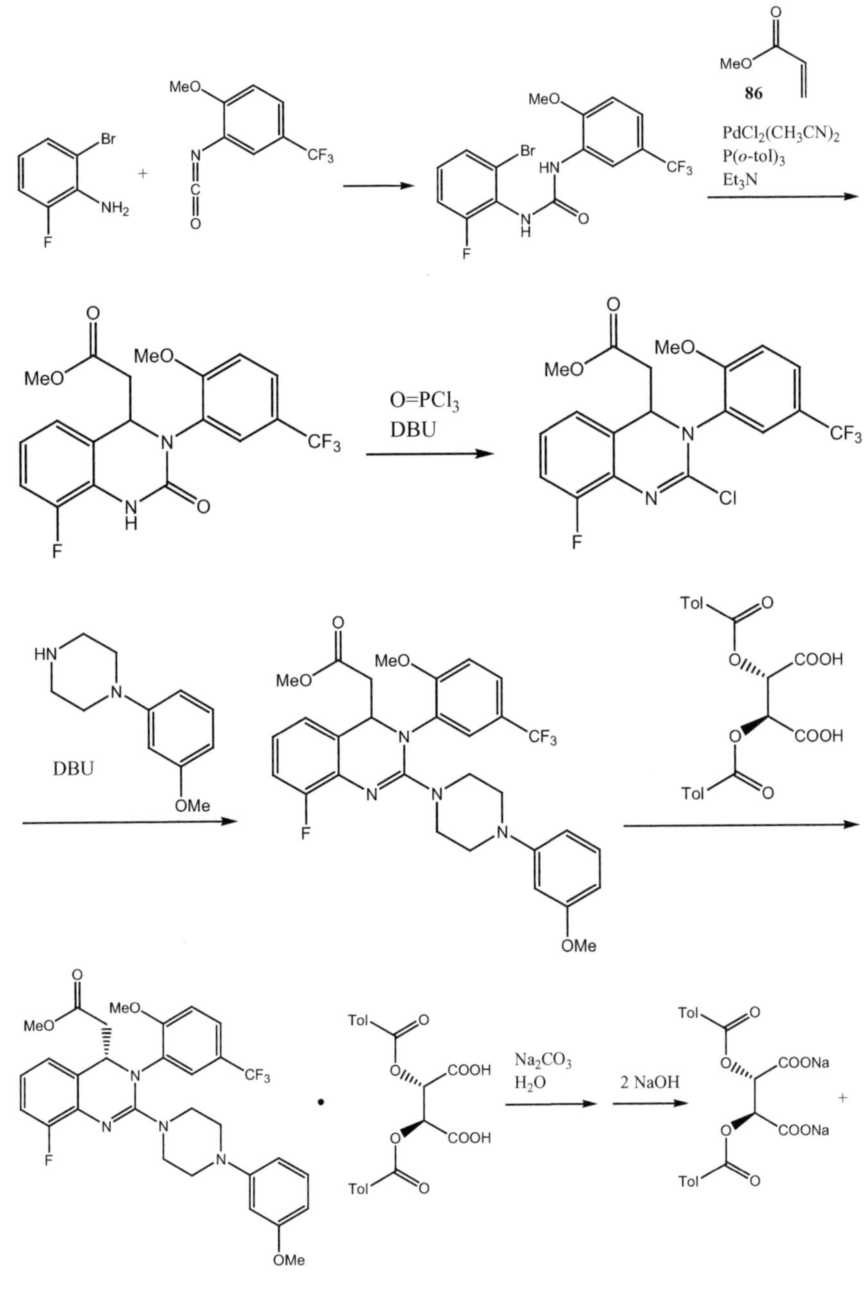

Examples from the Chemical Industry

SCHEME 7.14 AiCuris process synthesis for letermovir. (continued)

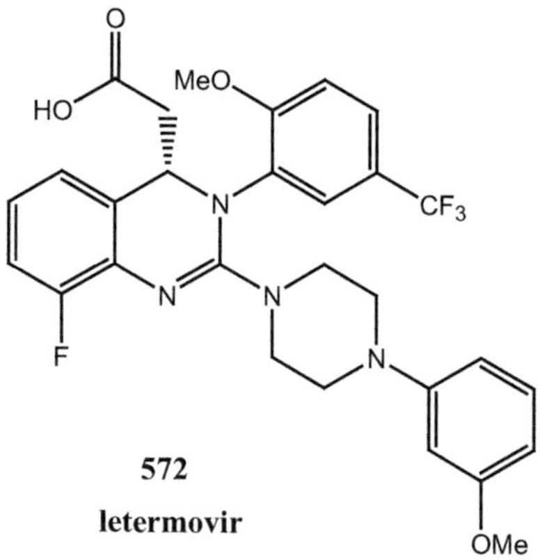

572
letermovir

SCHEME 7.14 (CONTINUED) AiCuris process synthesis for letermovir.

Part 1

From these data determine the % yield of reaction, % atom economy, stoichiometric factor, E-factor, reaction mass efficiency, and process mass intensity for the reaction. State any assumptions made in the calculations. Verify the wt% value of hydrogen peroxide.

Part 2

The patent states that after 100 h, 2.9 kg of methanolic solution of hydrogen peroxide is produced at a concentration of 4 wt%. Based on the results of Part 1 verify if this statement is true.

Part 3

A reported selectivity for hydrogen peroxide that is less than 100% suggests that the reaction produces side products. Suggest a plausible mechanism for the reaction that accounts for the production of hydrogen peroxide and other side products.

Part 4

Based on the experimental data given, determine the mass of side product identified in Part 3.

Part 5

Concentrations of hydrogen peroxide can be determined by a colorimetric redox titration method using cerium(IV)sulfate and ferroin as an indicator.

Write out a balanced chemical equation for the titration reaction. What atom is oxidized and what atom is reduced? Indicate the oxidation numbers of the relevant atoms.

PROBLEM 7.19[32]

Themes: balancing chemical equations, patent literature, material efficiency metrics.

Experimental details from a patent for the production of hydrogen peroxide directly from hydrogen gas and oxygen gas are given below.

Examples from the Chemical Industry

Catalyst preparation:

10 g activated charcoal
100 mL water
0.32 g sodium carbonate
10 mL solution containing 1 g Na_2PdCl_4 and 0.1 g H_2PtCl_6
0.85 g sodium formate in 10 mL water
1.2 g catalyst (dry) collected and used in hydrogen peroxide synthesis

Hydrogen peroxide synthesis:

Reactor volume = 400 mL
Reactor temperature = 8°C

Liquid medium:

150 g methanol-water (95:5) mixture
6 ppm HBr
300 ppm sulfuric acid
feed rate of liquid medium = 300 g per hour

Gaseous medium:

The autoclave is pressurized to 100 bars with a gas mixture consisting of 3.6% hydrogen, 11% oxygen, and 85.4% nitrogen by volume without stirring. Then, stirring is started and the pressure is maintained with 810 normal liters per hour (at 0°C and 1 atm) of the same gas mixture and 300 g per hour of the liquid medium is added at the same time.
After 800 h, 7.2 wt% hydrogen peroxide solution is obtained.
% selectivity for hydrogen peroxide = 73%

Part 1

From these data determine the % yield of reaction, % atom economy, stoichiometric factor, E-factor, reaction mass efficiency, and process mass intensity for the reaction. State any assumptions made in the calculations.

Part 2

If the % selectivity for hydrogen peroxide is 73%, what is the % conversion of hydrogen?

PROBLEM 7.20

Themes: balancing chemical equations, patent literature, material efficiency metrics.

Processes:

Sulfuric acid electrolysis[33]
Isopropanol oxidation[34]
Anthraquinone process[35]

Three industrial processes for the production of hydrogen peroxide are shown below along with experimental details.

Sulfuric acid electrolysis

$$2\ NH_3 + 2\ H_2SO_4 \longrightarrow 2\ [NH_4][HSO_4] \xrightarrow[-H_2]{\text{Pt electrode}} (NH_4)_2S_2O_8 \xrightarrow{2\ H_2O} H_2O_2 + 2\ [NH_4][HSO_4]$$

Input materials:

25 kg ammonia
30 kg sulfuric acid
1150 L water
8400 kg steam
Yield of hydrogen peroxide = 1000 kg of 70 wt% aqueous solution

Isopropyl alcohol oxidation

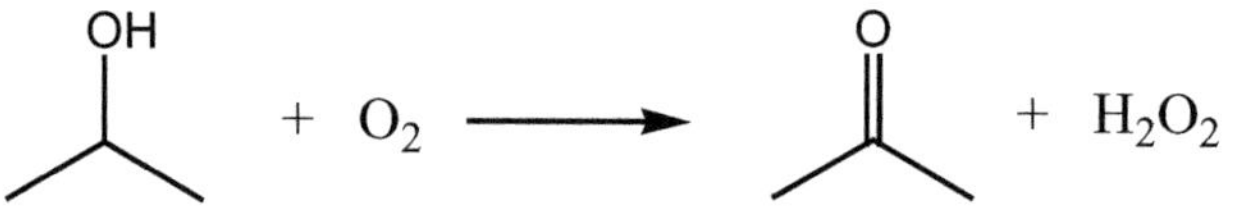

$$\text{(CH}_3\text{)}_2\text{CHOH} + O_2 \longrightarrow \text{(CH}_3\text{)}_2\text{C=O} + H_2O_2$$

Input materials:

500 kg isopropanol
190 cubic meters at STP of oxygen gas
Yield of hydrogen peroxide = 1000 kg of 25 wt% aqueous solution
Yield of acetone = 455 kg

Anthraquinone process

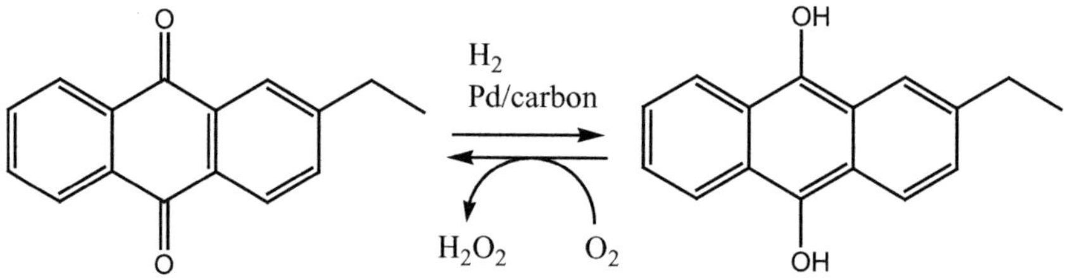

(ethylanthraquinone) + H₂ (Pd/carbon) ⇌ (ethylanthrahydroquinone); then + O₂ → (ethylanthraquinone) + H₂O₂

Input materials:

14 parts ethylanthraquinone
15 parts xylene
45 parts capryl alcohol
40 parts acetophenone

Pd on carbon (4 wt% with respect to ethylanthraquinone); Pd content is 5%
Hydrogenation is undertaken until 80% of the anthraquinone is converted to anthrahydroquinone
Oxidation is carried out with air.

Examples from the Chemical Industry

After 300 hydrogenation-oxidation cycles, a 1 wt% solution of hydrogen peroxide is obtained.

For each of these methods, determine the % yield of reaction, % atom economy, stoichiometric factor, E-factor, reaction mass efficiency, and process mass intensity. State any assumptions made in the calculations.

PROBLEM 7.21[36]

Themes: balancing chemical equations, synthesis plan analysis, synthesis tree diagrams, material efficiency metrics, energy metrics, process metrics.

Ciprofloxacin is a quinolone antibiotic that has been made by a sequential one-pot batch process according to Scheme 7.15. The intermediates are not isolated along the way. The experimental details for each step are also given.

Step 1

35.8 g 3-dimethylamino-acrylic acid ethyl ester and 27.3 g triethylamine are added to 50 mL toluene.

48.6 g 2,4,5-trifluoro-benzoyl chloride is added dropwise over 30 minutes at 50°C. Reaction was continued for 1 h at 50°C to 55°C.

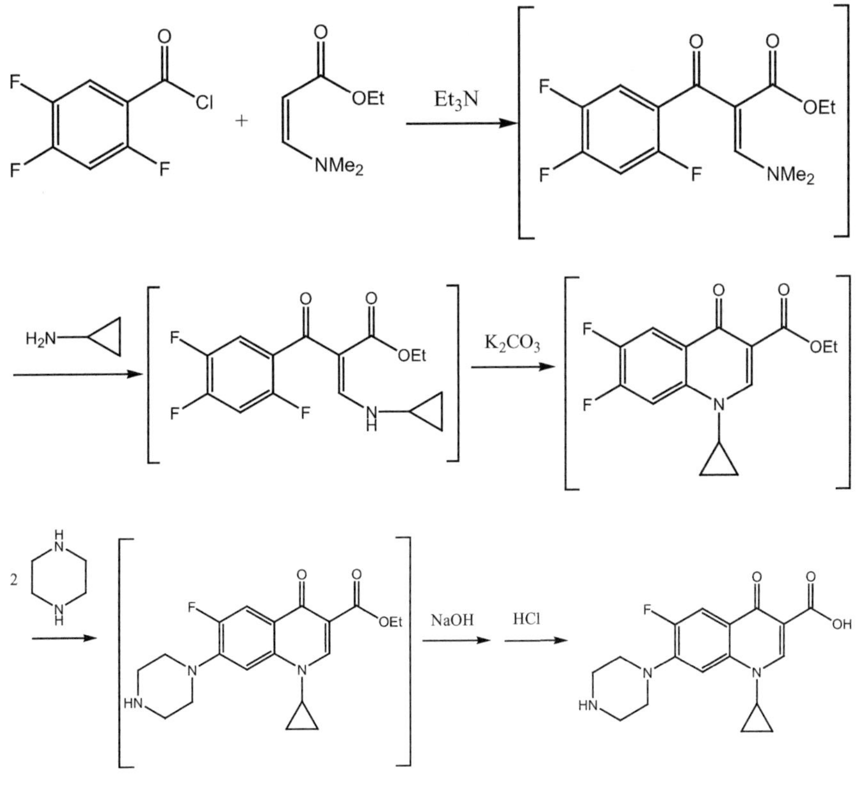

SCHEME 7.15 Process synthesis for ciprofloxacin.

Step 2

17.3 g acetic acid and 15.5 g cyclopropylamine are added dropwise at 30°C to 36°C.

After 1 h the salts were extracted with 100 mL water. The organic phase was then evaporated to yield 80.3 g of a crude oily residue.

Step 3

250 mL N-methylpyrrolidinone was added to the crude residue and 48.4 g potassium carbonated was added. The mixture was warmed to 80°C to 90°C for 1 h.

Step 4

After 1 h 86 g piperazine was added and the mixture heated for 1 h at 80°C to 90°C. Then it was diluted with 150 mL water.

Step 5

20 g sodium hydroxide was added and the mixture stirred for 1 h at 70°C. Then it was diluted with 500 mL water, filtered, and acidified with concentrated HCl solution to pH 7.5. After cooling the solution to 0 to 5°C for 2 h a precipitate formed and this was filtered by suction and washed with 2×200 mL water. Finally, the product was dried overnight under vacuum.

Yield of ciprofloxacin = 71.4 g

Part 1

Balance each chemical equation showing reaction by-products.

Part 2

Draw a synthesis tree for the plan

Part 3

Using the *template-REACTION.xls* and *template-SYNTHESIS.xls* spreadsheets determine the material efficiency metrics for the plan. State any assumptions made in the calculations.

Part 4

For the plan determine the following parameters:

Reaction volume
Process time
Throughput
Space-time-yield

Part 5

Using the *ENERGY-input-analysis-synthesis-plan.xls* spreadsheet and the thermodynamic data given below, determine the energy input required to produce 1 kg of ciprofloxacin by the synthesis scheme under MRT conditions. State any assumptions made in the calculations.

Examples from the Chemical Industry

PROBLEM 7.22[37]

Themes: balancing chemical equations, synthesis plan analysis, synthesis trees, material efficiency metrics, energy metrics, process metrics.

Ciprofloxacin is a quinolone-based antibiotic that has recently been made by a new sequential organic synthesis (SEQUOS) approach using microreactor technology (MRT). Instead of using conventional batch processes solutions containing reagents are flowed through microchannels at given flow rates and residence times in the heated microreactor. This allows for cleaner separation of desired intermediate products in each step and makes it possible to concatenate or telescope sequenced steps without having to isolate intermediates.

The synthetic sequence is shown below in Scheme 7.16 along with experimental data.

Step 1

1.8 g 3-dimethylamino-acrylic acid ethyl ester and 1.35 g triethylamine in 10 mL chloroform (flow rate = 0.12 mL/min)

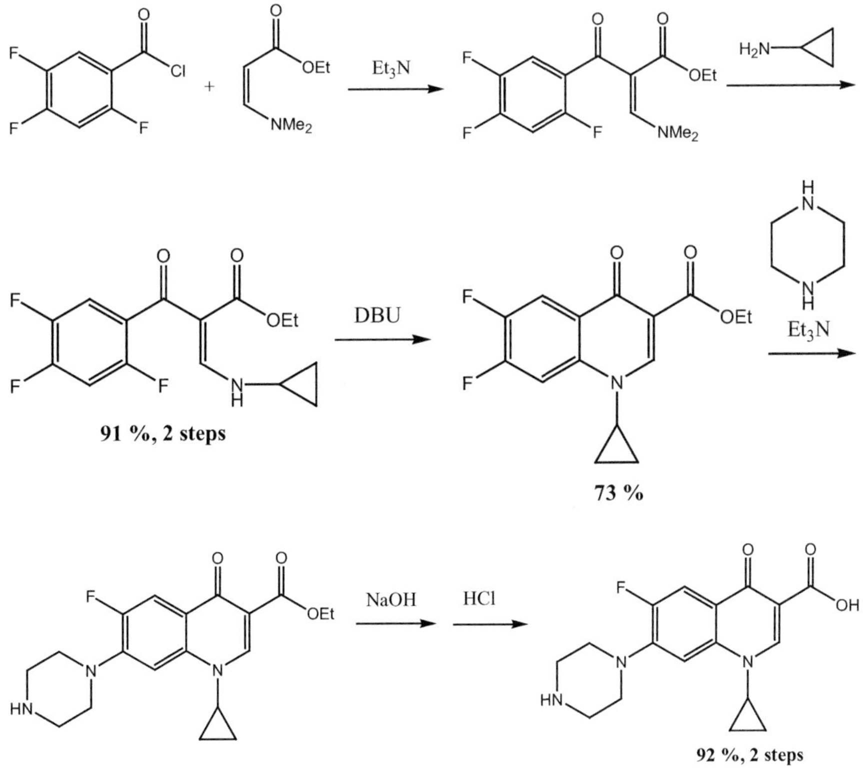

SCHEME 7.16 Synthesis of ciprofloxacin using microreactor technology (MRT).

2.65 g 2,4,5-trifluoro-benzoyl chloride in 10 mL chloroform (flow rate=0.12 mL/min)
Reaction temperature=55°C
50 mL chloroform purge

Step 2
3.18 g 3-dimethylamino-2-(2,4,5-trifluoro-benzoyl)-acrylic acid ethyl ester in 20 mL chloroform (flow rate=0.1 mL/min)
0.85 g cyclopropylamine and 0.9 g acetic acid in 10 mL chloroform (flow rate=0.1 mL/min)
Reaction temperature=35°C
Work-up: water
Yield of intermediate product after steps 1 and 2=3.6 g

Step 3
2 g 3-cyclopropylamino-2-(2,4,5-trifluoro-benzoyl)-acrylic acid ethyl ester in 10 N-methylpyrrolidinone (NMP) (flow rate=0.12 mL/min)
2.3 g DBU in 10 mL NMP (flow rate=0.08 mL/min)
Reaction temperature=120°C
Work-up: water, acetone, dichloromethane purge
Yield of intermediate product=1.37 g

Step 4
2.9 g 1-cyclopropyl-6,7-difluoro-4-oxo-1,4-dihydro-quinoline-3-carboxylic acid ethyl ester in 30 mL NMP (flow rate=0.24 mL/min)
4.0 g piperazine, 1.6 g triethylamine, and 8.0 g t-butanol in 20 mL NMP (flow rate=0.16 mL/min)
Reaction temperature=85°C
Work-up: 10 mL NMP, 50 mL dioxane, 50 mL methanol, and 50 mL dichloromethane purges

Step 5
Purge with 3 wt% sodium hydroxide aqueous solution
3.52 g 1-cyclopropyl-6-fluoro-4-oxo-7-piperazin-1-yl-1,4-dihydro-quinoline-3-carboxylic acid ethyl ester in 75 mL NMP (flow rate=0.76 mL/min)
1.5 g sodium hydroxide in 50 mL water (flow rate=0.48 mL/min)
Reaction temperature=90°C
2 g sodium hydrogen carbonate
concentrated 37 wt% HCl
Purification: washings with water and acetone
Yield of ciprofloxacin after steps 4 and 5=3.01 g

Part 1
Balance each chemical equation showing reaction by-products.

Part 2
Draw a synthesis tree for the plan.

Examples from the Chemical Industry

Part 3

Using the *template-REACTION.xls* and *template-SYNTHESIS.xls* spreadsheet determine the material efficiency metrics for the plan. State any assumptions made in the calculations.

Part 4

For each step determine the following parameters:

Reaction volume
Combined flow rate in microreactor
Process time
Throughput
Determine the overall throughput and overall space-time-yield for the process.

Part 5

Using the *ENERGY-input-analysis-synthesis-plan.xls* spreadsheet and the thermodynamic data given below determine the energy input required to produce 1 kg of ciprofloxacin by the synthesis scheme under MRT conditions. State any assumptions made in the calculations.

Part 6

Compare the performance of the flow MRT procedure with the one-pot batch process.[36]

PROBLEM 7.23[38]

Themes: balancing chemical equations, material efficiency metrics, energy metrics.

Four example reactions are shown below that were run on an industrial scale using continuous flow microwave irradiation. From the data given for each example determine the following parameters: % yield, atom economy, PMI, process time, space-time-yield, and energy consumption (kJ/kg product). State any assumptions made in the calculations.

Example 1

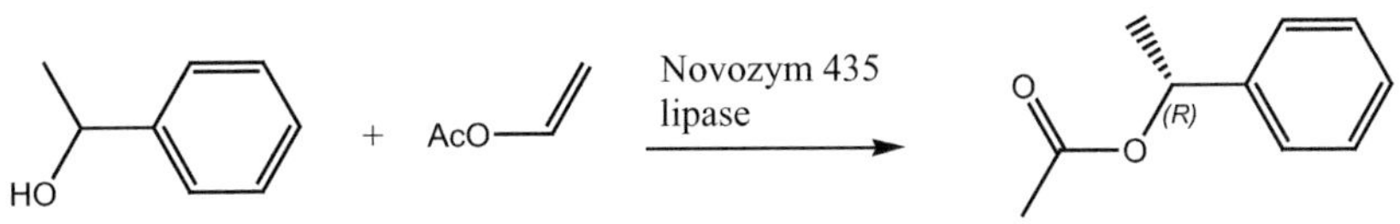

48 g (±)-1-phenylethanol
172.2 vinyl acetate
700 mL toluene
5 g Novozyme 435 lipase
Assumed yield of (R)-1-phenylethyl acetate = 45%
Reaction temperature = 70°C

Flow rate = 175 mL/min
Microwave power = 236 W
Residence time = 180 min

Example 2

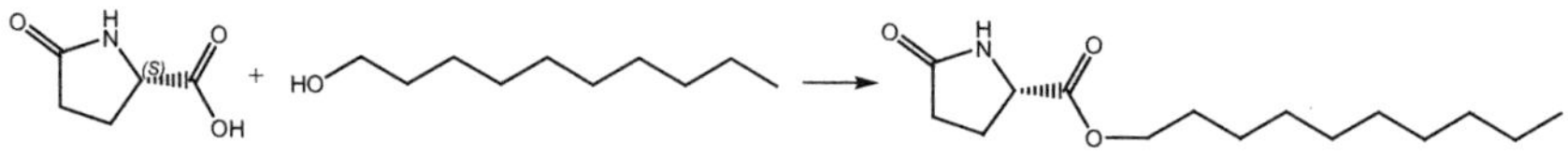

69.1 g (S)-pyroglutamic acid
622 g n-decanol
Assumed yield of laurydone = 80%
Reaction temperature = 150°C
Flow rate = 115 mL/min
Microwave power = 294 W
Residence time = 140 min

Example 3

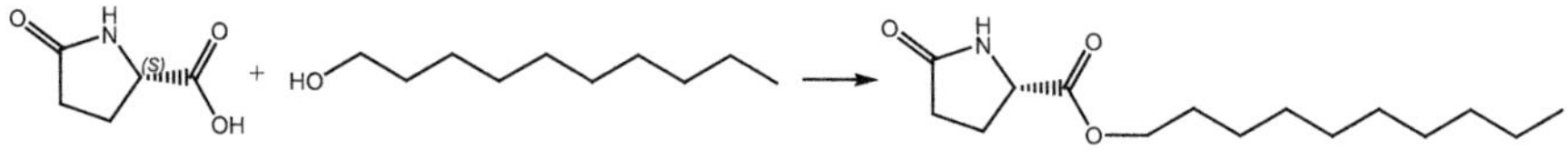

69 g (S)-pyroglutamic acid
622 g n-decanol
Assumed yield of laurydone = 80%
Reaction temperature = 150°C
Flow rate = 343 mL/min
Microwave power = 444 W
Residence time = 178 min

Example 4

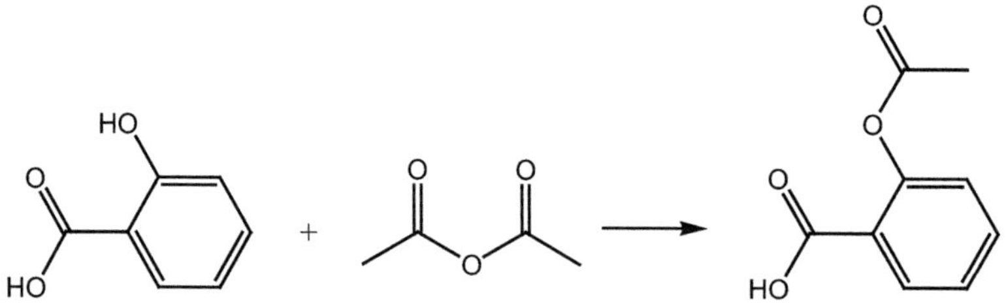

207 g salicylic acid
307 g acetic anhydride

1.12 L acetic acid
Assumed yield of aspirin = 80%
Reaction temperature = 120°C
Flow rate = 27 mL/min
Microwave power = 253 W
Residence time = 6.5 min

PROBLEM 7.24[39]
Themes: balancing chemical equations, synthesis plan analysis, synthesis tree analysis, material efficiency metrics, energy metrics, process metrics.

Meclinertant, developed by Sanofi Pasteur, is a neurotensin receptor antagonist used to treat conditions such as Parkinson's disease in which the neuropeptide, neurotensin, plays a role in its pathogenesis. A traditional batch process synthesis is shown below in Scheme 7.17.

Yields given are those reported by the authors.

Part 1
Balance each chemical equation showing reaction by-products.

Part 2
Draw a synthesis tree for the plan.

Part 3
Using the *template-REACTION.xls* and *template-SYNTHESIS.xls* spreadsheet and the experimental details given in the paper determine the material efficiency metrics for the plan. State any assumptions made in the calculations.

Part 4
For the plan determine the following parameters:

   Overall reaction volume
   Overall process time
   Overall throughput
   Overall space-time-yield

Part 5
Using the *ENERGY-input-analysis-synthesis-plan.xls* spreadsheet and the thermodynamic data given below determine the energy input required to produce 1 kg of meclinertant by the synthesis scheme under batch process conditions. State any assumptions made in the calculations.

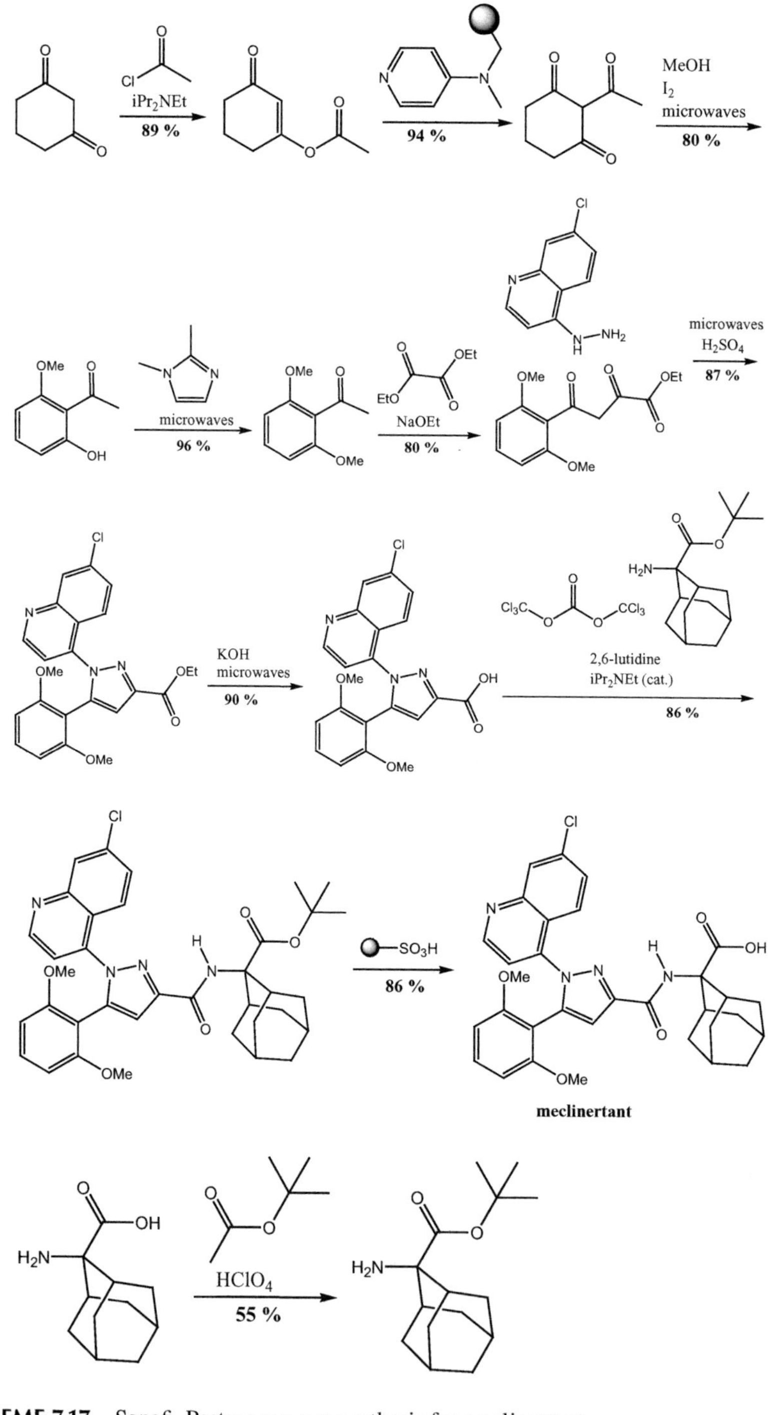

SCHEME 7.17 Sanofi–Pasteur process synthesis for meclinertant.

substance	Formula	MW (g/mol)	Tm (°C)	Tb (°C)	ΔH(fus) kJ/mol	ΔH(vap) kJ/mol
Acetyl chloride	C_2H_3ClO	78.45	−112	52	275.31	30.05
t-Butyl acetate	$C_6H_{12}O_2$	116	25	95	ND	32.58
1,3-Cyclohexanedione	$C_6H_8O_2$	112	103		ND	ND
Dichloromethane	CH_2Cl_2	84.9	−97	40	4.602	28.38
Diethyl oxalate	$C_6H_{10}O_4$	146	−41	185	ND	45.14
Diisopropylethylamine	$C_8H_{19}N$	129	−50	127	ND	ND
Dimethylcarbonate	$C_3H_6O_3$	90	2	90	12	33.72
Dimethylformamide	C_3H_7NO	73	−61	153	16.154	39.41
1,2-dimethylimidazole	$C_5H_8N_2$	96	37	204	ND	ND
Ethanol	C_2H_6O	46	−114.5	78	4.931	39.40
HCl	HCl	36.45		−85	1.998	17.43
Iodine	I_2	254	113	184	15.517	41.54
2,6-Lutidine	C_7H_9N	107	−6	144	10.043	38.35
Methanol	CH_4O	32	−98	64.7	3.205	35.14
Perchloric acid	$HClO_4$	100.45	−101	112	6.933	26.71
Potassium hydroxide	KOH	56	380	1320	8.619	99.37
Tetrahydrofuran	C_4H_8O	72	−108	65	8.54	30.26
Toluene	C_7H_8	93	−95	111	6.636	33.59
Triethylamine	$C_6H_{15}N$	101	−115	89	ND	30.89
Triphosgene	$C_3Cl_6O_3$	297.6	79	203	ND	ND
Water	H_2O	18	0	100	6.002	39.50

ND = no data

Heat capacity coefficients:

Substance	Phase	A	B	C	D	E	Tmin (K)	Tmax (K)
Acetyl chloride	Liquid	55.906	4.43E−01	−1.49E−03	2.24E−06		161	457
t-Butyl acetate	Solid	−34.758	8.08E−01				253	283
1,3-Cyclohexanedione	Solid	ND	ND	ND	ND	ND	ND	ND
Dichloromethane	Liquid	49.637	5.57E−01	−1.25E−03	1.17E−06		280	672
Diethyl oxalate	Liquid	85.853	1.2371	−3.20E−03	3.56E−06		234	581
Diisopropylethylamine	Liquid	ND	ND	ND	ND	ND	ND	ND
Dimethylcarbonate[a]	Liquid	170430					293	361
Dimethylcarbonate[a,b]	Gas	71662	189620	1423.2	122710	671.1	300	1500
Dimethylformamide	Liquid	63.727	6.07E−01	−1.62E−03	1.86E−06		214	582
1,2-dimethylimidazole	Liquid	ND	ND	ND	ND	ND	ND	ND
Ethanol	Liquid	59.342	3.64E−01	−1.22E−03	1.80E−06		160	465
Ethanol	Gas	27.091	1.11E−01	1.10E−04	−1.50E−07	4.66E−11	100	1500
HCl	Gas	29.244	−1.26E−03	1.12E−06	4.97E−09	−2.50E−12	50	1500
Iodine	Solid	45.389	9.60E−03	8.70E−05			100	387
2,6-lutidine[a]	Liquid	95834	298.91				267	417
Methanol	Liquid	40.152	3.10E−01	−1.03E−03	1.46E−06		176	461
Methanol	Gas	40.046	−3.83E−02	2.45E−04	−2.17E−07	5.99E−11	100	1500
Perchloric acid	Liquid	97.789	9.20E−02	−2.87E−04	7.55E−07		172	599
Potassium hydroxide	Solid	50.276	4.42E−02	5.35E−07			250	522
Tetrahydrofuran	Liquid	63.393	4.03E−01	−1.27E−03	1.83E−06		166	486
Tetrahydrofuran	Gas	32.887	2.46E−02	6.02E−04	−6.24E−07	1.85E−10	50	1500
Toluene	Liquid	83.703	5.17E−01	−1.49E−03	1.97E−06		179	533

(continued)

Examples from the Chemical Industry

Substance	Phase	A	B	C	D	E	Tmin (K)	Tmax (K)
Triethylamine	Liquid	114.243	7.06E−01	−2.26E−03	3.32E−06		159	482
Triphosgene	Solid	ND	ND	ND	ND	ND	ND	ND
Water	Liquid	92.053	−4.00E−02	−2.11E−04	5.35E−07		273	615
Water	Gas	33.933	−8.42E−03	2.99E−05	−1.78E−08	3.69E−12	100	1500

ND = No data

a Coefficients correspond to units of J/kmol deg K
b Functional form of temperature dependence of C_p for gas phase of dimethylcarbonate:

$$C_p = A + B\left[\frac{C/T}{\sinh(C/T)}\right]^2 + D\left[\frac{E/T}{\cosh(E/T)}\right]^2 \text{, J/kmol deg K}$$

Functional form of temperature dependence of C_p for all other substances:

$$C_p = A + BT + CT^2 + DT^3 + ET^4 \text{, J/mol deg K}$$

PROBLEM 7.25[39]

Themes: balancing chemical equations, synthesis plan analysis, synthesis tree analysis, material efficiency metrics, energy metrics, process metrics.

Meclinertant, developed by Sanofi Pasteur, is a neurotensin receptor antagonist used to treat conditions such as Parkinson's disease in which the neuropeptide, neurotensin, plays a role in its pathogenesis. Its synthesis by continuous flow microreactor technology (CFMRT) is shown below in Scheme 7.18. In this convergent plan two steps were performed under batch process conditions and three steps were telescoped using flow process conditions.

Yields given are those reported by the authors.

Part 1

Balance each chemical equation showing reaction by-products.

Part 2

Draw a synthesis tree for the plan.

Part 3

Using the *template-REACTION.xls* and *template-SYNTHESIS.xls* spreadsheet and the experimental details given in the paper determine the material efficiency metrics for the plan. State any assumptions made in the calculations.

Part 4

For the plan determine the following parameters:

    Overall reaction volume
    Overall process time
    Overall throughput
    Overall space-time-yield

Part 5

Using the *ENERGY-input-analysis-synthesis-plan.xls* spreadsheet and the thermodynamic data given below determine the energy input required to produce 1 kg of meclinertant by the synthesis scheme under MRT conditions. State any assumptions made in the calculations.

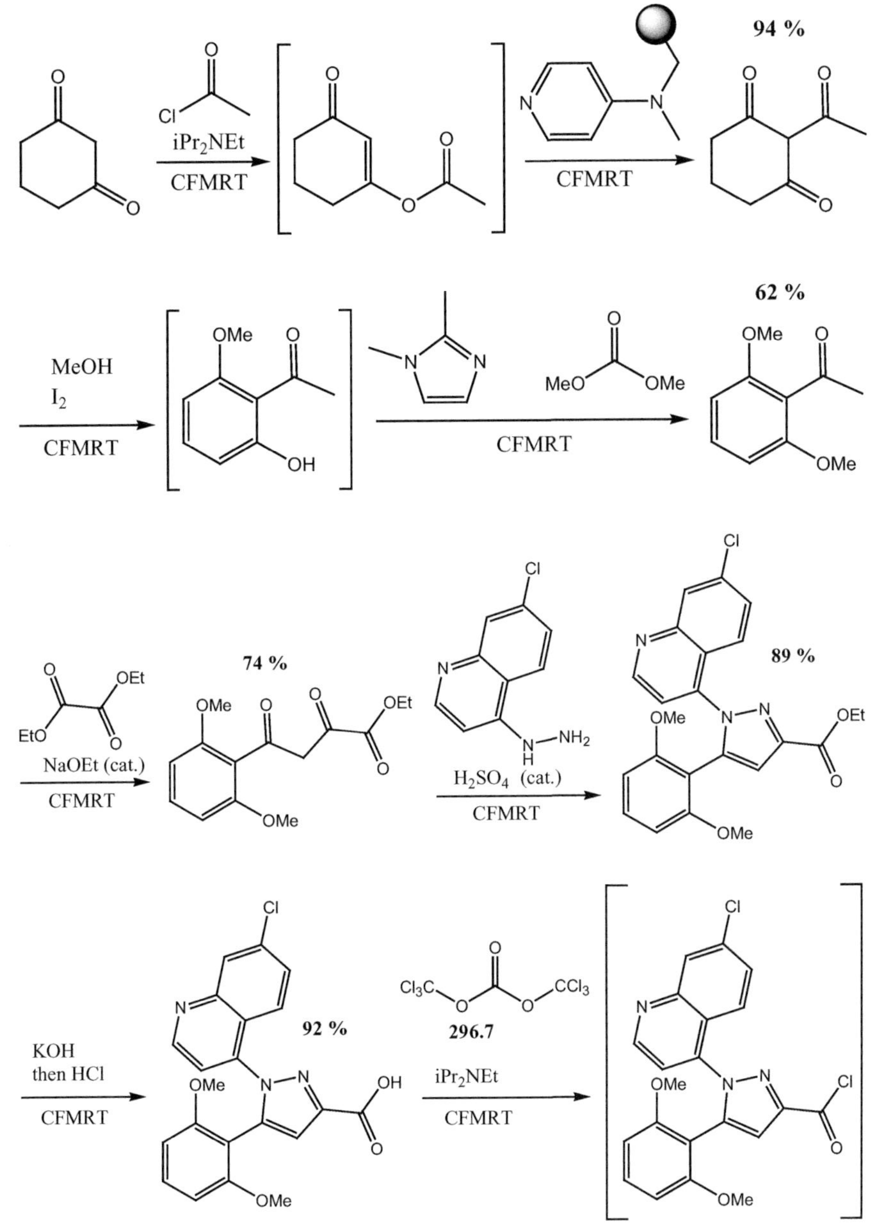

SCHEME 7.18 Synthesis of meclinertant using continuous flow microreactor technology. (CFMRT). (continued)

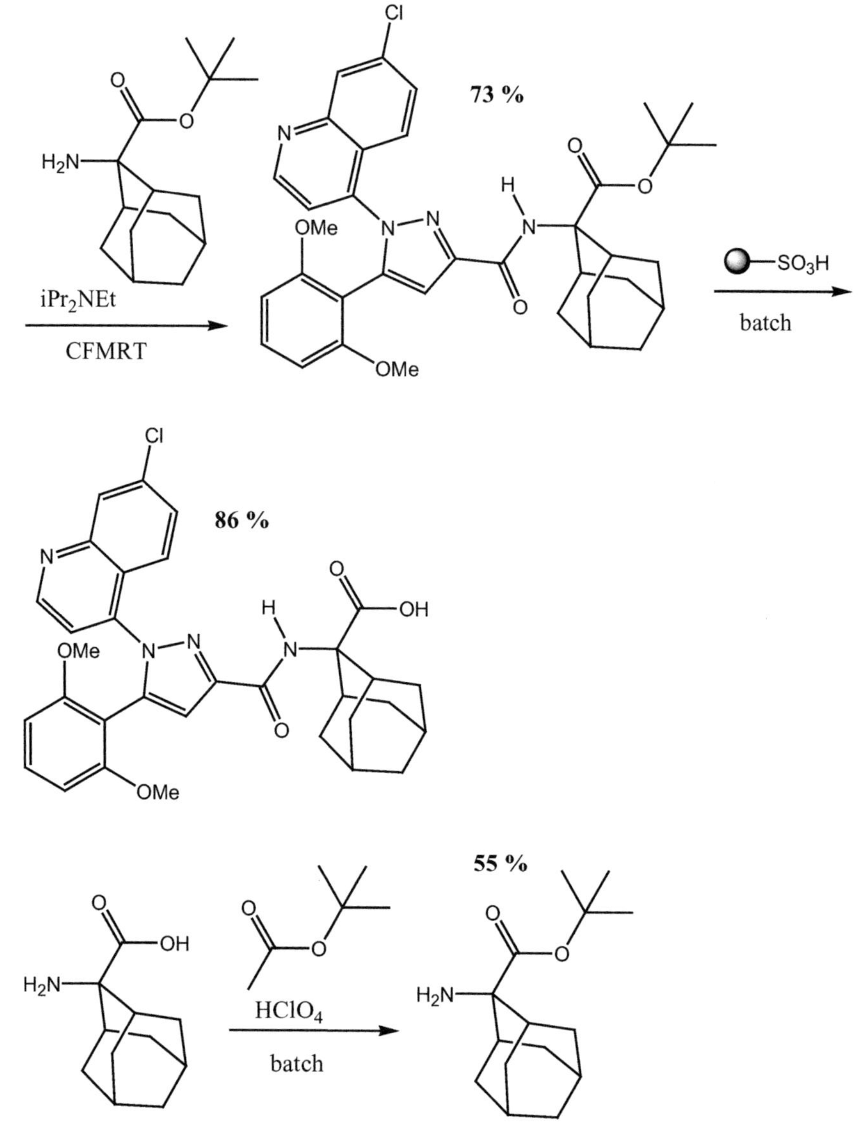

SCHEME 7.18 (CONTINUED) Synthesis of meclinertant using continuous flow microreactor technology. (CFMRT).

Examples from the Chemical Industry

Substance	Formula	MW (g/mol)	Tm (°C)	Tb (°C)	ΔH(fus) kJ/mol	ΔH(vap) kJ/mol
Acetyl chloride	C_2H_3ClO	78.45	−112	52	275.31	30.05
t-Butyl acetate	$C_6H_{12}O_2$	116	25	95	ND	32.58
1,3-Cyclohexanedione	$C_6H_8O_2$	112	103		ND	ND
Dichloromethane	CH_2Cl_2	84.9	−97	40	4.602	28.38
Diethyl oxalate	$C_6H_{10}O_4$	146	−41	185	ND	45.14
Diisopropylethylamine	$C_8H_{19}N$	129	−50	127	ND	ND
Dimethylcarbonate	$C_3H_6O_3$	90	2	90	12	33.72
Dimethylformamide	C_3H_7NO	73	−61	153	16.154	39.41
1,2-dimethylimidazole	$C_5H_8N_2$	96	37	204	ND	ND
Ethanol	C_2H_6O	46	−114.5	78	4.931	39.40
HCl	HCl	36.45		−85	1.998	17.43
Iodine	I_2	254	113	184	15.517	41.54
Methanol	CH_4O	32	−98	64.7	3.205	35.14
Perchloric acid	$HClO_4$	100.45	−101	112	6.933	26.71
Potassium hydroxide	KOH	56	380	1320	8.619	99.37
Sulfuric acid	H_2SO_4	98	10.5	337	10.711	33.28
Tetrahydrofuran	C_4H_8O	72	−108	65	8.54	30.26
Toluene	C_7H_8	93	−95	111	6.636	33.59
Triethylamine	$C_6H_{15}N$	101	−115	89	ND	30.89
Triphosgene	$C_3Cl_6O_3$	297.6	79	203	ND	ND
Water	H_2O	18	0	100	6.002	39.50

ND = no data

Heat capacity coefficients:

Substance	Phase	A	B	C	D	E	Tmin (K)	Tmax (K)
Acetyl chloride	Liquid	55.906	4.43E−01	−1.49E−03	2.24E−06		161	457
t-Butyl acetate	Solid	−34.758	8.08E−01				253	283
1,3-Cyclohexanedione	Solid	ND	ND	ND	ND	ND	ND	ND
Dichloromethane	Liquid	49.637	5.57E−01	−1.25E−03	1.17E−06		280	672
Dichloromethane	Gas	26.694	8.40E−02	8.97E−06	−5.09E−08	1.87E−11	100	1500
Diethyl oxalate	Liquid	85.853	1.2371	−3.20E−03	3.56E−06		234	581
Diisopropylethylamine	Liquid	ND	ND	ND	ND	ND	ND	ND
Dimethylcarbonate[a]	Liquid	170430					293	361
Dimethylcarbonate[a,b]	Gas	71662	189620	1423.2	122710	671.1	300	1500
Dimethylformamide	Liquid	63.727	6.07E−01	−1.62E−03	1.86E−06		214	582
1,2-dimethylimidazole	Liquid	ND	ND	ND	ND	ND	ND	ND
Ethanol	Liquid	59.342	3.64E−01	−1.22E−03	1.80E−06		160	465
Ethanol	Gas	27.091	1.11E−01	1.10E−04	−1.50E−07	4.66E−11	100	1500
HCl	Gas	29.244	−1.26E−03	1.12E−06	4.97E−09	−2.50E−12	50	1500
Iodine	Solid	45.389	9.60E−03	8.70E−05			100	387
Methanol	Liquid	40.152	3.10E−01	−1.03E−03	1.46E−06		176	461
Methanol	Gas	40.046	−3.83E−02	2.45E−04	−2.17E−07	5.99E−11	100	1500
Perchloric acid	Liquid	97.789	9.20E−02	−2.87E−04	7.55E−07		172	599
Potassium hydroxide	Solid	50.276	4.42E−02	5.35E−07			250	522
Sulfuric acid	Liquid	26.004	7.03E−01	−1.39E−03	1.03E−06		298	879
Tetrahydrofuran	Liquid	63.393	4.03E−01	−1.27E−03	1.83E−06		166	486
Tetrahydrofuran	Gas	32.887	2.46E−02	6.02E−04	−6.24E−07	1.85E−10	50	1500

(*continued*)

Examples from the Chemical Industry

Substance	Phase	A	B	C	D	E	Tmin (K)	Tmax (K)
Toluene	Liquid	83.703	5.17E−01	−1.49E−03	1.97E−06		179	533
Triethylamine	Liquid	114.243	7.06E−01	−2.26E−03	3.32E−06		159	482
Triphosgene	Solid	ND	ND	ND	ND	ND	ND	ND
Water	Liquid	92.053	−4.00E−02	−2.11E−04	5.35E−07		273	615
Water	Gas	33.933	−8.42E−03	2.99E−05	−1.78E−08	3.69E−12	100	1500

ND = No data

[a] Coefficients correspond to units of J/kmol deg K
[b] Functional form of temperature dependence of C_p for gas phase of dimethylcarbonate:

$$C_p = A + B\left[\frac{C/T}{\sinh(C/T)}\right]^2 + D\left[\frac{E/T}{\cosh(E/T)}\right]^2, \text{J/kmol deg K}$$

Functional form of temperature dependence of C_p for all other substances:

$$C_p = A + BT + CT^2 + DT^3 + ET^4, \text{J/mol deg K}$$

PROBLEM 7.26[40]
Themes: balancing chemical equations, catalytic cycles, mechanism, selectivity, conversion, reaction yield, turnover number.

Scheme 7.19 shows the conversion of ethanol to various higher alcohols.

Experimental Procedure:
[RuCl$_2$ (η^6-p-cymene)]$_2$ (0.184 g, 0.300 mmol, 0.05 mol%), ligand **L** (0.137 g, 0.599 mmol, 0.1 mol%) and sodium ethoxide (2.039 g, 29.97 mmol, 5 mol%) were added to a clean oven-dried fitted glass insert within a glovebox. The glass insert was then sealed within the autoclave, before removal from the glovebox. Ethanol (35 mL, 599 mmol) was then injected into the autoclave through an inlet against a flow of nitrogen. The autoclave was sealed and placed into the pre-heated (150°C) aluminium heating mantle for 4 h. After the reaction run time, the autoclave was cooled to room temperature in an ice-water bath. The autoclave was vented to remove any gas generated during the reaction. A liquid sample was removed, filtered through a short plug of Celite and analyzed by GC-MS (100 µL of sample, 25 µL of n-pentanol, 1 mL MeOH).

The percent conversion of starting ethanol is 31.4%.

Part 1
Write out a mechanism showing the catalytic cycle for the formation of butanol.

Part 2
Write out a balanced chemical equation for this transformation.

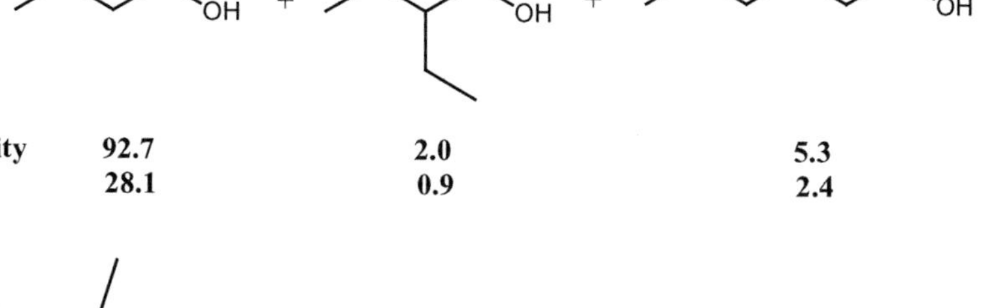

	butanol	2-ethyl-butanol	hexanol
% selectivity	92.7	2.0	5.3
% yield	28.1	0.9	2.4

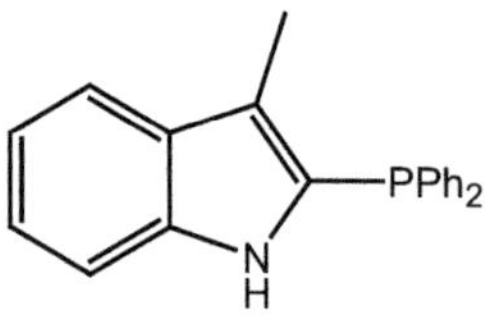

SCHEME 7.19 Synthesis of higher alcohols from ethanol.

Part 3
Write out a balanced chemical equation for the formation of 2-ethylbutanol.
  Write out a balanced chemical equation for the formation of n-hexanol.
  Suggest mechanistically how these products could have been formed.

Part 4
Using the *template-REACTION.xls* spreadsheet and the details of the experimental procedure, determine the material efficiency metrics for the production of each of the three products. State any assumptions made in the calculations.

Part 5
Determine the respective turnover numbers for the production of each of the three products.
  Verify the % conversion of ethanol and % selectivities of products.

PROBLEM 7.27[41]
Themes: balancing chemical equations, material efficiency metrics, experimental procedure errors.

An industrial procedure to manufacture adipic acid calls for the following list of materials to make 1 metric tonne of product. The chemical engineer authors claimed a yield of 90% with respect to cyclohexane.

Material	Mass (kg)
Cyclohexane	800
Nitric acid	1000
Air	600
Ammonium metavanadate	0.25
Copper	0.2
Cobalt naphthenate	0.1

The reaction takes place in two stages. The first stage is cobalt catalyzed air oxidation of cyclohexane to a 1:1 mixture of cyclohexanol and cyclohexanone. This mixture is then further catalytically oxidized to adipic acid in the presence of nitric acid.

Part 1
Suggest mechanisms for the nitric oxidation of cyclohexanol and cyclohexanone to adipic acid. What nitrogen oxide by-products are expected?

Part 2
Write out balanced chemical equations for both oxidation reactions and determine the overall balanced chemical equation for the entire process.

Part 3
Fill in the experimental quantities of reactants in the table below and determine the limiting reagent and the stoichiometric factor.

Reactant	MW (g/mol)	Stoich. coeff.	Mass (g)	Exp. Moles	Stoich. moles	Excess moles
Cyclohexane						
Oxygen						
Nitric acid						
Adipic acid						

Can the quoted % yield be verified? What can you deduce about the quantities of reactants quoted in the experimental procedure by these chemical engineers?

Part 4[42]

Another procedure given in a DuPont patent describing the same kind of chemistry gives the following quantities of materials.

In the first stage, 1997.1 parts cyclohexane, 6 parts cyclohexanone, 1.2 parts cobalt naphthenate are oxidized in a 1 gallon steel autoclave with air at p = 100 psi and T = 145°C for 77 minutes at an air space-velocity of 91 volumes (STP) per volume of charge per h. The reaction yielded 1748.48 parts of unreacted cyclohexane and 170.6 parts of 1:1 mixture of cyclohexanol and cyclohexanone. In the second stage, after stripping off the unreacted cyclohexane, 506.3 parts of this mixture was added gradually to 1490 parts nitric acid, 1010 parts water, 1.25 parts ammonium metavanadate, and 3.75 parts copper in a steel autoclave. After heating for 1 h at 114°C, 466.6 parts adipic acid were recovered.

STP = standard temperature and pressure taken as 0°C and 1 atm

Part A

Write out balanced chemical equations for the first oxidation stage with oxygen.
Write out a balanced chemical equation for the second oxidation stage with nitric acid.

Part B

Using the experimental quantities given fill out the following tables. Interpret the word "parts" as "grams." Determine the % yields of cyclohexanol and cyclohexane with respect to cyclohexane and with respect to oxygen. Determine the % conversion of cyclohexane to cyclohexanol and cyclohexanone. Determine the % yield of adipic acid with respect to cyclohexanol and to cyclohexanone.
1 US gallon = 3.785 L

Reactant	MW (g/mol)	Stoich. coeff.	Mass (g)	Exp. moles	Stoich. moles	Excess moles
Cyclohexane						
Oxygen						
Cyclohexanol						

Examples from the Chemical Industry

Reactant	MW (g/mol)	Stoich. coeff.	Mass (g)	Exp. moles	Stoich. moles	Excess moles
Cyclohexane						
Oxygen						
Cyclohexanone						

Reactant	MW (g/mol)	Stoich. coeff.	Mass (g)	Exp. moles	Stoich. moles	Excess moles
Cyclohexanol						
Cyclohexanone						
Nitric acid						
Adipic acid						

Part C
Determine the overall AE, overall yield, and overall PMI for producing adipic acid from cyclohexane.

PROBLEM 7.28
Themes: reaction mechanism, redox reactions.

In the nitric acid oxidation of cyclohexanol to adipic acid, nitrous acid (HNO_2) and hyponitrous acid (HNO) are reaction by-products (see Problem 7.27). Show, by means of reaction mechanisms, how these can be transformed to nitrogen dioxide (NO_2), nitric oxide (NO), and nitrous oxide (N_2O), respectively.

PROBLEM 7.29

Part 1

A 100 mL round bottom flask has a diameter of 62 mm. Suppose that the flask is half full of a solvent and that the inner wall of the flask is coated with a thin film of catalyst.

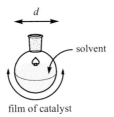

Determine the surface area to volume ratio for the region shown in the diagram. Assume the round bottom flask is a sphere. How does the surface area to volume ratio scale with the diameter of the flask?

Part 2

Suppose the same solvent is flowed through a hollow tube coated with a thin film of catalyst having the same inner diameter of 62 mm and length L mm. Determine the surface area to volume ratio. How does the surface area to volume ratio scale with the diameter of the tube?

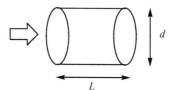

Part 3

If the same volume of liquid in the round bottom flask is flowed through a tube of the same diameter, what should be the length of the tube?

Part 4

If the same area coated with catalyst in the round bottom flask is placed in the tube of the same diameter, what should be the length of the tube?

Part 5

Suppose the 62 mm diameter round bottom flask (RBF) contains a 50 mL solution of reactants dissolved in ethyl acetate solvent. This solution is subjected to microwave irradiation for 10 minutes in order to produce 1 g of reaction product. If we were to flow this amount of solution through a hollow capillary microwave reactor tube of length 100 mm and inner diameter 1.5 mm at a rate of 0.2 mL/min, what should be the residence time of the liquid in the tube for the same reaction to take place with the same yield of product? Determine the space-time-yields for the batch process and the continuous flow process. If the power consumption for both the batch and continuous flow reactions is 100 W, determine the energy consumption of each method in kJ/kg product.

Part 6

Compare the results of the continuous flow single tube conditions in Part 5 with the following "numbering up" approach where the 50 mL of reaction solvent is partitioned and flowed through 10 tubes of identical dimension as before. Assume that these tubes are placed in the same microwave reactor run at the same power.

Part 7

Compare the surface area to volume ratios of the single tube continuous flow operation (Part 5) with the "numbering up" operation involving 10 tubes (Part 6) and the batch operation in a round bottom flask (Part 1).

Examples from the Chemical Industry

PROBLEM 7.30[43]

Themes: balancing chemical equations, synthesis tree diagram, materials efficiency metrics, target bond mapping, synthesis strategy.

The Thai medicinal plant *Pueraria mirifica* has been used in folk medicine in Asia and has been found to contain the phytoestrogen miroestrol. The only reported total synthesis of this compound is by Corey and coworkers using an enantioselective convergent strategy from the starting materials shown below in Scheme 7.20.

Part 1
Look up the synthesis plan in the literature and rewrite it showing balanced chemical equations for each step. Maintain the same structural aspect of starting materials, intermediates, and final product throughout the scheme.

Part 2
Draw a synthesis tree diagram for the plan showing the two convergent branches.

Part 3
Determine the overall kernel material efficiency metrics for the plan.

Part 4
Draw the final structure target bond map highlighting all synthesis bonds made in the plan. Determine the MW fraction of sacrificial reagents.

Part 5
Draw a target bond forming profile for the plan indicating the number of target bonds made per reaction step. Determine the fraction of steps producing no target bonds.

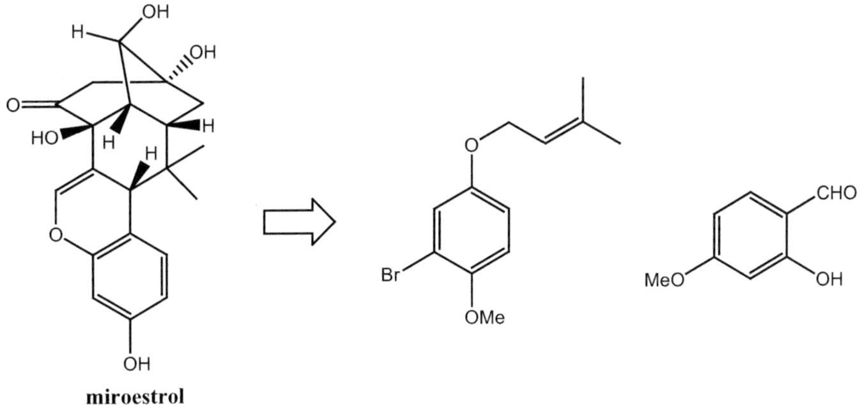

SCHEME 7.20 Corey synthesis of miroestrol.

Part 6

Suggest an alternative strategy to make the intermediate shown below. *Hint*: Apply the ring partitioning strategy on the non-aromatic ring six-membered ring.

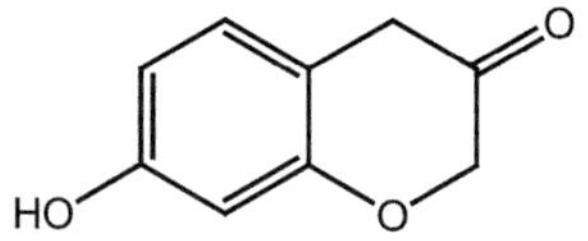

PROBLEM 7.31[44]

Themes: synthesis plan comparison, gate-to-gate, material efficiency, energy intensity, benign indices, safety-hazard indices.

Two industrial plans for the production of 1000 kg of acetic acid are compared.

Plan A

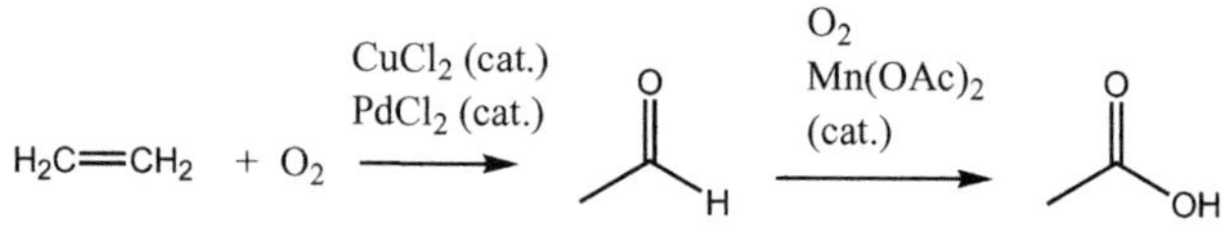

Step 1:

670 kg ethylene
290 cu m oxygen at STP (0°C and 1 atm)
1 mol% copper(II)chloride
1 mol% palladium(II)chloride
300 cu m cooling water
1300 kg steam at p = 1.4 MPa
300 kg steam at p = 460 kPa
1000 kg acetaldehyde
T (rxn) = 100°C
P (rxn) = 700 kPa

Step 2:

1100 kg acetaldehyde
3.3 kg manganese diacetate
1110 cu m air at reaction conditions
1000 kg acetic acid
T (rxn) = 65°C
P (rxn) = 520 kPa

Plan B

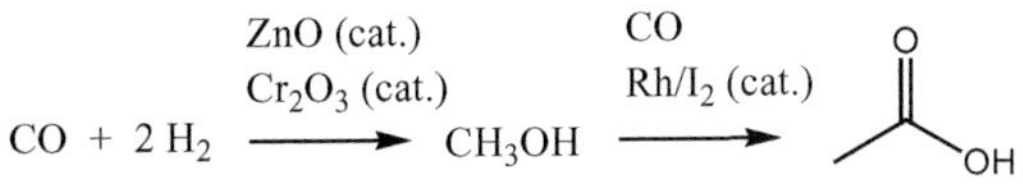

$$CO + 2H_2 \xrightarrow[Cr_2O_3 \text{ (cat.)}]{ZnO \text{ (cat.)}} CH_3OH \xrightarrow{CO, Rh/I_2 \text{ (cat.)}} CH_3COOH$$

Step 1:

1170 cu m carbon monoxide at STP (0°C and 1 atm)
2350 cu m hydrogen at STP (0°C and 1 atm)
1 mol% zinc oxide
1 mol% chromium oxide
1000 kg methanol
T (rxn) = 300°C
P (rxn) = 5000 psi

Step 2:

534 kg methanol
468 kg carbon monoxide
1 mol% rhodium
1 mol% iodine
1000 kg acetic acid
T (rxn) = 175°C
P (rxn) = 1 atm
1 atm = 101325 Pa = 14.696 psi

Part 1
Write out balanced chemical equations for each plan.

Part 2
Using the *template-REACTION.xls* and *template-SYNTHESIS.xls* spreadsheets determine the material efficiency metrics for the production of 1000 kg acetic acid for each plan. State any assumptions made in the calculations.

Part 3
For each plan write out a tree diagram showing the scaled masses of materials needed to produce 1000 kg of acetic acid. Summarize the results by showing a mass balance of input and output materials.

Part 4
For each plan determine the energy input requirements to produce 1000 kg of acetic acid using the *energy-input-synthesis-plans.xls* spreadsheet and the thermodynamic data given below.

Plan A

| | Phase | Tb (K) | Coefficients for Cp[a] | | | | | Tc(K) | Pc(atm) | Cp (J/mol K) |
			A	B	C	D	E			
Acetaldehyde	Liquid	292	152.99	598.64	−0.89481			455	54.97	
Acetaldehyde	Gas		48251	106650	1992.9	78851	912.78			
Oxygen	Gas	90.17	2.91E+04	1.00E+04	2.53E+03	9.36E+03	1.15E+03	154.58	49.77	
Mn(OAc)2	Solid									NA
Ethylene	Gas	169.5	3.34E+04	9.48E+04	1.60E+03	5.51E+04	7.41E+02	282.36	49.66	
CuCl2	Solid									71.9
PdCl2	Solid									NA
Water	Liquid	373	2.76E+05	−2.1E+03	8.13E+00	−1.41E−02	9.37E−06	647	218	
Water	Gas		3.34E+04	2.68E+04	2.61E+03	8.89E+03	1.17E+03			

Enthalpy of vaporization at boiling point for acetaldehyde = 23.99 kJ/mol
NA = not available
[a] Cp coefficients correspond to units of J/K kmol for heat capacity at constant pressure unless otherwise specified

Plan B

Coefficients for C_p[a]

	Phase	Tb (K)	A	B	C	D	E	Tc(K)	Pc(atm)	Cp (J/mol K)
Methanol	Liquid	337.9	1.08E+05	−3.8E+02	9.79E−01			512.58	79.9	
Methanol	Gas		3.93E+04	8.89E+04	1.96E+03	5.56E+04	9.10E+02			
CO	Gas	81.7	2.91E+04	8.77E+03	3.09E+03	8.46E+03	1.54E+03	132.92	34.53	
Rhodium[b]	Solid	3970	4.80E+01	−4.44E−03	8.78E−07	−3.54E−11				
Iodine	Solid	457	34122	134.16	−0.01997	−0.0018	3.8E−06	819	115.5	
Iodine[b]	Liquid		314.003	−1.3711	−2.57E−03	−1.5E−06				
Hydrogen	Gas	20.39	2.76E+04	9.56E+03	2.47E+03	3.76E+03	5.68E+02	132.92	34.53	
ZnO	Solid									40.3
Cr_2O_3	Solid									118.7

Enthalpy of vaporization at boiling point for methanol = 35.21 kJ/mol
Enthalpy of fusion at melting point for iodine = 15.517 kJ/mol; Tm = 387 K

[a] Cp coefficients correspond to units of J/K kmol for heat capacity at constant pressure unless otherwise specified
[b] Cp coefficients correspond to units of J/K mol for heat capacity at constant pressure

Redlich–Kwong equation of state parameters for gases:

Input Chemical	a (L^2 atm deg $K^{0.5}$/mol^2)	b (L/mol)
Carbon monoxide	16.9855	0.02739
Ethylene	77.6795	0.04046
Hydrogen	1.434	0.01848
Oxygen	17.1876	0.0221

Tait equation of state parameters for liquids:

Input Chemical	Acetaldehyde	Water
MW (g/mol)	44	18
ω	0.291	0.365
Tc (K)	455	647.13
Pc (Pa)	5.57E+06	2.21E+07
Lambda	8.11E–04	5.72E–04
m	–0.7224	–0.7143
A11	0.2806	0.3471
B21	0.25809	0.274
n	0.2776	0.28571
A12	87.3702	29.8605
B22	–3682.2	–3152.2
C32	–31.548	–7.3037
D42	2.01E–02	2.42E–09
E52	5.53E–13	1.81E–06

Part 5

Which plan has the advantage with respect to material efficiency metrics?
Which plan has the advantage with respect to energy intensity metrics?

Part 6

For each plan determine the benign indices for input and waste materials.
Which plan is comparatively more benign? Which chemicals have the most impact?

Examples from the Chemical Industry

Plan A

Input Materials:

Substance	Mass (g)	MW (g/mol)	No. Carbon Atoms	Alpha No. Acidic Hs	ABP	ODP	SFP
$CuCl_2$	3.54E+04	134.45	0	0	5.52		
Ethylene	7.37E+05	28	2	0	0		2.39
$Mn(OAc)_2$	3.30E+03	172.94	4	0	4.57		
Oxygen	1.84E+06	32	0	0	0		
$PdCl_2$	2.52E+04	177.3	0	0	0.72		
Water (cooling)	3.30E+08	18	0	0	0		
Water (steam)	1.76E+06	18	0	0	0		

Substance	LD50 (mg/kg)	LC50 (g/m³ 4 h)	H (atm m³/mol)	log Kow	ADP
$CuCl_2$	140		1.0E–100	–1E+100	1.94E–03
Ethylene		0.05	0.228	1.13	0
$Mn(OAc)_2$	2940		1.0E–100	–1E+100	1.38E–05
Oxygen			0.769	0.65	0
$PdCl_2$	2704		1.0E–100	–1E+100	0.323
Water (cooling)	90000		8.48E–09	–1.38	0
Water (steam)	90000		8.48E–09	–1.38	0

Output Materials:

Substance	Mass (g)	MW (g/mol)	No. Carbon Atoms	Alpha No. Acidic Hs	ABP	ODP	SFP
Acetaldehyde (unreacted)	3.67E+05	44	2	0	0		1.67
Acetic acid	1.00E+06	60	2	1	9.842		
$CuCl_2$	3.54E+04	134.45	0	0	5.52		
Ethylene (unreacted)	3.70E+04	28	2	0	0		2.39
$Mn(OAc)_2$	3.30E+03	172.94	4	0	4.57		
Oxygen (unreacted)	1.17E+06	32	0	0	0		
$PdCl_2$	2.52E+04	177.3	0	0	0.72		
Water (cooling)	3.30E+08	18	0	0	0		
Water (steam)	1.76E+06	18	0	0	0		

Substance	LD50 (mg/kg)	LC50 (g/m³ 4 h)	H (atm m³/mol)	log Kow	ADP
Acetaldehyde (unreacted)		23.4	6.67E–05	–0.34	0
Acetic acid	3310	39.3	1.00E–07	–0.17	0
CuCl$_2$	140		1.0E–100	–1E+100	1.94E–03
Ethylene (unreacted)		0.05	0.228	1.13	0
Mn(OAc)$_2$	2940		1.0E–100	–1E+100	1.38E–05
Oxygen (unreacted)			0.769	0.65	0
PdCl$_2$	2704		1.0E–100	–1E+100	0.323
Water (cooling)	90000		8.48E–09	–1.38	0
Water (steam)	90000		8.48E–09	–1.38	0

Plan B
Input Materials:

Substance	Mass (g)	MW (g/mol)	No. Carbon Atoms	Alpha No. Acidic Hs	ABP	ODP	SFP
Carbon monoxide	1.25E+06	28	1	0	0		0.029
Cr$_2$O$_3$	4.24E+04	151.99	0	0	24.08		
Hydrogen	1.12E+05	2	0	0	0		
Iodine	4.25E+04	254	0	0	0		
Rh	1.72E+04	102.91	0	0	0		
ZnO	2.27E+04	81.37	0	0	24		

Substance	LD50 (mg/kg)	LC50 (g/m³ 4 h)	H (atm m³/mol)	log Kow	ADP
Carbon monoxide		1.9	1.09	–1E+100	0
Cr$_2$O$_3$	621		1.0E–100	–1E+100	0.001716
Hydrogen			1.28	0.45	0
Iodine	14	0.36	3.13E–04	2.49	0.0427
Rh			1.0E–100	–1E+100	32.3
ZnO	8437	2.5	1.0E–100	–1E+100	9.92E–04

Examples from the Chemical Industry

Output Materials:

Substance	Mass (g)	MW (g/mol)	No. Carbon Atoms	Alpha No. Acidic Hs	ABP	ODP	SFP
Acetic acid	1.00E+06	60	2	1	9.842		
CO (unreacted)	3.15E+05	28	1	0	0		0.029
Cr_2O_3	4.24E+04	151.99	0	0	24.08		
H_2 (unreacted)	4.50E+04	2	0	0	0		
Iodine	4.25E+04	254	0	0	0		
Methanol (unreacted)	7.00E+02	32	1	0	−1.2		0.2
Rh	1.72E+04	102.91	0	0	0		
ZnO	2.27E+04	81.37	0	0	24		

Substance	LD50 (mg/kg)	LC50 (g/m³ 4 h)	H (atm m³/mol)	log Kow	ADP
Acetic acid	3310	39.3	1.00E−07	−0.17	0
CO (unreacted)		1.9	1.09	−1E+100	0
Cr_2O_3	621		1.0E−100	−1E+100	0.001716
H_2 (unreacted)			1.28	0.45	0
Iodine	14	0.36	3.13E−04	2.49	0.0427
Methanol (unreacted)	5600	84	4.55E−06	−0.77	0
Rh			1.0E−100	−1E+100	32.3
ZnO	8437	2.5	1.0E−100	−1E+100	9.92E−04

Part 7

For each plan determine the safety-hazard indices for input and waste materials. Which plan is comparatively safer? Which chemicals have the most impact?

Plan A

Input Materials:

Substance	Mass (g)	MW (g/mol)	O Consumption/Liberation rxn	No. O substrate	No. O products
$CuCl_2$	3.54E+04	134.45	$CuCl_2 = Cu + Cl_2$	0	0
Ethylene	7.37E+05	28	$C_2H_4 + 6\,[O] = 2\,CO_2 + 2\,H_2O$	0	6
$Mn(OAc)_2$	3.30E+03	172.94	$Mn(OAc)_2 + 7\,[O] = Mn + 4\,CO_2 + 3\,H_2O$	4	7
Oxygen	1.84E+06	32	$O_2 = O_2$	2	2
$PdCl_2$	2.52E+04	177.3	$PdCl_2 = Pd + Cl_2$	0	0
Water (cooling)	3.30E+08	18	$H_2O = H_2O$	1	1
Water (steam)	1.76E+06	18	$H_2O = H_2O$	1	1

Substance	LD50 Dermal (mg/kg)	LC50 (g/m³ 4 h)	FLP (°C)	Moles H₂ Generated	LEL	IS (N m)	LBT (cm³/10 g)	OEL (mmol/m³)
CuCl₂				0		1.0E+100	0	2.82E–06
Ethylene		0.05	–136.6	0	2.7	1.0E+100	0	8.2
Mn(OAc)₂				0		1.0E+100	0	6.06E–05
Oxygen			–52	0		1.0E+100	0	
PdCl₂				0		1.0E+100	0	
Water (cooling)				0		1.0E+100	0	
Water (steam)				0		1.0E+100	0	

Substance	SD (mg)	Q (R-Phrases)
CuCl₂	5.28E–01	6000
Ethylene	1.70E–01	48000
Mn(OAc)₂	0	300
Oxygen	2.31E–02	64000
PdCl₂	5.85E–03	15
Water (cooling)	0.00E+00	
Water (steam)		

Output Materials:

Substance	Mass (g)	MW (g/mol)	O Consumption/ Liberation rxn	No. O Substrate	No. O Products
Acetaldehyde (unreacted)	3.67E+05	44	$C_2H_4O + 5\,[O] = 2\,CO_2 + 2\,H_2O$	1	6
Acetic acid	1.00E+06	60	$C_2H_4O_2 + 4\,[O] = 2\,CO_2 + 2\,H_2O$	2	6
CuCl₂	3.54E+04	134.45	$CuCl_2 = CuCl_2$	0	0
Ethylene (unreacted)	3.70E+04	28	$C_2H_4 + 6\,[O] = 2\,CO_2 + 2\,H_2O$	0	6
Mn(OAc)2	3.30E+03	172.94	$Mn(OAc)_2 + 7\,[O] = Mn + 4\,CO_2 + 3\,H_2O$	4	7
Oxygen (unreacted)	1.17E+06	32	$O_2 = O_2$	2	2
PdCl2	2.52E+04	177.3	$PdCl_2 = Pd + Cl_2$	0	0
Water (cooling)	3.30E+08	18	$H_2O = H_2O$	1	1
Water (steam)	1.76E+06	18	$H_2O = H_2O$	1	1

Examples from the Chemical Industry

Substance	LD50 Dermal (mg/kg)	LC50 (g/m³) 4 h	FLP (°C)	Moles H$_2$ Generated	LEL	IS (N m)	LBT (cm³/10 g)	OEL (mmol/m³)
Acetaldehyde (unreacted)		23.4	−17	0	4	1.0E+100	0	8.16
Acetic acid		39.3	39	0	4	1.0E+100	0	0.42
CuCl$_2$				0		1.0E+100	0	0.000744
Ethylene (unreacted)		0.05	−136.6	0	2.7	1.0E+100	0	8.2
Mn(OAc)$_2$				0		1.0E+100	0	0.001156
Oxygen (unreacted)			−52	0		1.0E+100	0	
PdCl$_2$				0		1.0E+100	0	
Water (cooling)				0		1.0E+100	0	
Water (steam)				0		1.0E+100	0	

Substance	SD (mg)	Q (R-Phrases)
Acetaldehyde (unreacted)	1.01E+02	13500
Acetic acid	8.12E+01	300
CuCl$_2$	5.28E−01	6000
Ethylene (unreacted)	1.70E−01	48000
Mn(OAc)$_2$	0.00E+00	300
Oxygen (unreacted)	2.31E−02	64000
PdCl$_2$	5.85E−03	15
Water (cooling)	0.00E+00	
Water (steam)	0.00E+00	

Plan B
Input Materials:

Substance	Mass (g)	MW (g/mol)	O Consumption/ Liberation rxn	No. O Substrate	No. O Products
Carbon monoxide	1.25E+06	28	CO + [O] = CO$_2$	1	2
Cr$_2$O$_3$	4.24E+04	151.99	Cr$_2$O$_3$ = 2 Cr + 3 [O]	3	0
Hydrogen	1.12E+05	2	H$_2$ = H$_2$	0	0
Iodine	4.25E+04	254	I$_2$ = I$_2$	0	0
Rh	1.72E+04	102.91	Rh = Rh	0	0
ZnO	2.27E+04	81.37	ZnO = Zn + [O]	1	0

Substance	LD50 Dermal (mg/kg)	LC50 (g/m³ 4 h)	FLP (°C)	Moles H$_2$ Generated	LEL	IS (N m)	LBT (cm³/ 10 g)	OEL (mmol/ m³)
Carbon monoxide		1.9		0	6.3	1.0E+100	0	1.4
Cr$_2$O$_3$				0		1.0E+100	0	0.00329
Hydrogen				0	4	1.0E+100	0	0.41
Iodine		0.36		0		1.0E+100	0	0.004
Rh				0		1.0E+100	0	0.000972
ZnO		2.5		0		1.0E+100	0	0.061

Substance	SD (mg)	Q (R-Phrases)
Carbon monoxide	3.91E−05	14000000
Cr$_2$O$_3$	0	450
Hydrogen	3.90E−03	48000
Iodine	3.45E−02	300
Rh	0.00E+00	10
ZnO	0.00E+00	30

Output Materials:

Substance	Mass (g)	MW (g/mol)	O Consumption/ Liberation rxn	No. O Substrate	No. O Products
Acetic acid	1.00E+06	60	C$_2$H$_4$O$_2$+4 [O]=2 CO$_2$+2 H$_2$O	2	6
CO (unreacted)	3.15E+05	28	CO+[O]=CO$_2$	1	2
Cr$_2$O$_3$	4.24E+04	151.99	Cr$_2$O$_3$=2 Cr+3 [O]	3	0
H$_2$ (unreacted)	4.50E+04	2	H$_2$=H$_2$	0	0
Iodine	4.25E+04	254	I$_2$=I$_2$	0	0
Methanol (unreacted)	7.00E+02	32	CH$_4$O+3 [O]=CO$_2$+2 H$_2$O	1	4
Rh	1.72E+04	102.91	Rh=Rh	0	0
ZnO	2.27E+04	81.37	ZnO=Zn+[O]	1	0

Examples from the Chemical Industry

Substance	LD50 Dermal (mg/kg)	LC50 (g/m³ 4 h)	FLP (°C)	Moles H₂ Generated	LEL	IS (N m)	LBT (cm³/10 g)	OEL (mmol/m³)
Acetic acid		39.3	39	0	4	1.0E+100	0	0.42
CO (unreacted)		1.9		0	6.3	1.0E+100	0	1.4
Cr₂O₃				0		1.0E+100	0	0.00329
H₂ (unreacted)				0	4	1.0E+100	0	0.41
Iodine		0.36		0		1.0E+100	0	0.004
Methanol (unreacted)	15800	84	11	0	6	1.0E+100	0	3.1
Rh				0		1.0E+100	0	0.000972
ZnO		2.5		0		1.0E+100	0	0.061

Substance	SD (mg)	Q (R-Phrases)
Acetic acid	8.12E+01	300
CO (unreacted)	3.91E−05	14000000
Cr₂O₃	0	450
H₂ (unreacted)	3.90E−03	48000
Iodine	3.33E−02	300
Methanol (unreacted)	6.57E+01	612500
Rh	0.00E+00	10
ZnO	0.00E+00	30

PROBLEM 7.32[45]

Themes: synthesis plan comparison, gate-to-gate, material efficiency, energy intensity, benign indices, safety-hazard indices.

Two synthesis plans to manufacture 1000 kg of ethyl acrylate are compared.

Plan A (Reppe process)

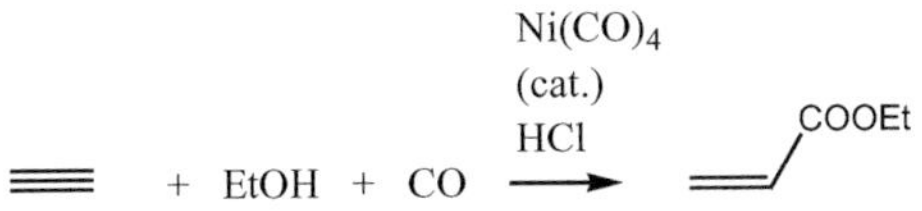

$$\equiv \;+\; EtOH \;+\; CO \xrightarrow[HCl]{Ni(CO)_4 \;(cat.)} \diagup\!\!\!= \text{COOEt}$$

260 kg acetylene
460 kg ethanol
560 kg carbon monoxide
85 kg nickel carbonyl
183 kg hydrogen chloride
800 kg ethyl acrylate
T (rxn) = 50°C
P (rxn) = 2 atm

Plan B (propylene oxidation process)

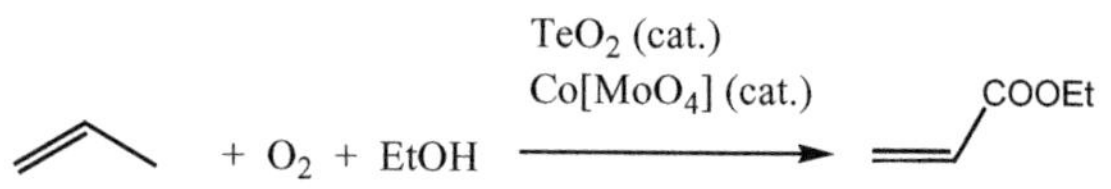

$$\text{CH}_2=\text{CHCH}_3 + \text{O}_2 + \text{EtOH} \xrightarrow[\text{Co[MoO}_4\text{] (cat.)}]{\text{TeO}_2 \text{ (cat.)}} \text{CH}_2=\text{CHCOOEt}$$

420 kg propylene
480 kg oxygen
460 kg ethanol
1 mol% tellurium(IV)oxide
1 mol% cobalt(II)molybdate
1000 ethyl acrylate
T (rxn) = 500°C
P (rxn) = 1 atm

Part 1
Write out balanced chemical equations for each plan.

Part 2
Using the *template-REACTION.xls* and *template-SYNTHESIS.xls* spreadsheets, determine the material efficiency metrics for the production of 1000 kg ethyl acrylate for each plan. State any assumptions made in the calculations.

Part 3
For each plan write out a tree diagram showing the scaled masses of materials needed to produce 1000 kg of ethyl acrylate. Summarize the results by showing a mass balance of input and output materials.

Part 4
For each plan determine the energy input requirements to produce 1000 kg of ethyl acrylate using the energy-input-synthesis-plans-plans spreadsheet and the thermodynamic data given below.

Plan A

	Phase	Tb (K)	Coefficients for Cp[a]					Tc(K)	Pc(atm)
			A	B	C	D	E		
Acetylene	Gas	189.15	3.68E+04	4.83E+04	1.80E+03	3.72E+04	7.08E+02	308.32	60.59
Ethanol	Liquid	351.44	102640	−139.63	−0.03034	0.002039		514	60.57
CO	Gas	81.7	2.91E+04	8.77E+03	3.09E+03	8.46E+03	1.54E+03	132.92	34.53
Ni(CO)$_4$	Liquid	316	734.18	−4.682	1.30E−02	−1.11E−05		508.4	32
Ni(CO)$_4$[b]	Gas		2.5804735	1.088934	−3.16E−03	5.02E−06	−4.34E−09		
HCl	Gas	188.15	2.92E+04	9.05E+03	2.09E+03	−1.1E+02	1.20E+02	324.65	82

Heat of vaporization at boiling point for nickel tetracarbonyl = 29.3 kJ/mol

[a] Cp coefficients correspond to units of J/K kmol for heat capacity at constant pressure unless otherwise specified
[b] Cp coefficients correspond to units of J/K mol for heat capacity at constant pressure; polynomial function used for nickel tetracarbonyl in gas phase

Plan B

	Phase	Tb (K)	Coefficients for Cp[a]					Tc(K)	Pc(atm)
			A	B	C	D	E		
Propylene	Gas	225.43	4.13E+04	1.53E+05	1.35E+03	7.44E+04	5.78E+02	364.76	45.52
Oxygen	Gas	90.17	2.91E+04	1.00E+04	2.53E+03	9.36E+03	1.15E+03	154.58	49.77
Ethanol	Liquid	351.44	102640	−139.63	−0.03034	0.002039		514	60.57
Ethanol	Gas		49200	145770	1662.8	93900	744.7		
TeO$_2$	Solid								
Co[MoO$_4$]	Solid								

Enthalpy of vaporization at boiling point for ethanol = 39.4 kJ/mol
Missing data for tellurium(IV)oxide and cobalt(II)molybdate.
[a] Cp coefficients correspond to units of J/K kmol for heat capacity at constant pressure unless otherwise specified

Examples from the Chemical Industry

Part 5
Which plan has the advantage with respect to material efficiency metrics?
Which plan has the advantage with respect to energy intensity metrics?

Part 6
For each plan determine the benign indices for input and waste materials.
Which plan is comparatively more benign? Which potential and chemicals have the most impact?

Plan A
Input Materials:

Substance	Mass (g)	MW (g/mol)	No. Carbon Atoms	Alpha No. Acidic Hs	ABP	ODP	SFP
Acetylene	3.25E+05	26	2	0	0		0.16
Carbon monoxide	7.00E+05	46	1	0	0		0.029
Ethanol	5.75E+05	28	2	0	−1.2		0.42
HCl	228750	170.7	0	1	20.851		
Ni(CO)$_4$	106250	36.45	4	0	0		

Substance	LD50 (mg/kg)	LC50 (g/m³ 4 h)	H (atm m³/mol)	log Kow	ADP
Acetylene		47.9	2.17E−02	0.37	0
Carbon monoxide		1.9	1.09	−1E+100	0
Ethanol	7060	94.1	5.00E−06	−0.3	0
HCl	900	0.41	5.26E−05	1.1	4.86E−08
Ni(CO)$_4$		0.0815	5.00E−01	5.23	1.08E−04

Output Materials:

Substance	Mass (g)	MW (g/mol)	No. Carbon Atoms	Alpha No. Acidic Hs	ABP	ODP	SFP
Acetylene (unreacted)	6.50E+04	26	2	0	0		0.16
CO (unreacted)	4.20E+05	28	1	0	0		0.029
Ethanol (unreacted)	1.15E+05	46	2	0	−1.2		0.42
Ethyl acrylate	1.00E+06	100	5	0	0		
HCl	228750	36.45	0	1	20.851		
Ni(CO)$_4$	106250	170.7	4	0	0		

Substance	LD50 (mg/kg)	LC50 (g/m³ 4 h)	H (atm m³/mol)	log Kow	ADP
Acetylene (unreacted)		47.9	2.17E–02	0.37	0
CO (unreacted)		1.9	1.09	–1E+100	0
Ethanol (unreacted)	7060	94.1	5.00E–06	–0.3	0
Ethyl acrylate	800	6	3.93E–04	1.32	0
HCl	900	0.41	5.26E–05	1.1	4.86E–08
Ni(CO)$_4$		0.0815	5.00E–01	5.23	1.08E–04

Plan B
Input Materials:

Substance	Mass (g)	MW (g/mol)	No. Carbon Atoms	Alpha No. Acidic Hs	ABP	ODP	SFP
Co[MoO$_4$]	3.28E+04	218.87	0	0	9.55		
Ethanol	4.60E+05	46	2	0	–1.2		0.42
Oxygen	4.80E+05	32	0	0	0		
Propylene	4.20E+05	42	3	0	0		3.03
TeO$_2$	1.60E+04	159.6	0	0			

Substance	LD50 (mg/kg)	LC50 (g/m³ 4 h)	H (atm m³/mol)	log Kow	ADP
Co[MoO$_4$]			1.0E–100	–1E+100	3.17E–02
Ethanol	7060	94.1	5.00E–06	–0.3	0
Oxygen			0.769	0.65	0
Propylene		18.04	0.196	1.77	0
TeO$_2$	5000		1.0E–100	–1E+100	52.8

Output Materials:

Substance	Mass (g)	MW (g/mol)	No. Carbon Atoms	Alpha No. Acidic Hs	ABP	ODP	SFP
Co[MoO$_4$]	3.28E+04	218.87	0	0	9.55		
Ethanol (unreacted)	0.00E+00	46	2	0	–1.2		0.42
Ethyl acrylate	1.00E+06	100	5	0	0		
Oxygen (unreacted)	0.00E+00	32	0	0	0		
Propylene (unreacted)	0.00E+00	42	3	0	0		3.03
TeO$_2$	1.60E+04	159.6	0	0			
Water	3.60E+05	18	0	0	0		

Examples from the Chemical Industry

Substance	LD50 (mg/kg)	LC50 (g/m³ 4 h)	H (atm m³/mol)	log Kow	ADP
Co[MoO$_4$]			1.0E−100	−1E+100	3.17E−02
Ethanol (unreacted)	7060	94.1	5.00E−06	−0.3	0
Ethyl acrylate	800	6	3.93E−04	1.32	0
Oxygen (unreacted)			0.769	0.65	0
Propylene (unreacted)		18.04	0.196	1.77	0
TeO$_2$	5000		1.0E−100	−1E+100	52.8
Water	90000		8.48E−09	−1.38	0

Part 7

For each plan determine the safety-hazard indices for input and waste materials.
Which plan is comparatively safer? Which potential and chemicals have the most impact?

Plan A
Input Materials:

Substance	Mass (g)	MW (g/mol)	O Consumption/ Liberation rxn	No. O Substrate	No. O Products
Acetylene	3.25E+05	26	$C_2H_2+5\ [O]=2\ CO_2+H_2O$	0	5
Carbon monoxide	7.00E+05	28	$CO+[O]=CO_2$	1	2
Ethanol	5.75E+05	46	$C_2H_6O+6\ [O]=2\ CO_2+3\ H_2O$	1	7
HCl	228750	36.45	$HCl+0.5\ [O]=0.5\ Cl_2+0.5\ H_2O$	0	0.5
Ni(CO)$_4$	106250	170.7	$Ni(CO)_4+4\ [O]=Ni+4\ CO_2$	4	8

Substance	LD50 dermal (mg/kg)	LC50 (g/m³ 4 h)	FLP (°C)	moles H$_2$ generated	LEL	IS (N m)	LBT (cm³/10 g)	OEL (mmol/m³)
Acetylene		47.9		0	2.5	1.0E+100	0	102.4
Carbon monoxide		1.9		0	6.3	1.0E+100	0	1.4
Ethanol		94.1	13	0	3.28	1.0E+100	0	41.2
HCl		0.41		0		1.0E+100	0	0.20
Ni(CO)$_4$	63	0.0815	−20	0	2	1.0E+100	0	0.000041

Substance	SD (mg)	Q (R-phrases)
Acetylene	5.27E–01	36000
Carbon monoxide	3.91E–05	14000000
Ethanol	1.00E+02	20
HCl	6.67E+02	9000000
$Ni(CO)_4$	1.15E+00	3.15E+09

Output Materials:

Substance	Mass (g)	MW (g/mol)	O Consumption/ Liberation rxn	No. O Substrate	No. O Products
Acetylene (unreacted)	6.50E+04	26	$C_2H_2 + 5\,[O] = 2\,CO_2 + H_2O$	0	5
CO (unreacted)	4.20E+05	28	$CO + [O] = CO_2$	1	2
Ethanol (unreacted)	1.15E+05	46	$C_2H_6O + 6\,[O] = 2\,CO_2 + 3\,H_2O$	1	7
Ethyl acrylate	1.00E+06	100	$C_5H_8O_2 + 12\,[O] = 5\,CO_2 + 4\,H_2O$	2	14
HCl	228750	36.45	$HCl + 0.5\,[O] = 0.5\,Cl_2 + 0.5\,H_2O$	0	0.5
$Ni(CO)_4$	106250	170.7	$Ni(CO)_4 + 4\,[O] = Ni + 4\,CO_2$	4	8

Substance	LD50 dermal (mg/kg)	LC50 (g/m³ 4 h)	FLP (°C)	Moles H_2 Generated	LEL	IS (N m)	LBT (cm³/10 g)	OEL (mmol/m³)
Acetylene (unreacted)		47.9		0	2.5	1.0E+100	0	102.4
CO (unreacted)		1.9		0	6.3	1.0E+100	0	1.4
Ethanol (unreacted)		94.1	13	0	3.28	1.0E+100	0	41.2
Ethyl acrylate	2997	6	10	0	1.4	1.0E+100	0	0.2
HCl		0.41		0		1.0E+100	0	0.20
$Ni(CO)_4$	63	0.0815	–20	0	2	1.0E+100	0	0.000041

Examples from the Chemical Industry

Substance	SD (mg)	Q (R-Phrases)
Acetylene (unreacted)	5.27E–01	36000
CO (unreacted)	3.91E–05	14000000
Ethanol (unreacted)	1.00E+02	20
Ethyl acrylate	3.81E+00	4500000
HCl	6.67E+02	9000000
$Ni(CO)_4$	1.15E+00	3.15E+09

Plan B
Input Materials:

Substance	Mass (g)	MW (g/mol)	O Consumption/Liberation rxn	No. O Substrate	No. O Products
$Co[MoO_4]$	3.28E+04	218.87	$Co[MoO_4]=Co+Mo+4\,[O]$	0	4
ethanol	4.60E+05	46	$C_2H_6O+6\,[O]=2\,CO_2+3\,H_2O$	1	7
oxygen	4.80E+05	32	$O_2=O_2$	2	2
propylene	4.20E+08	42	$C_3H_6+9\,[O]=3\,CO_2+3\,H_2O$	0	9
TeO_2	1.60E+04	159.6	$TeO_2=Te+2[O]$	2	0

Substance	LD50 dermal (mg/kg)	LC50 (g/m³ 4 h)	FLP (°C)	moles H_2 generated	LEL	IS (N m)	LBT (cm³/10 g)	OEL (mmol/m³)
$Co[MoO_4]$				0		1.0E+100	0	0.000228
Ethanol		94.1	13	0	3.28	1.0E+100	0	41.2
Oxygen			–52	0		1.0E+100	0	
Propylene		18.04	–108	0	2	1.0E+100	0	20.4
TeO_2				0		1.0E+100	0	0.000627

Substance	SD (mg)	Q (R-phrases)
$Co[MoO_4]$	0	750
Ethanol	1.00E+02	20
Oxygen	2.31E–02	64000
Propylene	3.75E–01	720000
TeO_2	0.00E+00	300

Output Materials:

Substance	Mass (g)	MW (g/mol)	O consumption/liberation rxn	No. O substrate	No. O products
Co[MoO$_4$]	3.28E+04	218.87	Co[MoO$_4$]=Co+Mo+4 [O]	0	4
Ethanol (unreacted)	0	46	C$_2$H$_6$O+6 [O]=2 CO$_2$+3 H$_2$O	1	7
Ethyl acrylate	1000000	100	C$_5$H$_8$O$_2$+12 [O]=5 CO$_2$+4 H$_2$O	2	14
Oxygen (unreacted)	0	32	O$_2$=O$_2$	2	2
Propylene (unreacted)	0	42	C$_3$H$_6$+9 [O]=3 CO$_2$+3 H$_2$O	0	9
TeO$_2$	1.60E+04	159.6	TeO$_2$=Te+2[O]	2	0
Water	360000	18	H$_2$O=H$_2$O	1	1

Substance	LD50 dermal (mg/kg)	LC50 (g/m^3 4 h)	FLP (°C)	Moles H$_2$ generated	LEL	IS (N m)	LBT (cm^3/10 g)	OEL (mmol/m^3)
Co[MoO$_4$]				0		1.0E+100	0	0.000228
Ethanol (unreacted)		94.1	13	0	3.28	1.0E+100	0	41.2
Ethyl acrylate		6	10	0	1.4	1.0E+100	0	0.2
Oxygen (unreacted)			−52	0		1.0E+100	0	
Propylene (unreacted)		18.04	−108	0	2	1.0E+100	0	20.4
TeO$_2$				0		1.0E+100	0	0.000627
Water				0		1.0E+100	0	

Substance	SD (mg)	Q (R-phrases)
Co[MoO$_4$]	0	750
Ethanol (unreacted)	1.00E+02	20
Ethyl acrylate	3.81	4500000
Oxygen (unreacted)	2.31E−02	64000
Propylene (unreacted)	3.75E−01	720000
TeO$_2$	0.00E+00	300
Water	0.00E+00	

PROBLEM 7.33[46]

Themes: synthesis plan comparison, gate-to-gate, material efficiency, energy intensity, benign indices, safety-hazard indices.

An industrial network to manufacture 1000 kg methyl methacrylate is shown below in Scheme 7.21.

Hydrogen (T = 925°C, 4.2 MPa):
    7 cu m propane at STP (0°C and 1 atm)
    163 kg water
    1 mol% each of NiO, Fe$_2$O$_3$, and Cr$_2$O$_3$
    Yield of hydrogen = 28 cu m at STP (0°C and 1 atm)

Ammonia (T = 925°C, p = 4.2 MPa):
stoichiometric amount of air to maintain 1:3 molar ratio of nitrogen to hydrogen
207.5 kg hydrogen
1229 kg air
1 mol% Fe$_2$O$_3$
Yield of ammonia: 1000 kg

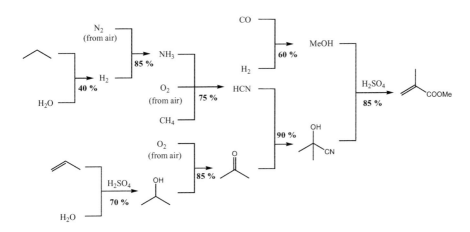

SCHEME 7.21 Industrial network to produce 1000 kg of methyl methacrylate.

Isopropanol (T = 27°C, p = 1 atm):

900 kg propylene
stoichiometric amount of water
12.5 kg 85 wt% sulfuric acid
Yield of isopropanol: 1000 kg

Acetone (T = 500°C, p = 350 kPa):

1218 kg isopropanol
10% excess oxygen from air
1 mol% copper
Yield of acetone: 1000 kg
Methanol (T = 300°C, p = 5000 psi):

1170 cu m carbon monoxide at STP (0°C and 1 atm)
2350 cu m hydrogen at STP (0°C and 1 atm)
Yield of methanol: 1000 kg

Hydrogen cyanide (T = 1090°C, p = 170 kPa):
830 kg ammonia
1155 cu m methane at STP (0°C and 1 atm)
7900 cu m air at STP (0°C and 1 atm)
725 kg sulfuric acid (sp. gr. 1.7)
Yield of hydrogen cyanide: 1000 kg

Acetone to acetone cyanohydrin (T = 25°C, p = 1 atm):
581 kg acetone
270 kg hydrogen cyanide
Yield of acetone cyanohydrin: 765 kg

Acetone cyanohydrin to methyl methacrylate (T = 90°C, 1 atm):
765 kg acetone cyanohydrin
320 kg methanol
981 kg 98 wt% sulfuric acid
Yield of methyl methacrylate: 765 kg

Part 1
Write out balanced chemical equations for each plan.

Examples from the Chemical Industry

Part 2

Using the *template-REACTION.xls* and *template-SYNTHESIS.xls* spreadsheets, determine the material efficiency metrics for the production of 1000 kg methyl methacrylate. State any assumptions made in the calculations.

Part 3

Write out a tree diagram showing the scaled masses of materials needed to produce 1000 kg of methyl methacrylate. Summarize the results by showing a mass balance of input and output materials.

Part 4

Determine the energy input requirements to produce 1000 kg of methyl methacrylate using the energy-input-synthesis-plans-plans spreadsheet and the thermodynamic data given below.

				Coefficients for Cp					
	Phase [1]	Tb (K)	A	B	C	D	E	Tc(K)	Pc(atm)
Acetone	Liquid	329.44	1.36E+05	−1.8E+02	2.84E−01	6.89E−04		508	46.4
ACH[2,3]	Liquid	463	55.844	1.34E+00	−3.44E−03	3.62E−06		647	41.94
CH_2	Gas	111.66	3.33E+04	8.03E+04	2.10E+03	4.21E+04	9.95E+02	191	45.44
CO	Gas	81.7	2.91E+04	8.77E+03	3.09E+03	8.46E+03	1.54E+03	133	34.53
Cu^3	Solid	2836	2.20E+01	8.84E−03	−1.01E−06				
H_2	Gas	20.39	2.76E+04	9.56E+03	2.47E+03	3.76E+03	5.68E+02	33.3	12.8
H_2SO_4	Liquid	610	5.98E+04	3.95E+02	−5.21E−01	3.12E−04	−7.06E−08	925	63.16
H_2SO_4	Gas		3.83E+04	1.12E+05	4.21E+02	−4.7E+04	5.48E+02		
HCN	Gas	298.85	3.01E+04	3.17E+04	1.61E+03	2.18E+04	6.26E+02	457	53.2
iPrOH	Liquid	355.3	4.66E+05	−4.1E+03	1.45E+01	−1.41E−02	4.66E+05	508	47.02
iPrOH	Gas		4.75E+04	1.94E+05	1.12E+03	9.38E+04	4.60E+02		
MeOH	Liquid	337.85	1.08E+05	−3.8E+02	9.79E−01			513	79.9
N_2	Gas	77.21	2.91E+04	8.61E+03	1.70E+03	1.03E+02	9.10E+02	126	33.5
NH_3	Gas	239.72	3.35E+04	4.82E+04	9.52E+02	−3.0E+04	1.06E+03	406	111.31
O_2	Gas	90.17	2.91E+04	1.00E+04	2.53E+03	9.36E+03	1.15E+03	155	49.77
Propane	Gas	231.11	59474	126610	844.31	86165	2482.7	370	41.92
Propylene	Gas	225.43	4.13E+04	1.53E+05	1.35E+03	7.44E+04	5.78E+02	365	45.52
Water	Liquid	373	2.76E+05	−2.1E+03	8.13E+00	−1.41E−02	9.37E−06	647	216.54
Water	Gas		3.34E+04	2.68E+04	2.61E+03	8.89E+03	1.17E+03		

[1] At T = 25°C and 1 atm
[2] ACH = acetone cyanohydrin
[3] Cp coefficients correspond to units of J/K mol for heat capacity at constant pressure; all other entries correspond to units of J/K kmol

Heat capacity of zinc oxide (solid) = 40.3 J/K mol
Heat capacity of chromium oxide (solid) = 118.7 J/K mol
Heat capacity of iron(III)oxide (solid) = 103.9 J/K mol
Heat capacity of nickel(II)oxide (solid) = 49.98 J/K mol
Heat of vaporization at boiling point of isopropyl alcohol = 39.85 kJ/mol
Heat of vaporization at boiling point of sulfuric acid = 50.20 kJ/mol
Heat of vaporization at boiling point of water = 40.65 kJ/mol
Redlich–Kwong equation of state parameters for gases:

Chemical	a (L^2 atm deg $K^{0.5}$/mol^2)	b (L/mol)
Acetone	361.3032	0.07793
ACH	730.962	0.10976
CH_4	31.7725	0.02984
CO	16.9855	0.02739
H_2	1.434	0.01848
H_2SO_4	1186.3039	0.1042
HCN	241.1855	0.06108
iPrOH	356.7337	0.07692
MeOH	214.3683	0.04565
N_2	15.3478	0.02678
NH_3	85.737	0.02593
O_2	17.1876	0.0221
Propane	180.6519	0.06277
Propylene	160.7288	0.05701
Water	140.7904	0.02113

Part 5

Summarize the metrics results in a table.

Part 6

For the synthesis plan determine the benign indices for input and waste materials. Which potential has the most impact? Which chemicals have the most impact?

Examples from the Chemical Industry

Input Materials:

Substance	Mass (g)	MW (g/mol)	No. carbon atoms	Alpha no. acidic Hs	ABP	ODP	SFP
Air	5.58E+06	28.56	0	0	0		
CO	6.12E+05	28	1	0	0		0.029
Copper (cat.)	9.80E+03	63.55	0	0	0		
Cr_2O_3 (cat.)	4.47E+04	151.99	0	0	24.08		
Fe_2O_3 (cat.)	6.06E+04	159.69	0	0	24.08		
H_2SO_4	1.55E+06	98	0	2	14.046		
Hydrogen	8.77E+04	2	0	0	0		
Methane	2.91E+05	16	1	0	0		0.005
NiO (cat.)	5.70E+03	74.71	0	0	24		
Propane	3.34E+05	44	3	0	0		0.27
Propylene	8.33E+05	42	3	0	0		3.03
Water	4.32E+06	18	0	0	0		
ZnO (cat.)	1.78E+04	81.37	0	0	24		

Substance	LD50 (mg/kg)	LC50 (g/m³ 4 h)	H (atm m³/mol)	log Kow	ADP
Air					0
CO		1.9	1.09	−1E+100	0
Copper (cat.)	300		1.0E−100	−1E+100	1.94E−03
Cr_2O_3 (cat.)	621		1.0E−100	−1E+100	0.001716
Fe_2O_3 (cat.)			1.0E−100	−1E+100	1.69E−07
H_2SO_4	2140	0.06	1.0E−100	−1.1	3.58E−04
Hydrogen			1.28	0.45	0
Methane		163	6.58E−01	1.09	0
NiO (cat.)			1.0E−100	−1E+100	1.08E−04
Propane		5760	7.07E−01	2.36	0
Propylene		18.04	0.196	1.77	0
Water	90000		8.48E−09	−1.38	0
ZnO (cat.)	8437	2.5	1.0E−100	−1E+100	9.92E−04

Output Materials:

Substance	Mass (g)	MW (g/mol)	No. carbon atoms	Alpha no. acidic Hs	ABP	ODP	SFP
Acetone (unreacted)	7.71E+04	85	3	0	2.851		
Acetone cyanohydrin (unreacted)	1.50E+05	58	4	0	0		0.13
Ammonia (unreacted)	7.07E+04	17	0	0	4.44		
CO (unreacted)	2.46E+05	28	1	0	0		0.029
CO_2 (by-product)	4.01E+05	44	1	0	7.95		
Copper (cat.)	9.80E+03	63.55	0	0	0		
Cr_2O_3 (cat.)	4.47E+04	151.99	0	0	24.08		
Fe_2O_3 (cat.)	6.06E+04	159.69	0	0	24.08		
H_2SO_4 (unreacted)	5.70E+05	98	0	2	14.046		
HCN (unreacted)	3.53E+04	27	1	1	4.901		
Hydrogen (unreacted)	4.46E+04	2	0	0	0		
^{i}PrOH (unreacted)	1.39E+05	60	3	0	−3.699		0.27
Methane (unreacted)	8.20E+04	16	1	0	0		0.005
Methanol (unreacted)	9.83E+04	32	1	0	−1.2		0.2
Methyl methacrylate	1.00E+06	100	5	0	0		
$NH_4[HSO_4]$ (by-product)	1.15E+06	115	0	1	4.14		
NiO (cat.)	5.70E+03	74.71	0	0	24		
Nitrogen	4.16E+06	28	0	0	0		
Oxygen	3.47E+05	32	0	0	0		
Propane (unreacted)	2.01E+05	44	3	0	0		0.27
Propylene (unreacted)	1.85E+05	42	3	0	0		3.03
Water (by-product)	4.66E+06	18	0	0	0		
ZnO (cat.)	1.78E+04	81.37	0	0	24		

Examples from the Chemical Industry

Substance	LD50 (mg/kg)	LC50 (g/m³ 4 h)	H (atm m³/mol)	log Kow	ADP
Acetone (unreacted)	5.9		1.97E–09	–0.03	0
Acetone cyanohydrin (unreacted)	3000	99.6	3.97E–05	–0.24	0
Ammonia (unreacted)	350	1.4	1.32E–05	–0.8	0
CO (unreacted)		1.9	1.09	–1E+100	0
CO_2 (by-product)		105.7	2.88E–02	–1.33	0
Copper (cat.)	300		1.0E–100	–1E+100	1.94E–03
Cr_2O_3 (cat.)	621		1.0E–100	–1E+100	0.001716
Fe_2O_3 (cat.)			1.0E–100	–1E+100	1.69E–07
H_2SO_4 (unreacted)	2140	0.06	1.0E–100	–1.1	3.58E–04
HCN (unreacted)	4	0.02	1.04E–04	–0.25	0
Hydrogen (unreacted)			1.28	0.45	0
iPrOH (unreacted)	5000	72.6	8.10E–06	0.05	0
Methane (unreacted)		163	6.58E–01	1.09	0
Methanol (unreacted)	5600	84	4.55E–06	–0.77	0
Methyl methacrylate	7872	31	3.37E–04	1.38	0
$NH_4[HSO_4]$ (by-product)			1.0E–100	–1E+100	3.58E–04
NiO (cat.)			1.0E–100	–1E+100	1.08E–04
Nitrogen			1.59	0.67	0
Oxygen			0.769	0.65	0
Propane (unreacted)		5760	7.07E–01	2.36	0
Propylene (unreacted)		18.04	0.196	1.77	0
Water (by-product)	90000		8.48E–09	–1.38	0
ZnO (cat.)	8437	2.5	1.0E–100	–1E+100	9.92E–04

Part 6

For the synthesis plan determine the safety-hazard indices for input and waste materials.

Which potential has the most impact? Which chemicals have the most impact?

Input Materials:

Substance	Mass (g)	MW (g/mol)	O consumption/liberation rxn	No. O substrate	No. O products
Air	5.58E+06	28.56			
CO	6.12E+05	28	$CO + [O] = CO_2$	1	2
Copper (cat.)	9.80E+03	63.55	$Cu = Cu$	0	0
Cr_2O_3 (cat.)	4.47E+04	151.99	$Cr_2O_3 = 2\ Cr + 3\ [O]$	3	0
Fe_2O_3 (cat.)	6.06E+04	159.69	$Fe_2O_3 = 2\ Fe + 3\ [O]$	3	0
H_2SO_4	1.55E+06	98	$H_2SO_4 = S + H_2O + 3\ [O]$	4	1
Hydrogen	8.77E+04	2	$H_2 = H_2$	0	0
Methane	2.91E+05	16	$CH_4 + 4\ [O] = CO_2 + 2\ H_2O$	0	4
NiO (cat.)	5.70E+03	74.71	$NiO = Ni + [O]$	1	0
Propane	3.34E+05	44	$C_3H_8 + 10\ [O] = 3\ CO_2 + 4\ H_2O$	0	10
Propylene	8.33E+05	42	$C_3H_6 + 9\ [O] = 3\ CO_2 + 3\ H_2O$	0	9
Water	4.32E+06	18	$H_2O = H_2O$	0	0
ZnO (cat.)	1.78E+04	81.37	$ZnO = Zn + [O]$	1	0

Substance	LD50 dermal (mg/kg)	LC50 (g/m³ 4 h)	FLP (°C)	Moles H2 generated	LEL	IS (N m)	LBT (cm³/10 g)	OEL (mmol/m³)
Air				0		1.0E+100	0	
CO		1.9		0	6.3	1.0E+100	0	1.4
Copper (cat.)	375			0		1.0E+100	0	0.016
Cr_2O_3 (cat.)				0		1.0E+100	0	0.00329
Fe_2O_3 (cat.)				0		1.0E+100	0	0.0313
H_2SO_4		0.06		0		1.0E+100	0	0.01
Hydrogen				0	4	1.0E+100	0	0.41
Methane		163	−188	0	5.3	1.0E+100	0	40.9
NiO (cat.)	50			0		1.0E+100	0	0.001339
Propane		5760	−104	0	2.37	1.0E+100	0	40.8
Propylene		18.04	−108	0	2	1.0E+100	0	20.4
Water				0		1.0E+100	0	
ZnO (cat.)		2.5		0		1.0E+100	0	0.061

Examples from the Chemical Industry

Substance	SD (mg)	Q (R-phrases)
Air	0	
CO	3.91E−05	14000000
Copper (cat.)	1.0E−106	15
Cr_2O_3 (cat.)	0	15
Fe_2O_3 (cat.)	0	
H_2SO_4	9.8	30
Hydrogen	3.90E−03	48000
Methane	5.04E−02	1440000
NiO (cat.)	0	50
Propane	2.42E−01	36000
Propylene	3.75E−01	720000
Water	0.00E+00	
ZnO (cat.)	0.00E+00	30

Output Materials:

Substance	Mass (g)	MW (g/mol)	O consumption/liberation rxn	No. O substrate	No. O products
Acetone (unreacted)	7.71E+04	85	$C_3H_6O + 8\ [O] = 3\ CO_2 + 3\ H_2O$	1	9
Acetone cyanohydrin (unreacted)	1.50E+05	58	$C_4H_7NO + 10.5\ [O] = 4\ CO_2 + 0.5\ N_2 + 3.5\ H_2O$	1	11.5
Ammonia (unreacted)	7.07E+04	17	$NH_3 + 1.5\ [O] = 0.5\ N_2 + 1.5\ H_2O$	0	1.5
CO (unreacted)	2.46E+05	28	$CO + [O] = CO_2$	1	2
CO_2 – by-product	4.01E+05	44	$CO_2 = CO_2$	2	2
Copper (cat.)	9.80E+03	63.55	$Cu = Cu$	0	0
Cr_2O_3 (cat.)	4.47E+04	151.99	$Cr_2O_3 = 2\ Cr + 3\ [O]$	3	0
Fe_2O_3 (cat.)	6.06E+04	159.69	$Fe_2O_3 = 2\ Fe + 3\ [O]$	3	0
H_2SO_4 (cat.)	5.70E+05	98	$H_2SO_4 = S + H_2O + 3\ [O]$	4	1
HCN (unreacted)	3.53E+04	27	$HCN + 2.5\ [O] = 0.5\ H_2O + CO_2 + 0.5\ N_2$	0	2.5
Hydrogen (unreacted)	4.46E+04	2	$H_2 = H_2$	0	0
iPrOH (unreacted)	1.39E+05	60	$C_3H_8O + 9\ [O] = 3\ CO_2 + 4\ H_2O$	1	10
Methane (unreacted)	8.20E+04	16	$CH_4 + 4\ [O] = CO_2 + 2\ H_2O$	0	4
Methanol (unreacted)	9.83E+04	32	$CH_4O + 3\ [O] = CO_2 + 2\ H_2O$	1	4
Methyl methacrylate	1.00E+06	100	$C_5H_8O_2 + 12\ [O] = 5\ CO_2 + 4\ H_2O$	2	14
$NH_4[HSO_4]$ (by-product)	1.15E+06	115	$H_5NSO_4 = 2.5\ H_2O + 0.5\ N_2 + S + 1.5\ [O]$	4	2.5
NiO (cat.)	5.70E+03	74.71	$NiO = Ni + [O]$	1	0
Nitrogen	4.16E+06	28	$N_2 = N_2$	0	0
Oxygen	3.47E+05	32	$O_2 = O_2$	2	2
Propane (unreacted)	2.01E+05	44	$C_3H_8 + 10\ [O] = 3\ CO_2 + 4\ H_2O$	0	10
Propylene (unreacted)	1.85E+05	42	$C_3H_6 + 9\ [O] = 3\ CO_2 + 3\ H_2O$	0	9
Water (by-product)	4.66E+06	18	$H_2O = H_2O$	1	1
ZnO (cat.)	1.78E+04	81.37	$ZnO = Zn + [O]$	1	0

Substance	LD50 dermal (mg/kg)	LC50 (g/m³ 4 h)	FLP (°C)	Moles H2 generated	LEL	IS (N m)
Acetone (unreacted)	8.5		74	0	2.2	1.0E+100
Acetone cyanohydrin (unreacted)	5000	99.6	−17.8	0	2.15	1.0E+100
Ammonia (unreacted)		1.4		0	16	1.0E+100
CO (unreacted)		1.9		0	6.3	1.0E+100
CO_2 (by-product)		105.7		0		1.0E+100
copper (cat.)	375			0		1.0E+100
Cr_2O_3 (cat.)				0		1.0E+100
Fe_2O_3 (cat.)				0		1.0E+100
H_2SO_4 (cat.)		0.06		0		1.0E+100
HCN (unreacted)	2.5	0.02	−18	0	5.6	1.0E+100
Hydrogen (unreacted)				0	4	1.0E+100
iPrOH (unreacted)	12800	72.6	12	0	2	1.0E+100
Methane (unreacted)		163	−188	0	5.3	1.0E+100
Methanol (unreacted)	15800	84	11	0	6	1.0E+100
Methyl methacrylate	7080	31	10	0	1.7	1.0E+100
$NH_4[HSO_4]$ (by-product)				0		1.0E+100
NiO (cat.)	50			0		1.0E+100
Nitrogen				0		1.0E+100
Oxygen			−52	0		1.0E+100
Propane (unreacted)		5760	−104	0	2.37	1.0E+100
Propylene (unreacted)		18.04	−108	0	2	1.0E+100
Water (by-product)				0		1.0E+100
ZnO (cat.)		2.5		0		1.0E+100

Examples from the Chemical Industry

Substance	LBT (cm³/10 g)	OEL (mmol/m³)	SD (mg)	Q (R-phrases)
Acetone (unreacted)	0	0.05	54.1	35
Acetone cyanohydrin (unreacted)	0	10.2	60.4	300
Ammonia (unreacted)	0	1.1	71.7	18000
CO (unreacted)	0	1.4	3.91E–05	14000000
CO_2 (by-product)	0	204.5	3.92E-02	40
Copper (cat.)	0	0.016	1.0E–106	15
Cr_2O_3 (cat.)	0	0.00329	0.00E+00	15
Fe_2O_3 (cat.)	0	0.0313	0.00E+00	
H_2SO_4 (cat.)	0	0.01	9.80E+00	30
HCN (unreacted)	0	0.2	2.05E+02	210000
Hydrogen (unreacted)	0	0.41	3.90E-03	48000
iPrOH (unreacted)	0	16.4	1.10E+02	300
Methane (unreacted)	0	40.9	5.04E-02	1440000
Methanol (unreacted)	0	3.1	6.57E+01	612500
Methyl methacrylate	0	0.5	4.14E+00	12000
NH_4 [HSO_4] (by-product)	0		7.42E-01	20
NiO (cat.)	0	0.001339	0.00E+00	50
Nitrogen	0		1.23E-02	1600
Oxygen	0		2.31E-02	64000
Propane (unreacted)	0	40.8	2.42E-01	36000
Propylene (unreacted)	0	20.4	3.75E-01	720000
Water (by-product)	0		0.00E+00	
ZnO (cat.)	0	0.061	0.00E+00	30

PROBLEM 7.34

Themes: balancing chemical equations, synthesis plan analysis, kernel material efficiency.

Synthesis routes:

G1 Medicinal chemistry route[47]
G2 Chemistry route[48]
G3 Chemistry route[49]
G4 One-pot chemistry route[49]

ABT-341 is a dipeptidyl peptidase IV (DPP4) inhibitor developed by Abbott Laboratories that is used to treat. Four synthetic routes are shown below for this pharmaceutical. G1 is the medicinal chemistry discovery route that employed traditional Diels–Alder chemistry to construct the key six-membered ring. G2 is a route that utilizes a tandem proline catalyzed Michael addition–Horner–Emmons–Wadsworth sequence in the first step to construct the same ring. G3 uses a similar strategy as G2 in a stepwise manner. G4 is a one-pot sequence following the same strategy as G3 where all six steps are telescoped into one sequential operation without isolation of intermediates.

Part 1

Using the *synthesis-plan-analysis-basis-materials-metrics.xls* spreadsheet determine the overall yield, overall AE, and overall kernel PMI for each plan.

G1: Medicinal chemistry route (Scheme 7.22)

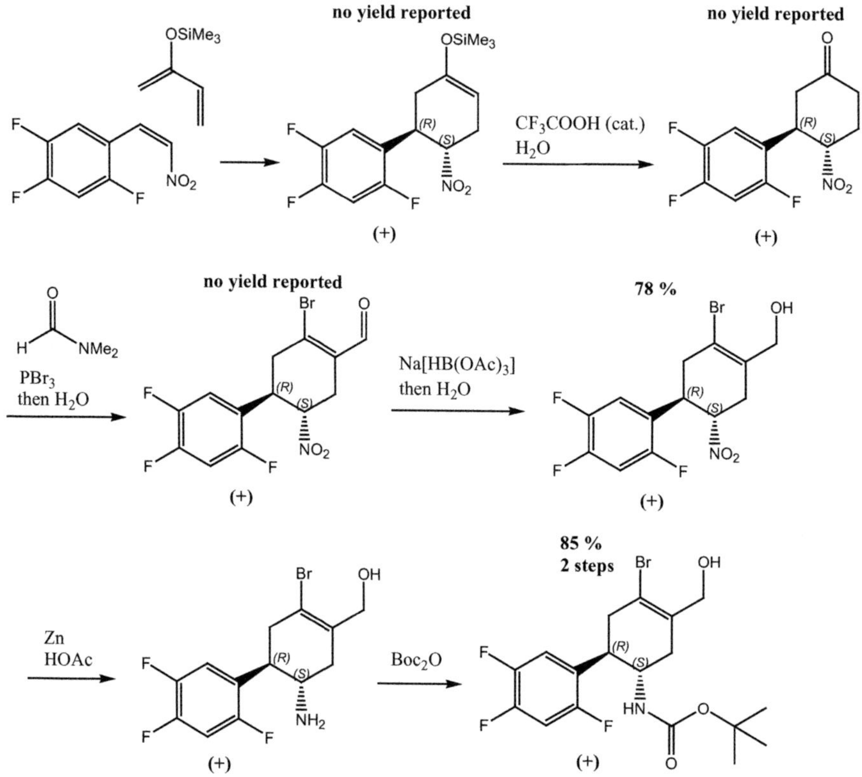

SCHEME 7.22 First generation medicinal chemistry route to ABT-341. (continued)

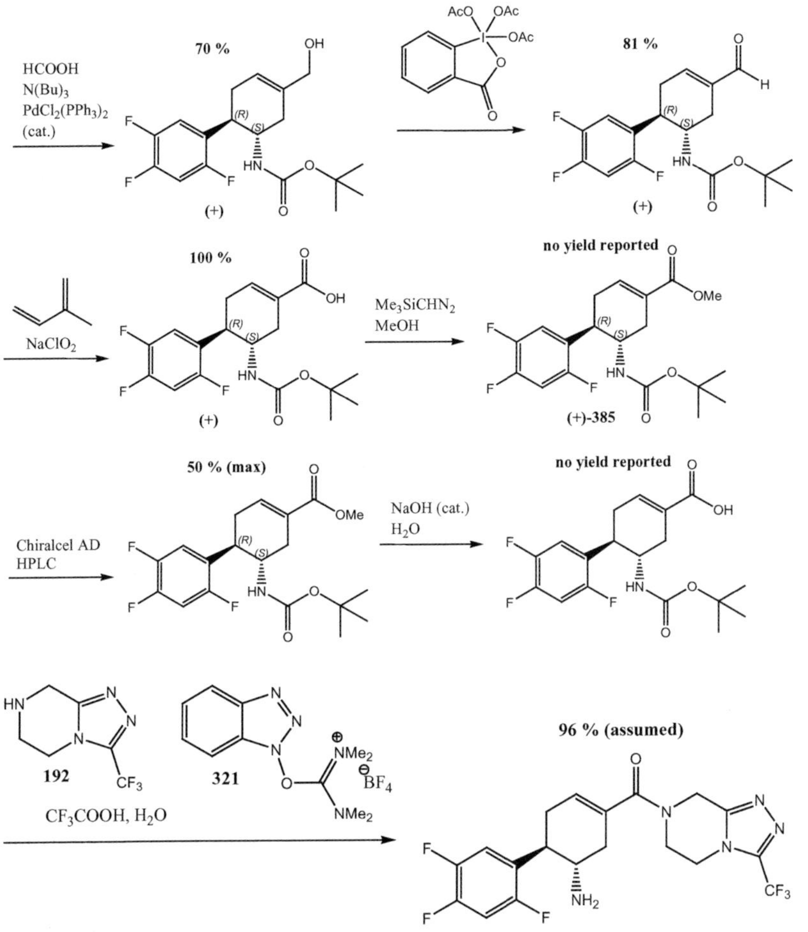

Examples from the Chemical Industry

SCHEME 7.22 (CONTINUED) First generation medicinal chemistry route to ABT-341.

G2: Chemistry route (Scheme 7.23)

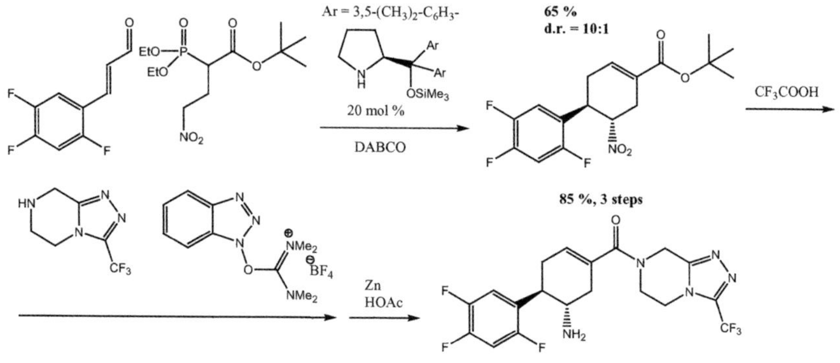

SCHEME 7.23 Second generation chemistry route to ABT-341.

G3: Chemistry route (Scheme 7.24)

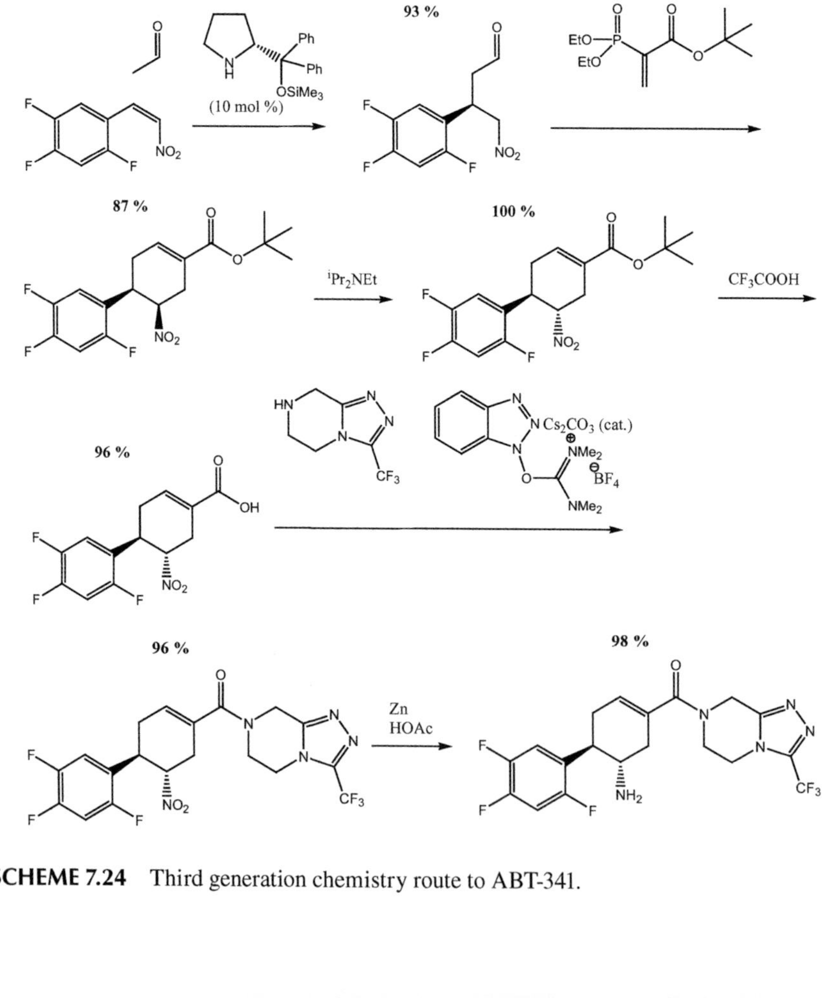

SCHEME 7.24 Third generation chemistry route to ABT-341.

Examples from the Chemical Industry

G4: One-pot chemistry route (Scheme 7.25)

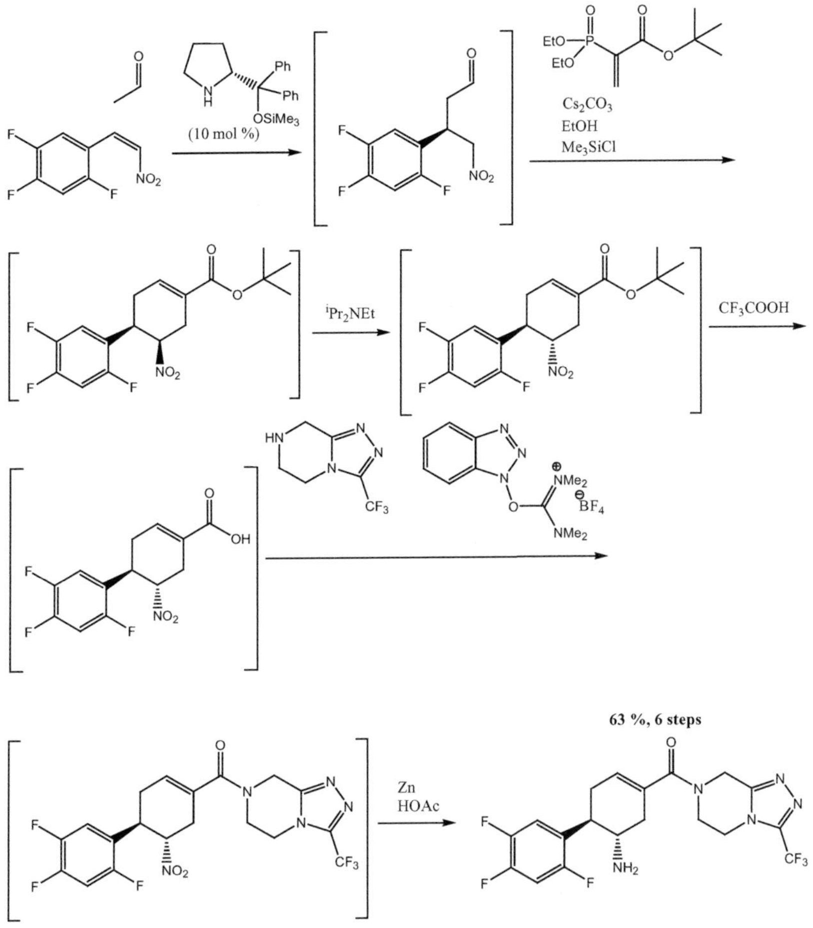

SCHEME 7.25 Fourth generation one-pot chemistry route to ABT-341.

Part 2
If the overall yield in the one-pot sequence in G4 is treated as separate steps, how does this change the overall kernel PMI value?

Part 3
Write out a catalytic cycle for the proline-based catalyzed Michael addition–Horner–Emmons–Wadsworth reaction used in G2.

Part 4
Compare the ring construction strategies employed in all four routes by highlighting the target bonds made.

PROBLEM 7.35
Themes: balancing chemical equations, synthesis plan analysis, kernel material efficiency metrics.

Synthesis routes:

Medicinal chemistry route G1[50]
Process chemistry route G2[51]
Process chemistry route G3[52]

Efavirenz is a non-nucleoside reverse transcriptase inhibitor developed by Merck as a treatment for HIV-1 (human immunodeficiency virus type 1). A medicinal chemistry discovery route (G1) and two process chemistry routes (G2 and G3) are shown below. All three routes begin from the same starting material, 1-(2-amino-5-chloro-phenyl)-2,2,2-trifluoro-ethanone. Merck claimed that the material efficiency for the synthesis of this pharmaceutical improved progressively from G1 to G3. The key challenge in the synthesis of this compound is the generation of the stereogenic quaternary center. The medicinal chemistry route makes the racemic compound which is then resolved using a diastereomeric discriminant. The process routes use a norephedrine analog as a chiral auxiliary.

G1: medicinal chemistry route (Scheme 7.26)

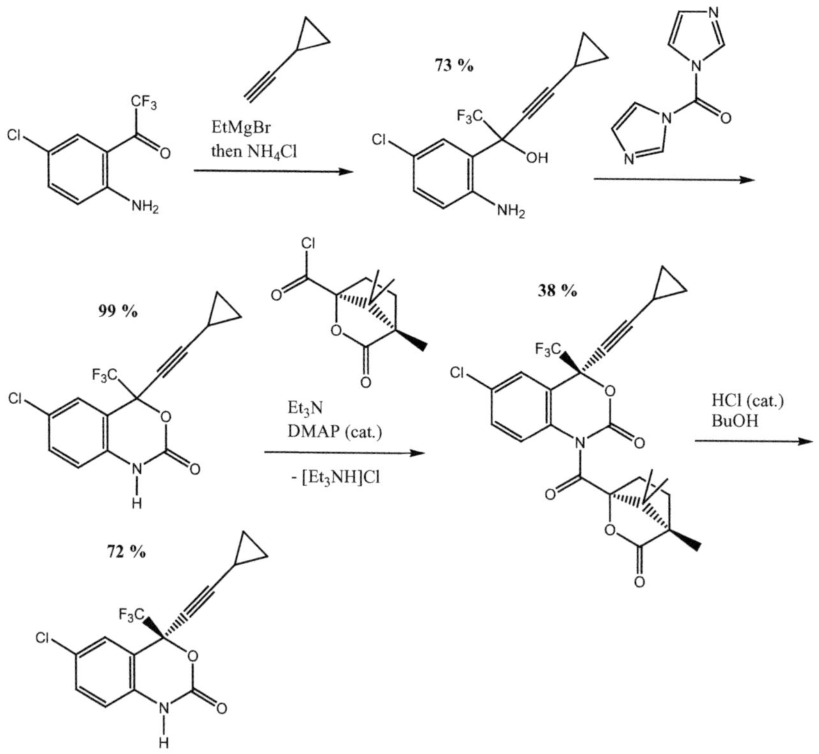

SCHEME 7.26 First generation medicinal chemistry route to efavirenz.

Examples from the Chemical Industry

G2: process chemistry route #1 (Scheme 7.27)

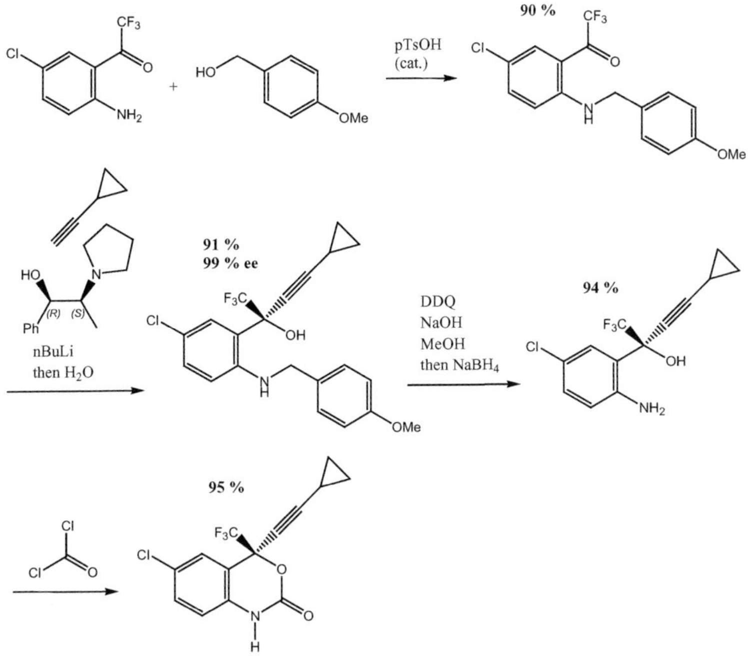

SCHEME 7.27 Second generation process chemistry route to efavirenz.

G3: process chemistry route #2 (Scheme 7.28)

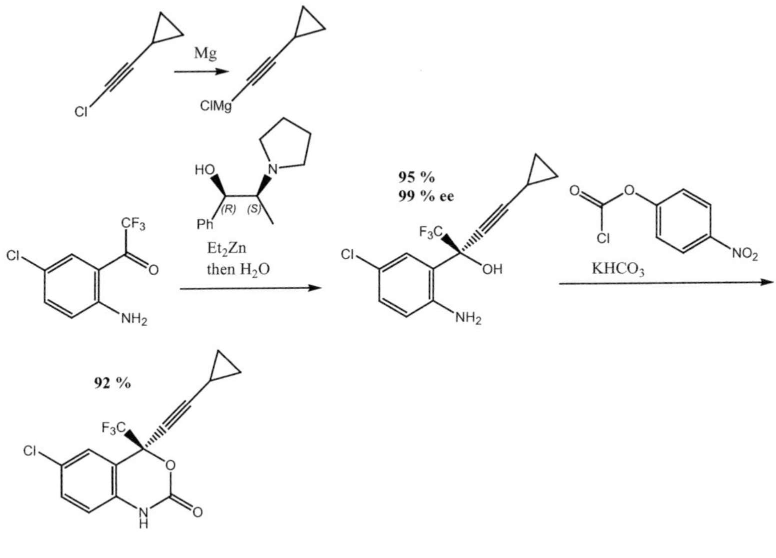

SCHEME 7.28 Third generation process chemistry route to efavirenz.

Using the *synthesis-plan-analysis-basis-materials-metrics.xls* spreadsheet determine the overall yield, overall AE, and overall kernel PMI for each plan. Verify whether or not the Merck claim is true at the synthesis strategy level.

PROBLEM 7.36[53]

Themes: balancing chemical equations, synthesis plan analysis, synthesis tree diagrams.

Merck developed the following process route to make the pharmaceutical efavirenz which is a non-nucleoside reverse transcriptase inhibitor used as a treatment for HIV-1 (Scheme 7.29).

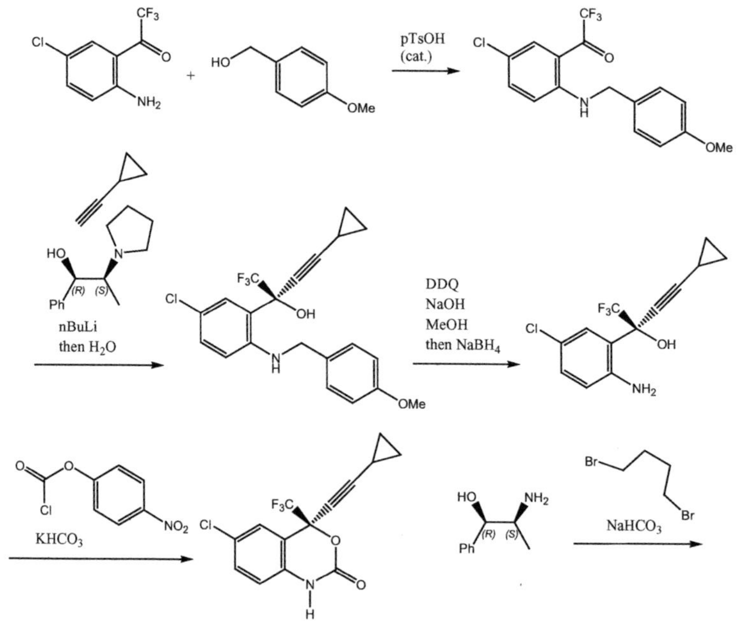

SCHEME 7.29 Merck process chemistry route to efavirenz.

Using the *template-REACTION.xls* spreadsheet and the details of the experimental procedure given in the reference, determine the material efficiency metrics for the synthesis of 1 kg of efavirenz. State any assumptions made in the calculations. Comment on any discrepancies found between the analysis and what the authors wrote in their paper regarding reaction yields and quantities of materials used. Draw a synthesis tree diagram for the plan. Summarize the metrics performances of each step. Identify the bottlenecks in this plan.

Examples from the Chemical Industry

PROBLEM 7.37

Themes: balancing chemical equations, synthesis plan analysis, kernel material efficiency metrics

Synthesis routes:

Medicinal chemistry route G1[54]
Process chemistry route G2[55]
Process chemistry route G3[56]

Raltegravir is an antiretroviral agent targeting integrase developed by Merck as a treatment for HIV-1. A medicinal chemistry discovery route (G1) and two process chemistry routes (G2 and G3) are shown below.

Part 1

All three routes begin from the same starting material, acetone cyanohydrin. Merck claimed that the material efficiency for the synthesis of this pharmaceutical improved progressively from G1 to G3. Using the *synthesis-plan-analysis-basis-materials-metrics.xls* spreadsheet determine the overall yield, overall AE, and overall kernel PMI for each plan. Verify whether or not the Merck claim is true at the synthesis strategy level.

G1: Medicinal chemistry route (Scheme 7.30)

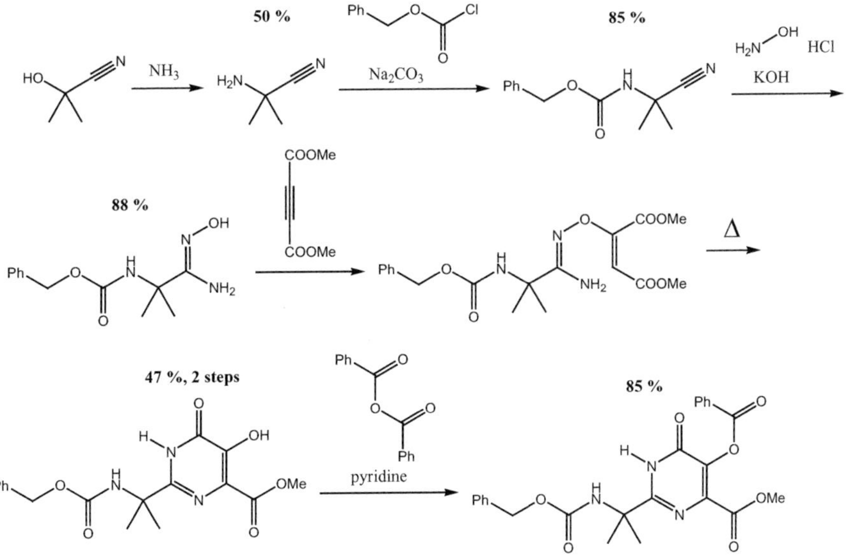

SCHEME 7.30 First generation medicinal chemistry route to raltegravir. (continued)

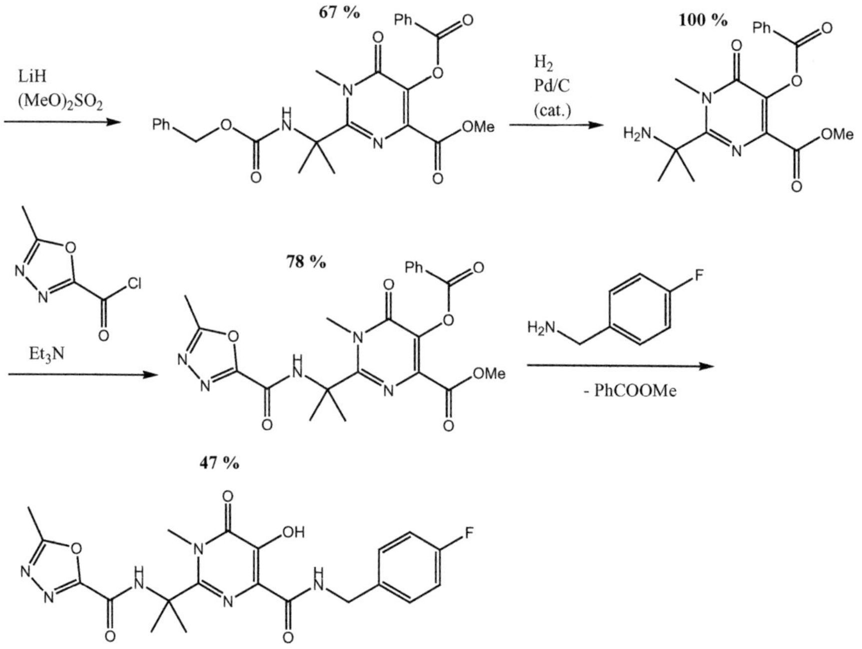

SCHEME 7.30 (CONTINUED) First generation medicinal chemistry route to raltegravir.

G2: Process chemistry route #1 (Scheme 7.31)

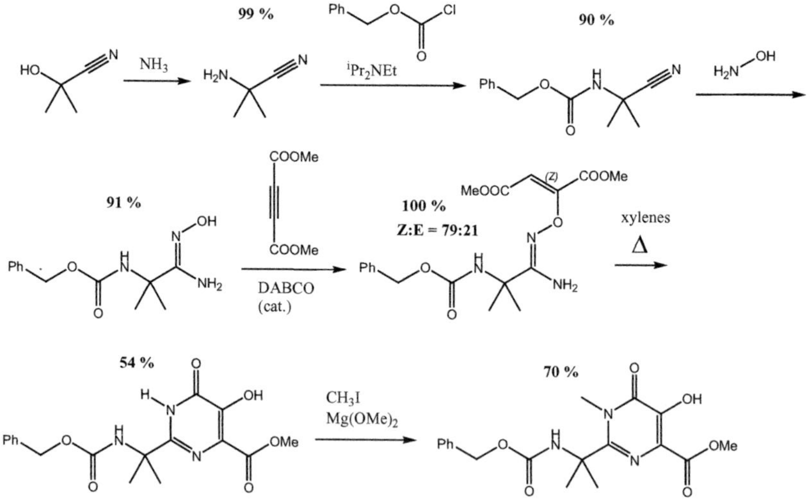

SCHEME 7.31 Second generation process chemistry route to raltegravir. (continued)

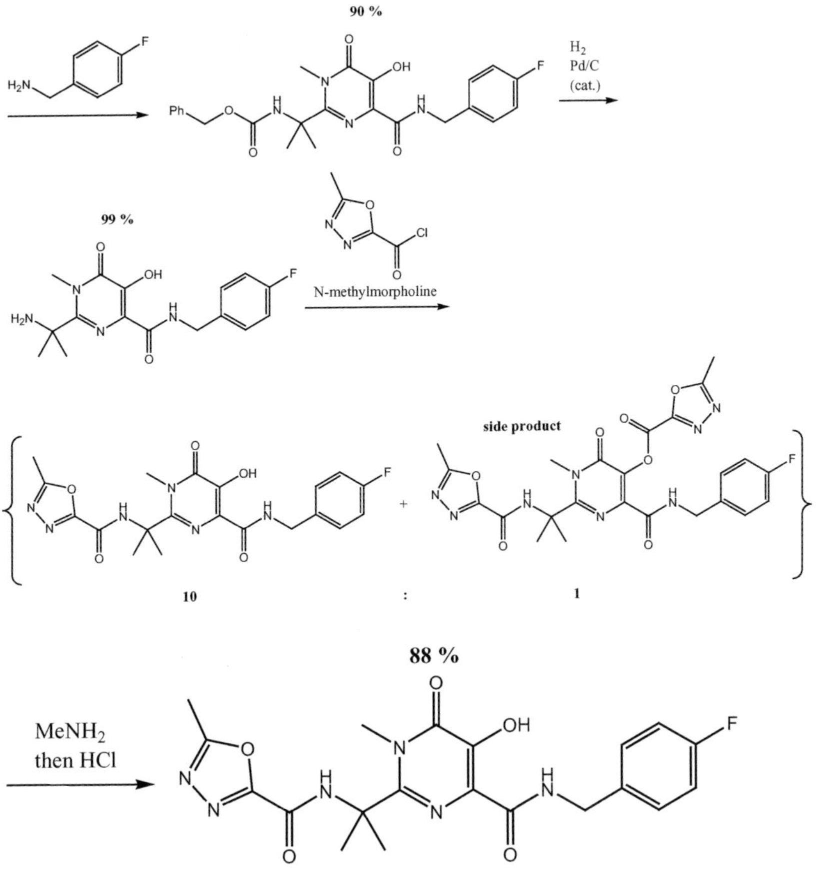

SCHEME 7.31 (CONTINUED) Second generation process chemistry route to raltegravir.

G3: Process chemistry route #3 (Scheme 7.32)

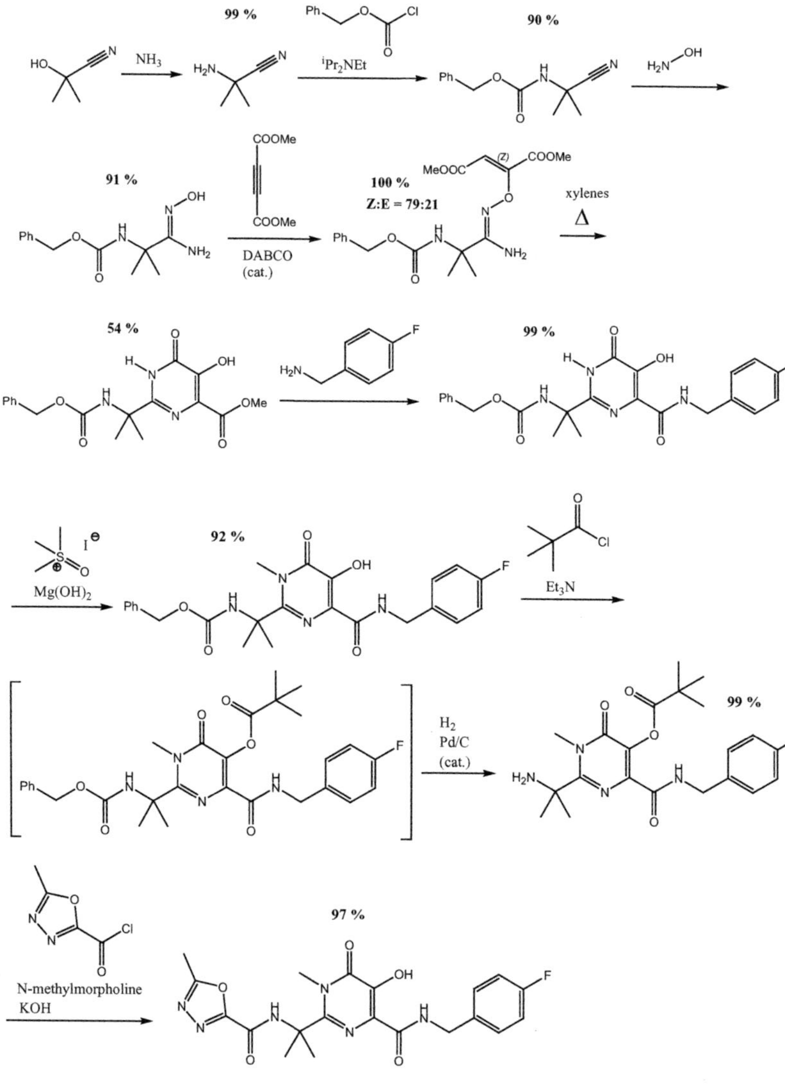

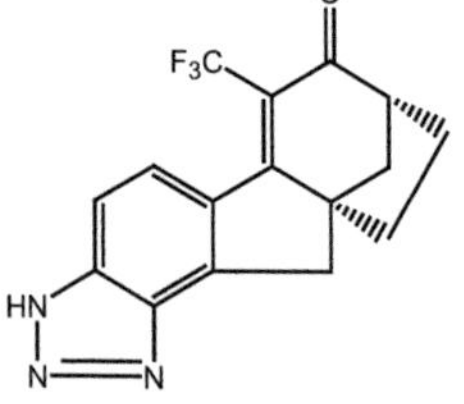

estrogen β-agonist

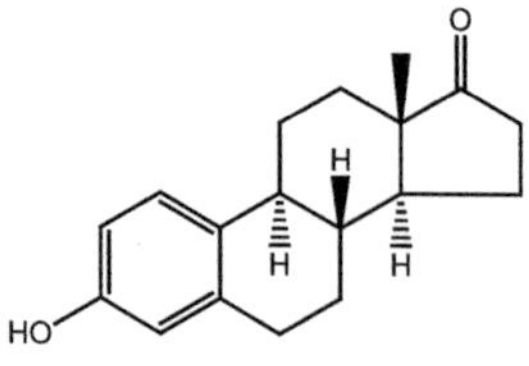

estrone

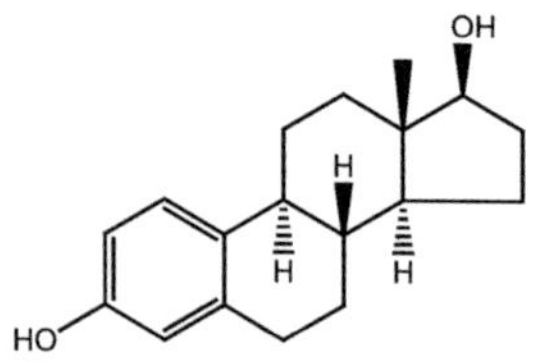

estradiol

SCHEME 7.32 Third generation process chemistry route to raltegravir.

Examples from the Chemical Industry

Part 2
Suggest a reaction mechanism for the thermal rearrangement shown below.

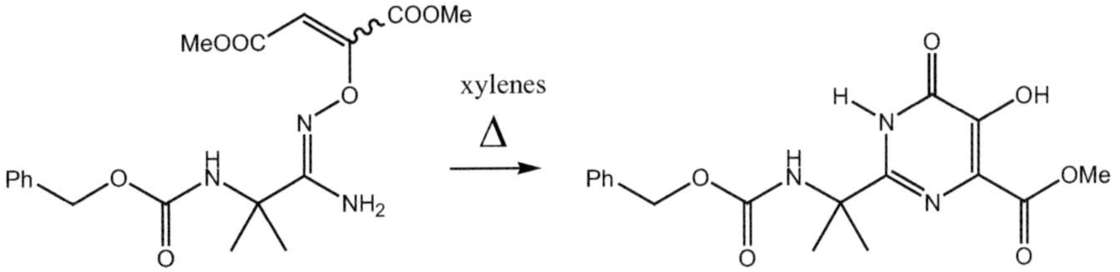

PROBLEM 7.38[57]
Themes: balancing chemical equations, synthesis plan comparison, % yields, % conversion, % selectivities, turnover numbers, process mass intensity.

Plan A (Guerbet reaction)

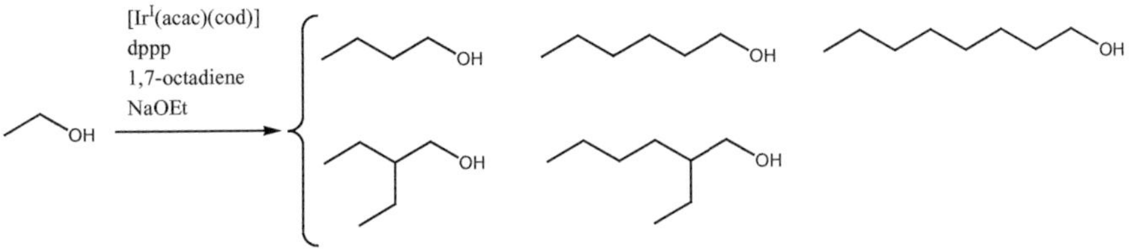

acac = acetylacetonato
cod = 1,5-cyclooctadiene
dppp = 1,3-bis(diphenylphosphino)propane

Product	% Selectivity	% Yield	TON
n-Butanol	51	21	1220
2-Ethylbutanol	15		464
n-Hexanol	15		464
2-Ethylhexanol	7		261
n-Octanol	2		87

Ethanol: 2 mL
Iridium complex: 0.01 mol%
dppp: 0.01 mol%
1,7-octadiene: 1 mol%
NaOEt: 5 mol%
Reaction temperature: 120°C
Reaction time: 15 h

Part 1
From these data, determine the following: the remaining % yields of products, the % conversion of ethanol to all products, the PMI for butanol production, and the

turnover numbers (TON) for producing each of the products based on the iridium complex used. How do the calculated TON compare with the authors reported values?

Part 2

Write out a mechanism for the production of *n*-butanol.

Part 3

Three other well known industrial processes to make n-butanol are shown below. One is based on a fermentation process and the other two utilize petrochemical feedstocks. Using the *template-REACTION.xls* and *template-SYNTHESIS.xls* spreadsheets and the experimental details given determine the corresponding % yield, % atom economy, and PMI values. State any assumptions made in the calculations. Compare the results with the Guerbet process. Which method is the most material efficient to produce *n*-butanol?

Plan B (fermentation)[58]

n-Butanol can be obtained by fermentation of molasses with *Clostridium saccharobutyl acetonicum liquefaciens*. In the process acetone, ethanol, carbon dioxide, and hydrogen gas are also produced as side products.

Starch → glucose → acetone + *n*-butanol + ethanol + carbon dioxide + hydrogen

In order to produce 1000 kg of *n*-butanol, 5425 L of molasses, 80 cubic meters of water, and 3.5 kg of nutrients are required. 1 L of molasses (0.8 kg) yields 174 to 200 g butanol, 48 to 78 g acetone, and 8 to 12 g ethanol.

Plan C (Wacker reaction – Aldol condensation – hydrogenation process)[59–61]

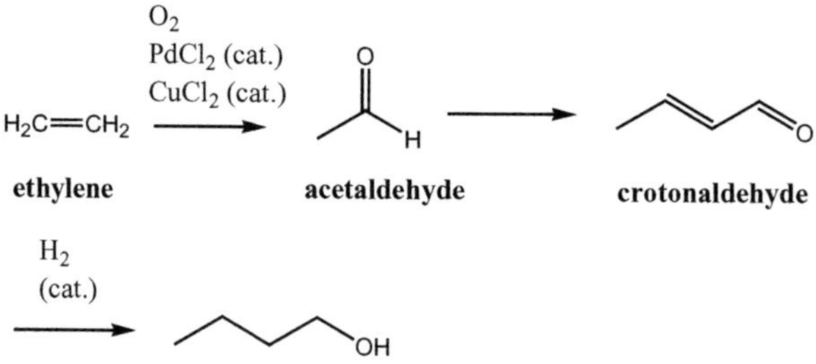

$H_2C=CH_2$ $\xrightarrow{\substack{O_2 \\ PdCl_2 \text{ (cat.)} \\ CuCl_2 \text{ (cat.)}}}$ acetaldehyde $\longrightarrow$ crotonaldehyde

ethylene

$\xrightarrow{\substack{H_2 \\ \text{(cat.)}}}$ ~~~~OH

Step 1

Ethylene: 670 kg

Oxygen: 290 cubic meters at STP (0°C and 1 atm)

Cooling water: 300 cubic meters

Demineralized water: 3 cubic meters

Steam: 1300 kg (1.4 MPa)

Steam: 300 kg (460 kPa)

Catalysts: small amount
Acetaldehyde: 1000 kg

Step 2
In a 1 L three-necked flask, equipped with a dropping funnel, mechanical stirrer and a thermometer and cooled in a bath of ice and salt, place 85 g. (109 mL.) of freshly distilled acetaldehyde. When the temperature has fallen to −5°C, add 25 mL of 15%, sodium sulfite solution, with stirring, during 1 h. The temperature rises to about + 10°C, and the liquid increases in viscosity and density. Aldol has a density of 1.103 g/mL, so that the progress of the condensation may be followed by a determination of the volume of the reaction mixture. Extract the aldol with two 75 mL portions of ether and remove the ether on a water bath. Transfer the residue to a Claisen flask with fractionating side arm, add 0.1 g. of iodine and distil slowly. A little water passes over first, followed by a constant boiling point mixture, containing 80%, of crotonaldehyde, at 84°C. The yield of the latter is 45 g. Pure crotonaldehyde boils at 102°C.

Step 3
Crotonaldehyde: 9900 g
Water: 747 g
Hydrogen: 70 mol per mol crotonaldehyde
Ni and Cr on silica: mole ratio of Ni to Cr = 97.8: 2.2
n-butanol: 9745 g
Plan D (Oxo Hydroformylation – Hydrogenation Process)[58,62]

propene + CO + H_2 $\xrightarrow{Rh^I(CO)_2(acac)\ L}$ butyraldehyde $\xrightarrow{H_2\ (cat.)}$ OH

L = 2,2'-ethylidene bis(4,6-di-t-butyl-phenyl)fluorophosphite

Step 1
A catalyst solution was prepared under nitrogen using a charge of 0.0375 g rhodium dicarbonyl acetonylacetate, (also known as rhodium dicarbonyl acac, 15 mg Rh),

2.12 g of 2,2'-ethylidene bis(4,6-di-t-butyl-phenyl)fluorophosphite (4.37 mmoles, [L]/[Rh]=30), and 190 mL of dioctylphthalate. The catalyst solution was charged to the reactor under an argon blanket and the reactor was sealed. The reactor was pressurized to 260 psig (18.9 Bar) with hydrogen, carbon monoxide, and nitrogen and heated to 115°C. Propylene feed was then started and the flows were adjusted to the following reported as liters/min at standard temperature and pressure (STP): hydrogen=3.70 L/min STP; carbon monoxide=3.70 l/min STP; nitrogen=1.12 L/min STP; and propylene=2.08 L/min STP. This is equivalent to having the following partial pressures in the feed to the reactor reported as pounds per square inch atmospheric (psia): hydrogen=96 psia (6.6 Bar); carbon monoxide=96 psia (6.6 Bar); nitrogen=29 psia (2.0 Bar); and propylene=54 psia (3.7 Bar). The reaction was carried out under the above flows for 53 h. The butyraldehyde production rate averaged 60.86 g/h for a catalyst activity of 4.06 kg butyraldehyde/gram of rhodium-hour. A total of 2.96 L of butyraldehyde was produced.

Step 2
butyraldehyde: 1030 kg
hydrogen gas: 328 cubic meters at 0°C and 1 atm (STP)
chromium and copper oxides: small amount
n-butanol: 1000 kg

PROBLEM 7.39[63]
Themes: balancing chemical equations, synthesis plan comparison, material efficiency metrics, target bond mapping, synthesis tree diagrams, fraction of sacrificial reagents.

Merck & Co. reported discovery and optimized scaled up process routes to an estrogen receptor β-agonist for the treatment symptoms associated with reduced estrogen levels in postmenopausal women. The structure of the agonist resembles those of estrone and estradiol.

estrogen β-agonist   estrone   estradiol

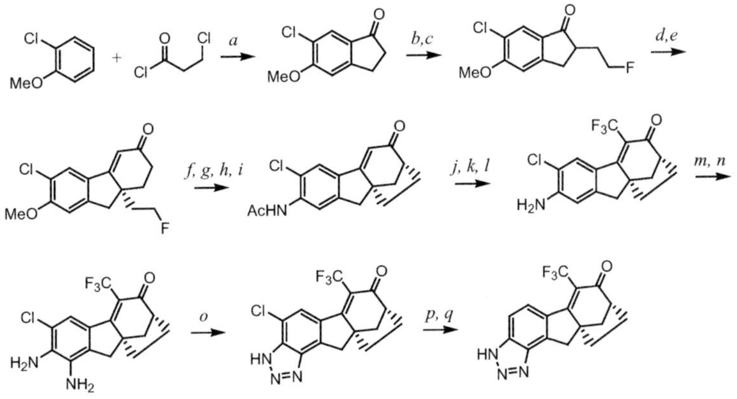

Examples from the Chemical Industry

SCHEME 7.33 Merck discovery route to an estrogen receptor β-Agonist. (a) AlMe₃, CH₂Cl₂, 0 °C then H₂SO₄, 100 °C, 51%; (b) LiHMDS, THF, −78 °C then NCCO₂Et, −78 °C to rt; (c) BrCH₂CH₂F, K₂CO₃, KI, DMA, 65 °C then 5 N NaOH, THF, H₂O, 0 °C, 58% over 2 steps; (d) MVK, N-[4-trifluoromethyl)benzyl]cinchoninium bromide, KOH, PhCH₃, 0°C, er = 2:1 favoring 2S enantiomer; (e) pyrrolidine, AcOH, PhCH₃, 95 °C, 65% over 2 steps; (f) LiCl, DMF, 150 °C, 67%; (g) Tf₂O, DIPEA, CH₂Cl₂, 0 °C, 73%; (h) MeC(O)NH₂, Pd₂(dba)₃, Xantphos, Cs₂CO₃, PhCH₃, 100 °C; (i) AcCl, DIPEA, CH₂Cl₂, rt, 73% over 2 steps; (j) NIS, DMF, 85 °C; (k) MFSDA, CuI, DMF, 75 °C, 79% over 2 steps; (l) 6 N HCl, AcOH, 80 °C, 86%; (m) 2,3,5,6-tetrabromo-4-methyl-4-nitrocyclohexa-2,5-dienone, TFA, rt; (n) H₂, Pd/C, KOAc, MeOH/EtOAc, rt, 69% over 2 steps; (o) NaNO₂, 12 N HCl, EtOH/H₂O, 0 °C; (p) H₂, Pd/CaCO₃, Pd(OH)₂/C, DMF, rt, 92% over 2 steps; (q) Chiral HPLC, 48%.

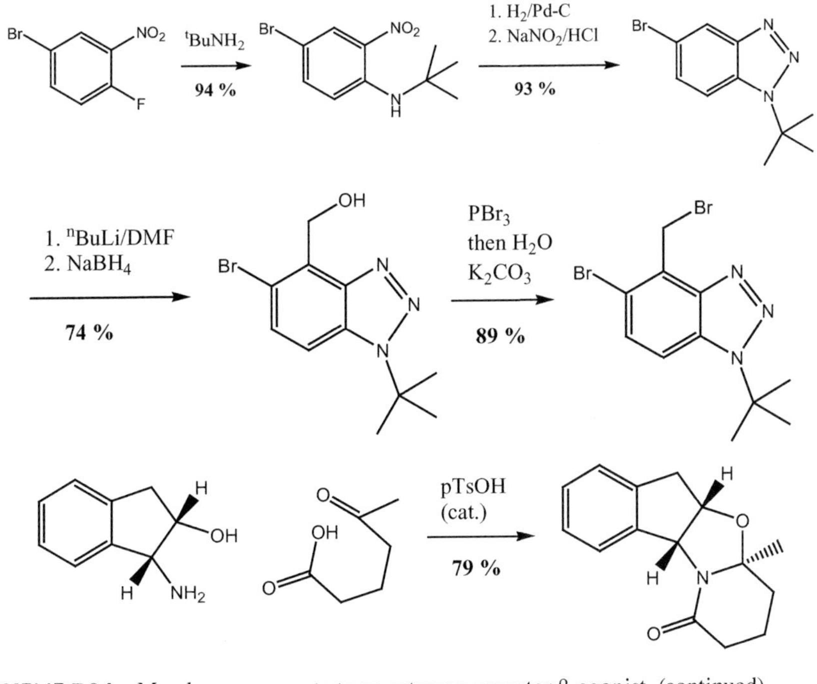

SCHEME 7.34 Merck process route to an estrogen receptor β-agonist. (continued)

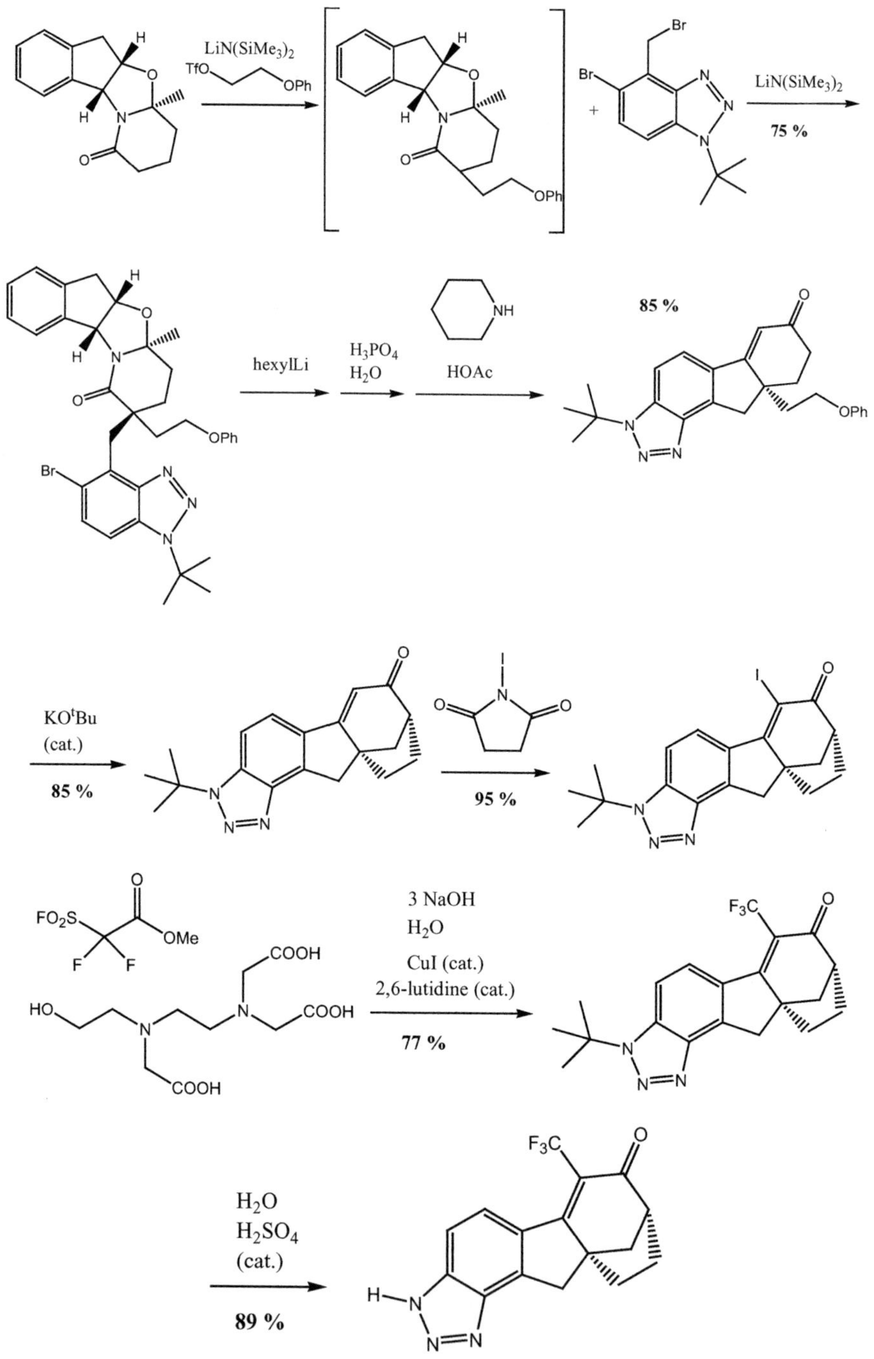

SCHEME 7.34 (CONTINUED) Merck process route to an estrogen receptor β-agonist.

Examples from the Chemical Industry

Part 1
Rewrite each route showing balanced chemical equations for each step.

Part 2
Draw synthesis tree diagrams for each route.

Part 3
Draw out the target bond maps for the product structure. Determine the MW fraction of sacrificial reagents for each route. How do the two routes compare from a synthesis strategy point of view?

Part 4
Based on the synthesis tree diagrams determine the corresponding overall yield, overall atom economy, and overall kernel PMI values. Which route is more material efficient?

Part 5
The authors disclosed a complete experimental procedure including all auxiliary materials used for the process route only. Based on those data determine the global PMI for that route using the *template-REACTION.xls* and *template-SYNTHESIS.xls* spreadsheets. State any assumptions made in the calculations. Compare the global PMI with the kernel PMI.

PROBLEM 7.40

Themes: balancing chemical equations, synthesis plan analysis, material efficiency metrics, energy metrics.

Vinyl chloride is a high volume industrial chemical that is primarily used to make polyvinyl chloride. Since World War II in Western countries, vinyl chloride is made via tandem chlorination of ethylene and dechlorination of 1,2-dichloroethane while in Eastern countries, India, and China it is made via hydrochlorination of acetylene using mercuric chloride as catalyst. Owing to its vast reserves of coal China is the leading producer of acetylene.

Below are 18 industrial routes to vinyl chloride grouped according to the starting materials used.

> acetylene + HCl ==> CH_2=CHCl
>
> acetylene + $ClCH_2CH_2Cl$ ==> CH_2=CHCl
>
> acetylene + 1,1,2,2-tetrachloroethane ==> CH_2=CHCl + trichloroethylene

$CH_2=CH_2 + Cl_2 ==> ClCH_2CH_2Cl ==> CH_2=CHCl + HCl$

$CH_2=CH_2 + HCl + O_2 ==> CH_2=CHCl + H_2O$

$CH_2=CH_2 + HCl + O_2 + ClCH_2CH_2Cl ==> CH_2=CHCl + H_2O + HCl$

$CH_2=CH_2 + Cl_2 + ClCH_2CH_2Cl ==> CH_2=CHCl + 2 HCl$

$CH_2=CH_2 + Cl_2 + O_2 ==> CH_2=CHCl + H_2O$

$CH_2=CH_2 + ClN=O ==> CH_2=ClCl + HCl + NO$

$CH_2=CH_2 + ClCH_2CH_2Cl ==> CH_2=CHCl + CH_3CH_2Cl$

$ClCH_2CH_2Cl ==> HCl + CH_2=CHCl$

1,1-dichloroethane $==> CH_2=CHCl + HCl$
1,1,1-trichloroethane $==> CH_2=CHCl + Cl_2$
1,1,2-trichloroethane $+ CH_2=CH_2 ==> ClCH_2CH_2Cl + CH_2=CHCl$

Part 1
Provide balanced chemical equations for all routes.

Part 2
Using the *template-REACTION.xls* and *ENERGY-input-analysis-synthesis-plans.xls* spreadsheets determine the PMI and thermodynamic energy consumption per ton of vinyl chloride produced for the following documented acetylene hydrochlorination routes. Assume all gases obey the ideal gas law. Use the enthalpy of vaporization and coefficients for temperature dependent heat capacities (J/K mol) as necessary. Identify the greenest routes.

$$acetylene + HCl ==> CH_2=CHCl$$

Catalyst	% Yield	T(rxn)/C	p(rxn)/atm	Reference
$HgCl_2$	97.3	250	1	64
$HgCl_2/CeCl_3$/activated carbon	98.5	100	1	65
$HgCl_2$/activated carbon	98.7	120	1	66
$HAuCl_4$/activated carbon/silicon carbide	20	185	0.5	67
Melamine/activated carbon	67.3	250	1.18	68
Au/Co(III)/SAC (spherical activated carbon)	91.5	150	1.18	69
$HgCl_2$/activated carbon	81	185	0.51	70
$HgCl_2/CeCl_3$/activated carbon	98.5	100	1	71

Examples from the Chemical Industry

Substance	Tb (K)	ΔH(vap)/ (kJ/mole)	A	B	C	D	E
Acetylene	189		19.36	1.15E−01	−1.24E−04	7.24E−08	−1.66E−11
Activated carbon			−0.832	3.48E−02	−1.32E−05		
$CeCl_3$			61.3	0.1157	−5.00E−05		
Chlorine	239		27.213	3.04E−02	−3.34E−05	1.60E−08	−2.70E−12
$Co(NH_3)_6Cl_3$		NA	NA	NA	NA	NA	NA
Gold	2933		21.992	1.30E−02	−5.23E−06		
$HAuCl_4$		NA	NA	NA	NA	NA	NA
$HgCl_2$	577		60.263	5.74E−02	−3.89E−05		
Hydrogen chloride	188		29.244	−1.26E−03	1.12E−06	4.97E−09	−2.50E−12
Melamine			15.006	0.2326	0.0016	−4.00E−06	2.00E−09
Nitrogen	77		29.342	−3.54E−03	1.01E−05	−4.31E−09	2.59E−13
silicon carbide			−4.1573	0.1433	−0.0001	7.00E−08	−1.00E−11

Part 3

Repeat Part 2 for the following ethane based routes to vinyl chloride.

$$CH_3CH_3 + HCl + O_2 \Longrightarrow CH_2=CHCl + 2H_2O$$

Catalyst	% Yield	T(rxn)/C	p(rxn)/atm	Reference
$KCl/CuCl/CuCl_2$	10.2	468	1	72
$CuCl_2/CeCl_3/LiCl/MnCl_2/Al_2O_3$	27.9	425	1	73
$Fe_2O_3/alpha\text{-}Al_2O_3$	23.1	550	1	74
$K_2O/Fe_2O_3/beta\text{-}Al_2O_3$	41.4	500	1	75
$CuCl_2/Al_2O_3/K_3PO_4$	74.5	600	1	76

$$CH_3CH_3 + 0.5Cl_2 + 0.75O_2 \Longrightarrow CH_2=CHCl + 1.5H_2O$$

Catalyst	% Yield	T(rxn)/C	p(rxn)/atm	Reference
$CuCl_2/ZnCl_2/KCl/celite$	70.4	500	1	77
$NaY\ zeolite/Ag_2O/MnO$	14.5	400	1	78

$$CH_3CH_3 + S + HCl \Longrightarrow CH_2=CHCl + H_2S + H_2$$

Catalyst	% Yield	T(rxn)/C	p(rxn)/atm	Reference
$Cr\text{-}Al_2O_3$	46.7	621	0.34	79

$$CH_3CH_3 + CH_2 = CH_2 + CH_3CH_2Cl + 4Cl_2 \Longrightarrow 3CH_2 = CHCl + 6HCl$$

Catalyst	% Yield	T(rxn)/C	p(rxn)/atm	Reference
CuCl/CuCl$_2$/KCl	40.6	474	1	80

Substance	Tb (K)	ΔH(vap)/ (kJ/mole)	A	B	C	D	E
Ag$_2$O			41.977	0.0965	−5.00E−05		
Al$_2$O$_3$			−8.121	3.87E−01	−3.16E−04		
CeCl$_3$			61.3	0.1157	−5.00E−05		
Celite (SiO$_2$)	2503		2.478	1.65E−01	−9.68E−05		
Chlorine	239		27.213	3.04E−02	−3.34E−05	1.60E−08	−2.70E−12
CuCl	1639		36.581	4.17E−02	−5.67E−06		
CuCl$_2$	1266		62.816	3.66E−02	−1.72E−05		
Ethane	184.4		28.146	4.34E−02	1.89E−04	−1.91E−07	5.33E−11
Ethyl chloride	285		35.946	5.23E−02	2.03E−04	−2.28E−07	6.91E−11
Fe$_2$O$_3$			65.237	0.1455	−4.00E−05		
Hydrogen chloride	188		29.244	−1.26E−03	1.12E−06	4.97E−09	−2.50E−12
KCl	1773		46.432	1.28E−02	7.04E−06		
K$_2$O			70.859	0.0472	−4.00E−06		
K$_3$PO$_4$		NA	NA	NA	NA	NA	NA
LiCl	1633		39.92	0.0285	−4.00E−06		
MnCl$_2$	1463		40.128	1.87E−01	−2.55E−04		
MnO			37.692	0.0253	−8.00E−06		
Na$_2$O Al$_2$O$_3$ 5.1 SiO$_2$		NA	NA	NA	NA	NA	NA
Nitrogen	77.2		29.342	−3.54E−03	1.01E−05	−4.31E−09	2.59E−13
Oxygen	90.03		29.526	−8.90E−03	3.81E−05	−3.26E−08	8.86E−12
Sulfur	718		2.003	1.20E−01	−1.62E−04		
ZnCl$_2$	1005		48.329	1.11E−01	−1.15E−04		

Part 4

Repeat Part 2 for the following ethylene based routes to vinyl chloride.

$$CH_2 = CH_2 + 1,1,2\text{-trichloroethane} \Longrightarrow ClCH_2CH_2Cl + CH_2 = CHCl$$

Catalyst	% Yield	T(rxn)/C	p(rxn)/atm	Reference
CuCl$_2$/ZrOCl$_2$/KCl/silica gel	17.8	350	1	81

$$CH_2 = CH_2 + 2ClN = O \Longrightarrow CH_2 = ClCl + HCl + 2NO$$

Examples from the Chemical Industry

Catalyst	% Yield	T(rxn)/C	p(rxn)/atm	Reference
Silica	17.5	480	1	82

$$CH_2 = CH_2 + Cl_2 \Longrightarrow ClCH_2CH_2Cl \Longrightarrow CH_2 = CHCl + HCl$$

Catalyst	% Yield	T(rxn)/C	p(rxn)/atm	Reference
None	95.8	380	0.34	83
None	80	451	1	84

$$CH_2 = CH_2 + HCl + 0.5\,O_2 \Longrightarrow CH_2 = CHCl + H_2O$$

Catalyst	% Yield	T(rxn)/C	p(rxn)/atm	Reference
KCl/CuCl/CuCl$_2$	36.2	420	1	85
CuO/firebrick	50	425	1	86

$$CH_2 = CH_2 + ClCH_2CH_2Cl \Longrightarrow CH_2 = CHCl + CH_3CH_2Cl$$

Catalyst	% Yield	T(rxn)/C	p(rxn)/atm	Reference
CaSO$_4$	97	260	20.4	87

$$CH_2 = CH_2 + HCl + 0.5\,O_2 + ClCH_2CH_2Cl \Longrightarrow 2\,CH_2 = CHCl + H_2O + HCl$$

Catalyst	% Yield	T(rxn)/C	p(rxn)/atm	Reference
FeCl$_2$/RhCl$_3$/ZnCl$_2$/LiCl/Al$_2$O$_3$	38.9	350	1	88

$$CH_2 = CH_2 + Cl_2 + ClCH_2CH_2Cl \Longrightarrow 2\,CH_2 = CHCl + 2\,HCl$$

Catalyst	% Yield	T(rxn)/C	p(rxn)/atm	Reference
none	95	540	27.2	89

$$CH_2 = CH_2 + 0.5\,Cl_2 + 0.25\,O_2 \Longrightarrow CH_2 = CHCl + 0.5\,H_2O$$

Catalyst	% Yield	T(rxn)/C	p(rxn)/atm	Reference
sand//CuCl$_2$/KCl/ThCl$_4$/silica gel	91.7	500; 400	3; 2.8	90

Substance	Tb (K)	ΔH(vap)/ (kJ/mole)	A	B	C	D	E
Al_2O_3			-8.121	3.87E-01	-3.16E-04		
$CaSO_4$			69.584	0.1021	-2.00E-06		
Chlorine	239		27.213	3.04E-02	-3.34E-05	1.60E-08	-2.70E-12
CuCl	1639		36.581	4.17E-02	-5.67E-06		
$CuCl_2$	1266		62.816	3.66E-02	-1.72E-05		
CuO			35.21	0.0304	-1.00E-05		
1,2-dichloroethane (liq)	357	32.16	26.31	7.76E-01	-2.23E-03	2.61E-06	
1,2-dichloroethane (g)			37.275	1.44E-01	1.04E-05	-7.83E-08	2.89E-11
Ethane	184		28.146	4.34E-02	1.89E-04	-1.91E-07	5.33E-11
Ethylene	169.3		32.083	-1.48E-02	2.48E-04	-2.38E-07	6.83E-11
Ethyl chloride	285		35.946	5.23E-02	2.03E-04	-2.28E-07	6.91E-11
$FeCl_2$			78.701	0.0098	2.00E-05		
Hydrogen chloride	188		29.244	-1.26E-03	1.12E-06	4.97E-09	-2.50E-12
KCl	1773		46.432	1.28E-02	7.04E-06		
LiCl	1633		39.92	0.0285	-4.00E-06		
Nitrogen	77.2		29.342	-3.54E-03	1.01E-05	-4.31E-09	2.59E-13
Nitrosyl chloride	268		28.551	7.59E-02	-9.44E-05	6.05E-08	-1.51E-11
Oxygen	90		29.526	-8.90E-03	3.81E-05	-3.26E-08	8.86E-12
$RhCl_3$	1073	NA	NA	NA	NA	NA	NA
Sand (sio_2)	2503		2.478	1.65E-01	-9.68E-05		
Silica gel (sio_2)	2503		2.478	1.65E-01	-9.68E-05		
$ThCl_4$	1194		106.03	0.0551	-2.00E-05		
1,1,2-trichloroethane (liq)	386.5	34.62	34.934	8.51E-01	-2.33E-03	2.65E-06	
1,1,2-trichloroethane (g)			28.881	2.49E-01	-1.76E-04	5.26E-08	-3.57E-12
$ZnCl_2$	1005		48.329	1.11E-01	-1.15E-04		
$ZrOCl_2$		NA	NA	NA	NA	NA	NA

Part 5

Repeat Part 2 for the following two-step routes to vinyl chloride.

Synthesis #1[91,98]

Step 1:
$CH_2=CH_2 + Cl_2 ==> ClCH_2CH_2Cl$ (T=85°C, p=1 atm; catalyst=1,2-dibromoethane)

Step 2:
$ClCH_2CH_2Cl ==> HCl + CH_2=CHCl$ (T=470°C, p=1 atm; catalyst: KCl/CuCl/$CuCl_2$)

Synthesis #2[92]

Step 1:
$CH_3CH_3 + Cl_2 + 0.5\ O_2 ==> ClCH_2CH_2Cl + H_2O$ (T=474°C, p=1 atm; catalyst: CuO/$CuCl_2$)

Examples from the Chemical Industry

Step 2:
$ClCH_2CH_2Cl \implies HCl + CH_2=CHCl$ (T = 454°C, p = 1 atm; catalyst: KCl/CuCl/CuCl$_2$)

Substance	Tb (K)	Δ H(vap)/ (kJ/mole)	A	B	C	D	E
Chlorine	238.94		27.213	3.04E–02	–3.34E–05	1.60E–08	–2.70E–12
CuCl	1639		36.581	4.17E–02	–5.67E–06		
CuCl$_2$	1266		62.816	3.66E–02	–1.72E–05		
CuO			35.21	0.0304	–1.00E–05		
1,2-dibromoethane (liq)	404.7		16.067	8.10E–01	–2.01E–03	1.99E–06	
1,2-dichloroethane (liq)	357	32.16	26.31	7.76E–01	–2.23E–03	2.61E–06	
1,2-dichloroethane (g)			37.275	1.44E–01	1.04E–05	–7.83E–08	2.89E–11
Ethane	184.4		28.146	4.34E–02	1.89E–04	–1.91E–07	5.33E–11
Ethylene	169.3		32.083	–1.48E–02	2.48E–04	–2.38E–07	6.83E–11
KCl	1773		46.432	1.28E–02	7.04E–06		
Oxygen	90.03		29.526	–8.90E–03	3.81E–05	–3.26E–08	8.86E–12

PROBLEM 7.41
Themes: synthesis plan comparison, materials efficiency metrics, green aspiration level (GAL), relative progress greenness (RPG)

Hoechst–Celanese plan[93,94]

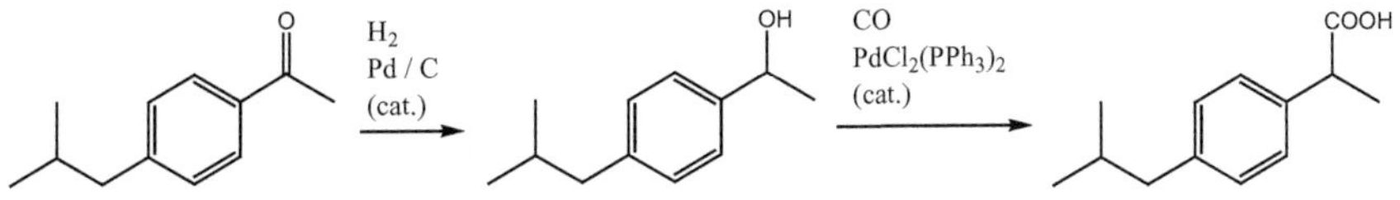

Boots plan[95]

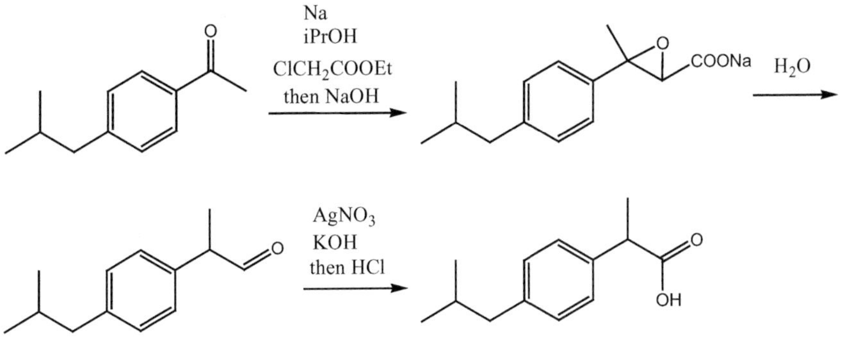

Upjohn plan[96]

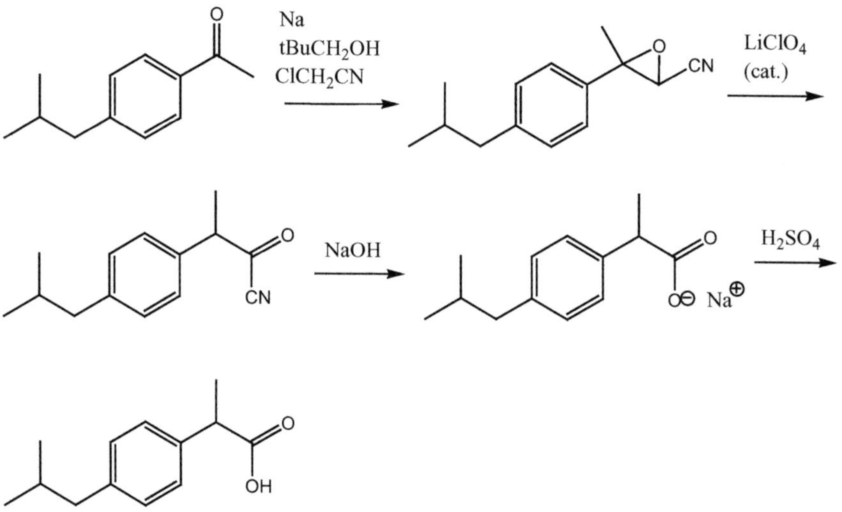

The relative progress greenness (RPG) and green aspiration level (GAL) metrics were developed by the pharmaceutical process industry as a rapid back-of-the-envelope method of gauging the relative greenness of various procedures to a given target molecule.[97] The steps in the calculations are given below.

Step 1: Determine the overall E-factor for a literature process by counting the number of reaction steps and multiply by 37.
Step 2: Count the number of construction steps in the synthesis plan, that is, those steps that involve forming target bonds in the final product structure.
Step 3: Determine GAL by multiplying the number of construction steps by 26.
Step 4: Determine RPG as the ratio between GAL (step 3) and overall E-factor (step 1).

Part 1

For each of the three industrial synthetic plans to make ibuprofen, determine GAL and RPG.

Part 2

Using the information given in the three procedures in the literature references and the *template-REACTION.xls* and *template-SYNTHESIS.xls* spreadsheets, determine the corresponding overall E-factors. State any assumptions made in the calculations. Compare the results with those found from step 1 in the calculation of RPG.

Part 3

Rank the three processes according to E-factor (according to Parts 1 and 2) and RPG. How does the back-of-the-envelope calculation measure up to the actual performances of the plans?

Part 4
For each plan add up the step E-factors and compare the results with the actual overall E-factors.

REFERENCES

1. Curzons, A.D., Constable, D.J.C., Mortimer, D.N., Cunningham, V.L. *Green Chem.* 2001, *3*, 1–6.
2. Jimenez-Gonzalez, C.; Constable, D.J.C. *Green Chemistry and Engineering: A Practical Design Approach*, Wiley: Hoboken, 2011, pp. 241–242.
3. Jimenez-Gonzalez, C.; Constable, D.J.C. *Green Chemistry and Engineering: A Practical Design Approach*, Wiley: Hoboken, 2011, pp. 269–272.
4. Jimenez-Gonzalez, C.; Constable, D.J.C. *Green Chemistry and Engineering: A Practical Design Approach*, Wiley: Hoboken, 2011, pp. 246–249.
5. Andraos, J.; Hent, A. *J. Chem. Educ.* 2015, *92*, 1831.
6. Fierz-David, H.E.; Blangey, L. *Fundamental Processes of Dye Chemistry*, Interscience Publishers: New York, 1949, pp. 228–229, 315–315.
7. Nair, M.G.; Dhananjayan, M.R.; Kron, M.A.; Milev, Y.P. US 2005267307 (Board of Trustees Michigan State University, 2005).
8. Jimenez-Gonzalez, C.; Constable, D.J.C. *Green Chemistry and Engineering: A Practical Design Approach*, Wiley: Hoboken, 2011, pp. 440–441.
9. Hessel, V.; Hofmann, C.; Löwe, H.; Meudt, A.; Scherer, S.; Schönfeld, F.; Werner, B. *Org. Proc. Res. Dev.* 2004, *8*, 511.
10. Kane, J.H.; Finlay A.C.; Amann, P.F. US 2385283 (Pfizer, 1945).
11. Chiusoli, G.P. US 3025320 (Montecatini Societa Generale per l'Industria Mineraria e Chimica, 1962).
12. Conant, J.B.; Tuttle, N. *Org. Synth. Coll.* 1941, *1*, 199.
13. Conant, J.B.; Tuttle, N. *Org. Synth. Coll.* 1941, *1*, 345.
14. Berg, R.G.; Hetzel, D.S. US 4100179 (Pfizer, 1978).
15. Linstead, R.P.; Mann, J.T.W. *J. Chem. Soc.* 1931, 726.
16. Shriner, R.L.; Ford, S.G.; Roll, L.J. *Org. Synth. Coll.* 1943, *2*, 368.
17. Horvath, I.T.; Rabai, J. *Science* 1994, *266*, 72.
18. Morschhauser, R.; Krull, M.; Kayser, C.; Boberski, C.; Bierbaum, R.; Puschner, P.A.; Glasnov, T.N.; Kappe, C.O. *Green Process Synth.* 2012, *1*, 281.
19. Bagley, M.C.; Jenkins, R.L.; Lubinu, M.C.; Mason, C.; Wood, R. *J. Org. Chem.* 2005, *70*, 7003.
20. McCaw, P.G.; Deadman, B.J.; Maguire, A.R.; Collins, S.G. *J. Flow Chem.* 2016, *6*, 226.
21. Nakagawa, M.; Sakai, Y.; Yasui, T. *J. Ferment. Technol.* 1984, *62*, 201.
22. Jimenez-Gonzalez, C.; Constable, D.J.C. *Green Chemistry and Engineering: A Practical Design Approach*, Wiley: Hoboken, 2011, pp. 646–647.
23. Jimenez-Gonzalez, C.; Constable, D.J.C. *Green Chemistry and Engineering: A Practical Design Approach*, Wiley: Hoboken, 2011, pp. 586–589.
24. Clarke, H.T.; Hartmann, W.W. *Org. Synth. Coll. Vol.* 1941, *1*, 233.
25. Jiang, J.A.; Chen, C.; Guo, Y.; Liao, D.H.; Pan, X.D.; Ji, Y.F. *Green Chem.* 2014, *16*, 2807.
26. Muheim, A.; Müller, B.; Münch, T.; Wetli, M. EP885968 (Givaudan-Roure, 1998).
27. Muheim, A.; Lerch, K. *Appl. Microbiol. Biotechnol.* 1999, *51*, 456.
28. Humphrey, G.R.; Dalby, S.M.; Andreani, T.; Xiang, B.; Luzung, M.R.; Song, Z.J.; Shevlin, M.; Christensen, M.; Belyk, K.M.; Tschaen, D.M. *Org. Process Res. Dev.* 2016, *20*, 1097.
29. Ritter, S.K. *Chem. Eng. News* 2017, *95*, 19–20.

30. Goossen, K.; Kuhn, O.; Berwe, M.; Krueger, J.; Militzer, H.C. US 2009221822 (AiCuris GmbH, 2009).
31. Fischer, M.; Kaibel, G.; Stammer, A.; Flick, K.; Quaiser, S.; Harder, W.; Massone, K. US 2001003578 (2001).
32. Paparatto, G.; Rivetti, F.; Andrigo, P.; De Albertini, G. US 2002025293 (ENI S.p.A., 2002).
33. Lowenheim, F.A.; Moran, M.K. *Faith, Keyes, & Clark's Industrial Chemicals*, 4th ed., Wiley: Hoboken, 1975, p. 487.
34. Lowenheim, F.A.; Moran, M.K. *Faith, Keyes, & Clark's Industrial Chemicals*, 4th ed., Wiley: Hoboken, 1975, p. 491.
35. Cosby, J.N.; Rissmiller, E.H. US 2902347 (Allied Chemical, 1959).
36. Zerbes, R.; Naab, P.; Franckowiak, G.; Diehl, H. EP 0657448 (Bayer AG, 1995).
37. Schwalbe, T.; Kadzimirisz, D.; Jas, G. *QSAR Comb. Sci.* 2005, *24*, 758.
38. Dressen, M.H.C.L.; van de Kruijs, B.H.P.; Meuldijk, J.; Vekemans, J.A.J.M.; Hulshof, L.A. *Org. Process Res. Dev.* 2010, *14*, 351.
39. Battilocchio, C.; Deadman, B.J.; Nikbin, N.; Kitching, M.O.; Baxendale, I.R.; Ley, S.V. *Chem. Eur. J.* 2013, *19*, 7917.
40. Wingad, R.L.; Gates, P.J.; Street, S.T.G.; Wass, D.F. *ACS Catal.* 2015, *5*, 5822.
41. Lowenheim, F.A.; Moran, M.K. *Faith, Keyes, and Clark's Industrial Chemicals*, 4th ed., Wiley: Hoboken, 1975, pp. 50–54.
42. Hamblet, C.H.; McAlvey, A. US 2557282 (DuPont, 1951).
43. Corey, E.J.; Wu, L.I. *J. Am. Chem. Soc.* 1993, *115*, 9327.
44. Lowenheim, F.A.; Moran, M.K. *Faith, Keyes, and Clark's Industrial Chemicals*, 4th ed., Wiley: Hoboken, 1975, pp. 1, 8, 10, 524.
45. Lowenheim, F.A.; Moran, M.K. *Faith, Keyes, and Clark's Industrial Chemicals*, 4th ed., Wiley: Hoboken, 1975, pp. 36, 40.
46. Lowenheim, F.A.; Moran, M.K. *Faith, Keyes, and Clark's Industrial Chemicals*, 4th ed., Wiley: Hoboken, 1975, pp. 21, 83, 482, 496, 524, 547.
47. Pei, Z.; Li, X.; von Geldern, T.W.; Madar, D.J.; Longenecker, K.; Yong, H.; Lubben, T.H. et al. *J. Med. Chem.* 2006, *49*, 6439.
48. Weng, J.; Li, J.M.; Li, F.Q.; Xie, Z.S.; Lu, G. *Adv. Synth. Catal.* 2012, *354*, 1961.
49. Ishikawa, H.; Honma, M.; Hayashi, Y. *Angew. Chem. Int. Ed.* 2011, *50*, 2824.
50. Yasuda, N.; Tan, L. in Yasuda, N. (ed.) *The Art of Process Chemistry*, Wiley-VCH: Weinheim, 2011, p. 19.
51. Yasuda, N.; Tan, L. in Yasuda, N. (ed.) *The Art of Process Chemistry*, Wiley-VCH: Weinheim, 2011, pp. 22, 26–27, 28, 29.
52. Yasuda, N.; Tan, L. in Yasuda, N. (ed.) *The Art of Process Chemistry*, Wiley-VCH: Weinheim, 2011, p. 34.
53. Pierce, M.E.; Parsons, R.L. Jr.; Radesca, L.A.; Lo, Y.S.; Silverman, S.; Moore, J.R.; Islam, Q. et al. *J. Org. Chem.* 1998, *63*, 8536.
54. Humphrey, G.R.; Zhong, Y.L. in Yasuda, N. (ed.) *The Art of Process Chemistry*, Wiley-VCH: Weinheim, 2011, p. 167.
55. Humphrey, G.R.; Zhong, Y.L. in Yasuda, N. (ed.) *The Art of Process Chemistry*, Wiley-VCH: Weinheim, 2011, pp. 172, 174, 176.
56. Humphrey, G.R.; Zhong, Y.L. in Yasuda, N. *The Art of Process Chemistry*, Wiley-VCH: Weinheim, 2011, p. 183.
57. Koda, K.; Matsura, T.; Obora, Y.; Ishii, Y. *Chem. Lett.* 2009, *38*, 838.
58. Lowenheim, F.A.; Moran, M.K. *Faith, Keyes, & Clark's Industrial Chemicals*, 4th ed., Wiley: Hoboken, 1975, p. 178.
59. Lowenheim, F.A.; Moran, M.K. *Faith, Keyes, & Clark's Industrial Chemicals*, 4th ed., Wiley: Hoboken, 1975, p. 1.

60. Vogel, A.I. *A Textbook Practical Organic Chemistry*, 3rd ed., Longman: Hoboken 1956, p. 460.
61. Young, C.O. US 1966157 (Carbide & Carbon Chemicals Corp., 1934).
62. Tolleson, G.S.; Puckett, T.A. US 2004/54211 (2004).
63. Maddess, M.L.; Scott, J.P.; Alorati, A.; Baxter, C.; Bremeyer, N.; Brewer, S.; Campos, K. et al. *Org. Proc. Res. Dev.* 2014, *18*, 528.
64. Lowenheim, F.A.; Moran, M.K. *Faith, Keyes, and Clark's Industrial Chemicals*, 4th ed., Wiley: Hoboken, 1975, p. 869.
65. GB 600785 (Monsanto, 1948).
66. GB 769773 (BASF, 1957).
67. Davies, C. PhD thesis 2012, U Cardiff, p. 48.
68. Wang, X.; Dai, B.; Wang, Y.; Yu, F. *ChemCatChem* 2014, *6*, 2339.
69. Zhang, H.; Dai, B.; Wang, X.; Li, W.; Han, Y.; Gua, J.; Zhang, J. *Green Chem.* 2013, *15*, 829.
70. Anderson, I.F.; Datin, R.C. US 2816148 (Allied Chemical & Dye Corp., 1957).
71. Boyd, T. US 2446123, US 2446124 (Monsanto, 1948).
72. Riegel, H. US 3557229 (Lummus Corp., 1971).
73. Beard, W.Q. Jr. US 3629354 (Ethyl Corp., 1971).
74. Magistro, A.J. US 4100211 (BF Goodrich, 1978).
75. Kroenke, W.J.; Nicholas, P.P. US 4119570 (BF Goodrich, 1978).
76. Riegel, H. US 3937744 (Lummus Corp., 1976).
77. GB 996323 (PPG Corp., 1965).
78. Pyke, D.R.; Reid, R. GB 2095245 (ICI, 1982).
79. Schuman, S.C. US 3377396 (1968).
80. Riegel, H.; Schindler, H.D.; Sze, M.C.; Brooks, M.E. US 3920764 (Lummus Corp., 1975).
81. Nielsen, D.R. GB 2121793 (PPG Ind., 1984).
82. FR 2064406 (ICI, 1971).
83. Tsutsumi, S. GB 956657 (1964).
84. Miller, F.D.; Jenks, D.P. US 2896000 (National Distillers & Chemical Corp., 1959).
85. Ludwig, G.H. GB 1213402 (BP, 1970).
86. Cass, O.W. US 2308489 (duPont, 1943).
87. Trotter, P.W. US 2681372 (Ethyl Corp., 1959).
88. Lemanski, M.F.; Leiter, F.C.; Vinson, C.G. Jr. US 4115323 (Diamond Shamrock, 1978).
89. GB 1177971 (Pullman Inc., 1970).
90. Otsuka, E.; Takahashi, T.; Fujisawa, K.; Abe, T. US 3291846 (Toyo Koatsu Industries, 1966).
91. Riegel, H. US 3935288 (Lummus Corp., 1976).
92. Riegel, H. US 3937744 (Lummus Corp., 1976).
93. Elango, V.; Murphy, M.A.; Smith, B.L.; Davenport, K.G.; Mott, G.N.; Moss, G.L. EP 284310 (Hoechst–Celanese Corp., USA, 1988).
94. Elango, V.; Davenport, K.G.; Murphy, M.A.; Mott, G.N.; Zey, E.G.; Smith, B.L.; Moss, G.L. EP 400892 (Hoechst Celanese Corp, USA, 1990).
95. FR 1545270 (Boots Pure Drug Co. Ltd., 1968).
96. White, D.R. US 4021478 (The Upjohn Company, USA, 1977).
97. Roschangar, F.; Sheldon, R.A.; Senanayake, C.H. *Green Chem.* 2015, *17*, 752.
98. Lowenheim, F.A.; Moran, M.K. *Faith, Keyes, and Clark's Industrial Chemicals*, 4th ed., Wiley: Hoboken, 1975, p. 392.

Appendix: Other Terminologies

Burden Shifting

This term is used in life cycle assessment (LCA) literature that refers to the situation where attempts to reduce environmental burdens at one stage of the life cycle create impacts at other stages. An example is when a process beginning with more environmentally friendly starting materials may produce greater life cycle impacts when considering how those starting materials are themselves manufactured.

ChemSpider[1]

This is a free online database of chemical substances sponsored by the Royal Society of Chemistry that includes various experimental and calculated physical and environmental property data that is important for carrying out life cycle assessments. It is useful since it incorporates the results of the EPA suite algorithm for predicting properties such as octanol–water partition coefficient, Henry law constant, water solubility, and persistence parameters in air, water, soil, and sediment.

Classification, Labeling and Packaging (CLP) Regulation[2]

A European Union set of rules that came into effect in 2009 that governs classification and labeling regulations to protect workers, consumers, and the environment by providing appropriate labeling that reflects a chemical's possible hazards. Companies in Europe are required to appropriately classify, label, and package their substances and mixtures before placing them on the market.

Regulation (EC) No 1272/2008 of the European Parliament and of the Council of December 16, 2008 on classification, labeling and packaging of substances and mixtures, amending and repealing Directives 67/548/EEC and 1999/45/EC, and amending Regulation (EC) No 1907/2006.

See *Globally Harmonizing System (GHS); Registration, Evaluation, Authorization, and Restriction of Chemicals (REACH)*.

Dangerous Substances Classification and Labeling (DSCL)

This system is part of the Dangerous Substances Directive law concerning chemical safety that has jurisdiction in the European Union.

Annex III of the directive defines standard phrases relating to the *Nature of special risks attributed to dangerous substances and preparations*, often referred to as R-phrases. The appropriate standard phrases must appear on the packaging and label of the product and on its MSDS. The R-phrases are the basis of the risk phrase potential used in life cycle assessment.

Design Institute for Physical Property Data (DIPPR 801)[3]

The DIPPR 801 project is affiliated with the American Institute of Chemical Engineers (AIChE). The database collates reliable thermophysical property data on 2424 industrial chemicals including 34 constant and 15 temperature dependent properties. Among these, the following are important in estimating input energy consumption in carrying out chemical reactions: acentric factor, normal boiling point, critical pressure, critical temperature, heat capacity of ideal gas, heat capacity of liquid, heat capacity of solid, heat of vaporization, liquid density, solid density, and vapor pressure of liquid.

EcoScale[4]

EcoScale is a semi-quantitative tool suitable for introductory undergraduate education on green chemistry that is used to assess environmental and hazard impacts of chemicals used in a chemical reaction. It uses an arbitrary penalty point system out of an ideal value of 100 covering the following categories: reaction yield, cost of reaction components (based on producing 10 mmol of final product), safety of reaction components, technical setup (type of equipment used), reaction temperature and reaction time, and workup and purification components. Greener procedures have high EcoScale values. The algorithm applies only to single reactions, not synthesis plans, and only to reaction input materials. It has limited coverage of actual toxicity and hazard parameters and is heavily weighted toward simplified WHMIS and NFPA-704 labeling systems and qualitative information found in MSDS sheets. There is no visual display and the EcoScale does not account for relative masses of input or waste materials in the assignment of penalty points. For example, 1 g of mercury used as reagent is assigned the same penalty points as if 100 g were used. The algorithm also does not consider waste reaction by-products. EcoScale is not recommended for work beyond teaching purposes at the introductory level.

Fast Life Cycle Assessment of Synthetic Chemistry (FLASC)™ [5]

FLASC is a trademark name of a life cycle assessment algorithm developed by GlaxoSmithKline (GSK) for the analysis of syntheses of pharmaceutical products.

Fire and Explosion Index[6]

The Dow fire and explosion index (FEI) is composed of general and specific process hazards factors and a material factor. The material factor covers flammability. The general process hazard factor covers exothermicity, endothermicity, material handling and transfer, accessibility of equipment, process unit design, and drainage and spillage control. The specific process hazard factor covers material toxicity, reaction pressure, quantities of flammable or unstable materials in storage, operation in or near the flammable range of a chemical, reaction temperature, corrosion and erosion, leakage from joints in equipment, fire equipment requirements, use of hot-oil heat exchange systems, and rotating equipment requirements. Although the Dow FEI is comprehensive, as the name suggests, it focuses mainly on flammability and explosiveness hazards associated with chemicals used and equipment requirements and wear and tear.

Globally Harmonizing System (GHS)[7-9]

GHS is an internationally agreed upon system of labeling chemical hazards managed by the United Nations that harmonizes descriptions and pictograms used by the European Union Classification, Labeling, and Packaging (CLP) Regulation and the United States Occupational Safety and Health and Administration (OSHA) standards into a single unified standard. Two broad categories of hazard are considered: physical, health, and environmental. Among physical hazards are the following nine sub-categories: explosives, gases, flammable liquids, flammable solids, oxidizing substances and organic peroxides, toxic and infectious substances, radioactive substances, substances corrosive to metals, and miscellaneous. Among the health hazards are the following 12 sub-categories: acute toxicity, skin corrosion, skin irritation, serious eye damage, eye irritation, respiratory sensitizer, skin sensitizer, germ cell mutagenicity, carcinogenicity, reproductive toxicity, specific target organ toxicity, and aspiration hazard. Among the environmental hazards are the following two sub-categories: acute aquatic toxicity and chronic aquatic toxicity. GHS label components include hazard pictograms, signal words such as "danger" or "warning", hazard statements, precautionary statements, product identifier (ingredient disclosure), supplier identification, and supplemental information.

See *Classification, Labeling and Packaging (CLP) Regulation, Registration Evaluation Authorization and Restriction of Chemicals (Reach).*

NATIONAL FIRE PROTECTION ASSOCIATION 704 LABELING SYSTEM[10]

The National Fire Protection Association (NFPA) is a United States trade association, that creates and maintains private, copyrighted standards and codes pertaining to fire protection for usage and adoption by local governments. The 704 labeling system refers to the Standard System for the Identification of the Hazards of Materials for Emergency Response (four-color hazard diamond symbol).

Liquid Class	FP (°C)	BP (°C)	NFPA Rating	Example
IA	FP < 23	BP < 38	4	Diethyl ether
IB	FP < 23	BP > 38	3	Acetone
IC	23 < FP < 33		3	Xylenes
II	38 < FP < 60		2	Acetic acid
IIIA	60 < FP < 93		2	*o*-Dichloro-benzene
IIIB	FP > 93		1	Dimethyl sulfoxide

National Institute for Occupational Safety and Health[11]

The National Institute for Occupational Safety and Health (NIOSH), established in 1970, is the United States federal agency responsible for conducting research and making recommendations for the prevention of work-related injury and illness. It compiles lists of occupational exposure limits for various industrial chemicals.

Registration Evaluation Authorization and Restriction of Chemicals (REACH)[12,13]

Registration, Evaluation, Authorization and Restriction of Chemicals (REACH) is a European Union regulation dated December 18, 2006 and came into force on June 1, 2007. REACH addresses the production and use of chemical substances, and their potential impacts on both human health and the environment.

Registry of Toxic Effects of Chemical Substances[14,15]

Registry of Toxic Effects of Chemical Substances (RTECS) is a database of toxicity information compiled from the open scientific literature without reference to the validity or usefulness of the studies reported. Until 2001 it was maintained by NIOSH as a freely available publication. It is now maintained by the private company Symyx Technologies and is available only for a fee or by subscription. The RTECS database is also available by subscription via the Canadian Centre for Occupational Health and Safety (CCOHS).

RTECS contains LD50(oral), LD50(dermal), and LD50(inhal) data.

Workplace Hazardous Materials Information System[16,17]

The Workplace Hazardous Materials Information System (WHMIS), (known as SIMDUT, Système d'information sur les matières dangereuses utilisées au travail in French) is Canada's national workplace hazard communication standard. The key elements of the system, which came into effect on October 31, 1988, are cautionary labeling of containers of WHMIS controlled products, the provision of material safety data sheets (MSDSs) and worker education and site-specific training programs.

References

1. www.chemspider.com
2. Chemical Glossary: CLP Regulation (www.reagent.co.uk/chemical-glossary/#clpregs)
3. www.aiche.org/dippr/projects/801
4. Van Aken, K.; Strekowski, L.; Patiny, L. *Beilstein J. Org. Chem.* 2006, 2, 3.
5. Curzons, A. D.; Jiménez-González, C.; Duncan, A. L.; Constable, D. J. C.; Cunningham, V. L. *Int. J. Life Cycle Assess.* 2007, *12*, 272.
6. *Dow's Fire and Explosion Index Hazard Classification Guide, 7th ed.,* American Institute of Chemical Engineers: New York, 1994.

Appendix: Other Terminologies

7. A Guide to the Globally Harmonized System of Classification and Labeling of Chemicals (GHS), United Nations, October 2015.
8. Globally Harmonized System of Classification and Labeling of Chemicals (GHS), United Nations, "The Purple Book."
9. Globally Harmonized System of Classification and Labeling of Chemicals (GHS), United Nations, 4th ed., 2011, ST/SG/AC.10/30/Rev.4.
10. NFPA 704: Standard System for the Identification of the Hazards of Materials for Emergency Response, National Fire Prevention Association: Quincy, MA, 2007 (http://law.resource.org/pub/us/cfr/ibr/004/nfpa.704.2007.pdf).
11. NIOSH Pocket Guide to Chemical Hazards; National Institute for Occupational Safety and Health, Centers for Disease Control and Prevention: Atlanta, GA; September Publication No. 2005-149, 2007 (www.cdc.gov/niosh).
12. https://en.wikipedia.org/w/index.php?title=Registration,_Evaluation,_Authorisation_and_Restriction_of_Chemicals&oldid=776015218
13. https://echa.europa.eu/regulations/reach
14. https://en.wikipedia.org/w/index.php?title=Registry_of_Toxic_Effects_of_Chemical_Substances&oldid=739902659
15. http://ccinfoweb.ccohs.ca/rtecs/search.html
16. https://en.wikipedia.org/wiki/Workplace_Hazardous_Materials_Information_System?oldid=785790304
17. www.hc-sc.gc.ca/ewh-semt/occup-travail/whmis-simdut/ghs-sgh/index-eng.php

Index

β-Carotene, 137–139
β-Lactam, 129–130
1-Methylbenzimidazole, 326–330
2-Methyl-2-propanol, 280–281
2-Methyltetrahydrofuran, 249–250
2,2-Diethoxy-1-isocyanoethane, 387
2,3,7,8,12,13,17,18-Octaethylporphyrin, 87
2,4,5-Trichlorophenoxyacetic acid, 90
2,6-Dimethylpyridine, 283
3-Chlorothietane 1,1-dioxide, 388–389
5-(Hydroxymethyl)furfural (5-HMF), 85

A

ABT-341, 473–477
Acentric factor, 255, 274–276
Acetaldehyde, 202–206, 275, 279, 442–451, 486
Acetic acid, 442–453
Adiabatic process, 255
Adipic acid, 96–97, 437–439
Agent Orange, 90
Alizarin, 395–396
Amantadine, 125
Andraos algorithm, 370
Aniline, 36–38, 174–175, 181–182, 192–196, 201–202, 240–241, 243–247, 249, 250, 345–346, 355–357, 360–362, 365–366, 368–370, 374–377, 385–386
Aniline yellow, 34
Apixaban, 53–58
Arens–von Dorp–Isler alkylation, 100
Aromoadendranediol, 68–71

B

Benign index, 191–196
Benzothiophene, 278
Benzyl cyanide, 388–389
Biphenyl, 83–84
Bohlmann–Rahtz synthesis of pyridines, 402
Branches, 7, 15, 19–28
Burden shifting, 503
Butanol-*n*, 321–322, 379, 381, 384, 436, 485–488

C

Carbofuran, 136–137
Carpanone, 66–67
Carvone, 125–126
Centroid, 6

ChemSpider, 503
Chiral rhodium catalyst, 13–15
Chlorobenzene, 284
Ciprofloxacin, 419–420
Classification, Labelling and Packaging regulation (CLP), 503
Clausius–Clapeyron equation, 275
Computational toxicology, 154, 199
Concession step, 45
Concise synthesis, 45
Construction step, 45
Continuous flow microreactor technology (CFMRT), 430
Cradle-to-cradle, 155
Cradle-to-gate, 155
Cradle-to-grave, 155
Creatine, 32–34
Cresol-*p*, 281, 409–411
Critical constants, 255–256
Cubic equation, 254, 266, 286–287
Cumulative energy demand, 319
Cumulative material efficiency metrics, 45–52
Cumulative yield, 58
Curtate LCA, 157
Cyclohexanol, 96–97, 437–439
Cyclooctatetrene, 130–131

D

Dangerous Substances Classification and Labelling (DSCL), 503
Dapoxetine, 16–18
Degree of asymmetry, 5
Degree of convergence, 5
Dendrimer, 142–145
Density, 256
Design Institute for Physical Property Data (DIPPR 801), 256, 504
Dicyclohexyl urea, 337
Diels–Alder reaction, 331
Diethyl oxalate, 352, 388–389
Di-isopropanolamine, 282–283
Diosgenin, 110–115

E

EATOS, 345–357
EcoScale, 355
Edmister formula, 274–276
Edwards–Lawrence algorithm, 357–361

Efavirenz, 478–480
Energy efficiency, 319
Energy index, 319
Energy input (absolute), 320
Energy input per mass of target product, 320
Energy intensity, 320
Enthalpy, 256–257
Enthalpy of formation (heat of formation), 257
Enthalpy of fusion (heat of fusion), 258
Enthalpy of reaction (heat of reaction), 258
Enthalpy of vaporization (heat of vaporization), 258
Environmental impact, 157
Environmental index, 158
Environmental quotient, 158
Epichlorohydrin, 119–120
Equation of state, 258–259
    ideal gas, 259
    Peng–Robinson, 259–260
    Redlich–Kwong, 259
    van der Waals, 259
Estrogen receptor, 488–490
Ethanol, 436–437
Ethyl acrylate, 453–463
Ethyl cinnamate, 100–101
Ethyl phenylcyanopyruvate, 389
Ethylene, 38–41
Ethylene glycol, 39–41, 206–207
Eugenol, 42–43

F

Fast Life Cycle Assessment of Synthetic Chemistry (FLASC), 504
Fire and explosion index (Dow Chemical), 504
First moment building up parameter, 58–60
Flash point, 215
Fluoxetine, 16–18

G

Gantt diagram, 61–64
Gate-to-gate, 158
Gate-to-grave, 158
Global atom economy, 64
Global E-factor, 64–65
Global Harmonized System (GHS), 505
Global mass efficiency, 65
Global material economy, 65
Global process mass intensity, 65
Global reaction mass efficiency, 65
Glycerol, 44–45
Gomberg-Bachmann reaction, 84
Green Aspiration Level, 497–498
GreenStar, 361–365
GreenStar area index, 362
Guaiacol, 42, 411

H

Hazard, 215
Hazard waste index, 215
Heat capacity at constant pressure, 261–262
Henry law constant, 154, 250
Hess's law, 262–263
Hodge–Sterner toxicity scale, 161–162
Hub intermediate, 65–66
Hydrogen cyanide, 340–341
Hydrogen peroxide, 414, 416–419
Hypochlorous acid, 203, 205–206, 393–394

I

Ibuprofen, 497
    Boots plan, 497
    Hoechst-Celanese plan, 497
    Upjohn plan, 498
Ideality, 67
Impact, 161
Indigo, 126–128
    Heumann–Fierz–David synthesis, 127
    Yamamoto synthesis, 127
Inherent safety index, 217
Inventory analysis, 67
Isobaric, 265
Isobutanol, 322
Isobutyraldehyde, 307–309
Isobutyric acid, 341
Isochoric, 265
Isometric, 265
Isoprene, 280
Isothermal, 265
Itaconic acid, 397–400, 406

K

Kappogenin, 112–113

L

L-DOPA, 11–12
LC50 (inhalation), 161
LD50(dermal), 218
LD50(oral), 162
Letermovir, 411–414
Life cycle, 163
Life cycle analysis or assessment (LCA), 153–155, 163
Life cycle costing, 163
Life cycle inventory, 163
Life cycle management, 502
Limonene, 104–105
Lipase, 16–18, 118, 423
Lombardo olefination, 108
Lower explosion limit, 216

Index

M

Marker degradation, 110
Materials Safety Data Sheets (MSDS), 164
Maximum molecular weight fraction of sacrificial reagents, 67
Maxwell equal area rule, 265–269
Meclinertant, 425–435
Menthol, 13–14, 104–105
Mercuric oxide, 7–8
Mescaline, 107–108
Methane, 264–265
Methanol, 338–339
Methyl isobutyrate, 313–319
Methyl methacrylate, 463–473
Methyl nitroacetate, 97–98
Meyer–Schuster rearrangement, 100
Microreactor technology (MRT), 421
Miroestrol, 441–443
Molar volume, 269
Molecular weight fraction of sacrificial reagents, 67–68
Multicompartment model (MCM), 166–169
Multivariate method (Queen's U), 362–365
Muscone, 102
Musk ketone, 102–103

N

National Fire Protection Association (NFPA) 704 labeling system, 505
National Institute for Occupational Safety and Health (NIOSH), 506
Nicotine, 94, 175–176, 200
  Craig synthesis, 94
Nitric acid, 96–97, 171–172, 249, 435–437
Nitrobenzene, 35, 192–196, 201, 243–247, 250, 345, 357, 382, 383
Nonane, 85–87
Normal boiling point, 269–270
Normal melting point, 270
Number of bonds made per step, 71
Number of branches, 71
Number of input materials, 71
Number of nonbonding steps per reaction stage, 71
Number of product nodes, 71
Number of reaction steps, 71

O

Octanol-water partition coefficient, 164–165
One-pot iterative assembly line strategy, 145–147
Overall atom economy, 71
Overall mass efficiency, 71
Overall process mass intensity, 71
Overall reaction mass efficiency, 71
Overall yield, 71
Oxygen balance, 218

P

Papaverine, 91–92
Persistence, 165, 187–189
Petasis titanocene carbene olefination, 109
Phase transition, 270
Phenylboronic acid, 396–397
(R)-1-Phenylethylamine, 89, 139–141
(S)-4-(Phenylmethyl)-2-oxazolidinone, 102–104
Plicamine, 75–81
Potential, 169
  abiotic resource depletion, 169–270
  acidification, 170
  acidification-basicification, 171–181
  bioaccumulation, 181
  bioconcentration, 181–182
  cancer potency, 182
  corrosiveness as a gas, 223
  corrosiveness as a liquid or solid, 223–224
  dermal absorption, 224
  endocrine disruption, 182–183
  explosive strength, 224–225
  explosive vapor, 225
  flammability, 225–226
  global warming, 183
  hydrogen gas generation, 226–228
  impact sensitivity, 228
  ingestion carcinogenicity, 184
  ingestion toxicity, 184
  inhalation carcinogenicity, 185
  inhalation toxicity, 185–187
  maximal allowable concentration, 228
  occupational exposure limit, 229–230
  oxygen balance, 230–234
  ozone depletion, 187
  persistence, 187–189
  photochemical ozone creation, 189
  risk phrase, 234–239
  skin dose, 239–241
  smog formation, 189–190
Predicting parameters
  critical constants (Joback method), 276–277
  enthalpy of formation (Domalski–Hearing method), 277
  enthalpy of fusion (Chickos method), 277–278
  enthalpy of vaporization (Vetere method), 279
  heat capacity of gases (Benson method), 279–280
  heat capacity of liquids (Ruzicka–Domalski method), 280–281
  heat capacity of solids (Goodman method), 281

normal boiling point (Nannoolal method), 282
normal melting point (Constantinou-Gani method), 283
vapor pressure of liquids (Lee-Kesler method), 337–338
vapor pressure of liquids (Riedel method), 283–284
Primary energy usage, 320
Principle of corresponding states, 270
Process solvent mass intensity, 391
Process time, 391
Process water mass intensity, 391
Process water use, 391
Progesterone, 110–115
Propanol, 82, 383, 384

R

Raltegravir, 481–483
Reaction network, 36–42
Reaction pressure hazard index, 218–219
Reaction stage, 72
Reaction step, 72
Reaction temperature hazard index, 219–220
Reduced constants, 272
Reduced digraph, 36, 39
Registration Evaluation Authorization and Restriction of Chemicals (REACH), 503
Registry of Toxic Effects of Chemical Substances (RTECS), 506
Residence time, 391
Resveratrol, 128–129
Risk, 165, 220
Risk index, 165, 220
Rowan solvent greenness index, 377–380

S

Sacrificial reagent, 72
Sacrificial step, 72
Safety-hazard impact, 241
Safety-hazard index, 241–247
Salvarsan, 133–134
Sarsasapogenin acetate, 110
Saturation temperature and pressure, 272
Scale out, 392, 408–409
Scaling factor, 72
Scaling up, 392, 408–409
Sea-nine, 132–133
Sequential organic synthesis (SEQUOS), 421
Sildenafil citrate, 380
Sobetirome, 98–100
Sodium thiosulfate, 35
Solvay process, 18–19
Solvent intensity, 391
Solvent recovery energy, 320

Space-time-yield (STY), 324
Standard state conditions, 272
Step atom economy, 72
Step economy, 72
Step E-factor, 73
Step process mass intensity, 73
Step reaction mass efficiency, 73
Suzuki coupling, 84
Suzuki reaction, 249
Synthesis
  convergent, 19–34
  divergent, 19
  linear, 15
  efficiency, 73
  strategy, 73
  tree, 7–19

T

Tait equation, 272–273
Target bond profile, 75–79
Target-bond forming reaction, 75
TD50, 165
Tebbe olefination, 108
Tetra(n-butyl)ammonium bromide, 392–393
Thermal expansion coefficient (isobaric), 273
Thietane 1,1-dioxide, 387–388
Thiete 1,1-dioxide, 387–388
Threshold limit value, 220–221
Thiocarbamates, 324–326
Throughput, 392
Time weighted average, 221
Toluidine-o, 277–278
Toluidine-p, 43
Toxicity, 165–166
Tributyltin oxide, 132–133
Triple point, 273
Tropinone, 120–122
  Robinson plan, 121
  Willstätter plan, 121–122

U

U Toronto Green Chemistry Initiative algorithm (UTGCI), 368–370
Ullmann coupling, 84
UM171, 52–53
Uncertainty in BI, 196
Uncertainty in SHI, 247
Vapor pressure of liquids, 273
Vapor pressure of solids, 274
VOC emission, 166

V

Vanillin, 11, 42–43, 92, 409–411
Vinyl chloride, 491–496

Index

Vinylformamide, 41–42
Vitamin C, 122–124

W

Waste treatment energy, 320
Water solubility, 221–223
Wittig reaction, 108

Work done on system, 254
Workplace Hazardous Materials Information
  System (WHMIS), 506
Wurtz–Fittig reaction, 84

X

Xylene-o, 382–383